GW00569171

Reports of the Research Committee of the Society of Antiquaries of London, No. 64

Landscape Plotted and Pieced

*Fyfield Down on 3 June 1934 (**see page vi for a descriptive analysis of this air photograph**)* (© Ashmolean Museum, University of Oxford*)*

Landscape Plotted and Pieced

Landscape History and Local Archaeology in Fyfield and Overton, Wiltshire

P J Fowler

The Society of Antiquaries of London

First published 2000
by
The Society of Antiquaries of London
Burlington House
Piccadilly
London W1J 0BE
for
English Heritage

ISBN 0 85431 276 5

ISSN 0953-7163

British Library Cataloguing in Publication Data
A CIP catalogue record for this book is available from the British Library

Edited for The Society of Antiquaries of London and English Heritage by Julie Gardiner

Original series design by Chuck Goodwin, London W2 5DA
Designed and laid out by Wenham Arts, Peterborough
Printed in Great Britain by BAS Printers Ltd

Dedication

To
Collin Bowen and the late Geoffrey Dimbleby
in admiration and with gratitude

and in memory of two fine institutions,
the Department of Extra-Mural Studies, University of Bristol (*ob* 1998)
and
the Royal Commission on the Historical Monuments of England (*ob* 1999)

Glory be to God for dappled things –
For skies of couple-colour as a brinded cow;
For rose-moles all in stipple upon trout that swim;

Fresh-firecoal chestnut-falls; finches' wings;
Landscape plotted and pieced – fold, fallow, and plough;
And áll trádes, their gear and tackle and trim.

from *Pied Beauty* by Gerard Manley Hopkins

Caption for frontispiece

The familiar oblique air photograph on page ii (taken by Major Allen in 1934) looks across Fyfield Down from the south east. It has been repeatedly published to illustrate well-preserved and typical 'Celtic' fields. In fact, it shows three major phases of pre-medieval land allotment on three discrete orientations, developed and used periodically during the two and a half thousand years up to about the middle of the first millennium AD. Various field types were incorporated. The arrangements, and individual fields, are not, in fact, well preserved in places because they were differentially modified by medieval cultivation. The ridge-and-furrow in the centre of the photograph is one such example.

The photograph shows much else besides. *Top left* is a Bronze Age ditch bounding the prehistoric arable, though the visible, markedly regular fields to its south east are of early Roman date. Across the *centre*, on land cultivated during the second and earlier part of the first millennium BC, are fields, defined by lynchets up to 4m high, which have finally accumulated against small, dry-stone walls built in the early Roman period. Overlying these early fields, and crossing their boundaries in places, is ridge-and-furrow arranged in butting furlongs. That in the centre of the photograph almost certainly represents the work in the mid-thirteenth century of a man called Richard of Raddun; his farm (*Raddun*) is *bottom left*, cut by the photograph's edge. Between the fields is a Romano-British track system with a contemporary settlement at the T-junction *right centre*. Passing along the track and through the settlement is a small, sharp bank and ditch continuing to the right: it is the nineteenth-century boundary of Overton Cow Down, as Fyfield Down was then called. It is smoothed over by what looks like an air-landing strip but which was in fact a racehorse-training gallop in 1934.

The prominent wood on the *left* is Wroughton Copse, a name derived from *Raddun*. The triangular enclosure in front of it is part of the medieval farm, with later medieval additions suggesting it may also be the site of a documented sheep-cote of 1490 and probably the site of a named barn, 'Rodden', in 1570. Further away on the left is a small but prominent beech plantation, also within an earthwork enclosure, and contemporary with the later eighteenth-century building alongside. One of only three houses on the down in modern times, Delling was the (game-)keeper's cottage. From *bottom right* to *left centre* runs a heavily rutted downland track, formerly the main London–Bath road via Marlborough and Avebury. Replaced by the turnpike (1743) along the Kennet valley, it was formally closed to through traffic in 1815, and now rejoices in a twentieth-century name, 'Green Street'.

Centre right is a low bank and ditch which is still the boundary of the straight north-east side of Fyfield parish. On it is 'Long Tom', a cut sarsen pillar which, as perhaps has been the case since at least Anglo-Saxon times when the area was called *Red Down*, marks where Fyfield ends.

Contents

Illustrations

Figures

Plates

Tables

Acknowledgements

A project that runs for thirty-nine years accumulates many debts as well as friends; it is impossible to list all of both. At the head of the 'Fyfod' list, nevertheless, must be two individuals who were not only there at the start and helped over many years but, far more than they probably realise, have been my inspiration throughout. Both Collin Bowen and the late Geoffrey Dimbleby accepted the dedication as a mark of my appreciation and long-term friendship.

Neither were, however, excavators. For the addition of archaeological excavation as a tool to a fieldworker's armoury, I acknowledge my debt to the late Professors Ian Richmond, Richard Atkinson and George Jobey, and to Charles Thomas and Nicholas Thomas, who introduced me to the delights of Wiltshire archaeology at Snail Down in 1955.

The project could not have happened, let alone been sustained, without the co-operation of the landowners of Fyfield and Overton: none has ever denied access, though perhaps they were not always given the opportunity so to do. We thank them all, in particular the Swanton family of North Farm, West Overton, the late George Todd of Manton, Robert Sangster of Manton, Count Konrad Goess Saurau of Temple Farm, John Bloomfield and English Farms, and the Forestry Commission.

Four institutions have principally afforded long-term support. They are the Department of Extra-Mural Studies, University of Bristol, which initiated the work in 1959 through its enthusiastic Resident Tutor in Wiltshire, the late Harry Ross. It then, under the knowing surveillance of its Director, Geoffrey Cunliffe, both sustained it in the field through the 1960s and facilitated post-excavation study through numerous classes in the 1970s; it even promoted one last small teaching excavation in 1996. The Royal Commission on Historical Monuments (England), subsequently the Royal Commission on the Historical Monuments of England, through numerous staff and in many professional ways, has been the main academic support throughout, 1959–98. From the start, and especially at the finish, 1995–8, the former Ancient Monuments Inspectorate in various government departments, now subsumed in English Heritage, has been supportive. The help in the later stages of another University, Newcastle upon Tyne, and especially of its Department of Agricultural and Environmental Science 1996–8, is also much appreciated.

Other institutions and organisations on whose long-term support and goodwill we have drawn include the Experimental Earthworks Committee of the British Association for the Advancement of Science; the former Nature Conservancy and English Nature; the Wiltshire and Swindon Record Office and the Archaeological Section with its Sites and Monuments Record at Wiltshire County Council, Trowbridge; Wiltshire Archaeological and Natural History Society and especially its Museum and Library at Devizes and its successive curators, Nicholas Thomas, Ken Annable and Paul Robinson; and the Society of Antiquaries of London for early financial support (1961, 1962 and 1964), for providing a platform for lectures on 30 November 1967 and 27 November 1997, and for publishing this work. The BBC has also, over the years, provided many opportunities to communicate on radio and television, and has incidentally helped financially. Other financial support from the Marc Fitch Fund is also acknowledged but, in revenue terms, the project was always very inexpensive – a few hundred pounds a year at most – until a six-figure grant from English Heritage to the University of Newcastle upon Tyne in 1995–6 enabled full-time preparation of publication and archive to proceed. The vision and personal interest in this matter of Dr Geoffrey Wainwright, formerly Chief Archaeologist at English Heritage, were crucial and are much appreciated.

I would also acknowledge a profound debt to a host of other individuals, not least the many hundreds who participated in the project over thirty-nine years. They have done the work. Among them are some whose commitment has been of two or more decades: I would thank by name Elizabeth Fowler, whose excavation records we now know to be effectively faultless, and Gillian Swanton (née Lawton) whose multi-faceted local support has been crucial. My parents, Phyllis and the late Bill Fowler, also provided vital help over many years.

Among those who made significant contributions in the field were three, then young, men who assisted with the surveying which produced many of the site plans: Nick Bradford (now Department of Archaeology, Southampton University), David Clarke (National Museum of Antiquities, Edinburgh) and Peter Drewett (Institute of Archaeology, University College London). Surveying of a sub-terrestrial sort was carried out before and during excavation by Dr, later Professor, Martin Aitken (Oxford University), Dr Collin Gill (Bristol University) and the late A J ('Tony') Clark. I would also like to thank all the site supervisors on the excavation programme: Delling enclosure: Anthony Witheridge, John Scantlebury; Down Barn enclosure: John Scantlebury, Gillian Swanton; Fyfield lynchet: Ann Woodward (née Heard); OD I–III: Jim Specht; OD X/XI: Peter Drewett, Dr Tania Dickinson (née Briscoe), Chris Gingell, Ann Woodward, Dr Isobel Smith; OD XII: Bill Fowler, Graham Thomas, David Fowler; TD I–III: F B Simpson and David Fowler; TD VIII: Andrew Tarr; TD IX: Brian Perry; WC: Joanna Close-Brooks, Bill Fowler and Chris Knowles. Among regular helpers on the excavations were: (the late) Norris Thompson, W G and Madge Drew, Nick and Sheila Pratchett, Mike and Steven Batt, Betty Bostock, Sue, Peter and John Dunstone, Diana Friendship, Jim Hancock, Philip Harding, Georgina Plowright, Arthur Selway, Alan Vince and Michael Wadhams. Others who helped significantly included E G H Kempson (documentary research), (now Professor) J G Evans (palaeo-environmental research) and Noel Grudgings (photography). The contributions of four Nature Conservancy staff were immense: Jim Hemsley (responsible for the Fyfield Down NNR and incidentally teaching this author about downland flora), and Professor A D Lacaille, Inigo Jones and Noel King (successive Wardens of the NNR). David Morgan Evans was a sympathetic Inspector of Ancient Monuments in those early decades, and he too has seen the project reach its final stages, not least with his successor many times removed, Amanda Chadburn, whose later support has been as much appreciated.

Contributions to this volume and the archive are acknowledged as appropriate *in situ*. It is proper, nevertheless, to thank the Trust for Wessex Archaeology (TWA) under the aegis of its sympathetic Director, Andrew Lawson. The work of its team of professionals was indispensable, and I would thank, in particular, Lorraine Mepham and her fellow specialists, notably Rachael Seager Smith, Tessa Machling and Dr Michael Allen. Other specialists who contributed included the late Dr Barbara Noddle, with pioneer work on animal bones, Juliet Rogers and (the late) Bob Everton on the human bones, (the late) Anne Everton on the flints, Professor Jenny Price (Roman glass), who produced a brilliant report and catalogue, D Penton on the OD XII Roman coins, and Dr Josh Pollard (the Beaker). In addition, thanks also go to the Tyne and Wear Museums Archaeology Department at Arbeia, South Shields, for its continuing expertise, especially Margaret Snape who, from the record alone drafted a report on a, to her, completely unknown site (OD XII), supported by the critical interest of Dr Paul Bidwell. Professor Mark Overton (Exeter University) reoriented the author in medieval agrarian studies in 1995, and Christopher Taylor and Desmond Bonney bravely tried to do likewise a little later in medieval settlement studies. Sylvia Young was of great help with local documentary research. Commissions to historians Dr John Hare (Overton manor) and Simon Yarrow (Knights Templar) promptly produced authoritative and distinctive contributions from original documentary research which encouraged the whole study to expand seriously into post-Roman landscapes. Only in the archive can the full substance of their work be appreciated. The whole downland landscape of all periods was exposed by the stunning aerial photographic cartography of Cathy Stoertz (RCHME), whose map provided the spatial and archaeological framework within which nearly four decades of exploration could generate some sort of understanding. David Renshaw's kindness and kind of landscape understanding have given a visual distinction to this volume's appearance, and I appreciate both greatly.

Curiously, there seemed little appropriate experience to draw on in arranging our archive in the way, using electronics, which was so obviously appropriate. Paul Robinson at Devizes Museum and Nigel Clubb at the National Monuments Record, Swindon, proved, however, helpful and constructive; and then, through Tim Williams at English Heritage, we were able to identify how the need could be met, and future users provided for, through the Archaeological Data Service, York University. We acknowledge the positive approach of Dr Julian Richards, Damian Robinson and Alicia Wise at York, and the technical skill of The Archaeological Practice, Newcastle University, in this matter.

Some of the graphics published here were prepared by Sandra Rowntree; most of her work is in the archive. Trevor Ermell, of Monochrome, Newcastle, prepared the prints of nearly all the ground photographs published here. Prints of air photographs, and other help, came from the National Monuments Record of England, the Ashmolean Museum and Cambridge University Committee for Aerial Photography. I would thank in particular Dr Robert Bewley and Roger Featherstone of the first, and David Wilson of the last, for their interest and assistance. Ann Clark first tackled editorially a recalcitrant text; Professor Richard Bradley and Christopher Taylor willingly accepted an invitation to be first peer-reviewers. The criticisms of all three were timely, constructive and ultimately helpful, leading directly to major revisions in the whole of the project's output. This revised entity subsequently went to anonymous reviewers.

Back in the engine-house at Newcastle, a small team, essentially part-time or on short-term contract, effectively brought the project to a tangible outcome, 1995–8. A tranche of half-a-dozen extremely able, recent graduates of the Department of Archaeology at Newcastle began to assemble the volume and archive, with significant contributions from Chris Lucas, Louisa Ward and Christine Martinez. Julie Watson proved the perfect amanuensis for an increasingly overwhelmed author and it was tragic that illness removed her from the project. Robert Johnston, in his first post-graduate year trying to keep his Professor up to the mark, rigorously addressed some of the major academic issues arising from the project. Graeme Stobbs and Kristian Strutt rescued a dire graphics situation in 1996, and both helped in the creation of this monograph and its archive far beyond the role of mere draughtsmen. From beginning to end of the creation of volume and archive, the author was indefatigably and percipiently supported by Ian Blackwell, initially his post-graduate student, then assiduous colleague, now co-author in a related book. He has literally kept the show on the road, drafted significant parts of the text and largely organised the archive now available at Devizes Museum, in the NMR and, through York, to the world at the touch of a button. At the very last stage of writing this volume, Mark Corney has brought a keen archaeological mind and a detailed knowledge of the Avebury area to bear on this monograph, to its great benefit. And at a final stage of not just editing this volume but integrating it and the entire output of the project into a single accessible whole, Dr Julie Gardiner (TWA) successfully completed on time a major piece of challenging and pioneering work. Kate Owen, Publications Manager at the Society of Antiquaries of London, ably saw the volume through the press.

The text was proofread by Sue Viccars and Kate Fielden. Elizabeth Fowler, as assiduous at the end as at the beginning, prepared the excellent Index.

Finally, I happily acknowledge that I would not be writing this passage were it not for the critical help of Gill Andrews. Officially manager for English Heritage of the 'post-excavation' phase, she never allowed 'the team', and certainly not this author, to believe other than that the publication and archival objectives were achievable. I did not always believe her, but never dared say so; my personal debt is irredeemable.

As author, I am profoundly conscious of an enormous debt accumulated through this project throughout a lifetime. I am, however, rather more conscious of, and deeply grateful for, friendship gained, as together we have plotted a landscape and pieced together a little understanding of it.

Peter Fowler
Clerkenwell, London
12 October 1999

Editorial notes

After an introduction to the study area, the project and some methodologies (Part I, Chapters 1–3), this volume's structure follows the well-tried formula of describing with some interpretation in Part II what was done (Chapters 4–13) and then discussing it (Part III, Chapters 14–17). Chapter 2, the bulk of the report (Part II, ten chapters) and the archive (*see below*) deal with the primary evidence and its acquisition. Chapter 2 is based on a major exercise in cartographic aerial photographic interpretation, and looks at the results from that point of view over the whole of the northern part of the study area, ie, essentially the downlands north of the Kennet valley. It provides a good introduction to the core of both the study area and of our approach, while suggesting some of the strands of thought to be considered later in Part III. Chapter 15 includes brief consideration of an equivalent map for the south of the study area.

Chapters 4 to 13 then explore the landscape of the whole study area. Four of those chapters, 4 to 7, examine in more detail parts of the northern landscape looked at as a whole in Chapter 2. Chapters 8 to 11 respectively examine aspects of the Kennet valley with particular reference to the villages, manors and estates, but always steering considerations to land-use and landscape change. The two remaining chapters (12–13) in Part II look at the rather different landscapes to the south, with their strong woodland flavour and the study area's one major monument, Wansdyke. All these chapters draw on a range of methodologies, appropriateness as often as availability of evidence indicating which particular one to use. Appropriateness itself was often influenced by types of land-use, categorised and explained in Chapter 1. Such types apart, the general progression of chapters is from north to south and west to east (Figure 3.4).

CONVENTIONS USED IN THE TEXT

1 Subsoil types with capital letters = Subsoil or bedrock. Thus 'Chalk' has a specific meaning, whereas 'chalk' could be used adjectivally, as in 'chalk rubble' or would need some explanation, eg, 'the post-hole was packed with chalk and cut into Chalk'.

2 Original field and site record book numbers were consistently used throughout the writing-up operation, 1995–7, and are extant in all the archive texts, ie, the original numbers are used for cuttings, excavated areas, context and all finds, etc, so that there is 1:1 correlation between the texts up to May 1997.

3 In the final preparation of the draft monograph (the 'July 1997' text, FWP 75: *see below* for explanation of these terms), we had to abandon such correlation and use a new set of layer numbers because too many of the layer numbers on the field drawings were just too nonsensical when transferred to the printed page and publishable graphic. A standardised graphic code and layer numbering system is used throughout this report and the excavation reports in FWPs 63–6. The key to both numbering and graphic code is allied with verbal descriptions in the Key, so it is often unnecessary to describe the appearance of a layer in the text: something like 'Figure 5.7, layer 32a' will usually suffice, provided a reader then looks up 'layer 32a' on the Key.

4 Measurements, dimensions and similarly factual metrical data are not provided as a matter of course in the text if they are present in a graphic, eg, the size of an excavated area, the width of a gully, the depth of a pit, the distance of one feature from another. This is both to save space and on the assumption that users will be able and willing to 'read a graphic' if they need such information. It is, however, usually spelt out in the archive texts and, of course, the site records as appropriate. Such data are, however, provided selectively, where they are not otherwise directly available in a published graphic, where they are seen to be significant, eg, the diameter of a structure, the depth of a find in a feature, or to help an argument or the flow of the text.

5 Similarly, spatial relationships are not necessarily spelt out if they are readily apparent in a graphic, though again significant ones will almost invariably have attention drawn to them, eg, ard-marks overlie many features on Site OD XI and such is indicated in several graphics; but the relationship is not mentioned in the text except where it is of general significance and where it is important to emphasise that a particular feature is earlier or later than a particular ard-mark.

6 To summarise with regard to excavations, it is policy NOT to include textually metrical data clearly contained in figures and plates where they are easily recoverable there from, for instance, dimensions of cuttings, heights/depths of layers, etc; nor of minutiae such as post-hole widths and depths, unless they are significant, where they are ordered and easily accessible in the archive. So there are two main filters in front of the excavation text in this monograph (and to a lesser extent in the excavation reports): data only appear if they are significant (for a host of reasons) and also contribute to the main story.

7 Material excavated was treated at the time as 'General finds' (GF) or 'Small finds' (SF). The latter were three-dimensionally recorded, but always within the context of a GF number which provided the general context and the GF and SF numbers of associated finds. Though scarcely used in this text, SF numbers to identify specific objects and GF numbers to identify associated material are used with decreasing density the later the date of drafts eventually building up to this monograph. In other words, the earlier work, 1995–6, on the project output was much more detailed than printed here; such detail tended to be progressively edited out 1996–8 as we struggled with the logistics of publication. There is, however, much more detail on finds in the excavation reports in the electronic archive, FWPs 63–5, and by and large even more in the FWPs behind them.

8 Notes published in the *Wiltshire Archaeological and Natural History Magazine* are referred to only by their volume and page number.

9 Except where otherwise stated, the photographs are by the author.

10 The numbered divisions around the margins of some of the maps are the 1km intervals of the Ordnance Survey National Grid.

11 The following figures are not printed in the body of the text but will be found in the pocket on the inside back cover: 2.1, 2.2, 2.6, 5.2, 6.5, 6.6, 6.7, 6.8, 6.9, 6.11, 6.20, 7.1, 7.3, 7.6, 7.7, 7.8, 10.2, 13.1 and 15.3.

Abbreviations used in the text

ADS	Archaeological Data Service, York University
aOD	above Ordnance Datum
CUCAP	Cambridge University Collection of Air Photographs
DB	*Domesday Book*
DM	Devizes Museum (Museum of the Wiltshire Archaeological and Natural History Society), Long Street, Devizes, Wiltshire
FWP	Fyfod Working Paper (*see* Appendix 1)
GPS	Global Positioning Satellite
IPM	*Inquisitiones Post-Mortem*
LPP	*Landscape Plotted and Pieced*
NMR	National Monuments Record, Swindon
NNR	National Nature Reserve
OS	Ordnance Survey
RCHME	Royal Commission on the Historical Monuments of England
SL	*The Land of Lettice Sweetapple* (Fowler and Blackwell 1998)
SMR	Wiltshire Sites and Monuments Record, Trowbridge
SSSI	Site of Special Scientific Interest
TWA	Trust for Wessex Archaeology
VCH	*Victoria County History*
WAM	*Wiltshire Archaeological and Natural History Magazine*
PNWilts	Gover *et al* 1939
SRO	Somerset Record Office, Taunton
WCL	Winchester Cathedral Library
WRO	Wiltshire Record Office, Trowbridge

Note on Appendix 1
and the Project Archive

The concept of an ordered and publicly accessible archive to the Fyfield and Overton Downs project has been fundamental to the intensive work that took place between 1995 and 1998 in attempting to complete the project satisfactorily. The two parts of the output, hard copy publication (this volume and Fowler and Blackwell 1998) and the archive, are conceived as an integrated product, each dependent on the other. Certainly, the nature of this volume is heavily conditioned by the existence, nature and accessibility of the archive.

The archive itself is in five media – material objects, writing, graphics, photographs and electronics – and physically is housed in two places: its primary depository in the museum of the Wiltshire Archaeological and Natural History Society at Devizes and also in the National Monuments Record, Swindon; and its electronic in another, the Archaeological Data Service (ADS), Department of Archaeology, University of York (*www.ads.ahds.ac.uk/catalogue/*).

An indicative catalogue of the archive can be found in Appendix 1 of this volume; catalogues of its main components are available electronically through the ADS. We would stress that, while this volume can in some ways stand alone, much of the evidence, and all of the primary evidence on which it is based, is in the archive or referenced from the archive.

In particular, this applies to all the excavations. Full, conventional, illustrated reports of these are available electronically. They are also available from the ADS in hard copy form, on request. Hard copies have also been placed in several libraries. For further explanation, *see* Appendix 1.

Summary

This monograph summarises and discusses a study (1959–98) of Fyfield and West Overton, two contiguous parishes in Wiltshire, England. They lie north to south across the upper Kennet valley between Avebury and Marlborough where the young River Kennet breaches the Chalk at the south-west corner of the Marlborough Downs. The two parishes embrace a large area of grass downland across their northern parts. Southwards are expanses of mainly historic permanent arable on the slopes north and south of the valley floor. The flood plain, much used in water management, is traditionally meadow. In all this the landscape is typical of much of the Wessex chalkland, but the parishes differ from many others in two respects, both concerning natural resources which have profoundly affected this landscape's history. To the north, much of Fyfield and Overton Downs is covered by the best remaining extent of Tertiary sandstone blocks ('sarsens'); while much of the parishes' southern reaches is covered by permanent woodland, a western outlier of Savernake Forest.

A combination of methods of study over thirty-nine years has been directed primarily to elucidating how and when the landscape came by its appearance. Four main factors have emerged in an answer. First, very little if any of this landscape is now 'natural': virtually all of it is an artefact. Second, the nature of the landscape artefact has been and continues to be strongly influenced by the natural characteristics of the study area, notably its geology (solid and drift), its hydrology, its soils and its climate. Third, land-use has both followed and, except for the climate, fashioned those natural characteristics in a long interaction which has nevertheless seen the present landscape's principal land-use features established at particular times in a sequence which was essentially over before the beginning of our era. Thus the post-glacial forest cover had been removed by 2000 BC at the latest, creating open downland, variously grassed, cultivated and under scrub, north of the river and probably to the south too. There the major landscape development was, however, the converse. Fyfield and Overton's permanent woodland, now called West Woods, was created through clearance around it to become a discrete feature and major component of the local landscape over 4,000 years. Despite long-term and probably even continuous management, it has not significantly changed its position, shape or size during that time.

The downland, in contrast, has enjoyed a chequered career, and most of it, though prehistoric, is not as old as the woodland. Most of the present grassland and at least some of the presently cultivated areas in the permanent arable were parts of an organised, axiometric landscape of enclosed fields, pasture, burial/ancestral lands, tracks and droveways in the mid/late second millennium BC, a downland landscape strangely lacking in settlements until circular, ditched and embanked ones appeared in the first half of the first millennium BC. Thereafter, the higher downs became and remained primarily unenclosed pasture, a land-use interrupted only by brief phases of spatially restricted cultivation in the first–second, fourth, tenth, thirteenth–fourteenth, nineteenth and mid–late twentieth centuries. Since their permanent establishment two and a half thousand years ago, they, like the woodlands, have seen only fitful habitation, notably in the first, fourth–fifth, thirteenth–fourteenth, sixteenth and nineteenth centuries, essentially marginal to a valley-based settlement pattern which has been dominant, despite locational and economic variations, since the early centuries AD.

Throughout, the two parishes have generally supported a resident population of under 1,000 people, while being both largely self-sufficient yet characteristically serving the interests of absent and distant landlords. Fourthly, therefore, theirs is a landscape of exploitation, as indicated by the absence of any major structure except East Wansdyke, locally unfinished. Local natural resources, such as chalk, clay, sarsen, wood, water, soil and grass, have been variously utilised in what has been basically an agricultural economy and landscape since farming communities developed in the area some 6,000 years ago. Yet the area has seldom been isolated and has, indeed, been characteristically easy of access receiving, for example, religious and architectural influences, building and household materials and artefacts such as pottery, metalwork and glass.

But the parishes are a place that people pass through rather than stay in. They lie at a natural, insular crossroads where a main west–east route across southern England meets a traditional north–south route between the English Channel and the English hinterland. The former perhaps initially ran through Avebury, across the embryonic downs in the third millennium BC and was then variously formalised as a Roman, turn-pike and twentieth-century trunk road along the Kennet valley; the latter, running across the valleys and along the ridges of the Wessex chalklands, has, by contrast, never become a metalled through road during its fitful existence and its use by local and transhumance traffic dating back to the last centuries of the prehistoric era.

Résumé

Cette monographie est une synthèse et un débat sur une étude (1959–98) de Fyfield et West Overton, deux communes contiguës du Wiltshire, en Angleterre. Elles s'étendent du nord au sud de la haute vallée de la Kennet, entre Avebury et Marlborough, là où la rivière naissante entaille la craie à l'angle sud-ouest des Marlborough Downs (collines crayeuses). La zone nord des deux communes comprend une large étendue de collines herbeuses. Au sud, sur les versants nord et sud du fond de la vallée, on trouve des étendues de terre arable principalement durant l'époque historique. La plaine d'inondation, particulièrement mise à profit dans la gestion de l'eau, est traditionnellement en prairie. Tout cela concoure à former le paysage typique de la majeure partie des Chalklands du Wessex, mais ces communes diffèrent de beaucoup d'autres sur deux aspects se rapportant chacun à des ressources naturelles et qui ont profondément affecté l'évolution du paysage. Au nord, les Downs (collines crayeuses) de Fyfield et Overton sont recouvertes par la principale étendue connue de blocs de grès tertiaires ('sarsen'), alors qu'au sud, les terrains communaux présentent un important couvert forestier constitué par une extension occidentale de la forêt permanente de Savernake.

Le but principal de l'étude conduite pendant près de trente-neuf ans, en combinant plusieurs méthodes d'analyse, a été de comprendre comment et à quel moment le paysage a pris cette apparence. Quatre principaux facteurs ont été isolés et permettent de répondre à cette question. En premier lieu, force est de constater que peu d'éléments sont aujourd'hui 'naturels' dans ce paysage: virtuellement, tout est artificiel. En second lieu, la nature de l'anthropisation du paysage a été et continue d'être fortement influencée par les caractéristiques naturelles de la zone d'étude, notamment par sa géologie (substratum et sédimentation), son hydrologie, ses sols et son climat. En troisième lieu, l'exploitation humaine a, à la fois, suivi mais aussi façonné ces caractéristiques naturelles, à l'exception du climat, au sein d'une longue interaction qui a aboutit à la mise en place des traits du paysage actuel au cours d'une séquence principalement située avant le début de notre ère. Ainsi, la couverture forestière post-glaciaire a été éliminée au plus tard vers 2000 BC, dénudant les collines qui ont été diversement mises en herbage, cultivées, abandonnées aux brousailles, au nord mais aussi probablement au sud de la rivière. Dans cette région, toutefois, le développement majeur du paysage a été inversé. La forêt permanente de Fyfield et Overton, aujourd'hui appelée West Woods, a été créée par un déboisement qui l'a circonscrite pour devenir un élément discret mais néanmoins majeur du paysage local il y a 4000 ans. Depuis lors, en dépit d'une gestion à long terme et probablement continue, la forêt n'a pas subi de changements significatifs dans sa position, sa forme et sa taille.

A l'opposé, le paysage des Downs a connu un sort plus diversifié et n'est pas aussi ancien que le paysage forestier même s'il est également préhistorique. La plupart des prairies actuelles et, pour le moins, certaines des étendues de terres arables permanentes aujourd'hui cultivées, s'intégraient, à la fin du deuxiéme millénaire BC, à un paysage axiométrique organisé en parcelles cloturées, en pâturages, en terres ancestrales d'inhumation, en allées et drailles, où faisaient étrangement défaut les installations humaines avant que n'apparaissent les habitations circulaires, fossées et avec un talus dans la première moitié du premier millénaire BC. Par la suite, les collines les plus élevées sont devenues et sont principalement restées des pâturages ouverts sauf pendant les brèves phases de mise en culture sur des superficies restreintes pendant les premier–deuxième, quatrième, dixième, treizième–quatorzième, dix-neuvième et fin du vingtième siècles. Depuis cette structuration, il y a 2,500 ans, elles n'ont connu, comme les terrains forestiers, que des occupations humaines épisodiques, en particulier pendant le premier, quatrième–cinquième, treizième–quatorzième, seizième et dix-neuvième siècles, en marge des occupations de base de fond de vallée, dominantes depuis le début de notre ère malgré les variations économiques et les changements de localisation.

Pendant tout ce temps, le nombre de résidents des deux communes a généralement été inférieur à 1,000 habitants et, bien qu'autonome économiquement, cette population servait, comme il était coutume, les intérêts des propriétaires terriens absents et lointains. En quatrième lieu, alors, l'absence de toute structure majeure, à l'exception d'East Wansdyke (inachevé sur ces communes), reflète un paysage d'exploitation. Les ressources locales naturelles comme la craie, l'argile, le grès, le bois, l'eau, les sols et les prairies ont été diversemment exploitées au sein de ce qui a été essentiellement une économie et un paysage d'agriculture depuis que des communautés de fermiers se sont installées dans cette region, il y a près de 6,000 ans. Cette région a, cependant, rarement été isolée et d'accès facile, elle en a reçu les influences caractéristiques telles que religieuses, architecturales, apport de matières premières de construction, d'objets domestiques ou encore d'éléments comme la poterie, le métal et le verre.

Mais ces communes sont surtout des lieux de passage et non de résidence. Elles sont situées à un croisement insulaire naturel par lequel la route principale ouest–est qui traverse l'Angleterre du sud rencontre la traditionnelle route nord–sud qui va de la Manche et remonte vers l'intérieur des terres. A l'origine, la première passait peut-être par Avebury à travers les Downs embryonnaires au cours du troisième millénaire BC et a été alors différemment utilisée par les Romains, comme route à péage et principale route le long de la vallée de la Kennet au vingtième siècle; la seconde, à travers les vallées et le long des crêtes des Chalklands du Wessex, n'est, au contraire, jamais devenue une route amenagée et a été emprunté épisodiquement pour assurer le traffic local et la transhumance depuis les derniers siècles de la préhistoire.

TRADUCTION: ALINE AVERBOUH

Zusammenfassung

Diese Monographie diskutiert und faßt die Ergebnisse einer Untersuchung zusammen, die von 1959 bis 1998 in den benachbarten Pfarrbezirken von Fyfield und West Overton in Wiltshire, England durchgeführt wurde. Die beiden Bezirke erstrecken sich von Norden nach Süden im oberen Kennet Tal zwischen Avebury und Marlborough, wo der kleine Fluß Kennet die Kreideformation an der südwestlichen Ecke der Marlborough Downs durchbricht. Ein großes Grashügelland erstreckt sich über ihre nördlichen Bereiche. In südlicher Richtung hingegen finden sich an den Hängen nördlich und südlich des Talbodens weite Flächen von in historischer Zeit dauerhaft benutztem Ackerland. Die Schwemmebene, die in starkem Maße in der Wasserhaushaltung genutzt wird, besteht traditionell aus Wiesen. Obwohl die Landschaft in allen diesen Merkmalen typisch für große Teile des Wessex Kreidelands ist, unterscheiden sich beide Pfarrbezirke von anderen vor allem in zweierlei Hinsicht, und zwar bezüglich der natürlichen Resourcen, die in der Geschichte der Landschaft eine wichtige Rolle gespielt haben. In nördlicher Richtung ist ein großer Teil der Fyfield und Overton Downs mit großen Resten von Blöcken tertiären Sarsengesteins bedeckt, und das südliche Gebiet wird überwiegend von dauerhaftem Wald, einem westlichen Ausläufer des Savernake Waldes, eingenommen.

Das Hauptziel der bereits 39 Jahre andauernden Untersuchung war es mit einer Kombination von Untersuchungsmethoden zu erforschen, wie und wann das heutige Aussehen der Landschaft entstanden ist. Vier Hauptfaktoren sind dabei deutlich geworden. Erstens, es kann heute sehr wenig, wenn überhaupt etwas von dieser Landschaft als 'natürlich' bezeichnet werden: praktisch alles ist Artefakt. Zweitens, die Natur des Landschaftsartefakts ist und wird weiterhin in starkem Maße von den natürlichen Gegebenheiten des Untersuchungsgebiets beeinflußt, besonders seiner Geologie (Felsgestein und Geschiebe), seiner Hydrologie, seiner Böden und seines Klimas. Drittens, zum einen ergab sich die Landnutzung aus diesen natürlichen Gegebenheiten, zum anderen aber, außer dem Klima natürlich, gestaltete sie sie auch in einer langfristigen Wechselwirkung, die dennoch die Hauptmerkmale der Landnutzung der gegenwärtigen Landschaft zu einer bestimmten Zeitperiode, die bereits vor dem Beginn unserer Ära abgeschlossen war, entstehen sah. Somit war der postglaziale Waldbewuchs bis spätestens um 2000 BC entfernt, was wiederum offenes Hügelland schuf, das nördlich des Flusses und wahrscheinlich auch in südlicher Richtung abwechselnd geweidet, bebaut und unter Gestrüpp gehalten wurde, wobei jedoch im Süden die Landschaftsentwicklung umgekehrt verlief. Das jetzt 'West Woods' genannte Land von Fyfield und Overton, das dauerhaften Waldbestand aufweist, erhielt seine Form erst durch die Rodung seines Umlands, und ist zu einem besonderen Merkmal und Hauptbestandteil der lokalen Landschaft der letzten 4,000 Jahre geworden. Dieses Land änderte trotz langfristiger und wahrscheinlich sogar kontinuierlicher Nutzung während dieser Zeit wenigstens nicht in signifikanter Weise seine Lage, Gestalt oder Größe.

Im Gegensatz dazu weist das Hügelland eine wechselvolle Geschichte auf, obwohl prähistorisch, ist der größte Teil davon nicht so alt ist wie das Waldland. Der größte Teil des gegenwärtigen Graslands und zumindest einige der jetzt kultivierten Gebiete im soliden Ackerland gehörten im mittleren/späten zweiten Jahrtausend BC zu einer aus eingefriedeten Feldern, Weiden, Begräbnis/Ahnen Land, Pfaden und Fahrwegen organisierten, axiometrischen Landschaft; eine Hügellandschaft, die sonderbarerweise bis zum Auftauchen kreisförmiger Siedlungen mit Graben- und Dammanlagen in der ersten Hälfte des ersten Jahrtausends BC keine Siedlungen aufweist. Danach wurde vor allem das höher gelegene Hügelland zu nicht umschlossenen Weiden, und blieb es auch, was eine Landnutzung darstellt, die nur durch kurze Phasen räumlich begrenzter Kultivierung im ersten–zweiten, vierten, zehnten, dreizehnten–vierzehnten, neunzehnten und von der Mitte bis zum späten zwanzigsten Jahrhundert unterbrochen wurde. Sie weisen seit ihrer dauerhaften Einrichtung vor zweieinhalb tausend Jahren, besonders aber im ersten, vierten–fünften, dreizehnten–vierzehnten, sechzehnten und neunzehnten Jahrhunderten, wie das Waldland nur eine sprunghafte und in der auf das Tal bezogenen Siedlungsstruktur geringfügige Besiedlung auf, die seit den ersten Jahrhunderten AD trotz örtlicher und ökonomischer Variationen vorherrschte.

Die beiden Pfarrbezirke haben im allgemeinen immer eine ansässige Bevölkerung von unter 1,000 Menschen ernährt. Obwohl beide Bezirke im Wesentlichen autark waren, dienten sie dennoch in charakteristischer Weise den Interessen abwesender und ferner Grundeigentümer. Es ist eine Landschaft der Ausbeutung, und dies wird durch das Fehlen jeglicher Struktur mit der Ausnahme von East Wansdyke (das in diesen Pfarrbezirken unvollendet war) angedeutet. In einer hauptsächlich auf Landwirtschaft fußenden Wirtschaftsweise und Landschaft, in der sich vor ungefähr 6,000 Jahren Bauerngesellschaften entwickelten, sind die örtlichen Naturschätze wie Kreide, Lehm, Sarsen, Holz, Wasser, Boden und Gras verschiedenartig genutzt worden. Dieses Gebiet war selten isoliert und immer offen für die Aufnahme von z.B. religiösen und architektonischen Einflüssen, Haus- und Haushaltsmaterial und Artefakten wie Keramik, Metallgeräte und Glas.

In den Pfarrbezirken haben wir jedoch eher Orte der Durchreise vor uns als einen Ort, an dem sich Menschen niederlassen. Sie liegen an einer natürlichen, inselartigen Straßenkreuzung, an der eine Hauptroute Englands in West–Ost Richtung eine traditionelle Nord–Süd Route zwischen dem Englischen Kanal und dem Britischen Hinterland trifft. Erstere verlief ursprünglich im dritten Jahrtausend BC wahrscheinlich durch das noch nicht entwickelte Hügelland durch Avebury, und wurde später verschiedenartig als gebührenpflichtige Schnellstraße in Römischer Zeit, und als Fernstraße entlang des Kennet Tals im 20. Jahrhundert formalisiert; im Gegensatz dazu wurde die letztere Straße, die durch die Täler und entlang der Kämme des Wessex Kreidelands verlief, niemals zu einer beschotterten Durchgangsstraße, sondern führte seit den letzten prähistorischen Jahrhunderten ein unbeständiges Dasein für den lokalen Verkehr und Transhumanz.

Übersetzung: Peter Biehl

Part I

The Evidence and its Assessment

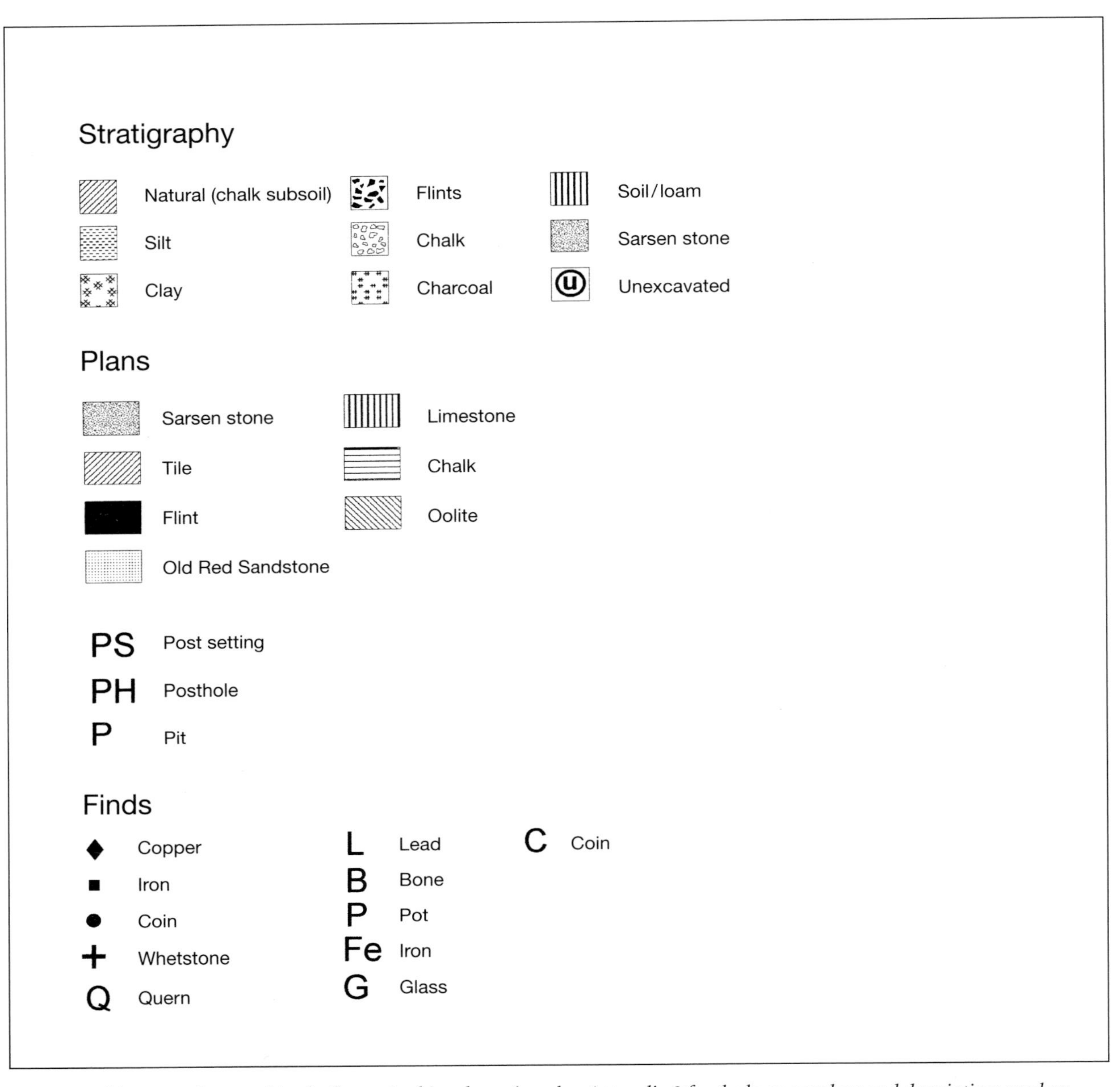

Key to graphic conventions used in the figures in this volume (see also *Appendix 2 for the layer numbers and descriptions used on excavation drawings)*

Chapter 1

A Landscape and its Setting

The facts of topography, soils and climate explain much, but beyond them lie purely historical facts like the laws of property and inheritance.

Hoskins 1955, 114

Fyfield and West Overton are two unremarkable villages in Wiltshire, roughly in the centre of southern England (Figure 1.1, 1–2). They lie beside a river called Kennet between the great prehistoric temple of Avebury, 4km to the west, and the medieval town of Marlborough 5km to the east (Figure 1.1, 3). Their area is pleasant, their history almost totally undistinguished. Not much of any significance ever seems to have happened there. No famous person has been born, lived or done anything there. The villages contain no major house or other architectural monument. Wansdyke is the only archaeological monument of individual distinction and it passes through many other parishes as well.

The archaeology of Fyfield and Overton might well appear to be limited to a famous air photograph of an ancient field system (*see* frontispiece) and one or two academically quite well-known sites and objects. Several relevant sites lie just beyond the parochial boundaries. The Late Neolithic/Early Bronze Age complex centred on 'The Sanctuary' on Overton (or Seven Barrow) Hill, presumably ceremonial, lies immediately over Overton's western boundary (Pollard 1992). Well-known megalithic tombs lie to the west – East and West Kennet long barrows (Piggott 1963) – and, immediately east of the Fyfield boundary, at Devil's Den. Many round barrows, mainly of the second millennium BC but including three of Roman date, dot the parochial lands, especially the downs, but nowhere is there a causewayed enclosure, a hillfort, an accredited Roman villa, a castle, a moated site or a religious house. What indeed is there to study?

The countryside hereabouts, not individual sites, is the subject of this volume. It comprises a deeply agrarian landscape witnessing the silent, largely anonymous doings of countless men and women over thousands of years. The great Wiltshire field archaeologist and local historian, Sir Richard Colt Hoare, recognised this some two centuries ago when he too set out to explore this self-same area: 'To whatever point we direct our steps, in the neighbourhood of Marlborough, we shall find objects either of British or Roman antiquity, to instigate our spirit of research, and to attract our inquiry' (1821, pt 2, 13, Iter III). We were not particularly searching for 'objects' and we are as much concerned with the most recent millennium and a half as with Roman and prehistoric times, but we were similarly motivated and the neighbourhood always led us on.

This volume is based on a historical study, periodic rather than continuous, so far spanning the years 1959 to 1998. The area studied encompasses the whole of the civil parishes of Fyfield and (now West) Overton. Together, Fyfield and Overton are of some 2,834ha (7,000 acres), disposed across the landscape roughly in the shape of a right-angled triangle (Figure 1.2). The 'upright' is on the west, now – and for how long? – the north–south line of The Ridgeway (Plate I). Their eastern boundaries stretch from adjacent north points high on the Marlborough Downs southwards for some 5km to the western edges of Savernake Forest. Both parishes are the subject of good modern historical summaries in volume XI of the *Victoria County History of Wiltshire (VCH)* (*see also* especially vols IV and X). We do not repeat those histories, not least because our approach is different. We nevertheless draw on them as suits our purposes, and follow them in one important

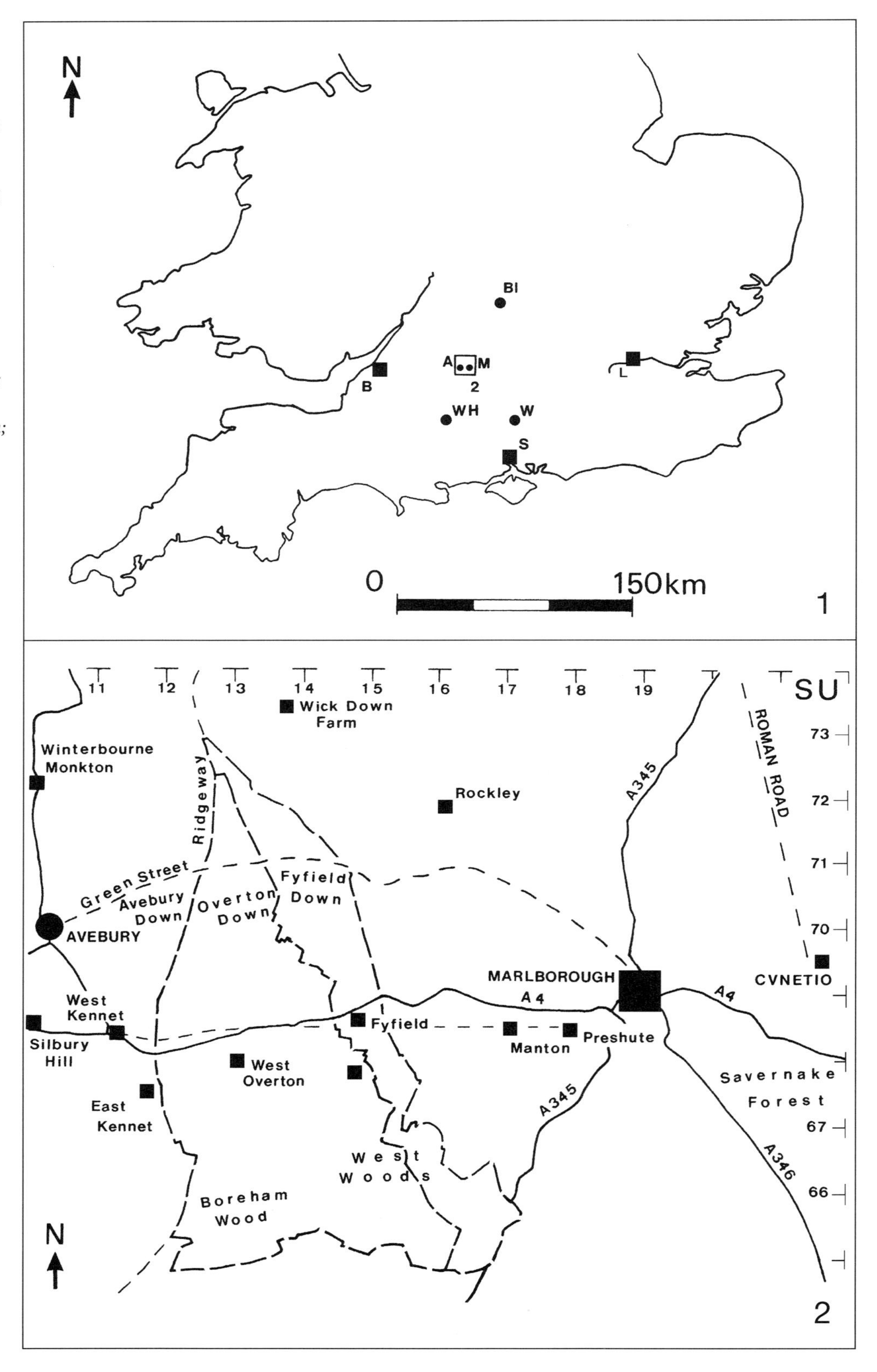

1.1 Location map: the study area in its English context (1 and 2), and West Overton and Fyfield in their parochial and topographical contexts on the Marlborough Downs (3, facing page)

B = Bristol; L = London; S = Southampton; A = Avebury; M = Marlborough; Bl = Blenheim; W = Winchester; WH = Wilton House

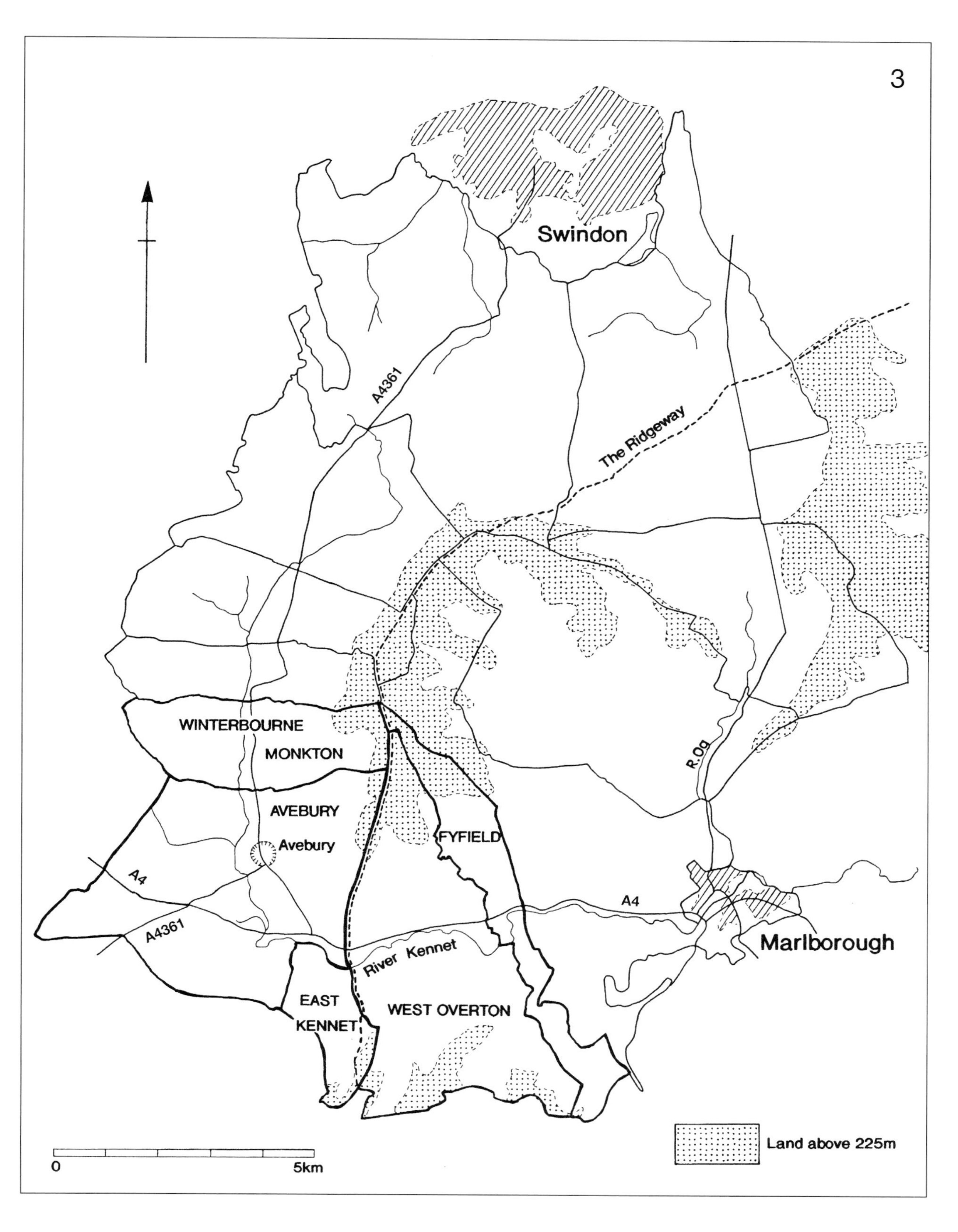
3
Swindon
A4361
The Ridgeway
R.Og
WINTERBOURNE
MONKTON
AVEBURY
Avebury
FYFIELD
A4
A4
A4361
River Kennet
Marlborough
EAST
KENNET
WEST OVERTON
0
5km
Land above 225m

Plate I *Crocker's 1821 view from near Wansdyke north along an unfenced Ridgeway across the Kennet valley to an unenclosed Overton Hill and Hackpen Down (from Hoare 1821)*

respect regarding nomenclature: 'West Overton' was the name of the new civil parish defined, as with Fyfield, in the later nineteenth century; the name had already been used of the 'poor-law parish' in the early nineteenth century (*VCH* XI, 181). Before then, however, there were two ecclesiastical estates, West and East Overton, each with its own tenurial history. We refer to them as appropriate, and use the single name 'Overton' (rather than 'West Overton') as a general descriptor of the area of the modern civil parish. Our main reason for so doing is that most of the modern West Overton consists of the historical East Overton. This can be confusing.

The Fyfield and Overton area occupies the south-western corner of the Marlborough Downs (Figure 1.3). These downs embrace *c* 50km^2 of rolling countryside cut to the south by the Kennet and the east by the River Og (Figure 1.1, 3), and marked to the north and west by prominent escarpments overlooking the geologically more complex area of the 'M4 corridor'. South of the Kennet the land rises through the deciduous and coniferous woodlands of West Woods (Figure 1.1, 2) to another escarpment, this time dropping steeply into the trench of the Vale of Pewsey, with Salisbury Plain beyond (Figure 1.3). Otherwise, the slopes are shallow and convex, rising to heights of just under 275m (900ft) above sea level. The underlying geology is a combination of Upper, Middle and Lower Chalk capped by rendzina soils and, in places, Clay-with-Flints. The valleys are a more complex mix of colluvial, alluvial and undisturbed deposits, often preserving beneath them a diverse and largely unexplored archaeology (Powell *et al* 1996; Evans *et al* 1993). The effect of time upon this geology has evolved a number of distinct landscape zones. These include the high plateau areas, asymmetrical valleys with dry, relatively flat bottoms, and broad valley floors. The overall visual effect is of a flowing, 'easy' landscape, which nevertheless peaks in relatively rough and marginal uplands (Plate II).

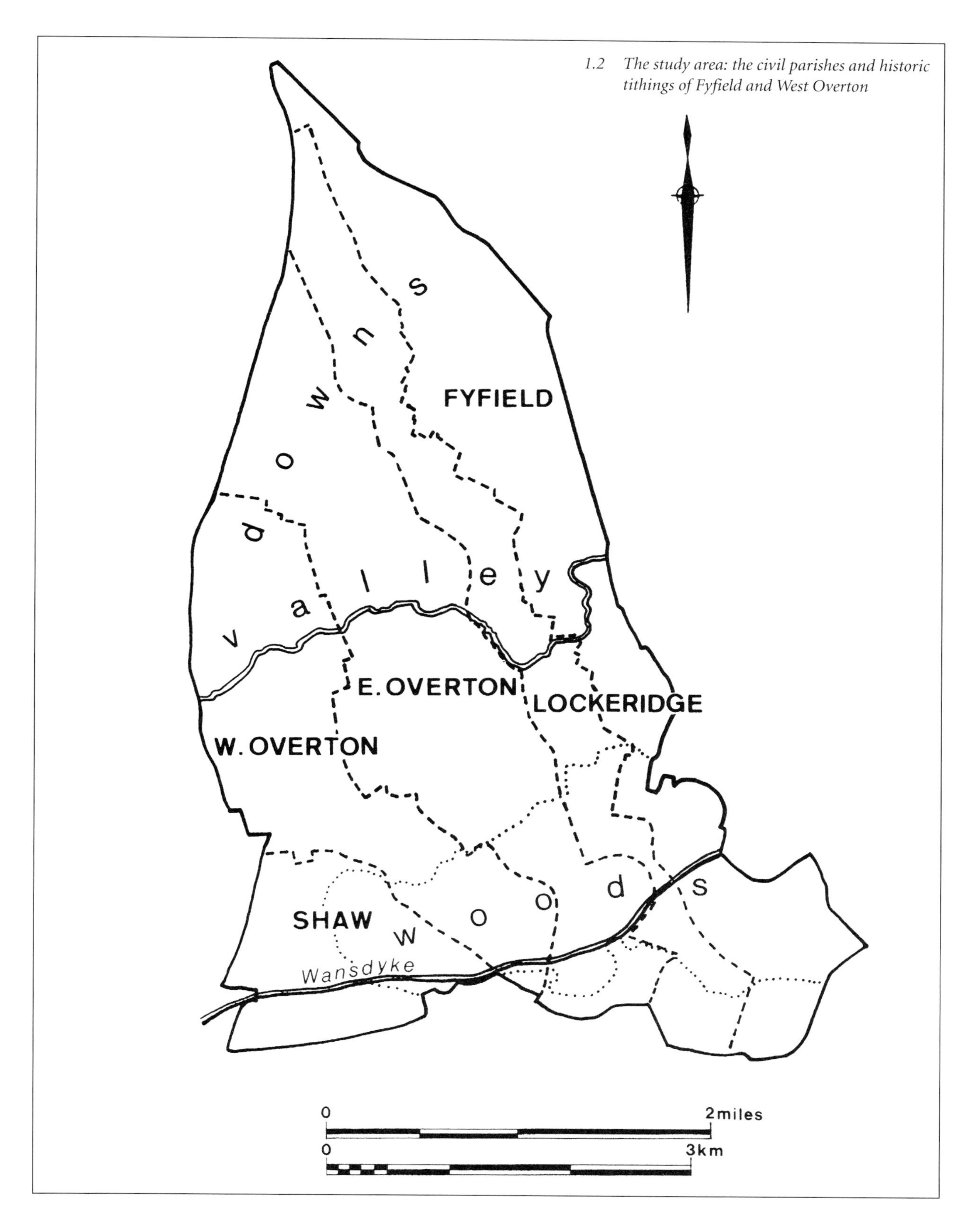

1.2 The study area: the civil parishes and historic tithings of Fyfield and West Overton

Both sides of the Kennet valley are broken by a number of dry valleys, markedly so on the north where Clatford Bottom and Pickledean (also known as Piggledean) incline north west and gently uphill far into the almost treeless downland. In contrast, their counterparts to the south reach through north-facing slopes up towards the remnants of Savernake Forest. Through and above the main Holocene deposits, on the downs, in the valley bottom, in the denes and in the woods, lies a spread of broken sarsen stone (Plate III). This is a material resulting from the breaking up of a much earlier (Tertiary) sandy crust that lay above the surviving land mass. The result was the deposition of the silcrete boulders across the reformed landscape. Such blocks of sandstone were incorporated into the great and well-known Neolithic–Early Bronze Age monuments at Avebury and Stonehenge, and it is often remarked that Fyfield and Overton could well have been the area from where such stones came. It is indeed a possibility, and who is to deny the place one of its rare brushes with great events beyond?

As is almost certainly the case down the Avon valley and around Stonehenge, the extent of the sarsen cover in the Fyfield/Overton area has been reduced drastically by local quarrying and use, not entirely locally, of the stone as building material. This has occurred, not continuously but continually, from at least the fourth millennium BC until early in the twentieth century (Plate IV; King 1968). Some areas still contain concentrations of sarsens, most notably in the Valley of Stones (Plate III) at the core of the Fyfield Down National Nature Reserve. The best of all survivals is a small area of densely close, and often large, sarsens in and on the north side of Delling Wood as the dry valley begins to run out between the north end of Overton Down and Totterdown (Figure 5.1, below). There, one can still enjoy the once-common experience (*see* opening quotation, Chapter 7) of walking across the

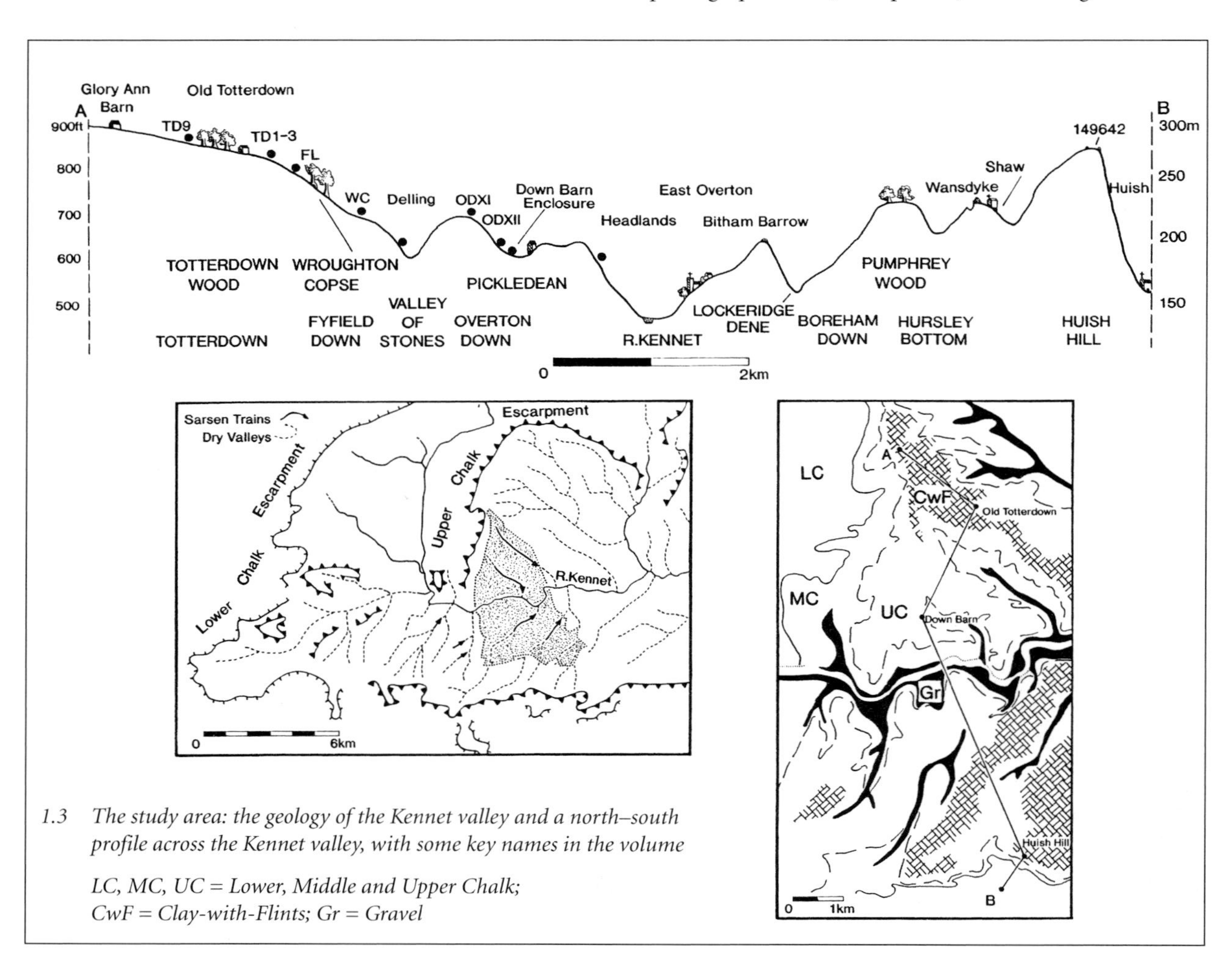

1.3 The study area: the geology of the Kennet valley and a north–south profile across the Kennet valley, with some key names in the volume

LC, MC, UC = Lower, Middle and Upper Chalk;
CwF = Clay-with-Flints; Gr = Gravel

Plate II The view south from the top of Fyfield Down towards West Woods and Lurkeley Hill

Plate III Valley of Stones showing sarsen train, looking north west to Delling Wood

Plate IV Sarsen stone split into blocks as if for removal, behind Delling, Fyfield Down

Plate V View within West Woods, with a typical boundary earthwork

landscape without touching the ground. Two National Trust properties, respectively much visited south of the Kennet in Lockeridge Dene and hidden at the south end of Pickledean, were acquired locally precisely because of the survival of the stones. Other good sarsen 'trains' still exist further south in West Woods. Individual stones have been variously used or moved or are *in situ* (eg, Plates XVII, XXIII and XLVIII; Figures 6.12 and 9.1).

The present downland landscape is visually dominated by openness – vast skies and large expanses of arable fields apparently devoid of boundaries and distinguishing marks. The occasional round barrow can be distinguished as a skyline hump or a small island of trees within this sea of arable. The core of the study area on the downs, hidden away from the passer-through and off-limits to vehicles, is in stark contrast to this cultivated zone. Fyfield and Overton Downs are now largely a National Nature Reserve; Overton Down, but not Fyfield Down (*contra* English Heritage 1998), forms a significant part of the Avebury World Heritage Site (*see below*, Chapter 17, for discussion). Both designations originally stemmed from the outstanding survival of spreads of sarsen stones. Now this downland is also one of the largest remaining tracts of 'old' chalk grassland in southern England outside Salisbury Plain. An archaeology is visibly preserved as earthworks, disturbed only by the grazing of sheep and rabbits and by racehorse training. Here, lines of sarsens, lynchets, banks, ditches, tracks and settlement enclosures witness former human action. Trees still exist in small pockets of woodland and along a few old hedgerows, but they stand out significantly because of their relative scarcity in an open landscape.

Southwards and across the Kennet, woodland is dominant in the equivalent zone in the landscape (Plate V). Modern commercial softwoods march with older deciduous plantations in a mosaic further enhanced by clearings documented since early medieval times and doubtless sometimes of earlier date. Some woodland is a recolonisation of such clearings, even former arable (Chapter 12). Permanent arable reaches up to these West Woods from Lockeridge and Overton villages, with diversity in land-use as well as land-history being maintained by two marked contrasts immediately beyond the woods. Southwards is a plateau of historically recent arable on Clay-with-Flints, historically 'heath', while adjacent to the south west is botanically rich old grassland (also a National Nature Reserve) on and around the peaks of Golden Ball Hill, Knap Hill, with its causewayed enclosure, and Walker's Hill, with Adam's Grave long barrow.

In summary, topographical/land-use zones within the study area can be identified as:

High (*c* 250m aOD): old grassland/woodland with sarsens on Upper Chalk and Clay-with-Flints (Plates II and XIX)
Medium altitude: permanent grassland with sarsens mainly on Upper Chalk (Plates VIII and XXIII)
Interface: area of old grassland and permanent arable around a sarsen-littered dene (Plates III and XXIX)
Valley slope: permanent arable on Middle Chalk (Plate XV)
Bottomlands: with permanent historic settlements, meadow and water-management (Plates XLVII and LXIII)

Each of these zones is represented in a series of examinations conducted in Part II, Chapters 4 to 13. There will be found key details, summaries and integrated results of the project's investigations, each chapter concerned with one small area and each complementing the other as we move towards more general issues in Part III. The presentation of this work in this way is dependent on the existence and availability of the archive (*see* Appendix 1).

The changing form of the landscape is directly reflected in contrasting archaeologies. The long, narrow nature of the parishes ensured that all of the locally available varieties of resources have been included in the strategies of exploitation used by people formerly living in them. Whether or not the parishes emerged to meet such a need on a rational, equitable basis, or have other origins, is a matter to which we return in Chapter 16. That the parishes reflect human uses of the landscape, natural and increasingly acculturated, we can see in the evidence of documents and old maps as well as in the archaeology on and in the ground (Chapter 3). Such sources are independently valid but from time to time they happily merge, particularly over specifics like a boundary (Chapter 3: Plate X) or a track (Plate VI). The sheer amount of archaeology present in these two otherwise fairly anonymous parishes would doubtless have surprised earlier generations but our belief is that they are not in fact remarkable, except perhaps in the good archaeological preservation of some of their landscape and our own long contemplation of it all.

RESEARCH BACKGROUND: A SUMMARY

Research into the archaeology of the Marlborough Downs has been dominated by the position of Avebury at their western foot. Few involved with that great

Plate VI Hollow-ways of the Old Bath Road (Green Street) climbing Overton Down towards Avebury (1959)

monument have, however, lifted their eyes to the hills to the east – Avebury, Overton, Fyfield and Manton Downs – though many have of course looked south to Overton Hill and 'The Sanctuary', West Kennet long barrow and Silbury Hill; west to Avebury Trusloe, South Street and the Beckhampton 'Avenue'; and north to Windmill Hill (Ashbee *et al* 1979; Burl 1979; Crawford and Keiller 1928; Cunnington 1931; Piggott 1963; Pollard 1992; Smith 1965; Ucko *et al* 1991; Whittle 1993; 1994; 1997; Whittle and Thomas 1986). Previously, as mentioned above, Colt Hoare had cast his extraordinarily percipient eye over our study area. He rode across and commented on Fyfield Down, causing a small excavation to be executed one morning at Rowden Mead into what is almost certainly our site 'Wroughton Copse' (1821, 45; Chapter 7; FWP 65). Over sixty years passed before the next systematic study (Smith 1885). Crawford (1922) followed up with some typically casual, perceptive primary field observations and comment but no further systematic work was published until Piggott (1942; 1950).

The Piggott model was very much in mind at the start of this project in 1959. It was still a major stimulant when Gingell, a site supervisor on OD XI in the 1960s (Chapter 6), began his own project on the Marlborough Downs in the 1970s. Gingell's work (1992) produced a substantial body of primary evidence from the downs which bears directly on this report. A more recent long-term project in the Kennet valley (Evans *et al* 1993) is even more important in the sense that its *locus* was in the valley and not on the downs. Furthermore, it was led by palaeo-environmental considerations, not archaeological sites, acquiring its data from non-monumental contexts and attempting to place them in a broad temporal context. It did not seek, however, seriously to consider downland matters, environmental or archaeological. In this it perhaps unconsciously perpetuated the tradition that, because the archaeologies of downland and bottomland are different in their appearance and method of record and acquisition, the two *locales* are functionally as well as environmentally separate.

Shades of the same dichotomy appear in another major study reporting work over the same generation and also centred on the bottomlands, notably Silbury Hill but also on other major monuments which are not quite so visible in today's landscape (Whittle 1997). The opportunity is therefore taken here to attempt to relate in some respects two approaches and two different aspects of the same area at the south-west corner of the Marlborough Downs (Chapters 14 to 16).

THE FYFIELD AND OVERTON PROJECT

The project developed from that earlier work, of course, but from its inception it tried to move away from individual monuments, and from big ones in particular. The immediate inspiration behind this approach was Collin Bowen, but just behind him were the influential writings of Crawford (especially 1953), the teaching and books of Hoskins (especially 1955), and, from two generations earlier, the combination of excavation with fieldwork which had enabled Pitt Rivers (1887–98) to extract knowledge from lowly humps and bumps on Cranborne Chase. The project deliberately set out to explore the potential of the ordinary, the inconspicuous, the accidental, the plentiful, the common, the inconsequential – fields, lynchets, tracks, boundaries, settlements, even archaeologically empty spaces – rather than those archaeologically ever-attractive funerary, ceremonial and military monuments (cf this author on archaeology in the intellectual climate at the time when the Experimental Earthworks Project also began; Bell *et al* 1996, xxiv–xxvi).

From such specifics, the project sought to move into considerations of landscape and of process in the two dimensions of local space and long-term time. As it happens, time through a lifetime has now lent a certain understanding to the view, not least in seeing the effects of the accumulation of many small landscape changes over a generation. The approach has of course subsequently become routine, and indeed this project has been overtaken methodologically. It became weak as it progressed because its resources did not provide for the adequate publication of the excavations, which were an integral part of its methodology in acquiring primary evidence. It remains weak here in its lack of radiocarbon estimates, its non-use of statistical methods and Geographical Information System (GIS) technology, and in the poverty of its palaeo-environmental evidence. On the other hand, especially in comparison with earlier work in Wessex, it employed not only wide spatial and temporal frameworks but also a range of methodologies and types of evidence, some pushed further than others. If the project has strengths, they lie in the quality of its fieldwork and closely associated use of air photographic interpretation and cartography, both for their own sake as lines of enquiry and also in setting the agenda for excavation; in its use of excavation as a question-specific tool; and in its attempts to integrate, during investigation and now, its interpretations with several different strands of evidence such as those from natural history and documentary sources.

In 1959 the project's initial aim was to investigate the history of the landscape of Overton and Fyfield Downs. This aim has been continually refined and expanded, not only in terms of area but also to allow for changing theoretical frameworks and the quality and quantity of the useable data-set. A concept of 'the history of the downland landscape' quickly became 'and the river valley'; whereupon it was realised, of course, that neither was intrinsically more significant historically than the wooded high ground to the south, and that none of the triad could be understood functionally without the others. The result has been to suggest an outline landscape history through study of the changing prehistoric and historic dimensions within the study area. This has identified the importance of change through time as an influence on past societies; it has also recognised the importance of change through time on the evolution of this landscape and on different spatial structures making up the various landscapes at different times (Chapter 16). We have come to realise the importance of not just how we see these landscapes but rather of how people in the past saw their landscapes – a sort of palaeo-phenomenology, to adapt a fashionable term of the 1990s (Tilley 1994).

By moving on early from a contemplation of the downs alone to an embrace of the whole of the parochial territories, the project took account of how the range of resources disposed non-randomly across the landscape was exploited in the past. It also simultaneously related itself to units of land and processes within them that were 'real' rather than theoretical or evidenced merely as the result of archaeological survival. 'Real', deliberately used as a dangerous word, here of course means that the parochial units provided a spatial framework for many people living and working in Fyfield and Overton in the past, certainly from over a thousand years ago and arguably longer. It is therefore believed that some of their activities and the processes in which they were involved can be usefully, though not exclusively, studied in relation to that framework.

Overall, early acquaintance with the area suggested that quite long periods of time were not represented on the downs by any evidence on the ground, ie, by earthworks. This led to the suggestion of breaks in the continuity of local land-use (Fowler 1975a, 121). Given the project's objectives, it also emphasised the importance of looking at as wide a range of evidence as possible and not just relying on either archaeological evidence or archaeological approaches. Place- and field-names, for example, and field botany, were sources of evidence brought to bear in the Overton/Fyfield study area in the 1960s and now routinely invoked (Figure 1.4). The issue of continuity itself needed to be defined first and then examined; now it needs to be re-examined carefully (Chapters 15 and 16). It appears at the moment that continuity on the downs, from the Neolithic onwards, is in only general terms of community or agrarian regime and that any unbroken occupation of one site is doubtful (ibid, 123). It is possible, therefore, that ruptures and discontinuities are more characteristic than continuities in several lines of enquiry, and some points are further discussed in Chapter 16.

Part III of this volume should, conventionally, present a conclusion, and it does indeed offer responses to the questions above (though not always directly); but we do not believe any specific conclusion is either desirable or likely to be correct in any absolute or permanent sense. Ideological doubts apart, for one thing the study has hardly begun. Part III is, in any case, not so much about conclusions as about interpretations, and interpretation as a dialogue between peoples, place and time. It is about ideas: about archaeology and research, about the place and history, and about abstractions such as resource, stewardship and World Heritage. It might give pause for thought, about their commonality for example, that the study of a parish pump ends with such a global, and noble, concept.

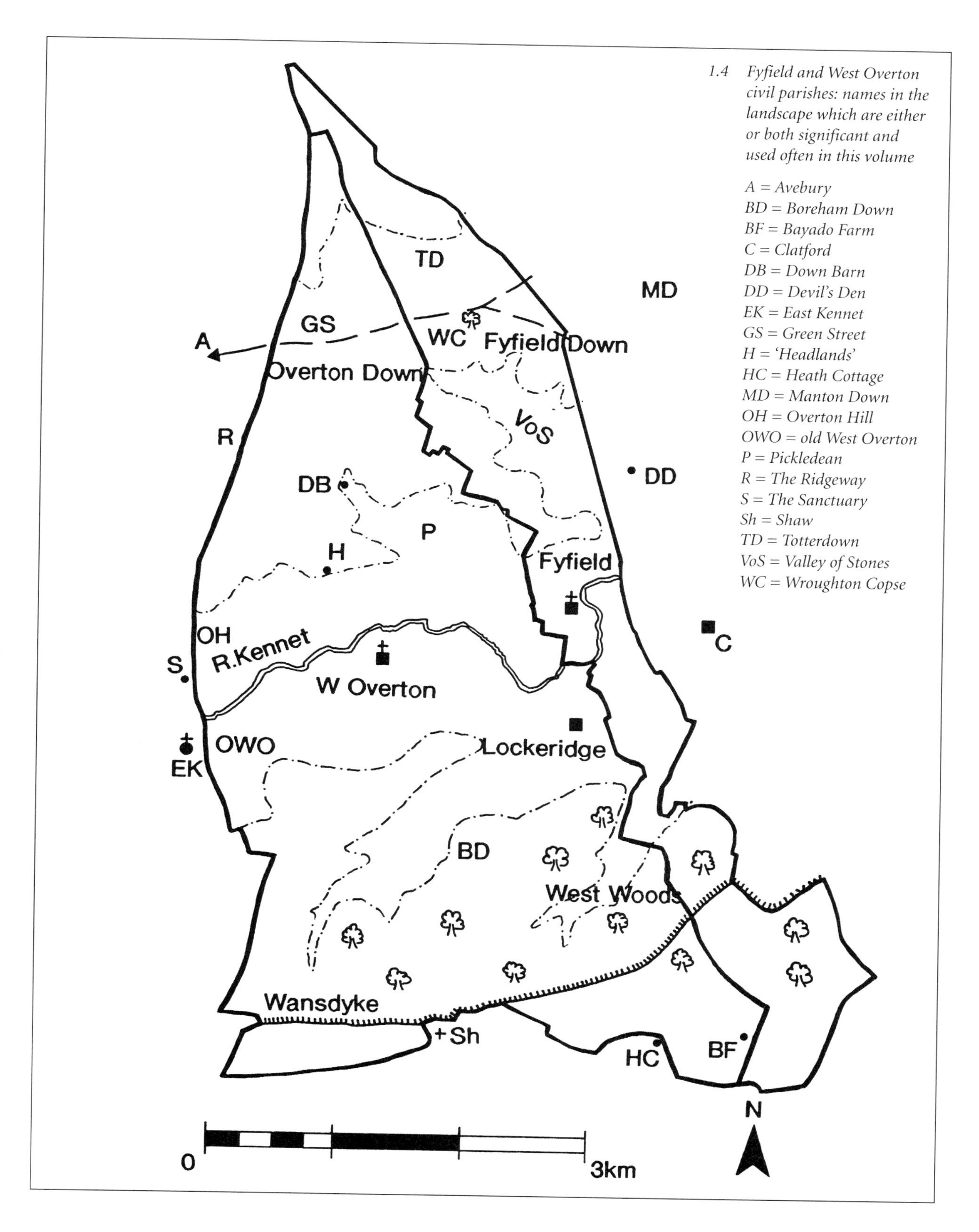

1.4 Fyfield and West Overton civil parishes: names in the landscape which are either or both significant and used often in this volume

A = Avebury
BD = Boreham Down
BF = Bayado Farm
C = Clatford
DB = Down Barn
DD = Devil's Den
EK = East Kennet
GS = Green Street
H = 'Headlands'
HC = Heath Cottage
MD = Manton Down
OH = Overton Hill
OWO = old West Overton
P = Pickledean
R = The Ridgeway
S = The Sanctuary
Sh = Shaw
TD = Totterdown
VoS = Valley of Stones
WC = Wroughton Copse

Chapter 2

Aerial Photography and Cartography

To complement the field investigations, the Royal Commission on the Historical Monuments of England (RCHME; now English Heritage) was commissioned to produce an archaeological air photographic transcript of Fyfield and Overton Downs at a scale of 1:10,000. Created with great interpretative and technical skill by Cathy Stoertz, the resulting map incorporated all available oblique and vertical air cover, exemplified by Plates VII and VIII, up to and including June 1995. The map (Figure 2.1) represents the final investigative component of the project and provides the broader landscape context.

The transcription is published here at a scale of 1:20,000, as received in digital form from the RCHME in a slightly amended version in 2000. It is overlain on the relevant OS base map (Figure 2.1). The cartography is also used as a base to show simplified outline field groups and other land divisions, the distribution of ridge-and-furrow and the spatial relationship of ancient fields and barrow groups (Figures 2.2 to 2.4). The Ridgeway (Plate I), Green Street (Plate VI), the present A4 road and the four existing woods on the downs (without prejudice as to their historicity) are also depicted. The map is an eloquent statement of the field archaeology of the project area in its own right and this chapter merely adds a guiding commentary to points of special interest and importance to the project.

The following text is our own analysis and interpretation based on the RCHME transcription and our knowledge of the mapped area. Seven areas within this map are further examined, at a greater level of detail, in Chapters 4 to 7. The full RCHME account is deposited in the project archive.

CARTOGRAPHIC BOUNDARIES

The aim of the air photographic cartography was to place the well-preserved field systems and other remains on the old grassland of the project core in their wider archaeological context. It was anticipated that it would be possible to build up links between earthworks east of The Ridgeway and cropmarks in the more or less continuous plough-zone to the west. This was achieved and a continuum of evidence, irrespective of its visibility on the ground, can now been seen. This extends from the foot of the western scarp, across The Ridgeway and over the downs beyond the eastern parish boundary of Fyfield to Clatford and Manton Downs and into Clatford Bottom (Valley of Stones) in Preshute parish.

The northernmost point technically required of the map was at the kink in The Ridgeway marking the northern tip of Fyfield parish (Figure 2.1). But an extension northwards was made to include Wick Farm and its surrounding earthworks to highlight the significance of this isolated area in relation to Lockeridge and the Templars (Chapter 10). The cartographic 'blanks' to east and west of Wick Farm merely represent areas not examined. North of Totterdown Wood the high land is capped with Clay-with-Flints and therefore unlikely to be particularly revealing of cropmarks. Equally, such a subsoil may well have inhibited activity there in the past though, at an altitude of 250m+ and exposed, the land's constraint now is the 'chill factor'.

The southern boundary of the map was quite deliberately drawn at the modern A4 road where it runs along the Kennet valley. This line marks a general change in the geomorphology and land-use, from arable

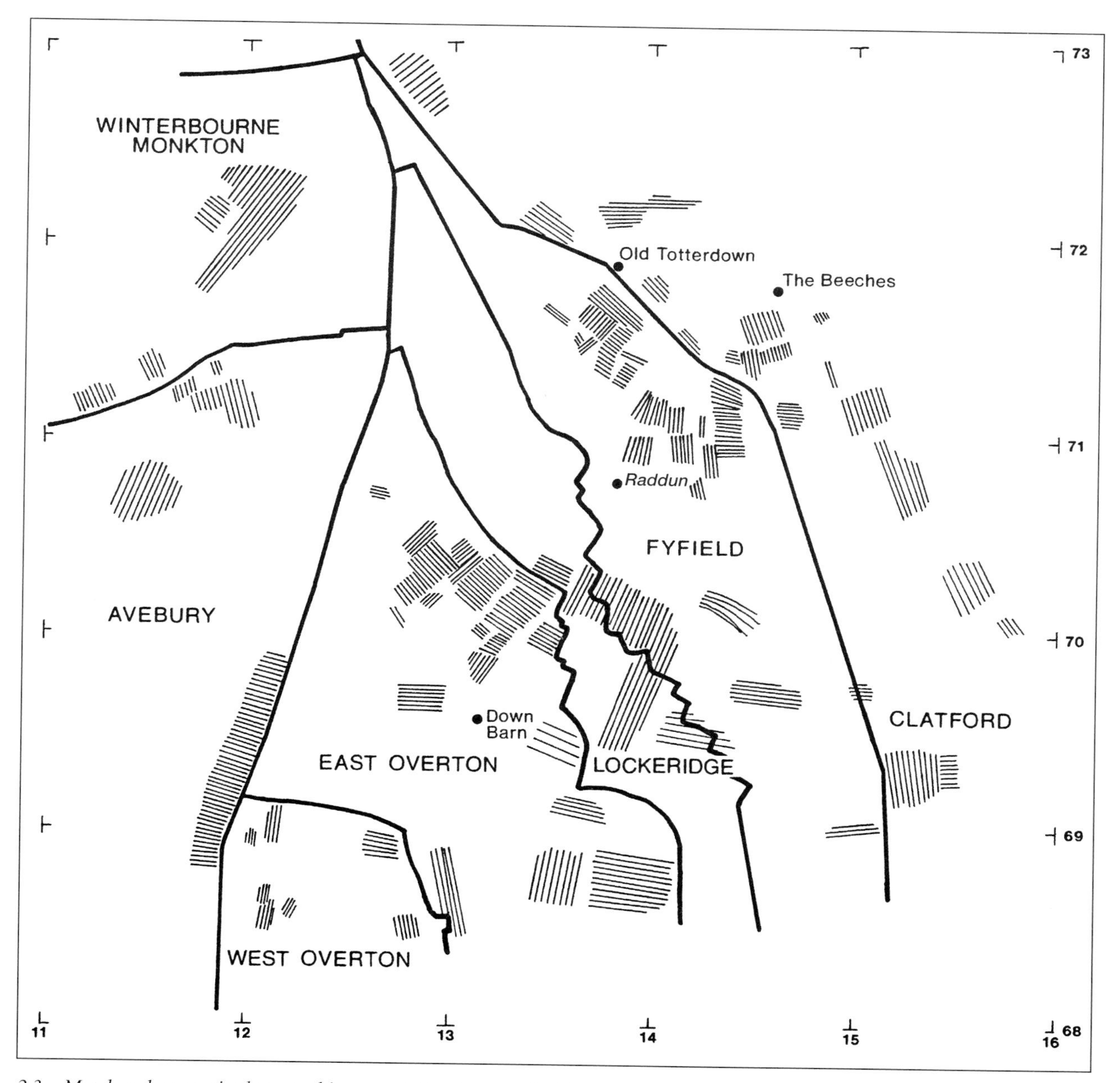

2.3 Map based on an air photographic transcription by the RCHME of Avebury, Overton, Fyfield and Manton Downs showing the incidence of ridge-and-furrow, here plotted against the tithing boundaries of West and East Overton, Lockeridge and Fyfield

on its north side to pasture over colluvium and alluvium to the south on the valley bottom. The modern valley surface is relatively recent (post-medieval) as well as unresponsive in air photographic terms.

THE ARCHAEOLOGY AS SHOWN ON THE AIR PHOTOGRAPHIC MAP

DISTRIBUTION

The overall impression is of an extensive enclosed landscape made up of small parcels, presumably fields (Figure 2.1). This landscape of ancient and fairly comprehensive enclosure stretches north–south for over 4.5km from Monkton Down to Seven Barrow (or Overton) Hill (*see* Figure 2.4) and from Green Street at the foot of the Marlborough Downs 4km eastwards to Clatford Down. The area involved is approximately 18km^2 (*c* 4,500 acres, or 7 sq miles). As recorded, the 'ancient' landscape is not continuous. The triangular-shaped south-east zone, where the linear remains are fragmentary, coincides with the medieval and modern permanent arable of the tithings of East Overton, Lockeridge (in West Overton civil parish), Fyfield and Clatford (Preshute civil parish). This area is a 'zone of destruction' in terms of its pre-medieval earthworks and as such the continuous ploughing has led to it being almost completely devoid of cropmarks (*SL*, figure 30). The cropmarks that do show are almost entirely of ring ditches, presumably of round barrows. Some of the linear features probably result from medieval cultivation so, in terms of the aerial archaeology in relation to earlier patterns, the record is fragmentary.

Despite the localised constraints of medieval and modern ploughing, a core block of parcels ('fields'), 'open' spaces, linear features and funereal monuments can be observed over an area some 14km square. The characteristic features of this zone are now considered on a thematic basis.

THE 'BLANK' AREAS

Three major areas of 'negative' evidence on the transcription appear to be archaeologically valid.

In the south east, two coombes penetrate the downland in a north-westerly direction from the Kennet valley. The westernmost, Pickledean (Figure 2.4), is mostly pasture, still with sarsens and a long history central to the Fyfield and Overton Downs story (Chapters 7 and 11 to 13); barrow groups L and M define the entrance to the more easterly, Clatford Bottom becoming the Valley of Stones (Figure 2.4). Neither of these coombes contains substantial earthworks, although they all have individual features and 'Celtic' fields occur among the sarsen spreads south of Delling Cottage (Figure 7.1). Both coombes have a long history as communication routes and have been extensively quarried for stones.

North of linear Ditch F4 (Figures 2.1 and 5.1, also labelled F4 in Bowen and Fowler 1962, 107) across Lockeridge Down (the northern part of Overton Down, Chapter 5) and Totterdown, an area of Clay-with-Flints, no earthworks exist and no buried archaeology is known. In particular, there are no traces of ancient field systems and the few cropmarks recorded are likely, at least in part, to be of medieval origin. The interpretation of the ditch bounding an arable field system to its south (Figure 5.1, Field Block 7) continues to be preferred. A slight earthwork heading north north east and parallel to the west side of Totterdown Wood was plotted, however, and subsequently found to exist. It is interpreted as the bank of an oval enclosure, overlying Ditch F4 and otherwise represented by a large lynchet, presumably later, curving away to the south and east. It is known locally as 'The Jousting Ground', a popular perception which emphasises its relative size in a complex of low field boundaries (Plate XIX, top left; *SL*, figure 25, top right; FWP 66).

The third significant archaeological 'gap' on the ground is faithfully echoed by the air photography on the grassland between Fyfield and Preshute Downs. The complexity of the former (Chapters 6 and 7, Figure 7.1) is underscored by the absence of evidence for intensive land-use on the latter.

The remaining blank areas are very small and probably reflect hostile land-use or deficient air photographic cover rather than real gaps. The blank area between Delling and Wroughton Copse, for example, is occupied by ancient fields overlain by the Delling enclosure (Figures 7.1 and 7.15), while the one in the middle of Totterdown, around the beech clump, is also largely illusory, though Bowen and Fowler (1962, 101), noting a small uncultivated area here surrounded by fields, are correct. The Overton Down experimental earthwork (SU 13007065) also stands in a small, field-less area (FWP 66).

LINEAR FEATURES: DITCHES AND TRACKS

Visually, the map seems to be held together by a complex of linear features. Essentially, these are of two types, ditches and tracks, although not necessarily as functionally distinct as that statement implies. Ditches were sometimes also used as tracks and much-used tracks tended to become hollowed out and can look like ditches.

Ditches

There are two major linear ditches in the study area. Located in the north-west quadrant, they may have a common point of origin west of the low rise occupied by barrow group E (Figures 2.2 and 2.4), and could define units running from the valley of the Winterbourne to the high downland.

Ditch 1 (Figure 2.2) runs in a north-easterly direction along the north side of a dry coombe, dividing fields on either side, and on to Monkton Down. At the western end two, probably successive, lines are visible. The more southerly arc on the coombe bottom appears to be the earlier. At the south-west end it becomes very pronounced, probably because it has also been used as a track which is referred to in an Anglo-Saxon charter (Chapter 13).

Ditch F4, now ploughed over west of The Ridgeway, runs for at least 3km from its junction with Ditch 1, crossing Lockeridge Down and Totterdown as far as Totterdown Wood from which it emerges as a hollow-way heading south-eastwards (*see also* Figures 16.6 and 16.7; Plates VIII and IX; *SL*, figure 25). It has served as a

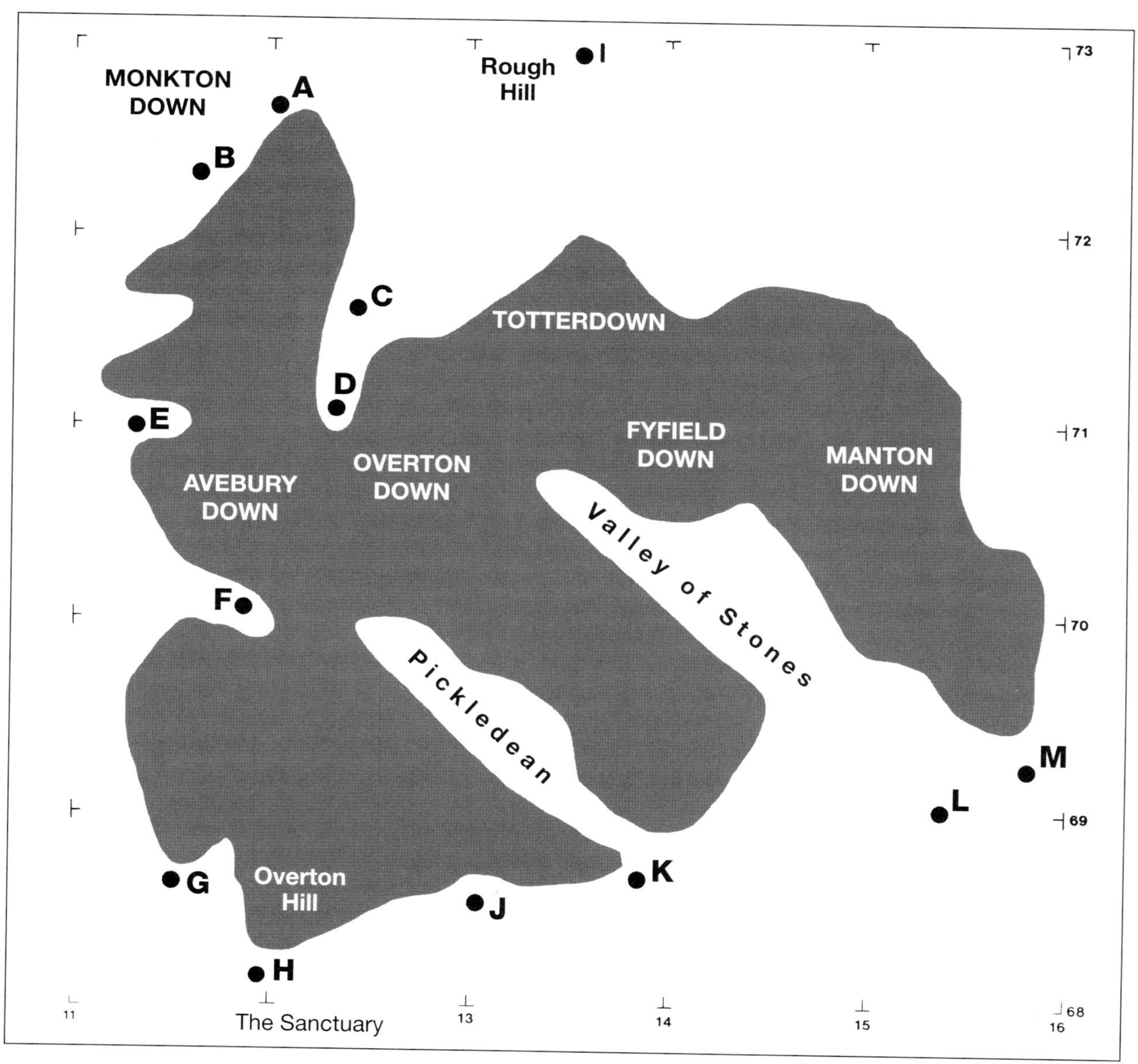

2.4 *Diagram to enumerate the round barrow groups (A–M), illustrate the extent of ancient fields on the downs (shown by the use of tone) and demonstrate the spatial relationship between barrows and fields*

Plate VII Avebury and Overton Down: vertical photograph taken on 1 December 1952, under light snow (NMR 540/958 3158, © Crown copyright/MoD*)*

Plate VIII Overton and Fyfield Downs: overlapping vertical photograph taken on 1 December 1952, under light snow (NMR 540/958 3160, © Crown copyright/MoD*)*

track both at its west end, where it feeds into Ditch 1, and over much of its length east of The Ridgeway. It divides, cuts and is overlain by fields, suggesting a considerable period of use and modification. Immediately east of The Ridgeway the ditch does not relate to any fields at all. On Totterdown, fields are laid off the south side (Group 8) and then a bank interpreted as the arc of an (Early Iron Age) enclosure overlies its line (Figure 5.1).

Tracks

The transcription is laced with man-made tracks of a variety of form, function and date. They are fundamental to any attempt to understand how the various landscapes represented on the map have functioned at different times. The treatment here is broadly chronological, working backwards from the most recent (*see also* Figure 16.6).

The Ridgeway This is the only major track running north–south and is demonstrably later than the landscape(s) it overlies and cuts. The Ridgeway simply could not have existed along its present line when that 'ancient' landscape was in use. In its enclosed form, as shown here, the route is very late (Chapter 15). In its unenclosed form it veered west and east of its present line, with earthworks of wheel ruts and hollow-ways still visible either side of its present boundaries. These are especially clear close to the junction with Green Street (Plate VII, Figure 2.1). The slightly curved line on the map across the angle south west of that junction is also a former course of The Ridgeway, interestingly running along a line of field boundaries belonging to the 'ancient' system.

Although the map does not provide us with an absolute date for The Ridgeway, the *Herepath* of tenth-century land charters follows the same line. When considered with other sources one can postulate a chronological horizon for The Ridgeway as a north–south throughway. It developed between the abandonment of the 'ancient' landscape depicted on the transcription – suggested below to be in the fifth–sixth centuries AD – and the early decades of the tenth century. The point is discussed further below (Chapters 8, 15 and 16, and in *SL* and Fowler 1998. *See also* FWP 30).

The A4 In contrast to The Ridgeway, this line provides for west–east/east–west traffic. It passes along the Kennet valley, preferring the northern edge of the floodplain. Of little local purpose – a network of lanes mostly on the south side of the river does that – its main function was and is as a through-road. It was part of the main mid-twentieth-century main road from London to Bath and the West Country, superseded in the early 1970s by the M4 motorway to the north. It was already 'the Great Road' as early as 1705, according to a Terrier of 11 January in that year, and became an engineered toll road in 1743. Within the study area the A4 lies in part on the line of the Roman road from *Londinium* to *Aquae Sulis* (*see below*, Margary [1967] 53).

Green Street (Old Bath Road) The downland route from Avebury to Marlborough formally ceased to be a through-road at the time of the Enclosure Award in 1815, when the valley toll road superseded it. Although it cuts across the earthworks of the ancient field systems, Green Street (Figure 2.1) does link to one of the 'old' trackways running through the field system immediately south of the junction with The Ridgeway. That Green Street was much used is indicated by an impressive group of hollow-ways on the east-facing slope of Overton Down (Plate VI). The route bifurcates on the eastern incline across Fyfield Down. The two continuations of the route are clearly marked on the ground by rutting and cuts over lynchets, partly picked up on the aerial cartography of east Fyfield, Clatford and Manton Downs. The main route, to the south, trends south-eastwards towards Manton House (SU 157709) and into Barton Coombe and Marlborough. An alternative northerly route, at least latterly called Green Street and in part now surfaced, runs from Delling Copse to Manton Down just east of The Beeches (D) (Figure 5.4) and then south east across Barton Down and Marlborough Common.

Roman road Forming the southern boundary of the transcription (Figures 2.1 and 2.2), the line of the Roman road (Margary 53) either side of The Ridgeway has long been known, surviving into the early 1960s as an unploughed *agger* on Overton Hill (Plate XIII). East of Overton Hill, the line is well established by air photography to North Farm. Here it coincides with the modern A4 for a distance of 500m, probably changing from an *agger* to a stone-revetted causeway across redeposited material on the valley floor (Evans *et al* 1993). Beyond the eastern edge of Figure 2.1 (SU 13856850), the air photographic evidence is ambiguous. The faint traces suggesting that the road continued uphill north of the A4 are unconvincing and were dismissed as the likely line. Far more suggestive was an imaginary line south of the A4 making straight for a large ditch or hollow-way south of Fyfield church (shown as a double broken line on Figure 2.1) which significantly coincided with the boundary between Fyfield and Lockeridge tithings (Chapter 10). A low linear earthwork on this line on the bottomlands between the A4 and the north bank of the River Kennet was dramatically emphasised by a strong parchmark in the dry, hot conditions of August 1996 (*SL*, figure 67). This was sectioned in 1997–8 and shown to cover a complicated sequence of road structures, undoubtedly of Roman date (G Swanton, pers comm).

Otherwise, the landscape is full of shorter stretches of now abandoned trackway, most of which are of considerable antiquity and relate to the prehistoric and Romano-British landscape. Many are integral with, or fit into, the axial field system (*see below*) and are suggestive of a localised 'grid'. For example, a 250m-length of trackway aligned south east–north west running between The Ridgeway and Green Street, *c* 500m south west of their junction, is parallel to a *c* 1km-long stretch of hollow-way heading south east from within Totterdown Wood to Clatford Down, some 2km distant to the north east. Between these two, a similar track (*hric weges*, Figure 6.11) runs the length of Overton Down on approximately the same axis. Starting beside – or coming out of – linear Ditch F4, and following the spine of Overton Down, it passes a Romano-British settlement (B3, Figure 6.1), before turning south west to run into the large Romano-British settlement, ODS (B2, Figure 6.13; Chapter 6). It probably linked with its counterpart on the north of Fyfield Down, passing north east across the Valley of Stones, up the west side of Wroughton Copse and then, still very clear as an earthwork, towards Romano-British settlement B4 (south of the T-junction at SU 141714, Figure 2.1; Fowler 1966, fig 8).

A pair of tracks cut by The Ridgeway appear as cropmarks north of Overton Hill (SU 120690). Both lie essentially west–east, quite markedly different from the others so far noted, even though each one curves slightly to the west north west with the lie of the land west of The Ridgeway. They appear to be associated with an extensive settlement and the more northerly track connects with a bank running east to a ditch associated with an Early Iron Age enclosed settlement, referred to in this volume as 'Headlands' (Chapter 4; Figure 4.2).

Fields

The transcription essentially depicts an ancient landscape, or landscapes, characterised by a network of small, enclosed fields that pre-date the present land divisions. Wherever there is a relationship with broad rig, or ridge-and-furrow (*see below*), the latter is always on top of and/or within the boundaries of the 'ancient' fields.

We are assuming – in the sense that this is not the place to argue the case in detail – that, in general, these fields were cultivated, though clearly at any one time many would have been fallow, pasture or for folding stock. The plentiful evidence of ploughsoils in, and lynchetting at the edges of, the fields and the excavated evidence of ard-marks in at least five different fields on both Totterdown and Overton Down, give credibility to the assumption.

Two analyses are attempted. The first seeks to identify the components of the 'flat' spread of fields across the transcription by defining eleven cohesive groups of fields, here called 'Blocks'. The second analysis, effected quite simply with plan, drawing board and T-square, suggests that axial arrangements underlie the non-random pattern and identifies the remains of three such arrangements. They are placed in a relative and absolute chronology and the three analyses are briefly compared (*see below*).

Field blocks

The eleven blocks are shown on Figure 2.2, with barrow groups shown on Figure 2.4. Whether or not these blocks are accepted, the exercise makes the point that considerable morphological variety exists within the transcription area.

Block 1: large rectangular fields, *c* 180 x 80m, but with subdivisions making smaller units, covering the south/south-west-facing sides and bottom of Monkton Down and coombe in the north west of the study area, generally related to Ditch 1 (*see above*). The northern, north-west, south-west and eastern limits are marked by barrow groups (A, B, E and C; Figure 2.4), with Ditch F4 providing a firm south-western boundary. It may once have extended further to the south east, perhaps as far as the eastern continuation of Ditch F4 towards The Ridgeway, but the mapped evidence suggests that it is overlain in that area by fields of Block 4.

Block 2: on the west-facing slopes of Avebury Down and extending beyond the western limit of the map, these fields are probably the western fringe of an originally more extensive system, subsequently overlain by part of Block 4. The pattern of fields either side of Green Street indicates two phases, with long rectilinear fields similar to those in Block 1 overlying fragmentary smaller ones. The northern boundary is defined by Ditch F4 and barrow group E (Figure 2.4), the southern edge by barrow group F, with both barrow groups lying in uncultivated, 'reserved' areas. To the east, barrow group

D (later incorporated into Block 4) originally bounded Block 2 towards the ridge-top.

Block 3: separated from Block 2 to the north by barrow group F, this Block is bounded to the south west by the extensive West Kennet Farm barrow cemetery (group G). The Overton Hill barrow group (H) may have been the original southern boundary. The western and eastern limits are imprecise, the latter perhaps having been somewhere under the later Block 5. In plan, Block 3 is impressively cohesive, reflecting the topography as it curves on to Overton Hill but ultimately related to a long common boundary curving west across the contours. The rectilinear fields are characteristically *c* 70 x 30/40m, with a suspicion that a unit of measurement of *c* 10m was in use, at least in determining field width. Tracks through the fields give out on to unenclosed land to west and south west.

Block 4: a distinctive group of long, rectilinear fields stretching almost from Monkton Down to central Overton Down and, if Block 5 is included, probably as far south as Overton Hill. The two Blocks, 4 and 5, essentially reflect the grain of the land along the ridge-top. Fields are *c* 70m long but only *c* 15m wide, a length to width ratio of around 1:5, making them morphologically distinct. This distinction conveys the cartographic impression, also gained on the ground, that this block overlies earlier arrangements, specifically field Blocks 1, 2 and 6. Except where it apparently follows the east boundary of Block 1 and respects barrow group C, Block 4 boundaries do not relate to barrow groups. This suggests that the group post-dates the 'barrow horizon'.

Block 5: the field remains are fragmentary but suggest two phases. Block 4 probably stretched north–south over this area but a separate number is given because Block 5 possesses a distinct morphological characteristic of associated settlements, one of which is probably later than Block 4 fields.

Block 6: the main block on Overton Down, stretching 1.5km north west–south east as preserved earthworks and a further 1km as cropmarks in modern arable to the south east. The latter area was permanent arable in medieval times, which partly accounts for the incomplete and strip-like nature of the southern extent of the plot. At least five landscape phases are distinguishable in the transcription: barrows/early fields, settlement, later fields/trackways, settlement and enclosures, and later (medieval) cultivation. Further consideration of this complexity is examined in Chapters 3, 6 and 7, including discussion of six excavations within the Block 6 area.

Block 7: small in area, it is laid off from linear Ditch F4 as it ascends the south-west-facing slope of Totterdown. The fields are *c* 80 x 40m, some of the smallest, complete fields to survive in the study area. The reason for their survival is twofold: they have boundaries of sarsens arranged in substantial lines, if not proper walls (an act which goes beyond mere clearance in an area thick with stones), and there has been no subsequent cultivation here. The north, west and south boundaries of the block are secure; it perhaps extended further east but is now overlain by Block 8 and possibly by Block 9.

Block 8: a morphologically distinct group of small rectilinear fields, integrated with a system of tracks running up and along the slope of Totterdown. Surveyed, excavated and published as 'late first/early second century' (Fowler 1966, 59; Fowler and Evans 1967, 291–2), that interpretation is now reinforced (Chapter 5, Plate XX; FWP 66), though the Block is no longer seen as an isolated patch of Roman arable.

Block 9: the main field block occupying much of Fyfield Down. It is of several phases and displays considerable structural complexity with field Block 8 being imposed upon it. It is bounded on the north west by linear Ditch F4 and, on the north east, by a deep hollowed trackway which may continue the ditch. As this ditch-cum-hollow-way approaches Clatford Down, between Blocks 9 and 10, it conveys an impression of being an established boundary zone – the parish boundary with Preshute lies just to its north – along the high, northern reaches of Fyfield Down. To the east, it incorporates two small, sarsen-littered dry valleys but then simply fades away on an area of open grassland. In contrast, the western and southern limits are marked by the floor of the sarsen-filled Valley of Stones with, on the south east, a pair of barrows, a rarity on Fyfield Down (Plate XXXVIII).

Block 10: facing north east and separated from Block 9 by a hollow-way along the ridge of Fyfield Down. Its other limits are uncertain. Unusually, it also faces north east. Cultivated in medieval times and still under intense cultivation, the block shows well on early post-war air photography. Discounting the medieval over-ploughing, the early fields seem to be rectilinear and large, perhaps as much as *c* 180 x 70m. It may be a northern extension of Block 11 but the cartographic link is tenuous and the latter is best considered separately.

Block 11: now hardly extant, this extensive group, here accurately delineated for the first time (and further delineated in Figure 5.4 and Plate XXII), is the Manton and Clatford Downs equivalent of the large blocks along The Ridgeway and Overton and Fyfield Downs. It stretches north north west–south south east with the grain of the land in a band up to 0.5km wide for almost 2km to a southern boundary in the vicinity of barrow group M. Unlike the blocks on Fyfield and Overton Downs, however, it is now almost entirely under plough. The western edge is delimited variously by a length of ditch, a single barrow, the parish boundary with Fyfield and the eastern rim of Clatford Bottom, below which lies Devil's Den (*see* Figure 5.4), originally a megalithic long barrow. With only minor adjustment, the line of field boundaries south west–north east across the block, on the axis immediately north of the ploughed-out round barrow at SU 14568964, could be made out to be aligned on the long barrow itself. The eastern edge is not well defined, partly due to modern development around Manton House. Manton long barrow, its correct position now relocated, lies near the northern edge (Figure 5.4).

It is possible to show, therefore, that at least eleven blocks of pre-medieval fields exist and, far from being contemporaneous, display considerable complexity and depth. The penultimate major episode is medieval ridge-and-furrow, which provides a *terminus ante quem* (Figure 2.3). Although only relative chronology has so far been observed, the outline of an absolute framework is beginning to be implicit in the relationships to other features, in particular the independently dated constructions such as barrows and the Roman road. Post-medieval land-use is considered elsewhere (Chapters 3, 7 to 11 and 16).

Axial analysis

Two clear orientations are revealed in the relict landscape of Figure 2.1. The first has its long axis aligned north west/south east, with corresponding shorter lines north east/south west (Figure 2.2). It is argued below that a third, less obvious axis is also present. A major inference is that the fields are not deployed across the downs at random but were organised in an axial arrangement as a sustained act of land management. Such observations had been central to the project from the early days when it was observed that the north-east/south-west orientation of earthworks across central Overton Down continued across the line of the uncultivated Valley of Stones. It required Fleming's work (1987, 1988, with references) on Dartmoor, however, to bring the point into focus.

The north-west/south-east axial field systems

The north-west/south-east axis (Figure 2.6) is within a degree of 45° west of true north (46° is actually used for calculations). Its orientation of 314/134° follows the grain of the terrain, whereas the other axis, south west–north east (224/44°), is 'topographically oblivious' (Fleming 1987). Allowing for very local topographical anomalies, the axes of fields in Block 1 relate to linear Ditch F4 in general and share the same orientation as fields on Totterdown and Fyfield Down (Block 9) and, most markedly, on Overton Down (Block 6). That the same axial arrangement continued further east is hinted at by the ploughed-out remains above Devil's Den on the south-west-facing slope at the south end of Block 11 (*see* Figure 5.4). There may indeed be an element of coincidence in all this, but the presence of an axial guiding orientation can hardly be doubted when it recurs over a landscape of unconformable, and in many cases non-intervisible, areas of terrain.

The south-west/north-east axis apparently operated not only north west–south east but also south west–north east. It seems to have conditioned alignments from the south end of Block 3 crossing three downland ridges and two dry valleys as far north as the south-west part of Block 10, a distance of 3.8km. It could well be argued that it was common sense, as well as good husbandry, to spread the fields across the warmer, south-west-facing slopes. On the other hand, a degree of order and control going beyond what can reasonably be explained away as 'natural' or coincidence is evident in the scale of the field disposition. The pattern, once established, was enduring. The fields and tracks of the early Roman period on Totterdown (Block 8) are exemplars. The point is made primarily in relation to structures of prehistoric and Roman date, the two and a half millennia during which most of the great spread on Figure 2.1 seems to have accreted; but also pertained in places in the medieval period (Chapters 6, 7 and 16).

The north/south axial field system

Within the prevailing south-west/north-east and north-west/south-east orientation the eye is drawn to an area of differing alignment, especially around the Green Street/Ridgeway intersection (Figures 2.5 and 2.6). Closer analysis defined a block of fields characterised by

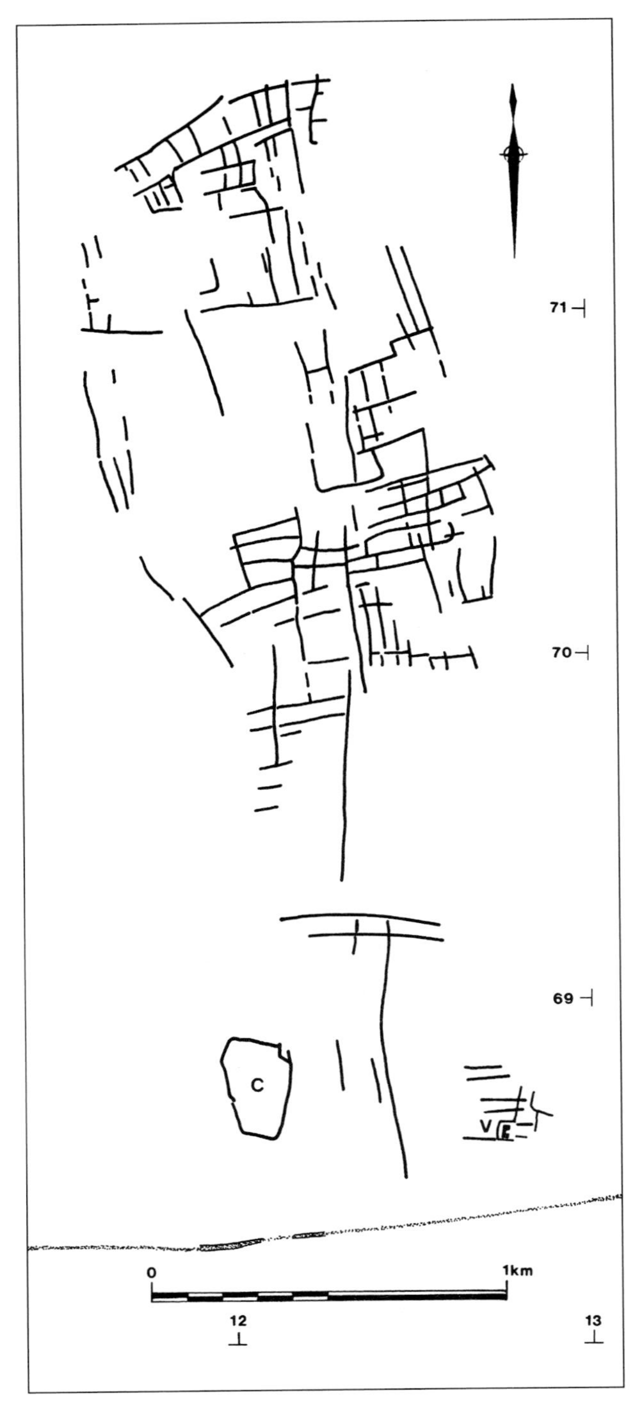

2.5 *Map of the Roman field system on Overton Hill as far north as modern Hackpen Hill, abstracted from Figure 2.1 largely on the basis of field morphology and the axis of the system at right angles to the Roman road across the bottom of the figure*

C = 'Crawford's complex'; V = Roman villa

their different orientation (7° west of true north) and distinctively rectilinear shape and size (Blocks 4 and 5). The fields tend to be 100m long or longer and 40–50m wide, not just larger than conventional 'Celtic' fields but of different shape and proportions, ie, roughly 2:1 or more, up to 5:1. These proportions are similar to the smaller rectilinear, 'Roman' fields on Totterdown (Block 8) and contrast sharply with the 'roughly 50m square' guideline for a typical 'Celtic' field on Wessex chalk.

This block can be fairly accurately defined taking the north–south axis and field morphology into account (Figures 2.1 and 2.2, Blocks 4 and 5 combined). On the northern side its edge seems to be at or just beyond a west–east field bank located *c* 200m west north west of barrow group C. The core of the system then sweeps south for 1.5km to a west–east, 300m-long, field division either side of The Ridgeway marking the boundary between Blocks 4 and 5 (the argument that, in reality, they are one is made below). The western edge is reasonably clear, marked by fields of a length to width ratio of 2:1 or more. The track heading north west towards Green Street (*see above*) may mark a boundary, especially if, as will be argued later (Chapter 15), it is part of the Romano-British landscape. Southwards from Green Street the difference in alignments is particularly well marked, the east edge of the system probably being on the west side of Pickledean above Down Barn. The dimensions of the area thus defined are: *c* 2.1km north–south by 0.6km (*c* 2,300 x 660yd); and the area of the field system is *c* 126ha (315 acres).

This system was, however, probably considerably larger. Its 600m width probably stretched for another 1.5km to the south as far as the Roman road, giving it a total length north–south of 3.6km, embracing 196 hectares (490 acres). The air photographic evidence for this suggestion is not strong, with only two lengths of north–south field bank on approximately the correct axis in Block 5. Overlying settlements and long-lived, permanent arable (Chapter 4, Figure 4.2; Plate XV) hampers further analysis (*SL*, figure 30).

The single most important piece of evidence about this field system, however, is that its axis is exactly at right angles to the Roman road across Overton Hill, supporting the suggestion that the field system came as far south as the road and was part of the same landscape arrangement. This geometrical association is viewed as a strong indication that the field system defined as Blocks 4 and 5 is of Roman date.

Mention has been made of a possible third axial arrangement. It lies 15½° west of true north, and mainly comprises field banks on, and at right angles to, that axis on eastern Fyfield Down and Manton Down (Figure 2.2, Blocks 9 east, 10 and 11; Figure 2.6). Block 11 appears to have been originally constructed on this axis, though later modified, and is the most convincing evidence for its existence. Much of Block 10, although over-ploughed, also accords with this axis. A further outlier on this alignment lies 0.5km distant, down slope on the east of Fyfield Down. This is located on the east side of a shallow, sarsen-filled re-entrant running northwards from the Valley of Stones, hinting that there may have been an overall 15½° alignment across the eastern part of the study area (Chapter 12). Such a layout is potentially early in the overall land-division scheme and could be related to an orientation towards barrow groups L and M.

Other, albeit fragmentary, hints of a 15½° axis are to be observed in field Blocks 2, 3 and 6. The impression conveyed is one of early field arrangements, closely related to round barrows, now peeping out from under and at the edges of later, superimposed field systems. Dating them would, therefore, be useful for our purposes.

Barrows

The area contains many barrows. The long barrows are well known but few in number, although their distribution is argued as being significant in landscape terms (Chapter 16). About 130 round barrows are recorded in the study area. In considering them the emphasis is placed on landscape and territory rather than individual structure or content. The transcription brings out three facets of the round barrows in the area. The first is their sheer number.

The second point is the number of groups and their siting. Thirteen groups (A–M) are identified on Figure 2.4, all consisting of three or more barrows. Some of the pairs and single barrows may, as further evidence accumulates, turn out to be groups. This has happened during the lifetime of this project with barrow groups J and K. Conversely, the singleton barrow at group C is so labelled because, although only one is visible on air photographs, three are shown by the OS 1:10,000 map. Thirteen is a very high number of barrow groups for an area of 16 sq km (6½ sq miles): exactly two per sq mile, comparable with the density of barrow groups along the south Dorset Ridgeway and in the Stonehenge area (RCHME 1970; 1979a; Woodward and Woodward 1996).

The third feature of the barrow groups is their distribution. With the exception of groups C and D, all are peripheral to the known ancient landscapes. Group C is also peripheral in that it is at the eastern limits of cultivation in its particular area. Otherwise the groups are peripheral specifically to the western and southern sides of this part of the Marlborough Downs. Conversely, barrow groups do not mark the northern and eastern edges of the ancient landscape. In part, local topographical considerations and the relationship with contemporary settlements and communications might explain this distribution. The groups either side of the entrance into Clatford Bottom (L and M), for example, are sited on local spurs and would have looked impressive when viewed from below. A monocausal model is unsatisfactory, however, and more complex reasons might be sought in the relationships between the barrow builders and users to each other and the surrounding landscapes.

Their distribution is not just peripheral; it is also remarkably regular. So much so that the pattern demands 'missing' groups to appear. For example, the two 'new' barrows at *c* SU 145692 may mark the emergence of a group, which would fill a gap between groups K and L (Chapter 16).

Individual barrow groups can also be seen as relating to specific field groups. Barrow groups A, B, F and probably H are sited just beyond the limits of cultivation, and, in the case of F, set in a reserved area between two field Blocks (2 and 3). In two cases (E and G), barrow groups are approached by tracks through the fields. Tracks are necessary in such contexts, not for through traffic, but specifically to lead stock to and from pasture without the risk of them encroaching on to crops.

The pattern is one of an integrated landscape with recognised and ordered arable, grazing and funereal areas. The functional pattern described here, as distinct from the geometric patterns suggested above, implies either that the fields could only stretch so far without trespassing on to the grounds of the ancients or that the burial grounds developed on the periphery of what was already regularly farmed land. They lie on the interface between arable and pasture, between enclosed fields and unenclosed grazing, or even between infield and outfield (Chapter 16).

Whatever the functional significance of the interaction between fields and barrows, the map clearly demonstrates the existence of a complex spatial relationship which is unlikely to be the result of chance. It also has clear chronological implications, and it is to this that attention is now turned.

CHRONOLOGY

The map, regarded as a primary source in its own right, provides a relative chronology in many instances. Some are examined in more detail in Chapters 4 to 7. Six areas of particular significance in terms of relative chronology are discussed here, beginning with the great zone of old grassland to the north. There the evidence is largely from standing earthworks and can, therefore, be checked on the ground. In the last two examples the evidence changes to plough-levelled archaeology revealed by aerial photography.

1 Linear Ditch F4 cuts through fields on the north-west/south-east alignment of Block 2 and acts as a boundary to presumably contemporary fields laid off from it on Totterdown, in Block 7. The ditch later becomes part of a track system related to Block 8, dated to the first–second centuries AD.

2 On Fyfield Down the aerial photography has revealed considerable evidence of a complex yet recoverable sequence. This is especially so at Wroughton (or Rowden) Mead, immediately east of Wroughton Copse. The density of features here reflects considerable medieval activity on top of Roman and earlier remains, all of which the cartography faithfully but unconsciously delineates.

3 Manton Down (Block 11) displays what should be quantifiable relationships between fields and other features, notably a long barrow, a prominent ditch and a small rectangular enclosure.

4 The great time-depth on Overton Down has long been known (eg, Crawford and Keiller 1928, 124–5). Suffice to note here that the transcription shows a complex layering of fields, overlaid by tracks, with differing settlement and enclosure types to be fitted into the sequence.

Plate IX *Fyfield Down from the north, on the evening of 24 May 1960, looking towards Wroughton Copse* (top centre*), from almost above a probable small Romano-British settlement at the T-junction of the 'Ridgeway route' (Figures 16.7 and 16.8). Otherwise, all the visible trackways run east–west and all, including those in use, are of medieval or earlier origin. The excavation of* Raddun *(site WC,* see *Chapter 7) is left of the Copse. The excavated lynchets and ridge-and-furrow (site FL,* see *Chapter 7) lie* left centre*, between the arc of a racehorse-training gallop and a small crater where World War II ammunition was detonated. The narrow ditch (*thin dark line, bottom left centre to right-hand corner of Wroughton Copse*) is the boundary of Overton's early nineteenth-century 'Cowdown'. The new fence running diagonally* bottom left to top right *represents another attempt, brief and abortive, to turn the down into a cattle ranch (AAU-79,* © Cambridge University Collection of Air Photographs*)*

5 The north–south field system (Blocks 4 and 5) overlies fields which were part of the general north-west/south-east orientation. Indeed, fragments of the earlier alignment only survive on Overton and Avebury Downs where they are peripheral to the later landscape. Thus we see the edges of the north-west/south-east system surviving on Monkton Down and along the scarp west of The Ridgeway between the 150m and 175m contours. In Block 5 we see as cropmarks a visually sharp juxtaposition of these successive alignments. A Roman date for the north–south system has been proposed.

6 The prominent complex of features along the east side of The Ridgeway in Block 5 is plotted as cropmarks, supplemented by air photographs taken before it was ploughed. A settlement appears to overlie fields. Nine hundred metres east, another settlement, 'Headlands', enclosed by a bank and ditch, was discovered as a cropmark. It is overlain by an Anglo-Saxon boundary (Chapter 4).

Ridge-and-furrow

The final point concerns the incidence of ridge-and-furrow and its importance as a chronological indicator. Indeed, early publication was almost obsessed by it, to the exclusion of other field matters (Bowen and Fowler 1962, 104; and cf Crawford and Keiller 1928, fig 24).

Eleven blocks of ridge-and-furrow were detected on the aerial photography (Figure 2.3). Here we are concerned primarily with this phenomenon surviving as earthworks specifically as a marker in local relative chronologies. The transcription also shows linear features derived from cropmarks in modern arable, in most cases best interpreted as evidence for medieval strip cultivation. Its overall distribution can only really be understood within the framework of parishes and tithings and its survival has much to do with post-medieval land-use.

Perhaps the most important points about the ridge-and-furrow are:

1 the demonstration of its existence on the high downland. Though this is now a commonplace, it was not even perceived as such when the project began and its early recognition was a significant contribution to the understanding of this landscape;

2 it is definitely not ubiquitous. Indeed, it is clearly disposed on the landscape in a non-random pattern. This suggests some accommodation might well be found with documentary evidence discussed elsewhere (Chapters 5, 9 and 15; FWP 43; Harrison 1995).

Absolute chronology

Clearly the transcription as such can say little about actual dates. Within the framework of relationships outlined above it is perfectly apparent that we are looking at a composite, complex and long-lived landscape. Monumentally, the earliest features depicted here are the long barrows, presumably of the fourth millennium BC though none is closely dated here by excavation. The round barrows, generally of the period 2500–1500 BC, may relate to the fragmentary traces of field system aligned north north west–south south east (15½°) and may represent an 'Early/Middle Bronze Age' landscape. However, in cultural terms the preference is for a 'Middle/Late Bronze Age' date of *c* 1500–900 BC. Linear Ditch F4 across the north of Overton Down and Totterdown should fit in here, if not earlier. The great spread of north-west/south-east (46°) landscape would conventionally be thought of as Late Bronze Age/Early Iron Age onwards (*c* 800 BC) up to and including the first–fourth centuries AD. Here, however, the evidence suggests otherwise, with an end to the use of fields for arable farming apparently occurring in the mid-first millennium BC.

The track system through the fields reached its developed form, not least to connect new settlements such as ODS, as a landscape phenomenon of the early Roman period. It was in the later first century AD too that the north–south (7°) field system ran north from the new Roman road on Overton Hill, with another block of contemporary fields on Totterdown. The final major phase of activity as illustrated by aerial photographic evidence was in medieval times when areas of downland, outside the permanent arable, were cultivated and partly settled. This is demonstrated by surviving ridge-and-furrow, at its best on Overton Down, where it is not independently dated, and on Fyfield Down where a context in the thirteenth century is firmly indicated by association with a settlement of that date (Figure 7.11). Later phases of land-use are neither very obvious nor widespread in terms of air photographic evidence, but farming continued and is indicated by a few sites such as the Delling Enclosure.

Unavoidably omitted, because there is little or no aerial cartographic evidence for them, are whole periods of time such as the third millennium BC, the last 500 years BC, and *c* AD 450–1200. These are serious gaps for an intended landscape history, and it is in part to them, together with the rather different problems provided by a plethora of evidence for other times and a host of activities, that we now turn, using a range of theory and practice in addition to aerial photography and cartography.

Chapter 3

Methodologies: Theory and Practice

... an intensive study here should help considerably in the solution of archaeological problems over much of Wessex ... ... Ultimately we hope to have large-scale plans and detailed descriptions incorporating every feature in the area which seems to be worth this treatment.

Bowen and Fowler 1962, 98

PERSPECTIVES AND POSSIBILITIES

Such were the optimistic aims of youth, in both the project and the author: the second aim has been largely achieved, but how far we are from solving what are now the problems of Wessex archaeology!

As stated in Chapter 1, the project's initial aim was to investigate the history of the landscape of Overton and Fyfield Downs. Such a simple statement belies not only the complexity of the landscape history itself as it was to be revealed, but also of the concepts and methodologies required to explore it. Moreover, in the more than forty years which have elapsed since conception of the project, British (and indeed world) archaeology has undergone many transformations in terms of theoretical frameworks, fieldwork and recording methods, and the application of a range of scientific and pseudo-scientific techniques unheard of in 1959. Perspectives, methods and possibilities, not surprisingly, have evolved along with the project and with the author himself.

From the outset, however, fieldwork was seen as the key to recording, and *integration* with other forms of recording – principally, as we have seen, aerial photography and cartography, combined with historical documentary research – as the means towards interpretation. As such, the project remains a model of innovation and, if the passage of time decrees that some of the methodologies employed now seem commonplace, even archaic, we may justifiably claim to have been among the first to employ them, at least on such a scale.

Work on the project was initially prompted by the extraordinary state of archaeological preservation in the area combined with the apparent opportunities, after appropriate fieldwork, to date phases of landscape development by *question-specific excavation.* To put it more fully, the initial objective, as stated unequivocally in the first interim report, was and remains:

> ... to establish ... the forms and sequence of human activity in the region. This involves a search for the areas of primary occupation, a consideration of settlement types and pattern and their development, of fields and pasture, boundaries, ritual and burial structures, of continuity and communications, and of the relationship of all to each other, to natural features and to conditions in different phases.
>
> Bowen and Fowler 1962, 98

In more contemporary terms, we can summarise this as follows:

The establishment of the main phases of human activity in the area in different periods and in relation to the natural environment overall, and especially in terms of landscape development with particular reference to questions of rupture, survival and continuity over four or more millennia.

All encompassing as this statement is, it is important to realise that the project was, from the outset, *structured* around the investigation of particular themes and by the posing of two principal questions:

How, why and when did the landscape, particularly the landscape of the downs, evolve into its twentieth-century form?

What types of economic activity were carried out in the study area and how were they distributed within it?

Implicit within these questions is a sub-set of further questions with which the project became involved, whether or not explicitly acknowledged at the time. These 'new' lines of enquiry could also be framed in terms of specific questions:

To what extent is the downland 'marginal':

- *in area?*
- *in terms of settlement pattern, through time and at different times?*
- *economically, tenurially and socially?*

Is it possible to define a history of the changing pattern of land use?

What was the chronology, extent and function of the 'Celtic' fields?

Do the types of settlement represented in the study area increase our understanding of settlement morphology?

Almost inevitably, further objectives emerged as the project developed and the strengths and weakness of both data and methodology became apparent. Field survey led not merely to a 'feel' for the landscape but to the rapid discovery of many new sites in a manner which has since become very familiar. In particular, it led to the recognition of phases of chronological development in the landscape (already described in outline in Chapter 2), demonstrable land-use 'zones' within the landscape (considered in Chapters 3, 8, 9 and 14), and to specific places where relative and possible absolute dating evidence could be obtained (Chapters 3 to 7). Feedback between air photography and survey, between both and documentary evidence, and between survey, excavation and documentary/cartographic evidence played a particularly important part in the conceptual development of *how* as much as *what* to investigate in the study area.

As a result, in its now evolved and mature form, the original project can be interrogated to consider yet further questions:

How has the environment of the area changed over time and what were/are/have been the consequences of such changes?

What, in chronological terms, were the main phases of human activity in the area?

How did the excavated sites function within the landscape, with particular reference to economic activity?

What can we infer about the nature of 'marginal' land? And if the downs (and forested area) were indeed 'marginal', how was this marginal land exploited at different times?

What does the evidence tells us about changing methods of farming and land-use?

What does the evidence tell us about changes in settlement pattern, settlement morphology and building types?

Are there constructive pointers to the use of evidence in interpretation in considering comparatively a rural settlement with, and rural settlements without, contemporary documentation? (For example, compare how we handle the evidence from, and what we make of, sites WC and OD XII.)

THEORY

The theory, perhaps more an assumption in the early years, was that the questions in mind could be appropriately addressed by what was, in the late 1950s and early 1960s, regarded as field survey. This concept was heavily influenced by the work and writings of O G S Crawford, as expressed in particular in his *Archaeology in the Field* (1953), his increasingly comprehensive notes for the Ordnance Survey (1951 onwards) and his editorship of *Antiquity*. The theory was simply that significant amounts of archaeological evidence lay on the ground, particularly in the form of earthworks, that this evidence could be recognised and recorded by going out, finding it and measuring and photographing it, and that it could then be interpreted in historical terms. This model was heavily influenced by generations of fieldworkers carrying out their surveys on southern English chalklands, though Crawford's own vision embraced other parts of the British Isles, and parts of Europe and north Africa (1953).

He had also been largely instrumental in bringing a new dimension to such approaches to what was still regarded as 'geography' rather than 'landscape' by his development of the application of aerial photography to archaeological field survey (1924; 1929; Crawford and Keiller 1928). This development was very much in mind as our project began and expanded, not least because one of the early investigators was simultaneously completing a book exploiting it in relation to early fields (Bowen 1961).

Overall, theory was initially implicit rather than explicit but there were good pragmatic grounds for believing that a project based on thorough and critical ground examination, backed up by aerial photography, would 'reveal' not only new sites and much archaeological evidence but also throw light on answers to the questions being asked. Weaknesses were that the theory rather assumed that earthworks and air photographic phenomena were adequately meaningful, and that such archaeological evidence as was visible could and would be perceived by the investigator(s).

METHOD

Field survey, excavation and documentary research, like air photography (Chapter 2), were integral to the whole project. Fieldwork was practised on the very first day in 1959 and was still in use in 1998. This approach included in particular fieldwork on foot – reconnaissance and metrical survey – characteristically closely linked to aerial photographic cartography.

Field survey involved seven methods of investigation:

- i Slow and reasonably thorough ground examination on foot.
- ii Metrical ground survey of selected sites and areas.
- iii Fieldwalking on arable land.
- iv Aerial reconnaissance.
- v Aerial photographic reconnaissance.
- vi Aerial photographic cartography.
- vii Sub-terrestrial survey:
  - (a) resistivity survey;
  - (b) magnetometer survey;
  - (c) dowsing.

In addition, a disproportionate amount of our effort went into another field methodology:

- viii Excavation.

Throughout, but especially in the 1990s, we put a lot of our resources into another method altogether:

- ix Documentary and cartographic research on both primary and printed sources.

COMMENTARY ON THE METHODS

Method i

Fieldwork on foot was, throughout, the principal means of investigation. The objective was to record as far as possible all man-made structures and disturbances. The whole of the two parishes has been walked at some time or other in this way, in the first place looking mainly for earthworks and stone structures in the countryside but later looking at roads, hedges, woods and, softer still, relationships, while extending the range from the conventionally archaeological to the present villages and other settlements, with their houses, buildings and other appurtenances. In general, the coverage was from east to west on the downs, beginning in Wroughton Mead and reaching The Ridgeway by the mid-1960s (with a backtracking to the east on to Manton Down in the 1990s). Sporadic forays to the south had already occurred in the 1960s, but in general the valley with its villages and the relatively large areas southwards in, to and through the woods were walked as a second stage, extending into 1998.

In practice, this method involved an overall visitation to the study area at the reconnaissance level, with many places and areas subsequently enjoying repeat visits ranging in number from one to dozens. Almost invariably, such repeat visits produced new information, not so much because of initial human fallibility – though of course we missed data on some visits – but as a result of variations in the conditions in which observations were made. Such variability embraced weather, light, vegetation, land-use and access. Increasing experience was also undoubtedly a factor. This method alone led to significant differences in the completeness and quality of the evidence at our disposal, depending largely on the thoroughness of investigation resulting from repeat visits but also produced by the quality of conditions on the ground when the visit or visits were made (Figures 3.1 and 3.2).

Method ii

Metrical ground survey of selected sites and areas was carried out at all the main sites individually, and of some areas judged to be of particular significance. These were selected rather more for their opportunities and challenges than for the impressiveness of their earthworks, for instance, on Totterdown and central Overton Down (Figures 5.2 and 6.1).

Most of this work was accomplished using a plane-table and/or a cross-head, always in conjunction with an artillery director for laying out the main survey axes when the site or area was more extensive than was appropriate for visual sighting alone. Field scales varied, depending on the area being recorded but also reflecting the degree of detail either available or required. They were characteristically 1:2,500, 1:1,250 or 1:500, but with some sites at much larger scales such as 1:32 (Wroughton Mead: archive drawing 208).

The whole of such work up to the early 1970s, which encompassed most of it, was, of course, implemented in

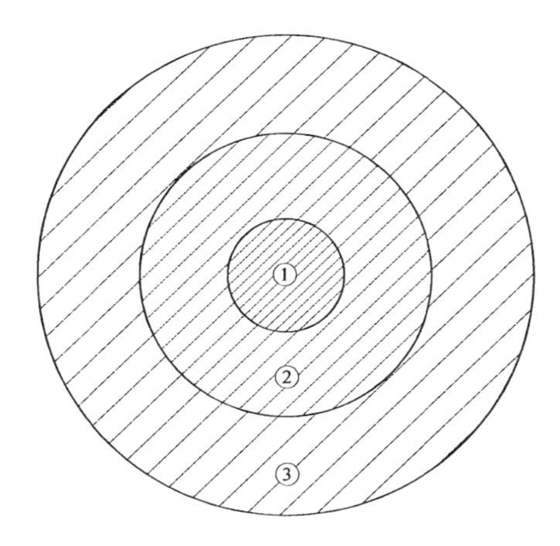

3.1 *Model of intensity of field investigation, showing three zones: ① intense ground cover, with widespread metrical survey and specific and large-scale excavation; ② comprehensive, extensive ground cover with specific metrical survey but no excavation; and ③ partial, selective ground cover, no surveys*

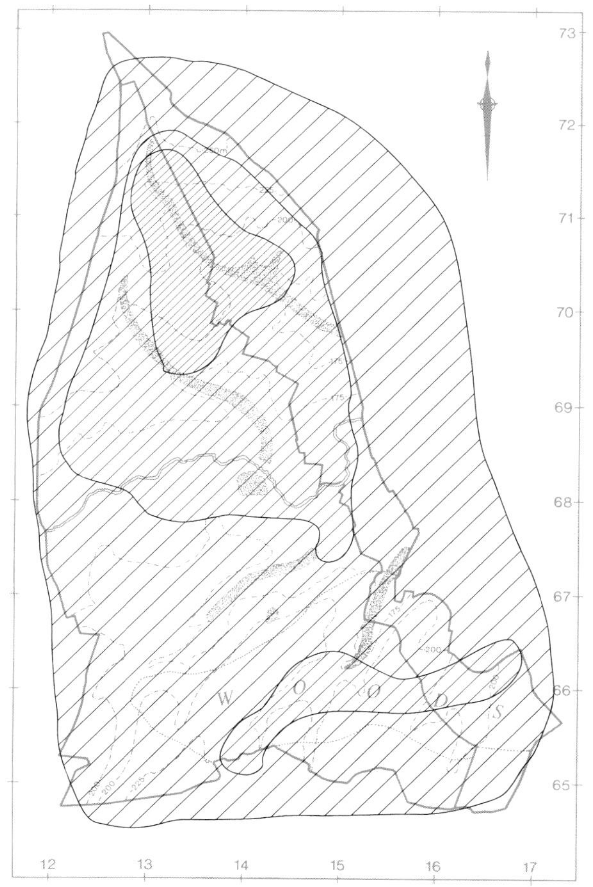

3.2 *Actual intensity of field investigation in Overton and Fyfield, 1959–98. The zones are as explained in Figure 3.1*

imperial measurements, using 100ft (equivalent to 30m) tapes in the field. All the measured field drawings were either done or supervised by this author, aided (and indeed taught) in many early examples by Collin Bowen. All the original field plans, pencil on either plastic or paper, are in the archive at Devizes Museum (*see* Appendix 1). The results of this work form a significant part of the line drawings used in this volume, the last one of which was surveyed in November 1998 (Figure 12.2).

Method iii

Fieldwalking was used at three known settlement sites but was not widely practised on the landscape. We differentiate the technique from other fieldwork in the sense of systematic, collective walking across arable fields searching for artefacts and other signs of human activity and simultaneously making a detailed locational record. This was not a serious part of the fieldworker's armoury in the 1950s and 1960s, at least as practised by the RCHME, which was very much mentor to the project's early years. Nor was the technique considered by either Atkinson (1953, 13–41) or Crawford (1953), though it was widely used by flint collectors, then and much earlier. That, however, was the point: scouring ploughed fields for objects was alright for collectors but not for serious archaeological fieldworkers who, in the great tradition canonised by Crawford, eschewed 'mere finds'. This project should have used the technique much more extensively, but its methodological priorities lay elsewhere.

Method iv

Aerial reconnaissance is distinguished from aerial photography because such operations played a significant part in the 1960s. Then, the author, usually piloted by the late Stanley Sharpe, was able to fly over the study area and its surrounds on numerous occasions. Aerial observation was also possible from helicopter and balloon flights primarily organised for media purposes (eg, *SL*, colour plates 6, 28, 30). All contributed to a growing familiarity with the area and its archaeological characteristics.

Method v

Aerial photographic reconnaissance began in the early days of archaeological air photography (Crawford and Keiller 1928; Plate XIX) and has continued ever since (FWP 85 indicates the archive of air photography now available in the NMR). The 1946–7 RAF vertical cover is excellent in itself for the whole study area and now provides a datum for post-war landscape change (eg,

Plates VII and LII). A particularly telling set of oblique air photographs was taken by Professor St Joseph on the evening of 24 May 1960: all have underpinned much of the subsequent work on the downs and four are reproduced here (Plates X, XVI, XX and XXXVIII; two of them, which have also been published elsewhere, are in *SL*, figures 25, 36). The Air Photography Unit of the RCHME (now English Heritage) has kept the area under review throughout the project's span, producing some particularly excellent results in the 1990s (eg, Plate XXII; *SL*, colour plate 25).

Method vi

Aerial photographic cartography (*see* Chapter 2) has been and remains fundamental to the methodology, mainly because of the extent of the visible archaeological evidence, particularly of medieval and earlier fields. Although two parishes is a small area in most perspectives, here they comprise too large an area to survey comprehensively at, say, 1:2,500 on the ground; or at least it was with this project's resources. Archaeological cartography from air photography was therefore the only feasible way to proceed. Crawford (and Keiller 1928, fig 24) began the process more than thirty years before our start, but from 1960 we were making various attempts to plot the archaeology of, first, the downs and then the whole study area basically from air photographs. Those attempts are in the archive at Devizes Museum (eg, drawing no. 409); some artwork versions were published (eg, Bowen and Fowler 1962, figure 1; Fowler 1963b, figure 33; Fowler 1969, maps A–C). Two poor versions based on what was then available in the SMR (Evans *et al* 1993, 2, and Powell *et al* 1996, figure 22) should now be disregarded as misleading; they have in any case been superseded by Figure 2.1 here and a new generation of derivatives based on it (eg, *SL*, figure 21; English Heritage 1998, figure 2).

This approach culminated in the aerial photographic map already discussed (Chapter 2, Figure 2.1). An equivalent map of the area south of the A4 was produced in-house (Figure 12.3). Both have subsequently been revised, and now a map of the whole study area has been derived from them and from the RCHME's independent mapping of a larger area in its Avebury Environs Project (Figure 15.3).

Method vii

Sub-terrestrial survey was used both experimentally and as a matter of routine, but its contribution overall was minimal compared with all the other methods. For one thing, the area of investigation was large in relation to those areas then realistically surveyable with what was still fairly cumbersome and, in the case of the proton magnetometer, pioneering equipment. Ground coverage was very slow in pre-computerised days, so in practice such methodology was only used in the hope of increasing information about specific sites already selected for excavation on other grounds. These were sites WC on Fyfield Down and OD X/XI on Overton Down.

Method viii

Archaeological excavation was a key part of the methodology used in the project. In concept, it was always the handmaiden of both the main objectives and other lines of enquiry, though at times its own prerogatives temporarily took over. Initially, however, each project excavation was planned to answer specific questions arising from fieldwork, mainly to elucidate sequence and function in trying to understand the workings of the landscape at various times, and therefore its evolution.

In the event, our own excavation was confined to the downland over the northern part of the two parishes (Figure 3.3). Those downs extend over some 26 sq km (*c* 2,590ha; 6,400 acres), of which less than a hectare (*c* 2 acres), that is about 0.03 per cent, was excavated. These excavations, including their code names, are listed in Tables 1 and 2. Hereafter they are referred to in the text by their code names.

Logistically, the main excavation effort went into fairly extensive work on three settlements, respectively of late prehistoric, late Roman and medieval date (Table 1). Together they provided a useful chronological and functional range across the landscape (Chapters 6 and 7; FWPs 63, 64, 65). We also conducted eight other excavations which, like the big ones, were conceived as small intrusions but, unlike the big ones, stayed small (Table 2; Chapters 5–7; FWP 66). They all provided, as was intended, critical evidence about phases of landscape development. Five of them were carried out directly by the project team; one was carried out by others in pursuing other but related objectives in the project's core area (involving this author in another guise, not least as director of excavations); and two were executed by others on sites suggested by us as a direct result of project fieldwork (Chapters 7 and 8; FWP 66).

The series of excavations produced a great deal of primary evidence. Some of it was absolutely crucial to the project's purposes, but much of it was superfluous from that point of view though, of course, interesting in other perspectives. Full, conventional excavation

Plate X *Methodology: fieldwork, excavation and air photography. Delling Enclosure in the background, independently discovered by our fieldwork and Cambridge air photography (1960), with* Raddun *in the foreground under excavation (Building 1) partly as a result of the air photographic rediscovery in 1954 (Plate XL) of a site discovered during reconnaissance fieldwork and dug into by Sir Richard Colt Hoare* c *1817 (AAU-82,* © Cambridge University Collection of Air Photographs*)*

reports have been prepared but they are not included in this volume. They are available in and from the archive, by Internet, on disc or as hard copy (*see* Appendix 1).

Method ix
Both primary documents and original maps comprised key sources of information for this project and support much of our interpretation for the most recent 1,500 years. We have also made extensive use of printed sources, some of them editions of primary sources, and of printed maps since *c* AD 1800, particularly those made by the Ordnance Survey. Such research has covered the whole study area more or less evenly and falls outside the 'levels of intensity' of investigation in the field indicated in Figures 3.1 and 3.2.

Table 1 The three main excavations

Code	*Description*
OD X and OD XI (Chapter 6)	Overton Down: *c* 30m^2 of one embanked and ditched Late Bronze Age/Early Iron Age settlement stratigraphically sandwiched between earlier and later fields (Figures 6.2–6.10; FWP 63)
OD XII (Chapter 6)	South end of Overton Down: the whole(?) of one late Romano-British farm on top of 'Celtic' fields (Figures 6.15–6.23; FWP 64)
WC (Chapter 7)	Wroughton Copse, Fyfield Down: much of a medieval farm overlying 'Celtic' fields (Figures 7.4–7.14; FWP 65)

Table 2 The smaller excavations (Chapters 5–7: all in FWP 66)

Code	*Description*
OD I (Chapter 5)	North end of Overton Down: trench across a Bronze Age linear bank and ditch which cut Beaker occupation (Figures 5.1, 5.2)
OD II (Chapter 5)	North end of Overton Down: small excavation around a Neolithic axe-sharpening stone (Figure 5.1)
OD III (Chapter 5)	North end of Overton Down: small excavation of a stone structure (Figure 5.1)
TD VIII and IX (Chapter 5)	Totterdown: two trenches across an eastward continuation of the linear ditch at OD I (Figures 5.1, 5.2)
FL 1, 2–5 (Chapter 7)	Fyfield Down: the major trench cut through a large 'Celtic' field lynchet; with subsidiary cuttings (Plate XXXIX, Figures 7.2, 7.3)
TD I, Ia, II, IIa, IIb and III (Chapter 5)	Totterdown: six cuttings to examine the boundaries and date of a rectilinear field system containing a cup-marked sarsen and overlying 'Celtic' fields (Plates XX and XXI; Figure 5.3)
Down Barn Enclosure (Chapter 6)	Trapezoidal and post-Roman earthwork enclosure, stratified above a long prehistoric and Roman sequence filling the formerly sarsen-filled and now dry coombe called Pickledean (Figure 6.14)
Delling Enclosure, Fyfield Down (Chapter 7)	Sharply defined, rectilinear earthwork and sarsen-stone enclosure overlying 'Celtic' fields, probably a sixteenth–seventeenth-century farmstead (Figure 7.15)

Most of the documents that we have consulted are curated in one of three places: Wiltshire County Record Office, Trowbridge (WRO); the Library of the Wiltshire Archaeological and Natural History Society (WANHS) at Devizes; and the muniment room of Winchester Cathedral, whose Bishop was ultimately the largest landlord in the two parishes in medieval times. The other main landlord was the Abbess of Wilton, but we have not researched in like vein her original archive. Nor have we pursued original research on the documents of their post-Dissolution successors, principally the Earl of Pembroke and the Duke of Marlborough.

That negative must, however, be qualified by the fact that our principal early modern source, the great 1567 *Pembroke Survey*, is available in a printed edition (Straton 1909) and in translation (FWP 76). With that, and above all with the several relevant volumes of the admirable *VCH Wiltshire*, we have studiously tried to avoid duplicating work already done and to build on the scholarly documentary basis already in place for the medieval and post-medieval periods. In particular, we have tried to give a landscape dimension to what is already known historically, both by fixing in a specific place a particular boundary or tenement and by extrapolating from such detail to generalities about the landscape at any one time and about its changes through time. The Fyfield and Overton study area is not, however, particularly distinguished in its documentary and cartographic evidence. It lacks, for example, any sixteenth- or seventeenth-century maps, and the material which is available is fairly typical of the sort surviving for many places in southern England.

From the point of view of landscape history, however, the area proved to possess a most valuable asset: its landlords found a welcome necessity to survey their estates every so often. As a result, our investigations have been blessed with two Saxon charters of adjacent estates, in addition to the *Domesday* survey and a mid-twelfth-century survey of the lands of the Knights Templar. There is plentiful landscape evidence during the rest of the medieval period, not least from the descriptions in custumals and property transfers. After the 1567 *Survey*, the focus of our approach shifts to maps rather than words alone. In fact, cartography was a process rather than a single event in a particular year so, though our interpretative use of it is similar to that of the earlier written surveys, the evidence itself is somewhat different.

We also used documentary and cartographic sources for palaeo-environmental purposes, but we must make an important caveat. We have not pursued systematic research on the voluminous post-Conquest documentary evidence for the study area from an environmental point of view: to do so would be to embark on a project in its own right. Similarly, the intensive research on the area's documentation by the *VCH*, Kempson (1962), Yarrow (FWP 18) and Hare (FWP 43) was not directed to elucidating environmental matters. Nevertheless, we are well aware that medieval and later documentation contains a wealth of evidence bearing on the environment and changes in it, a potential we have to some extent selectively illustrated in Chapters 8 to 13. Overall, however, the project methodology was not directed towards producing a dramatic improvement in the environmental record from documentary evidence since it was for a long time thought that archaeology could answer many of the questions that were being asked. This proved to be correct up to a point, and uniquely for all time up to the late Saxon period; but we now know that much palaeo-environmental evidence also lies on parchment and paper (Chapter 14).

Documentary evidence

Main sources

1. S449: charter of East Overton of AD 939 which details the boundary of the fifteen-hide estate granted by King Æthelstan to Wulfswyth, a nun. A detailed analysis of this charter can be found in FWP 11 (Grundy 1919, 240–4; Brentnall 1938a).
2. S784: charter of West Overton of AD 972 which details the boundary of the ten-hide estate granted by King Edgar to a lady Ælflæd. A detailed analysis of this charter can be found in FWP 68 (Grundy 1919, 245–7; Brentnall 1938a).

The translations for both charters (S449 and S784), with minor alterations by us, are based on Brentnall's, who in turn used Grundy's. Use of these two charters was made in FWPs 44–48 while working on the history of the villages and the area south of the Kennet, as well as in FWPs 51–55 which looked at smaller areas of study called 'windows' in drafts of this monograph up to 1997. They are much the same as the areas looked at in Chapters 4 to 13 (Figure 3.4).

Other charters used include S547, a charter from AD 949 with West Kennet Farm at its centre, and charters S348 and S424, dating from AD 934, which delineate the boundary of Oare and North Newnton (Grundy 1919, 320–1; FWP 68). Brief studies were also made of charters S272, S1403, S1507 and S1513 (Alton Priors) and S341, S399, S543 and S668 (Winterbourne).

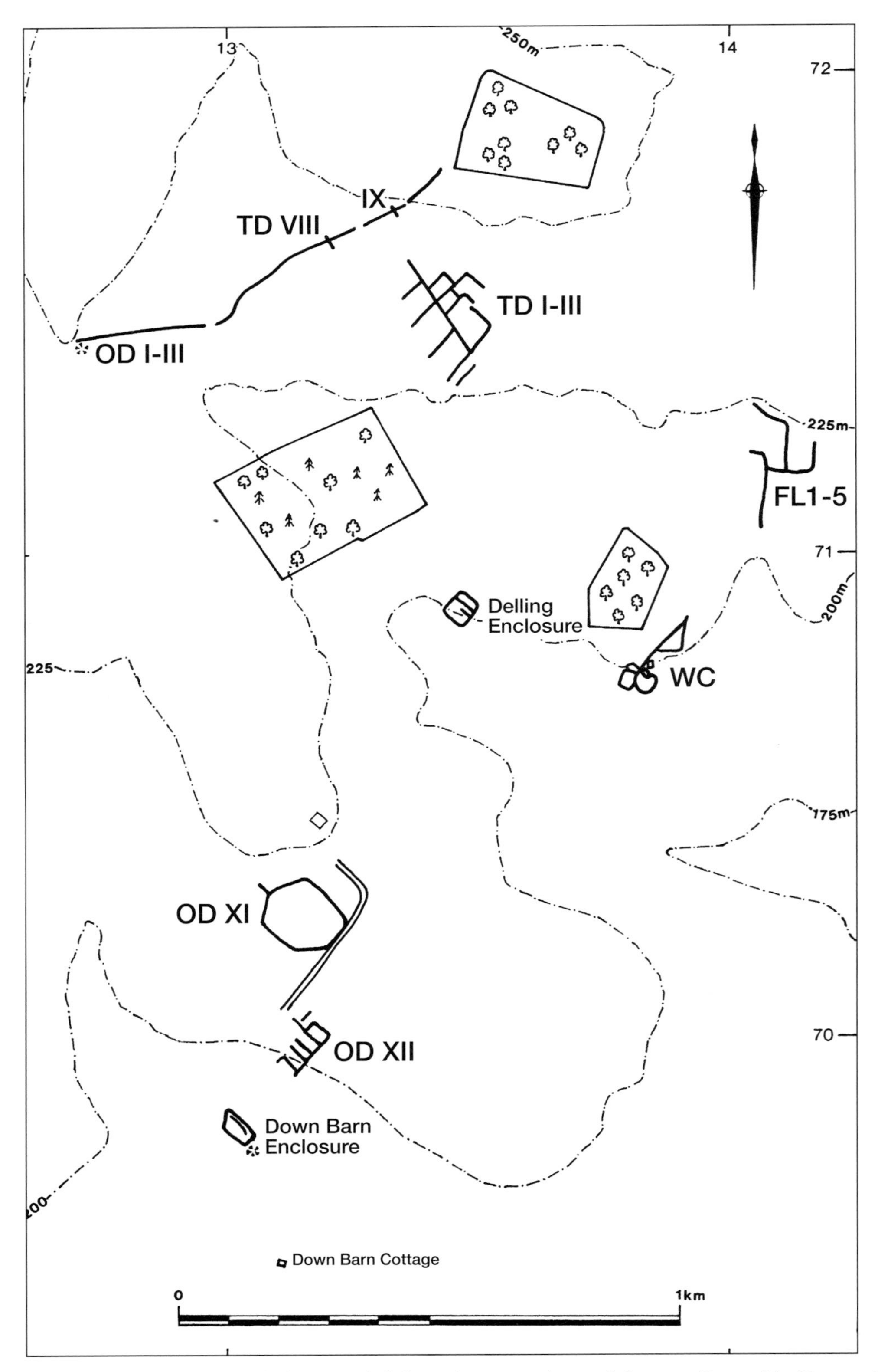

3.3 *Map of the northern downs, showing the location of all the project excavations and the areas discussed in Chapters 5–7*

Information from all eleven of these supplementary charters pertinent to our study is included in the FWPs mentioned above, as well as in The Ridgeway Report (FWP 30).

3 *VCH Wiltshire*: mainly volumes II (*Domesday Book*) and XI, 104–203 (the history of the parishes of West Overton and Fyfield), though many other volumes were consulted (*see* Bibliography).

4 'Lockeridge, Knights Templar and Rockley', by Simon Yarrow (FWP 18), is an original study commissioned by the project in May 1995. Yarrow transcribed leases or donations of land at Lockeridge to the Knights Templar at Rockley into modern English. In addition, he produced a brief history of Lockeridge, Fyfield Down and Rockley during the time of the Knights Templar, noting any information that was potentially relevant to the project's landscape emphasis. The central part of this thesis is the Templar survey of 1185. FWP 18 also contains all the known donations, in Latin and English, from Rockley and Lockeridge to the Knights Templar.

5 'Copies of Terriers and other ancient Records relating to this Vicarage' (Overton with Alton Priors and Fyfield). These hand-written and unpublished notes by the Revd E H Goddard are his literal transcription (in 1936) of 'a small note book in the custody of the Vicar of Overton' containing copies of the records associated with the parish of Overton dating from between 1290 and 1704. References in our text to the *Nona Inquisitions* (1342), the *Parliamentary Survey of 1649–50* and to Terriers come from this source. Goddard's notes are held in the SRO, accession unknown.

6 'Agriculture and Land Use on the Manor of Overton 1248–1539', by John Hare (FWP 43), was also commissioned by, and carried out specifically for, the project. Hare completed a thorough examination of as many documents as possible, especially those in the archives at Winchester Cathedral, relating to the manor of East Overton and Fyfield, with particular emphasis on place-names, specifically *Raddon/Raddun*. In addition, FWP 43 considers the manorial economy of East Overton and Fyfield in the medieval period. Dr Hare's essay is used extensively, often verbatim, in Chapters 7 and 9 and his research significantly underpins this volume.

7 *Survey of the Lands of William First Earl of Pembroke* (Straton 1909; referred to here as the '1567 Survey'). For each entry the manorial holdings, the names of tenants, their holdings and how these are held, and, to a varying extent, the boundaries of pasture and woodland, are detailed. The survey for the West Overton entry was actually carried out in 1566 and that for East Overton and Fyfield in 1567 (*Wiltshire Archaeological Magazine*, 32, 291–2). A translation of the Survey into modern English was commissioned from Dr Chris Grocock (FWP 76). *Surveys of the Manors of Philip, First Earl of Pembroke and Montgomery 1631–2*, edited by Eric Kerridge (1953 WAHNS Records Branch, vol IX), was also used.

Cartographic sources

1 *A Map of Shaw Farm within the Parish of Overton ...* surveyed by John Walker in 1734 (WRO 1553/109). Shaw Farm sits astride Wansdyke with the farmland predominantly to the north and west up to the boundary with West Overton and East Kennet.

2 *Andrews and Dury's Map of Wiltshire, 1773* (WANHS 1952).

3 *The Manors of East Overton, Lockeridge in Overton, Fyfield and Clatford in Preshute* (WRO/open access). Late eighteenth-century, referred to as 'late eighteenth-century map' in the text.

4 *West Overton in the County of Wilts*, 1783 (WRO 2203). Two versions of this map exist: one, a 'rougher' version dated 1783, the other 1794 (referred to respectively as '1783 map' and '1794 map').

5 *A Plan of the Manor of West Overton in the County of Wilts*, 1802 (WRO EA61). This map was the final stage of the 1783 and 1794 maps. It was drawn for the Parliamentary Inclosure Commissioners of West Overton (25 March 1802). Referred to as '1802 map'.

6 *A Plan of an Estate belonging to The Rev.ᵈ F. C. Fowle with the Lands adjoining comprehending the Manor of Fifield and East Overton in the County of Wilts* by A Dymock, 1811 (WRO/628/49/4). West Overton does not appear on this map (except 'Weylands'). Its delineation of the Fyfield manorial boundary shows the pre-Enclosure bounds of Fyfield, especially north of the A4. It excludes Clatford Park and Overton Heath, then not part of Fyfield

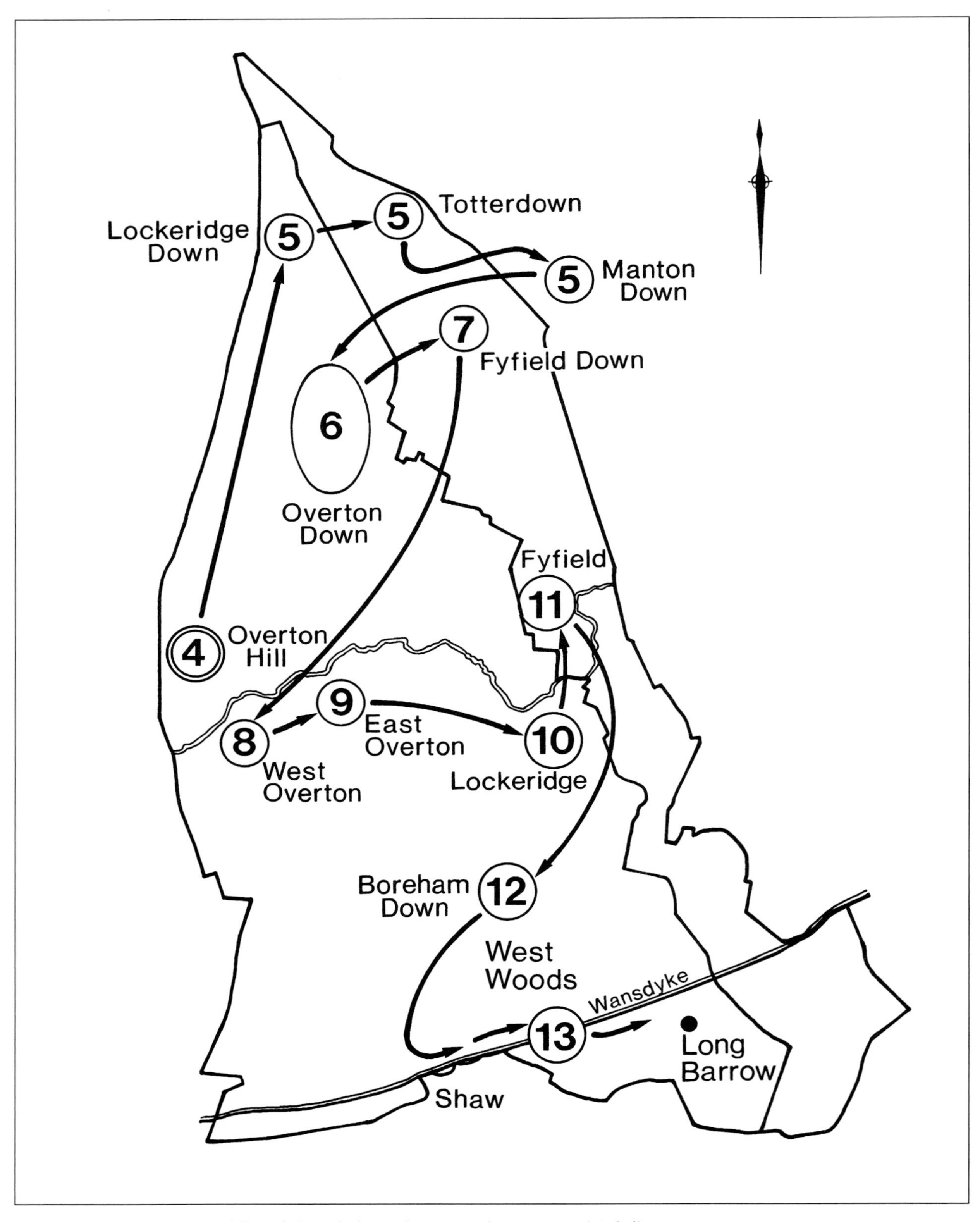

3.4 Map showing the route followed through the study area in Chapters 4–13 (circled)

parish. Exists on microfilm. Referred to as '1811 map' (Plate XI).

7 *A Map of East Overton, West Overton, Lockridge and Fifield, made in the Inclosure. 1815 and 1816* (WRO/EA117). This details the size of plots, land allotments and land exchanges. A full report of the proceedings concerning the Enclosure Award to East Overton (1821) accompanies the map at WRO. Referred to as 'Inclosure map' or '1815/16 map'.

8 *A Plan of East Overton, West Overton, Shaw, Lockeridge and Fifield in the County of Wilts* by Abraham Dymock, 1819 (WRO/778/2). Dymock replaced Decimus Godson as surveyor and cartographer for the Enclosure commissioners in East Overton and Fyfield (*see also* 1811 map). This multi-coloured map relied heavily on the previous maps of the area (especially WRO/628/49/4 and WRO/EA117), though it is far more detailed, marking every gate and even noting crops. Exists on microfilm. Referred to as '1819 map'.

9 Ordnance Survey maps at all available scales from the early nineteenth century onwards were consulted, especially OS two inches to the mile drawings from 1814–15, the first edition OS maps at one inch to the mile (*c* 1820), the 1887, 1900 and 1924 twenty-five inches to the mile maps, and the 1889 six-inch maps.

10 *Map of West Overton 1862*, British Library (Plate XII).

11 Smith 1885 contains some excellent, detailed maps.

12 Olympia Agricultural Co Ltd Map, 1922 (WRO 2444/3), shows all our study area except the tithing of West Overton. Each plot and field is numbered.

General significance of these sources Documents and maps alter the character of the interpretation. Though material remains can be made to tell a story, the addition of documentary and cartographic evidence often supports such stories and indeed suggests others. The significant contribution of documents and maps in this study was mainly, therefore, to complement and supplement the archaeological evidence, by:

(a) indicating land use;
(b) giving an insight into the natural environment;
(c) 'peopling' the landscape and villages;
(d) naming known sites;
(e) helping to explain place-names;
(f) locating 'new' features, including boundaries and settlements;
(g) underlining the antiquity of some features, especially boundaries;
(h) suggesting areas of further research.

Palaeo-environmental investigation

The environmental dimension was present in the thinking about the study area from the start, and illuminating it was pursued in several ways. The methodology of the environmental programme employed, however, largely reflects the approaches current in the 1960s when the bulk of the evidence was collected. In particular, assumptions made then about the relationship between excavation and environment are now glaringly anachronistic. No systematic programme of environmental sampling was pursued and there was no on-site sieving or flotation. Indeed, flotation was not a serious option. On the other hand, those involved were well aware of the palaeo-environmental dimension and many environmental materials were visually collected, almost all from carefully specified contexts. Animal bones in particular were collected by the thousand, and carefully bagged, washed, marked and roughly sorted. They were examined in the 1970s by Dr Barbara Noddle, who then disposed of them; her report was revised in 1995.

Numerous 'soil samples' were also taken (mostly discarded as uninformative in 1995) and, critically, in several places sequences of stratified soil samples were taken 'blind' (ie, without specific questions in mind). The most informative sequence, from OD X/15, is discussed below in Chapter 6.

The project drew heavily upon the advice in two books basic to the project's approach (Cornwall 1956; 1958), and Professor G W Dimbleby encouraged and advised on the environmental dimension of the investigation in its early days. Indeed, he himself undertook standard pollen analysis of a sequence of samples through FD 1 (Chapter 7); the FD 1 series examined in 1995 was a duplicate set of samples. Dimbleby demonstrated that pollen did not survive in the chalk environment of Fyfield Down, so no further pollen analytical work was pursued. Instead, J G Evans (now Professor), Dimbleby's post-graduate student, applied and developed his pioneering techniques in molluscan analysis, notably on Overton Down (OD XI). This was the way in which an environmental dynamic, so vital to the project's objective of landscape

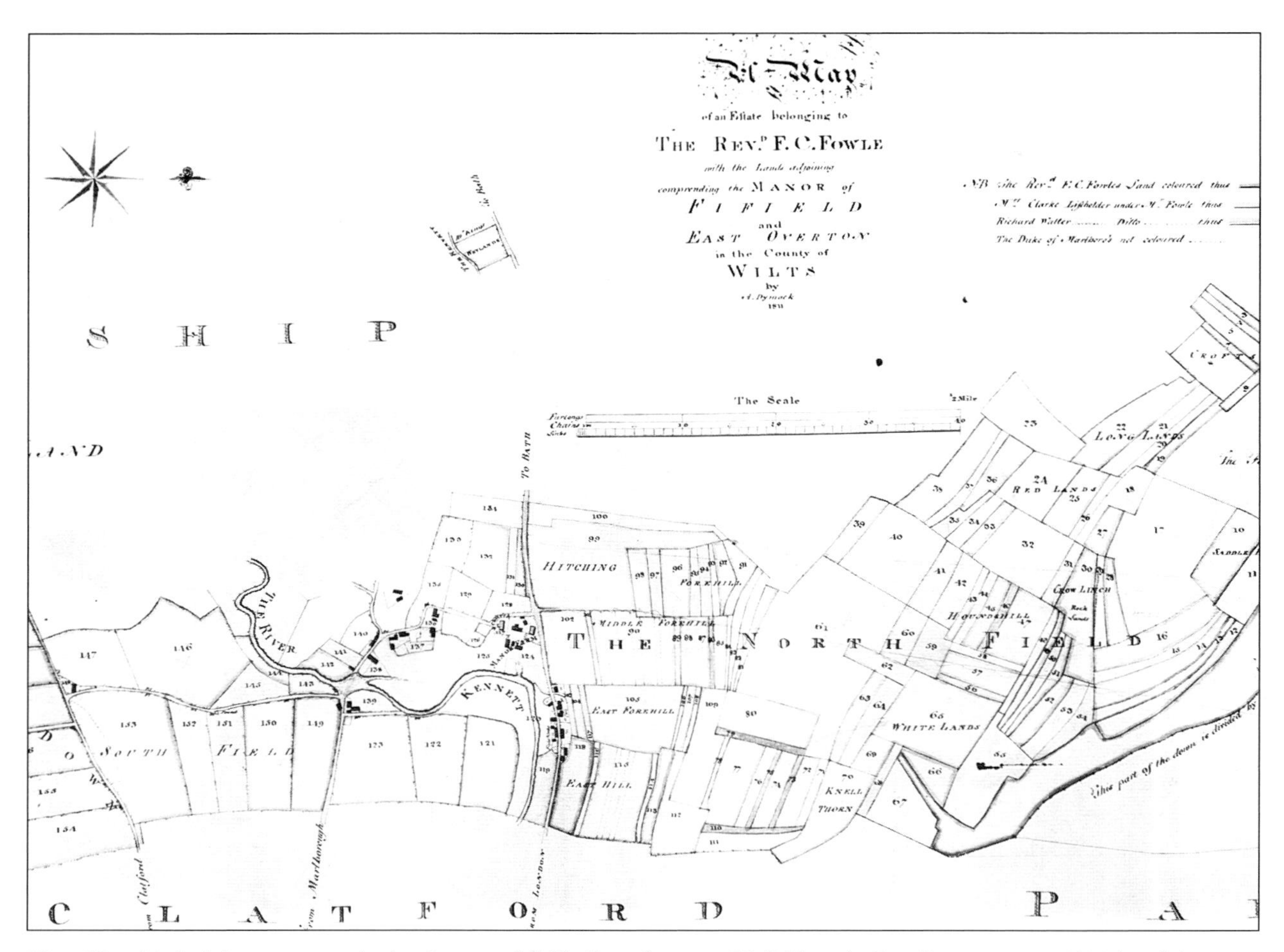

Plate XI Methodology: cartography, landscape and fields. Part of a map of Fyfield on the East Overton estate, 1811 (north is to the right) (reproduced by permission of the WRO)

change, was introduced into local considerations. It was also the technique that brought new environmental evidence to bear.

Some of this was given primary publication at the time (eg, Fowler and Evans 1967; Evans 1968; 1970; 1972, 380–1; Dimbleby and Evans 1974). When Evans subsequently incorporated his Fyfield/Overton work in synthesising publications, its significance percolated far beyond the parish boundary (Evans 1972, 316–21; 1975, 152–3; 1978, 121–2). The environmental dimension was also subsumed in other syntheses (eg, Fowler 1975a; 1981a; 1983a; Jones 1986, 13–15). One result of this environmental input from the start was that interpretation of the study area was always open to new evidence and interpretations, and has constantly been modified by them. Evans has continued his environmental investigations in the area, recently providing an overall assessment of the upper Kennet valley (Evans *et al* 1993). Other new evidence (Powell *et al* 1996; Swanton, pers comm), as the following account shows, was absorbed during the first half of 1996. Discoveries published then from along the line of the Kennet valley sewer pipeline in 1993 (Powell *et al* 1996) included environmental evidence interesting in itself and additionally so when related to data already obtained from the study area. The new evidence complemented that from the downs by including, in particular, information on valley in-filling and about second-millennium BC flora south of the river at Pound Field, West Overton.

Four points should be made in relation to the project methodology and the subsequent environmental evidence:

1 The reliance overall on visual collection during 1960s-style excavation, and the absence of sieving or

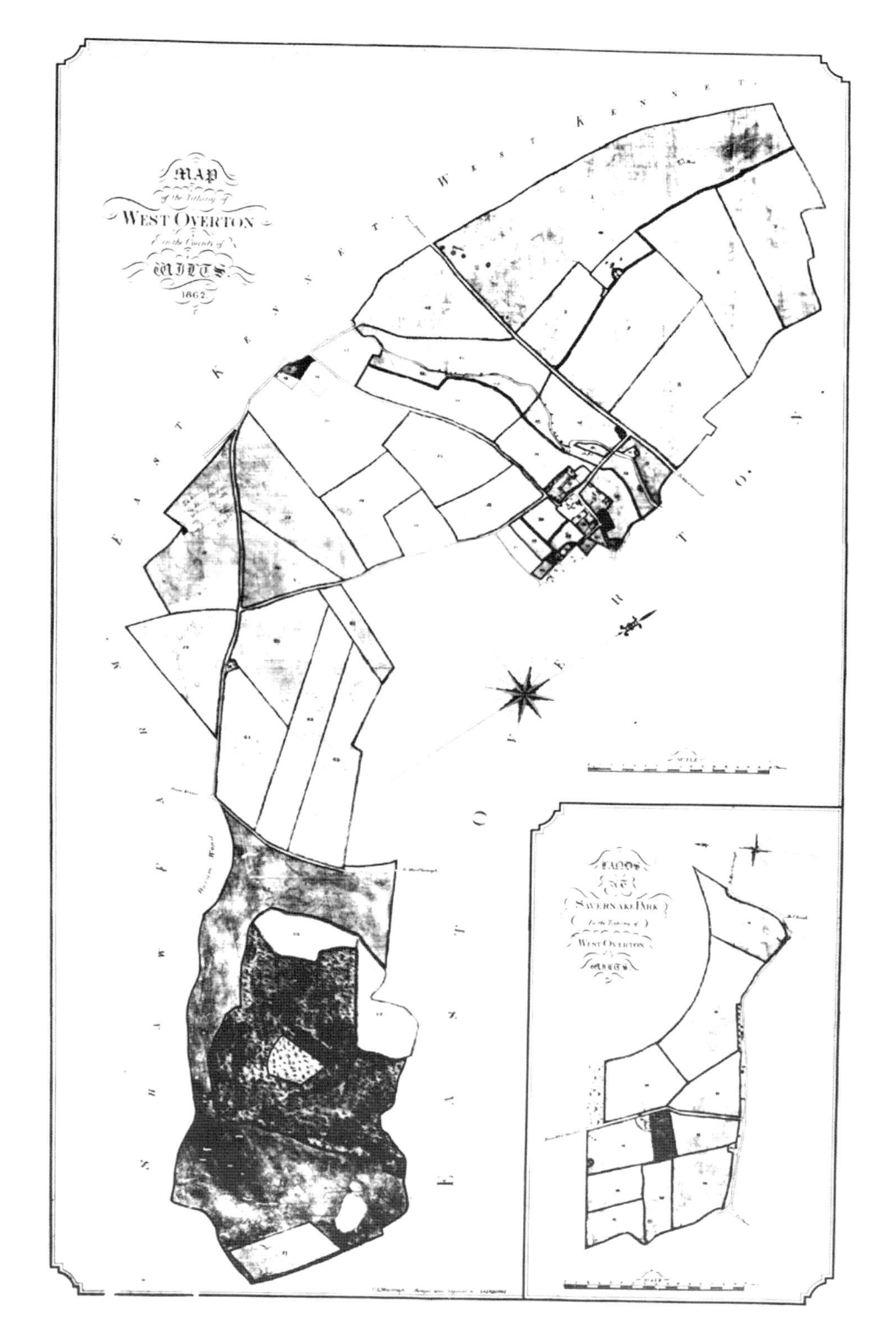

Plate XII Methodology: cartography and village morphology. Part of a map of West Overton, 1862 (reproduced by permission of the British Library)

flotation, have resulted in differential recovery of animal bones and charred remains, and no means of checking whether the absence of other organic remains was misleading or otherwise.

2 Soil profiles sampled for Mollusca occurred only on archaeological excavations; no off-site examination by coring, for example, was carried out. The main sequence, on OD X/15, was highly suggestive but can present only a partial picture of one portion of one site.

3 The methodology used for animal bone analysis has recovered less information than would be expected in a more recent study.

4 The charred remains and animal bones, although individually contextualised on site when recovered, have been reported on by species, not context, except for those associated with the pit deposits on OD XI. It is therefore difficult to make anything more than broad inferences about the nature of agrarian activity and impossible from this data to detect short-term change within main occupation phases.

These limitations affect the possible level of analysis, not the validity of the record as such or the inferences from it. These last are discussed in Part III with reference to other local sites, and a number of conclusions are presented.

IMPLEMENTATION AND EXAMPLES

To return to our initial, principal questions:

How, why and when did the landscape, particularly the landscape of the downs, evolve into its twentieth-century form?

and

What types of economic activity were carried out in the study area and how were they distributed within it?

The project enabled us to begin to look at not just abandoned settlements but at *patterns* of settlement through time and at different times, economically, tenurially and socially. We began to ask whether it was possible to define for pre-medieval as well as for medieval and modern times a history of the changing pattern of land-use while, at the other end of the range of ambition, tackling in detail questions about the chronology, extent and function of 'Celtic' fields. Our changing perceptions, both of the archaeology of our chosen study area and of the methods and theoretical constructs by which it might be examined, have subtly (or not) transformed these initial questions into the likes of:

How has the environment of the area changed over time and what were/are/have been the consequences of such changes? (Chapter 14)

and

What, in chronological terms, were the main phases of human activity in the area?

The fact that three quite large settlement excavations were conducted led, obviously but not in a way at first in mind, to consideration of how these settlements functioned within the landscape, with particular reference to economic activity (Chapters 14 to 16). The comparability, so we came to think, of the excavated settlements archaeologically encouraged thoughts, not here developed, about interpretative differences in considering small rural settlements with and without documentary evidence (cf Chapters 6 and 7).

The early application of ground reconnaissance followed by selective metrical survey led to a considerable increase in the amount of detail on Fyfield Down itself, an area where people seemed to have admired Allen's air photography (frontispiece) but not examined its subject matter. The site identified as WC and indeed the whole of Wroughton Mead (Chapter 7), for example, were given detailed examination and measured survey in 1959, the latter being revised and significantly supplemented during a window of opportunity presented by drought conditions and over-grazing in 1995 (Figure 7.5).

From such detail, different and better understanding flowed from an early stage, first on Fyfield Down, then on Overton Down and much of the rest of the downland, and later along the valley and across the lands to the south. We were led to certain sites on Overton Down, for example, by Crawford's air photography and cartographic analysis but detailed ground work and selective measured survey both improved the record and shed new light on the landscape history there (Figures 6.1, 6.2, 6.13 and 6.11). We were led to the detail of Totterdown by St Joseph's air photography (Plate XX) and, there again, measured survey was needed to help elucidation of that local landscape. Air photography, ground reconnaissance and measured survey were complementary and, as discussed elsewhere, in the three cases of Fyfield and Overton Downs and Totterdown were then, in turn, further complemented by excavation and, later, documentary research.

Much of this early fieldwork was published during the 1960s, in interim reports (Bowen and Fowler 1962; Fowler 1963a; 1967) and in a synthesis limited to the downland Roman material (Fowler 1966). But other surveys had also been carried out by the end of the 1960s, on sites and areas of other periods and situated off the downs, and such fieldwork continued sporadically into the 1990s. The well-preserved earthworks of the deserted part of the former East Overton village, for example, were both discovered during reconnaissance and planned as a discrete 'site' (Figure 9.2), though here they are also used in their settlement context (Plate XLVII). A re-examination of Wansdyke in Overton and Fyfield forty years after the definitive work of the Foxes (Fox and Fox 1958), and thirty years after our initial reconnaissance, led to further fieldwork, survey and re-interpretation (Chapter 13).

As a result, the record published here includes examples of plans already published (eg, Figure 6.14), sites already published but here with revised plans (eg, Figure 5.1), new plans of known sites (eg, Figure 13.1), plans of known but hitherto unplanned sites (eg, Figures 6.12 and 12.2) and plans of hitherto unpublished sites (eg, Figures 6.1 and 7.4). The original fieldwork may have occurred at any time from 1959 to 1998, but in any case what is published here will almost certainly have involved at least a revisit in the 1990s. Such fieldwork continues, and new observations continue to be made, for example, the slight traces of water-meadow ridges cut by and to the right of the abandoned leat on Plate LXIIIb, recorded in ideal conditions of sheep-cropped grass and strong cross-light on the morning of 14 November 1998.

That is exemplified south of the River Kennet. The archaeological record for that area is graphically displayed in Figure 12 .1, a map derived from the county SMR. The record is typical of many other places, comprising an assortment of miscellaneous observations, antiquarian record, chance finds and inexpert air photographic cartography and interpretation. This project has added little to that record in terms of archaeological specifics, but it has produced, for the first time, reasonable plans from measured survey and/or air photographic cartography, of West Woods long barrow (Figure 12.2), Shaw deserted medieval village (Figure 13.1) and Ring Close medieval deserted settlement area in West Overton village (Figure 9.2), and the prehistoric field systems on Boreham Down and adjacent area (Figure 12.3). Air photographic inspection and new RCHME photography plus field reconnaissance have also added, especially during 1995–7, a number of ploughed-out barrows, including the new long barrow above Lockeridge Dene with its nearby group of circular cropmarks and rectilinear enclosure (*SL*, colour plate 25). Another small rectilinear enclosure (at *c* SU 133667) is morphologically more akin to Middle Bronze Age ones elsewhere on the Marlborough Downs. Fragmentary air photographic traces of a larger, more circular enclosure on Lurkeley Hill suggest a site akin to 'Headlands' and Overton Down XI, with traces of probably prehistoric fields to its north and east in the area subsequently cultivated in the furlongs identified in relation to Crooked Crab, Hollow Snap and Alton Way (Figure 8.3b).

While fieldwalking was not seriously practised by archaeologists in the 1960s, clearly by the 1970s it had become an option (Shennan 1985; though not recognised as such in Fowler 1977). The decision was taken in the 1980s not to pursue it at Fyfield and Overton. Already the preparations for publication were protracted and daunting, and there were doubts about our ability to address the task properly (doubts which, with the advantage of hindsight, can now be seen to have been justified). So, rather than add another uncompleted set of data to the record, the extensive areas of arable in the two parishes were not systematically searched. Had they been, there can be little doubt that the record of the whole landscape would be different, and probably significantly so (cf Russett undated).

Mesolithic material, for example, is recorded in only four places: at North Farm, where it was sought in research excavations by Professor J G Evans; at Down Barn where it serendipitously occurred beneath an excavated earthwork (Chapter 6); at Shaw where it occurred in a ploughed field which was deliberately searched (Chapter 13); and near Bayardo Farm where the farmers have amassed a considerable flint collection from unsystematic but eagle-eyed collection over decades (Chapter 12). A clear implication is that systematic fieldwalking would alter the 'Mesolithic map' of the area, and the same is probably true of later, flint-using periods too. Here is a clear opportunity for someone to take up.

Two other ploughed areas were deliberately fieldwalked: once over the 'Headlands' settlement complex (Chapter 4, Figure 4.2), and many times over the eastern part of ODS, a Romano-British settlement (Chapter 6). The arable over the former west of the tithing boundary between the two Overtons was searched to test the hypothesis suggested by Plate XV that the circular enclosure, a possible villa and possible timber buildings (Chapter 4) were successively of the Early Iron Age, Roman and post-Roman periods. Potsherds identical to those from site OD XI were found over the first but nowhere else; late Roman potsherds and a few pieces of building debris were found over the second; and nothing was found over the last. East of the fence which splits ODS, the permanent arable always produces a few potsherds, characteristically of earlier rather than late Roman type; similar material occasionally comes from molehills on the western strip of the settlement in grassland.

Methods v and vi, aerial photographic reconnaissance and cartography, are also continuing, and they too continue to add data and, nearly always, further understanding of the study area. This approach culminated for project purposes in the production of the overall map of the northern part of the study area (Figure 2.1) rather than in the discovery of particular major sites. With such a map, for the first time it was possible to work from a reliable base, for although the map was limited to what could be mapped from aerial photography as distinct from the total archaeology of the area plotted, it linked earthworks to evidence in the extensive arable around all but the eastern side of Fyfield and Overton Downs. This was particularly helpful in understanding relationships between Overton Down and Avebury Down to its west, across a visual divide marked by The Ridgeway which could immediately be seen to be relatively recent (Chapter 2). To the south too, where numerous new sites were apparent in modern and indeed historic arable, landscape relationships of medieval times came into focus (Chapters 8 and 10). Cartography based on air photographs of the area south of the Kennet similarly brought to light suggestive relationships between forest, fields and tracks (Chapter 12).

Within the aerial cartographic approach to the landscape, air photography also provided what it is exceptionally good at, that is, the discovery of 'new' sites and, equally important, new contexts. Examples from arable which significantly affected interpretation of the study area were the soilmarks of prehistoric fields, not previously noted, providing a 'new' context for the well-known Manton long barrow (Chapter 5, Figure 5.3); a parchmark complex of round barrows and rectilinear enclosures close to a hitherto undiscovered long barrow above Lockeridge Dean (Chapter 5: *SL*, colour plate 25); and another settlement/burial complex named by us 'Headlands' north west of North Farm, West Overton, initially discovered as cropmarks by St Joseph (Plate XV) and then expanded in later soil and cropmark records by others (Chapter 4). This last in particular is a key site in the development of this landscape – and, indeed, of our understanding of it.

In comparison to aerial cartography, resistivity and magnetometer survey (methods viia and b) were insignificant; they were only used on two sites, WC in Wroughton Mead and Site XI on Overton Down, at both as an aid to excavation management (*SL*, colour plate 3). They were supplemented by a little *ad-hoc* dowsing on the latter (some data, including plots, in the Devizes archive). The Wroughton Copse site (WC, Chapter 7) was on Clay-with-Flints, which seemed to affect the efficiency of both methods; in particular, neither indicated features cut into the subsoil and then refilled with it. But, while the excavation was largely led by the visible remains of what were taken to be the sites of buildings, the complementary geophysical indications influenced decisions about placing cuttings in Enclosures A and B and, more importantly, about proceeding no further with excavations in them in the absence of visible and sub-terrestrial evidence of occupation (FWP 65).

Both methods proved effective on the Upper Chalk of Overton Down Site OD XI in locating some, though not all, chalk-cut features subsequently excavated (Chapter 6). The first phase of excavation there was earthwork-led, the results of which indicated the need for geophysics. So, on the flattened interior of a settlement with no visible, contemporary features, the excavation was guided in its second phase by geophysical results. Small test-pits were dug at all geophysical 'hot-spots', some being unexplained and some proving to be geological or man-made features. Among the latter were military features of the 1940s, including a rubbish pit, and settlement features of the first millennium BC. The foundation gully of Building 4, for example, was initially intersected from a magnetometer indication. On the other hand, geophysics and dowsing proved hopeless, and indeed misleading, in trying to locate the ditch surrounding the settlement. Given that its north-western arc was indicated by the slightest of earthworks, the whole of the rest of its course was eventually determined entirely by excavation around an area roughly twice as large as that indicated by sub-terrestrial data (Figure 6.2).*

Method viii, excavation, was clearly highly selective in this project, not only in its size, absolute and relative, but more particularly in its targets. It was targeted specifically to date, and to examine the structure of, fields, a key relict element of the downland landscape;

and to investigate the nature of some other features related to fields and land-use. The excavations are listed in Tables 1 and 2, summarised in Chapters 5–7, and reported on in FWPs 63–66.

Several other archaeological excavations have been carried out in the area and its immediate vicinity, both before and since 1959. Here they are tabulated and briefly summarised, selecting landscape data from a fuller treatment in FWP 66.

SUMMARY OVERVIEW OF METHODOLOGY

This suite of methods in the field 'worked' in the sense that it produced rational results amenable to checking by others and interpretable by us. Three weaknesses, however, stand out. This approach produces no finality; the data-set, the evidential base, is never complete, and in that sense interpretation is always vulnerable. Here, it is unlikely that major new earthwork complexes have been missed from ground survey but any day could produce significant new evidence. Ground disturbance, during farming, gardening or grave-digging, will extend the record qualitatively as well as quantitatively, and air photography will discover not just individual new sites but whole complexes. Such has been the experience while writing up the project (1995–8), with a view to 'finishing' it, and there is every reason to believe that such will continue. And another site like 'Headlands' could change the perception of this study area overnight.

That first weakness applies to many areas where the past exists in a working landscape. Our biggest weakness was that no systematic fieldwalking was undertaken on a regular basis. This reflects both the origins of the project, with its emphasis on earthworks and air photography, and a deliberate decision in the 1980s not to pursue the technique here.

The same was true of the grassland downs in respect of 'finds'. Numerous artefacts have been noted over the years, mainly in molehills and rabbit scrapes, but the fieldwork across the grasslands and through the woods has not involved a systematic search for artefacts. This means that the record is particularly weak in its information about 'flint sites' and almost certainly the near-absence of such information exposes a lacuna in our understanding of the landscape evolution in the study area.

A third weakness is obvious from our comments on geophysics. That whole dimension in this landscape has barely been touched. With modern equipment capable of covering larger areas, much more could readily be examined. For example, a couple of days' work could probably resolve at least some of the queries relating to the settlement complex of ODS and Sites OD XII and XIII (Figure 6.11).*

A fourth weakness only became apparent with time, and is not a mistake that would be made now by this author or anyone else starting such a project. The fact is that, in 1959–60, no one had any idea either that the work would take so long to reach a satisfactory point at which to stop or that so long would be spent on the project. By 1966, nevertheless, we were referring to 'the 10-year project …' (Fowler 1967, 16), meaning 1969, by which time all project excavation had in fact ceased. All the same, in general no long-term controls or standards were built into the methodology at the beginning: how we did what we decided to do developed empirically. The notable exception was, however, excavation. A fairly standardised excavation recording procedure was established for all project excavations in 1960 and, though its details developed for the better over the next three years, it was thereafter followed almost automatically throughout the big excavations of OD XI and XII and their numerous small associates. It was the record of the fieldwork which was unsystematic, with no overall ground-control, such as would now be inserted from GPS, no standard scales or conventions (though generally following the RCHME's practices), and no established routine method of recording minor features or surface material. The same applied in general to documentary and cartographic work until a modern, disciplined approach was brought to bear from 1995 onwards. The absence of such discipline in the compilation of the record throughout the project is very definitely a weakness, right up to the present. Preparing this volume and the archive would have been much simpler if overall project management, in the best modern sense, had been applied from the start.

Three lessons from this approach and its implementation are clear. The work cannot ever be completed, either conceptually or even at the practical level of finding all that there is to be found on and in the ground. In the 1990s, the point is a truism; so it is probably worth stating that such a relativist view was not part of the intellectual framework in the late 1950s when the project began. A paradigmatic change came with new thinking about the nature and interpretation of archaeological evidence (eg, Clarke 1968), the 'quantitative revolution' (Fowler 1972a, 105) in the landscape from air photography, and a significant increase in the size of the archaeological response to the threats to it in the 1960s and 1970s. As we finished our intensive, systematic field surveys in the early 1970s, we realised we could never 'finish', a realisation which, twenty-five years later, has so far proved correct. Meanwhile, the relativist concept has been accepted (Fowler 1981c) as the basis for the official curation of archaeological data at county and national levels. There, Sites and Monuments Records and the National Monuments Record are regarded merely as the best set of data available at any one moment. Our records and thoughts about Fyfield and Overton were overtaken (yet again) as we prepared this monograph.

Secondly, we learnt to adapt our methodology to current land-use, and so came to realise that past land-use was fundamental to what we were able to find. Again, the point is well understood now, and has been since Taylor's (1972) adumbration of the 'zone of destruction' phenomenon. Beginning with our focus very much on the excellently preserved palimpsest of earthworks on Fyfield and Overton Downs, we realised very early on that, to understand the downs, we had to extend our survey south into the valley and beyond to the woods (Bowen and Fowler 1962, fig 1; 1966). The realisation was triggered by the recognition of a medieval settlement on Fyfield Down (Chapter 7) and extensive patches of ridge-and-furrow, phenomena which clearly could only be understood within manorial and parochial frameworks (Figure 2.3). That thought encouraged questions about the possibilities of similar tenurial arrangements before the thirteenth century AD. But we also realised earlier on, not least because of comparable survey by the RCHME in Dorset, that to look at the earthworks of the two downs as entities was quite inadequate. We had to look at the areas immediately outside the old grassland of the National Nature Reserve, to the west into Avebury parish as well as south towards Fyfield and Overton villages. Just because those areas were under plough – and had been in places for at least a thousand years, as we discovered later – did not mean they could be ignored. It took a long time to pull an overview of these areas together but when it eventually arrived (Figure 2.1), we were able to appreciate our downland earthworks in a meaningful context.

A third point is of a different nature. Perhaps our experience overall leads to the placing of a large question mark about the reliability and therefore validity of much archaeological field survey as practised in Britain in the 1990s. In so far as much of it is based on the single visit at a time often dictated by non-archaeological considerations, it probably gives rise in numerous cases to incomplete or misleading data and perhaps, therefore, wrong inferences. Archaeological information obtainable from field survey is often not necessarily self-evident, sufficiently secure nor complete enough on a single visit, to support weighty judgements.

* A gradiometer survey was carried out over the eastern part of the 'Headlands' enclosure in two days in October 1999 (*see* Plate XV and Figure 4.2). It was immensely successful in locating details of the perimeter, third entrance and over the interior (Hamilton, M A, Dennis, I and Swanton, G, nd but 2000, *A Geophysical Survey at Snail Creep Field, North Farm, Wiltshire*, 1999, Cardiff Studies in Archaeology, Specialist Rep 15, Cardiff University).

Part II

A Landscape and its Interpretation

We now move on to an exploration of the study area and begin to interpret it. We look at it in three parts: the northern downland (Chapters 4 to 7), essentially the area shown on Figure 2.1; the valley of the River Kennet (Chapters 8 to 11; Figure 8.1); and the southern upland characterised above all by woodland between the valley's southern slopes and the southern boundary of the parishes (Chapters 12 and 13; Figure 12.1). To handle the evidence, and to begin to tell a story, we break up each part into chapters, topographically arranged. In this first section, the northern downland, the focus of each necessarily multi-period chapter, leads the arrangement from earliest to latest. Our route is indicated in Figure 3.4.

Chapter 4

The Northern Downland: Overton Hill, The Ridgeway and 'Headlands'

On the western edge of the study area the southern part of the northern downland abuts The Ridgeway, stretching south west from Pickledean to Overton Hill and the prehistoric monument known as 'The Sanctuary' (Figure 4.1, Plate XIII). Thence, this first area to be examined in more detail keeps north of the Roman road and stretches eastwards over both arable and grass downland to an eastern boundary marked by a curving hedge-line that is the historic division between Saxon West and East Overton. 'Headlands' were referred to at the time, and we use that word as an identifier in our landscape (Figures 4.1 and 4.2).

Apparently almost featureless at first glance, the area primarily selects and defines itself archaeologically (Figure 4.2). Indeed, it could be argued that this particular local landscape, incorporating The Sanctuary, barrow group H (*see* Figures 2.1 and 2.4) and the area of Figure 4.2, was a focal zone for the whole of the study area throughout prehistory and into the middle of the first millennium AD.

Hedgeless and under plough, the southern half of this exemplary 'window' into the landscape lies on a gentle south-facing slope dropping towards the Kennet. Easily mistaken as the product of modern grant-led arable farming, its open character today is precisely because it was not enclosed (1802, 1821). In medieval times the area was the whole or part of 'North Field' (1631); it might have been part of Anglo-Saxon West Overton's limited downland pasture five hundred years earlier (*see* Chapter 8). The manorial map of 1783 shows it divided by a north–south boundary. West of this was 'Farm Down' and 'The Cow Down', or, as the earlier, rougher version of the map calls it, 'Common Cow Down', presumably all pasture. To the east lay 'Upper', 'Middle' and 'Lower' 'Fields', presumably all arable. Farm Down was separated from Cow Down by an open boundary containing two standing stones and the barrow near The Ridgeway. The fields were divided by open boundaries, as was Middle Field from its Upper and Lower counterparts. The lower half of the central boundary, however, was demarcated by 'pales'. The permanent pasture on its west was ploughed up in 1960 and restored to grass in 1993.

The northern part of this 'window' into the landscape lies beyond a hedge, still a tenurial division, which perpetuates the tenth-century estate boundary between West and East Overton (Figures 4.1 and 4.2). Beyond are apparently unremarkable hedged fields beside The Ridgeway, which almost certainly reflect a previously unrecognised post-medieval park (Figure 4.1; *see below*).

From west to east this area (Figure 4.2) stretches over 1km between boundaries defined on Saxon charters (S449 and S784). The western limit is marked by the *Herepath*, on the line of the present Ridgeway, and the eastern by the sinuous boundary with East Overton, passing through the 'Headlands' complex (*see below*). This line persists today as a bridle path running to a ford

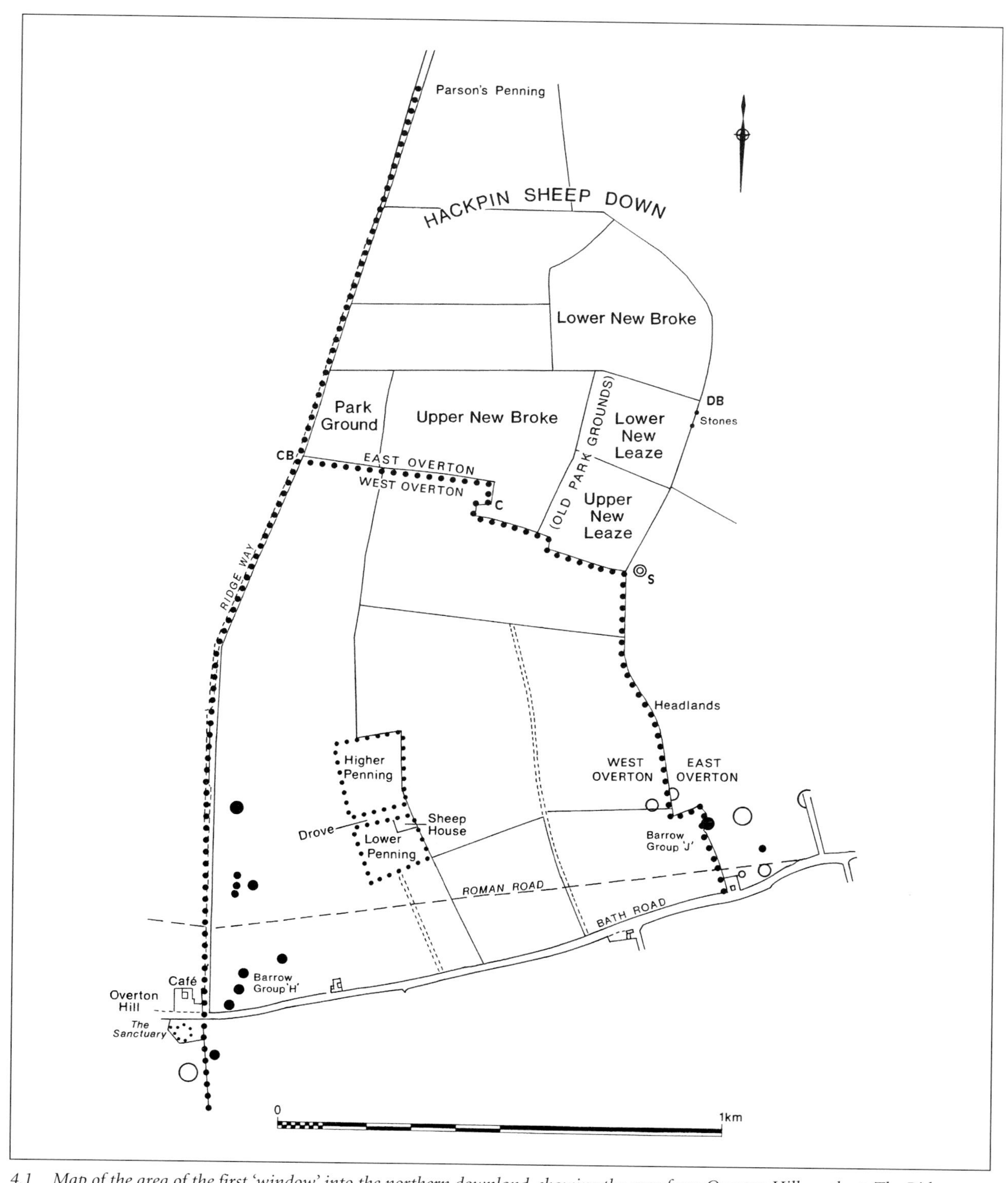

4.1 *Map of the area of the first 'window' into the northern downland, showing the area from Overton Hill north up The Ridgeway to parts of Overton Down, respectively formerly a park and called 'Hackpin', and east across to 'Headlands' on the boundary between the Anglo-Saxon estates of West and East Overton; cf Plate XIV*

DB = Down Barn; CB = Colta's Barrow; C = 'Crundel'; S = 'Scropes pyt'

Plate XIII *Overton Hill with The Sanctuary newly outlined in concrete, the 'Seven Barrows' and Roman road under grass and the transport café in embryo (photograph taken by Major Allen in the early 1930s; NMR ACA 7094 915,* © Ashmolean Museum, University of Oxford*)*

across the River Kennet, whence it ascends to Hursley Bottom in West Woods (Chapter 16). Lines of communication and settlement are a major component of this particular area. Two barrow groups (G, J) lie just outside the west and east extremes (Figure 4.1), but the Seven Barrow group (H) and The Sanctuary are included.

We begin with a brief description and interpretation of the prehistoric–early medieval archaeology in the southern half of this area, review the Anglo-Saxon charter evidence and some documentation, and end with the putative park to the north.

THE ARCHAEOLOGY

The main archaeological features are disposed essentially in two clusters (Figures 4.2 and 4.3), respectively on the west and east of the area. Presented in gazetteer form, followed by a discussion and interpretation, the first section commences with the western grouping around Overton Hill. The second section deals with the 'Headlands' complex on the eastern edge.

The Sanctuary

SMR no. 159. Plate XIII. Cunnington 1931; Pollard 1992, and discussion below.

The site has an added interest now as an early (*c* 1930) example of an attempt to display for public edification the structural components of a mainly timber site (Chapter 17).

Seven Barrow (or Overton) Hill barrow group (Group H)

SMR nos 657, 658, 659, 660, 661, 662, 663. Figures 2.1, 2.2; Plate XIII.

The seven barrows lie immediately east north east of The Sanctuary and are roughly aligned south–north, excepting 663, which is offset to the east. South of The Sanctuary is barrow 610, marked as 'site of' by the OS, but actually surviving as a visible swelling in the arable. Best viewed as a southern outlier of the 'Seven Barrow' group, its westerly offset position mirrors that of 633 to the north. North of The Sanctuary was a pair of barrows, 611 and 612. The name 'Seven Barrows' has been in use for at least a millennium; in AD 972 it was called *seofon beorgas* (S784) and in the 1670s it was 'Seven burrowes Hill … near London way' (*WAM* 7, 227).

Roman road, *Aquae Sulis ad Cunetio*

SMR no. 160.

Now reverted to pasture after more than thirty years of cultivation, the road (Margary 53) is still visible as a low spread *agger*. In 1884 the causeway measured 5ft (1.5m) in height, comparable to Ackling Dyke in Cranborne Chase (Bowen 1991), and was 18ft to 20ft (*c* 5.5m) across, with distinct ditches on either side (*WAM* 33, 326). Overton Hill marks an alignment change in the route, aiming for a point in front of North Farm where the modern A4 road takes up the Roman line. Its first cartographic depiction is in the *Antonine Itinerary* (Rivet 1970, *Iter* XIV), reappearing some 1,500 years later on Hoare's map of 1819. A charter of AD 949, relating to the bounds of a four-hide estate at West Kennet (called, confusingly, *Ofærtune*; S547), follows the Roman road (*stræt*) from Silbury Hill to the West Overton boundary, suggesting the road was still in use, even if only as an estate boundary, in the tenth century.

Barrow G6b

SU 11966835. Excavated in 1962. Smith and Simpson 1966.

Situated north of the Roman road and best regarded as an outlier of the Seven Barrows, this round barrow is unusual within the project area in having been well excavated and published. The pre-barrow sequence is crucial to our landscape interest: a 'quantity of pottery' suggested 'several phases of fairly intensive Neolithic activity' supposedly 'from successive short-lived settlements' spanning the third and reappearing in the earlier second millennia BC, ceasing a short time before burial began (Smith and Simpson 1966; FWP 66). The first funerary structure consisted of a flint and sarsen ring-cairn around a pit-burial, subsequently covered by a central turf stack *c* 6m in diameter within a mound *c* 60m across and at most *c* 3.30m high, with no surrounding ditch. A central pit contained a primary inhumation burial with a Beaker and bronze awl of the sixteeenth century BC. Five secondary inhumation and six inurned cremation burials occurred in the mound. A grey clay sealing the primary burial pit contained snail shells 'of land species indicative of dry calcareous grassland' (Smith and Simpson 1966, 142. For Saxon inhumations, *see below*).

Roman burials

SU 118683. Excavated in 1962. Smith and Simpson 1964.

Three small, low mounds, G6, G6a and G7, west of barrow G6b, lay in a row aligned south–north, north of and roughly at right angles to the line of the Roman road (Figure 4.3). All were under plough between 1962 and 1993 but their sites have now been restored to grass.

A pit on the south-west side of G6a, *c* 30m west of the prehistoric G6b, contained late Neolithic pottery comparable to that from beneath G6b (Smith and Simpson 1964, 83).

G7, the most northerly mound, was 0.60m high with a surrounding palisade trench of *c* 7m diameter enclosing a pit containing sherds of Roman pottery. A Saxon burial cut the trench on the north east. G6a was similar to G7 but only *c* 4.5m in diameter, with remains of cremated bone and potsherds in the mound material. G6 had similar dimensions to G6a but was much disturbed.

The traces of close-spaced wooden posts in the ditches of G7 and G6a suggest these structures took the form of wooden-sided cylinders or timber-revetted earthen mounds enclosing one or more cremation burials. The pottery recovered, especially that from barrow G6b, suggests 'an early 2nd-century date may fairly be inferred for the whole assemblage' (ibid, 79–81).

Anglo-Saxon cemetery

SU 119683. Excavated in 1962. Smith and Simpson 1964; Eagles 1986.

The inhumation burials of a fifth-century AD adult female, a sixth-century warrior and two children were excavated in the barrow group, suggesting sub-Roman/early Saxon settlement in the area (Eagles 1986). Four graves were found in G6b, a woman and a male and two child inhumations. A shield boss and spearheads suggested a sixth-century warrior burial. Another child inhumation lay at the edge of G7 and further Saxon burials around the Roman tombs were suspected (Smith and Simpson 1964). The *Anglo-Saxon Chronicle* records battles at Barbury Castle, 10km north of Overton Hill, in AD 556 and 592, and at *Wodnesbeorg*, 5km to the south, in AD 715 (Eagles 1986).

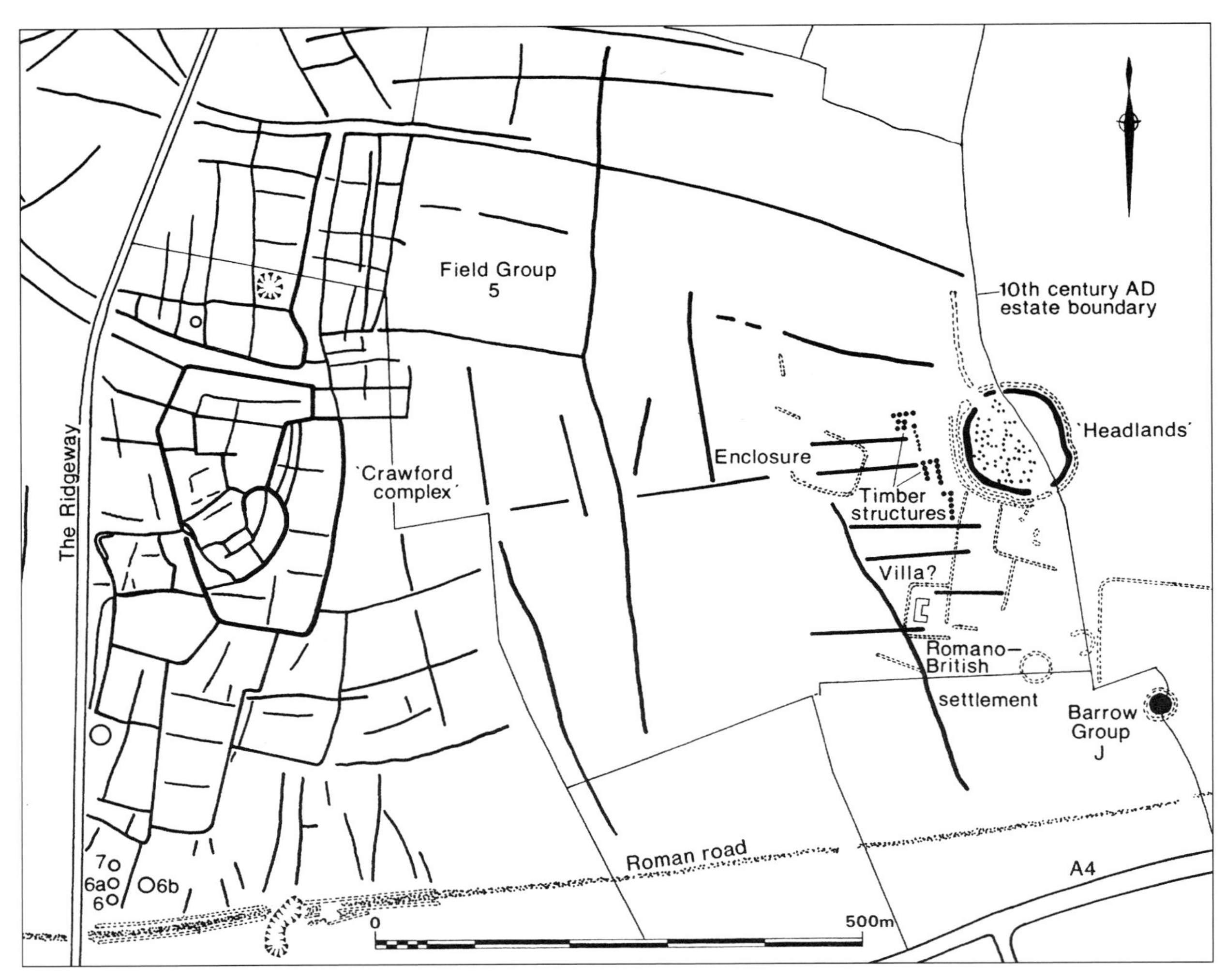

4.2 Map of the area from the 'Crawford complex' on Overton Hill to 'Headlands' in the northern part of the former Anglo-Saxon estate of West Overton, showing a field archaeology of earthwork and cropmark evidence from prehistoric to early medieval times built up from many air photographs and maps

Settlement and field complex

Centred at SU 125683, partly SMR 688. Figures 4.2, 4.3; Plate XIV.

This settlement complex, immediately east of the modern Ridgeway, was discovered and photographed from the air by O G S Crawford in 1924. It is referred to hereafter as 'Crawford's complex'. The area was at that time still pasture and the whole complex and its environs survived as earthworks (Plate XIV). Its significance has not previously been fully appreciated because Crawford's photograph was not published and consequently the complex, having no record in antiquarian literature, went unnoticed. It appears perfectly preserved on RAF air photographs dating from 1952 and, by implication, continued in that state until the ploughing of 1962 (Smith and Simpson 1964, 68). At this date our project's ground coverage had not yet reached Overton Hill and the complex never, therefore, enjoyed ground survey. As a result, no serious protest was made when the area was ploughed in 1962, though that year saw the, at that time, customary reaction to agricultural threat: the excavation of the visible barrows (Smith and Simpson 1964; 1966). Bronze Age barrow, Roman tombs and Anglo-Saxon cemetery were, however, treated and depicted as being in a landscape void (Smith and Simpson 1964, figure 1, top right). Most of the settlement area remains under cultivation.

Figure 4.2 is an attempt to reconstitute the site and

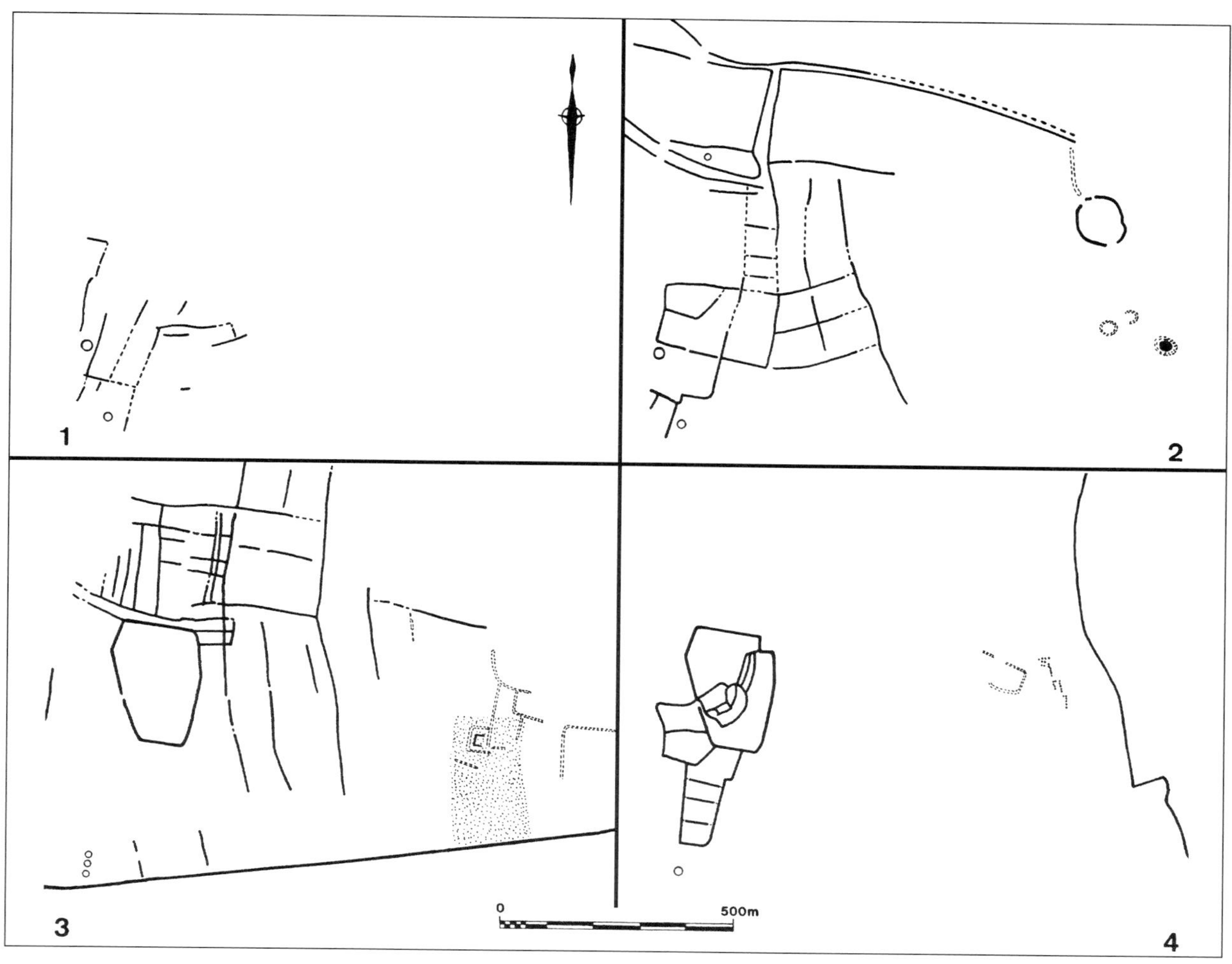

4.3 *Interpretative diagram based on the evidence displayed in Figure 4.2 and proposing four major chronological phases in the use of the landscape:* **1** *fragments of Early/Middle Bronze Age field system related to Early Bronze Age round barrows;* **2** *fragmentary Middle/Late Bronze Age 'landscape of enclosure' with ancestral round barrows and peripheral pasture, settlement, fields and tracks;* **3** *Roman landscape with new road and rectilinear fields, ditched enclosures and a villa, with a large hexagonal embanked enclosure incorporating earlier field boundaries to the west;* **4** *fifth–seventh-century* AD *landscape with small enclosed settlement and small fields imposed on late Roman earthwork enclosure, angular ditched enclosure and large rectangular timber structures in echelon-like arrangement parallel to an estate boundary which might well have been shown on diagram 3*

its environs from air photographs. The plan, essentially mapped at 1:10,000 and then prepared for publication at 1:5,000, was composed from many different air photographs taken between 1924 and the mid-1980s. Although most of the data were manifest as crop and soilmarks after 1962, their elucidation was signally aided by the availability of earlier air photographs of the complex as earthworks. To complete the record, the RCHME plot of the same settlement complex (Figure 2.1) was integrated into our independently created air photographic plan to produce a single site plan. This was then amalgamated with the RCHME cartography of the area east to 'Headlands' to produce the slightly schematised but reasonably accurate Figure 4.2.

'Crawford's complex'
Figures 4.2, 4.3; Plate XIV.

The most substantial enclosure is the seven-sided example enclosing *c* 4.5ha with an entrance on the west. It is clearly superimposed on the Bronze Age fields. Within it was a 'kidney-shaped' enclosure of *c* 0.8ha. Some air photographs indicate the probability of a rectangular building within it, with small plots fitted into the outlines of the earlier fields immediately to its north west and still within the large rectilinear enclosure. Although such downland enclosures are notoriously difficult to date (Fowler *et al* 1965), a Roman or later date is suggested by analogy with similar, but larger, enclosures in Cranborne Chase, notably Rockbourne Down and Soldier's Ring, Hampshire (Bowen 1990, 94).

'Headlands' Early Iron Age settlement enclosure
SMR no. 203/674/675; AM 822, now SM 21763. Figures 4.2, 4.3; Plate XV.
Settlement complex so-named (by us) from the reference to this area in the West Overton Saxon charter as west *heafde* ('west headlands'; S449; *see below*). Situated 0.5km to the east of 'Crawford's complex', it is known entirely from aerial photography and geophysics, supplemented by limited fieldwalking. The principal elements are:

203 A slightly ovoid ditched enclosure of 'Little Woodbury type' encompassing *c* 1.5 ha. (3.7 acres), morphologically similar to OD X/XI (Chapter 6). Entrance and 'antennae' are unusual in facing to the north west (cf Bowen and Fowler 1966, figure 1), though recent air photographs strongly suggest further anomaly in the form of a southern entrance too. Air photographs also show many pits and other dark features on the interior (Plate XV). Early Iron Age pottery, burnt flint and sarsen fragments, similar to the main phase on OD XI, have been collected from the interior and its immediate vicinity. The enclosure is divided almost equally north north west–south south east by the tithing boundary, itself on the tenth-century estate boundary between West and East Overton. It can be argued that this boundary swerves north west as it climbs from the Roman road in order to dissect, perhaps even bisect, the Early Iron Age settlement enclosure.

674 Three sides of rectangular ditched enclosure known only from air photographs.

675 Rectangular ditched enclosure, probably around a small, winged Romano-British villa which shows reasonably clearly on Plate XV; not accepted as such by the RCHME. Romano-British pottery has been collected from ploughsoil here and further south. A ditch clearly runs between 203 and 675.

Arranged in a staggered line to the north west on Plate XV are three possible rectangular features defined by pit-like blotches, arguably large post-holes. Not recognised as such by the RCHME, which nevertheless plots the features (Figure 2.1); the air photographic evidence hints at large timber buildings. Fieldwalking produced no artefacts. The site requires sub-surface investigation to establish its nature. One possibility is a sub/post-Roman successor to the villa (Figure 4.3, 4).

303 A cross on the SMR map, probably the find-spot of Romano-British material recorded in *VCH* I, 121, appears to be within a settlement area, partly visible on air photographs including Plate XV, stretching perhaps between SMR 675 and the Roman road (cf Bowen and Fowler 1962, 101, B1).

Barrow group J

Figures 2.1, 4.1, 4.2, 16.8.
Included because of its relationship to the Anglo-Saxon estate boundary.

665 Here refined as 665a and 665b for two barrows, both known only as cropmarks. Barrow 665a lies on the east side of the Saxon boundary at a point where it 'dog-legs' before resuming its south south east–north north west course. An obvious inference is that the boundary deviated to respect an existing landmark(s). The West Overton charter refers to two barrows hereabouts and the boundary clearly passes between them. The recent recognition of 665b (*c* 110m west north west of 665a and just to the west of the boundary) provides a pleasing concordance between documentary and archaeological sources.

666, 667, 668, 670 Round barrows which, together with barrows 665a and b, make up an irregular linear cemetery west and north of North Farm. The group occupies a somewhat unusual topographic position, strung out along the 150m contour close to the valley floor of the River Kennet. Barrow 667 is right under the north-west corner of the farm's domestic surrounds and 668 is in an almost identical situation at the north-east corner of the farm *enceinte*. Even though the present buildings and layout do not appear to pre-date 1801, the evidence hints at possible long-lasting influences of prehistoric land-markers, especially where there could be an argument for North Farm as the eastern tenurial inheritor of the 'Headlands' complex (*see below*).

669 A round barrow, probably to be regarded as part of the same group, but out of line and further down the slope, south of the Roman road and right on the northern edge of the present A4.

Plate XIV *Overton Hill: 'Crawford's complex' and related features as recorded on a previously unpublished air photograph taken in December 1952 (cf Figure 4.1). North is to the top; cf Figures 4.1 and 4.2 (NMR 540/958,* © Crown copyright. NMR*)*

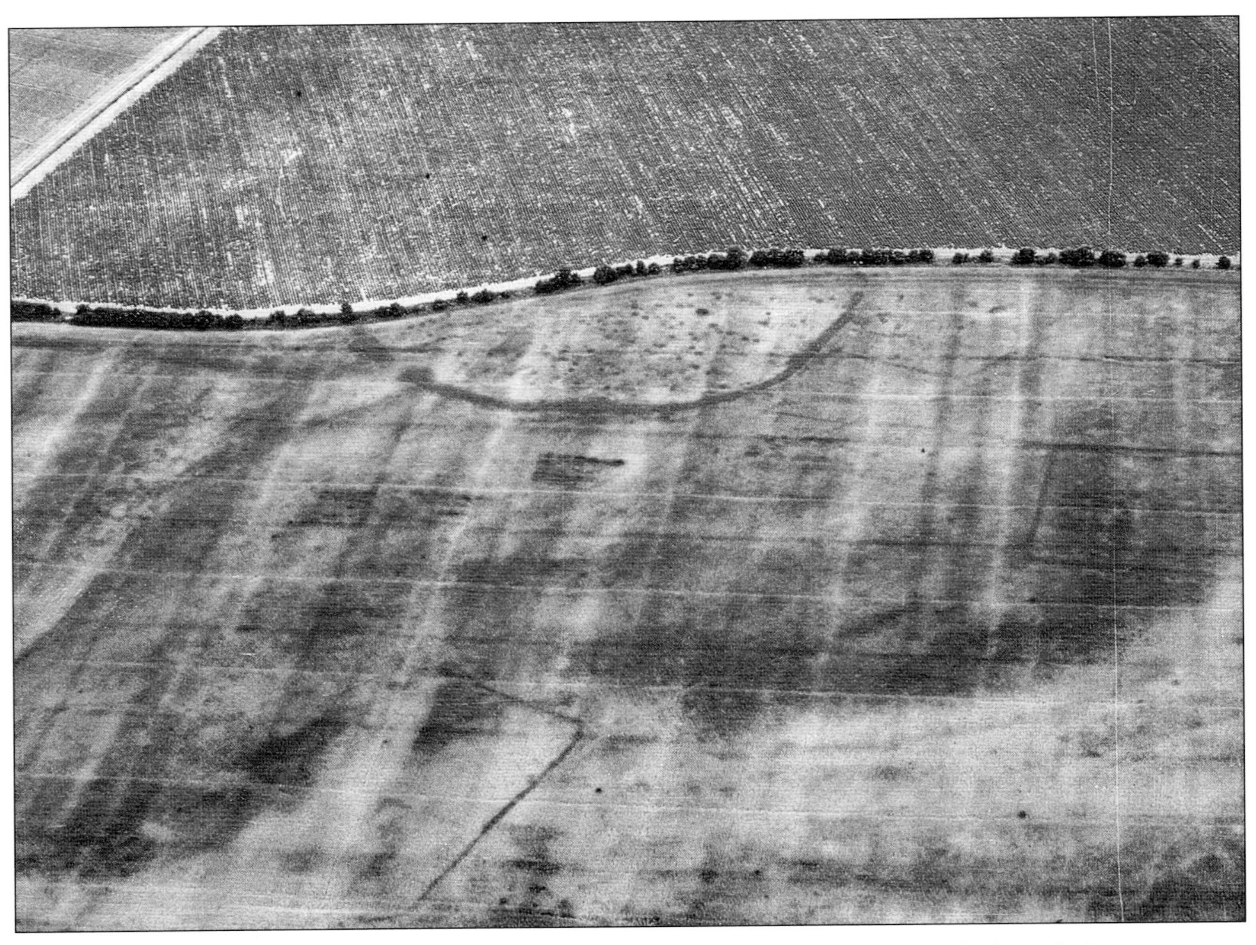

Plate XV *'Headlands' from the west, showing the Late Bronze Age/Early Iron Age enclosure overlain by the Anglo-Saxon estate boundary with various cropmarks, including a possible villa and large rectangular timber buildings (AW-27,* © Cambridge University Collection of Air Photographs*)*

INTERPRETATION

The location of the area in Figure 4.2 is vital to its appreciation (cf Figures 2.1 and 4.1). The Ridgeway is largely irrelevant as it post-dates everything on the map except the A4 road and some modern field boundaries. The map's southern edge is the *Londinium–Aquae Sulis* Roman road. Immediately off the south-west corner are The Sanctuary and the Seven Barrow group of round barrows. Much of the history of the study area, or at least some of its main phases, could well be represented in this figure, especially considered in context (Figure 4.2).

The Sanctuary itself has recently enjoyed a re-examination (Pollard 1992). It was preceded, as were other monuments in this locality, by 'pre-monument activity' represented by Neolithic material. Here, such material was considered to be residual and, unlike the excavators' interpretation at barrows G6a and b (*see above*), was suggested to 'reflect activity of a special nature or simply the middening of ceramics away from areas of flint working and habitation' (Pollard 1992, 219). Presumably Pollard means the manuring of farmland, for arable or pasture. This, rather than some 'special activity' anticipating The Sanctuary's construction, would fit a picture of a landscape being increasingly worked for farming from non-monumental habitations in the earlier third millennium. When the monument came to be built around the middle of that millennium, and used over perhaps five hundred years, its landscape significance appears to have been expressed in two ways. In general, it was part of what can reasonably be recognised as a 'sacred geography' in the Avebury district, one in which there was 'little division between areas of the landscape involved in routine, day-to-day, and special activities' (Pollard 1992, 225). More immediately, Figures 4.2 and 4.3 provide hints as to possible functions and physical elements in that landscape, both during and after The Sanctuary's heyday.

Largely on structural grounds, The Sanctuary has understandably been seen as west-facing, relating to the West Kennet Avenue and Avebury. The depositional patterning within the structure, and access to it, so clearly elucidated by Pollard (1992, figures 5, 6, 8), strongly points to an eastern and northern orientation of movement and activity. This provides an alternative 'landscape context' where The Sanctuary faced and perhaps welcomed people coming down the long spine of Hackpen Hill or along the Kennet valley. After its demise as a destination, a lingering tradition of sanctity may be implied by the continuation of funerary activities. This occurred not on the western or 'official' side, but on the east, where an arc of round barrows (group H) developed and where, coincidentally or otherwise, people were still being buried two and a half thousand years later.

There may be unrecognised elements of the third millennium BC on Figure 4.3. Excluding the material from under barrows G6b and G7, the earliest features seem to be fragments of the field system which sweeps round from the north west and barrow group G. The two double-banked droves (Figure 4.3, 2; Chapter 2) may be part of that mid-second-millennium landscape: the southernmost approaches 'Crawford's complex' and the more northerly the 'Headlands' complex. They are certainly integral with a more extensive and cohesive pattern of fields which overlies the earliest fragments. Given the association with the Early Iron Age 'Headlands' enclosure, this phase could be contemporary with Phase 3b on site OD XI, Overton Down, dated to the first half of the first millennium BC (Chapter 6).

That landscape was in turn modified by the creation of longer, rectilinear fields (although the southern drove is retained). Probably of first–second-century AD date (Figure 4.3, 3; Chapter 2), this Romano-British landscape included the Roman road, the second-century tombs (barrows 6, 6a and 7) on Overton Hill and a possible villa with associated settlement stretching down the south-facing slope from 'Headlands'. The evidence for the villa is suggestive rather than definitive. Within the small enclosure, 675, air photography (Plate XV) seems to outline a small corridor building with short eastern wings at both north and south ends. Another ditched enclosure, *c* 150m north west, is undated but seems to relate to the 'Roman' layout. A villa here would also provide a context for the tombs on Overton Hill. These would have been visible from the 'Headlands' complex and can be broadly compared to the hilltop location of the temple-mausoleum associated with the villa at Bancroft, Buckinghamshire (Williams and Zeepvat 1994).

The large polygonal enclosure within 'Crawford's complex' would fit comfortably into a Roman landscape, perhaps late in the period and contemporary with OD XII (Chapter 6) and, arguably, a villa. Its size and shape suggest a function associated with livestock.

The most striking feature of this landscape is the 'kidney-shaped enclosure' within the polygonal one (Plate XIV; Figure 4.3). It looks like a small, enclosed settlement with 'gardens' or 'paddocks'. It could be earlier than the polygonal enclosure, but interpretation of the air photography, coupled with the non-conformity of its alignment in relation to surrounding features, strongly suggest that it may have been inserted into an existing but 'empty' embanked enclosure. A further, and slighter, rectilinear enclosure on the south west of the polygonal one may also be associated. This, along with the kidney-shaped and polygonal enclosures, clearly overlies 'Celtic' fields.

The important point is, whatever the detail of the sequence, that there appear to be two phases of earthworks on top of Bronze Age fields. Overton Hill, despite its flattened appearance now, is archaeologically on a par with other areas of the northern downland in having a long structural sequence, both earlier and later than the prehistoric 'horizon' provided by the multi-phased 'Celtic' field systems.

'HEADLANDS'

This complex, in conjunction with barrow group J, is of great importance as a local centre. Plate XV shows very clearly part of the earlier first millennium BC enclosed settlement and a possible Roman villa. It also shows three clusters of dark 'blobs' arranged in linear fashion immediately west of the enclosure (slightly exaggerated in size on Figure 4.3, 4 so as to be visible). The southernmost and central rectangular arrangements, the northern single line, and the northern line of the west–east rectangular arrangement are all *c* 20m long. The widths of the three arrangements are all *c* 10m–11m. Such consistency clearly indicates deliberation, leading to the suggestion that the features represent either structured pit-digging or timber structures with large post-holes. The individual 'blobs' seem to be in the 1m–2m diameter range.

On balance, it is suggested these phenomena represent three or four rectangular timber buildings. If so, they appear to be double square in plan and of the order of 20m by 10m (*c* 66ft by 33ft). Their date is unknown, though they do not appear to relate spatially to the fragmentary but reasonably certain Roman pattern. Fieldwalking has failed to collect any diagnostic material over their position, a result in sharp contrast to the Early Iron Age potsherds to the east and the Romano-British material slightly to the south. The shallow soil and exposed position in long-term arable should preclude any realistic possibility of their being Neolithic. Similarly, large rectangular structures are not a recognised component of later prehistoric settlement in Wessex. Therefore the possibility exists, if buildings they be, that they might be sub- or post-Roman, *sensu* fifth–seventh century. The 'Saxon' burials adjacent to the Roman tombs on Overton Hill take on a particular significance in this respect.

The 'building' dimensions, as plotted from the air photographs, are well within the mean for structures of this period (James *et al* 1984). If so, an attractive model presents itself of 'Headlands' as a local *caput* over some 1,500 years, with a sequence of residences, commencing with a presumed timber round building in an enclosure, followed by a villa and a cluster of timber 'halls'. Such a sequence here, if it is to be chronologically continuous, requires a residential claimant for the last centuries BC (cf Chapter 6).

The proximity of the settlements and the tenth-century boundary also begs the question of whether the former were placed at a territorial margin, or whether the tenth-century documentation reflects the subdivision of an earlier, prehistoric and Roman land-unit. Perhaps the two later settlements, if substantiated, continued to occupy the traditional habitation site, even though the lands which they oversaw were now much-reduced, emerging into history as the tithing of West Overton. The interpretation could be pushed further by arguing that a new *sub-caput* would also be required east of the new boundary, the equivalent of the 'Headlands' Roman and sub-Roman settlements. Such may have been at ODS (Figure 6.13; with or without OD XII), at an as yet undiscovered site close to North Farm, or across the river adjacent to medieval East Overton. In addition an as yet unlocated settlement may be anticipated within the 'nodal place' of Down Barn (cf Chapters 6, 15 and 16).

How old is the tithing boundary at 'Headlands'? Presumably the Early Iron Age enclosure, like barrows 665a and b, was visible as an earthwork when it was crossed. Or was some concept of tenure and property still alive when the boundary sought out its line across an already flattened ancient marker? Conversely, is it conceivable that the Early Iron Age enclosure was placed on this spot precisely because it was on a boundary, representing either the colonisation of a sort of 'no-mans' land' or the uniting of two previously separate estates? On balance, the seemingly deliberate association of settlement and boundary suggests estate fragmentation, perhaps in the first half of the first millennium AD. Another interesting aspect, especially given the mirror-like quality of 'Headlands' enclosure compared to OD XI (itself subdivided by a Roman boundary, Figure 6.2), is that here too is an example of a Late Bronze Age/Early Iron Age site being abandoned but retaining an influence on adjacent land arrangements, in this case for over 2,000 years. The Anglo-Saxon boundary across it still survives as a property boundary and visually stands out as a line of bushes in an ocean of arable (Plate XV). The only reason for its survival is history and tenure.

The Ridgeway and boundaries

The present line of The Ridgeway dutifully follows, as the name suggests, the ridge of the Marlborough Downs, forming part of the boundary of several parishes including the western boundary of both Overtons and, higher to the north, of Fyfield. The name is clearly derived from the OE *rigte weye* ('straight way') or *hric wege* ('ridge way'). There are, or were, however, many such 'ridge ways' throughout Wessex (*see* Andrews and Dury 1773), not all of them following ridges. Though a relatively late arrival on this scene, the present Ridgeway is certainly old, if an age of 1,000 years counts as old in this landscape. Its oft-quoted status as 'the oldest road in England/the world' (eg, Anderson and Godwin 1982) is assumed rather than fact. It remains to be determined, however, whether The Ridgeway was established as a fixed line firm enough to form the common boundary for estates or land units when they were being formed in the tenth century and earlier. Or is it early medieval in origin, keeping to the edges of recently formed estates to reduce the necessity of directly crossing land belonging to different owners? Its counterpart along the boundary between West and East Overton should perhaps be borne in mind.

The Saxon charter tells us that on the northern edge of West Overton the boundary with East Overton ran from the middle of the *crundel* to *coltan beorh oth thaene herpoth an hacan penne* – from the quarry to Colta's Barrow as far as the *herepath* on Hackpen – (S449; Grundy 1919, 242). The East Overton boundary then follows the *herepath* to the prehistoric ditch known to us as ditch F4 (Chapter 2; Figures 2.1 and 5.1). Clearly, The Ridgeway was a *herepath* in the tenth century. A charter of AD 922 or 972 (S668) records that the east side of Winterbourne continued up to the *mearce*, or common boundary; further south, just beyond Wansdyke, *an ealdan herepathe* ('old army-path') was noted in the Alton Priors charter of AD 825 (S272), both seemingly following a roughly similar line to that of today's Ridgeway. If the *mearce*, *herepathe* and *herpoth* are reflected in the line of the modern Ridgeway, which seems likely, then a tenth-century path passed through a broad strip of land much as it did between at least 1821 and the mid-twentieth century (Plate I). To an extent, modern traffic still oscillates similarly between the fences 40ft (13m) apart, which Enclosure defined as the width of The Ridgeway in the nineteenth century.

Perhaps the significant point is the reiteration in the three charters (S272, S668 and S449) of the description 'path' rather than *wege*, a term which implies more traffic and a firmer imprint on the landscape. Brentnall notes (1938a, 124) that 'herepaths led to the meeting-places of the various hundreds where the levies gathered when the army was mobilised'. The Herepath/Ridgeway was the nearest way from Overton to the point on the Marlborough–Broad Hinton road called Man's Head (SU 140739) which was probably the meeting-place of the Hundred of Selkley in Saxon times (cf Manshead, Beds). This again hints at periodic military use rather than daily agrarian through-traffic, but, as Costen has argued (1994, 105), when the charters were written this 'path' was most likely already a 'highway' mainly for non-military use or simply the place, being a major route, where one might encounter a war band. Heavy use of The Ridgeway during the mid- to late tenth century led to it being metalled at the West Overton ford (*strætford*).

The Ridgeway was called *Ryggeweye* in the late twelfth century, at least south of the river. *Ryggeweye* was, however, a popular name for routes in the vicinity (Brentnall 1941, 394, 395), and Andrews and Dury (1773) show an unnamed track running approximately along the line of the present Ridgeway without distinguishing it from many other downland tracks in the area. Other late eighteenth- and early nineteenth-century maps do not show The Ridgeway, while depicting other routes. The maps do, however, show three standing stones marking the boundary of West Overton with Avebury along the east side of the line of the modern Ridgeway. These would have stood out starkly on the open downland. No track is noted along this boundary. The Ridgeway had at this period simply ceased to serve as a major route, being replaced by other north–south arterial routes and by several local east–west routes. In short, it had been replaced by other tracks deemed more suitable. Needless to say, this is why, over time, tracks come and go out of use, and why this landscape is laced with remains of earlier routes (Chapter 16).

The Ridgeway enjoys another aspect: its emergence as a specifically 'British trackway', ie, of prehistoric origin (Anderson and Godwin 1982). This is discussed elsewhere (Fowler 1998). Here we need but note that it appears in that guise in the works of neither Aubrey nor Stukeley, that Colt Hoare may possibly have been thinking of it (1821, 32) and that the phrase apparently appears first in Long (1862, facing title page; 1). By 1885, 'British Trackway' had become embedded in the cartographic record through the Ordnance Survey and, since 1889, 'The Ridgeway' has nearly always appeared in the Gothic style reserved for non-Roman antiquities. It is, however, just 'Ridgeway' on the very latest OS production (*Explorer* 157, summer 1998).

Boundary markers on The Ridgeway

Fieldwork recorded a number of stones along the length of The Ridgeway. Pre-dating the concrete markers of the 1980 National Trail, they are believed to be the boundary stones recorded on the early nineteenth-century maps. Indeed, their locations, restricted to the parish boundary line, or west side of The Ridgeway, correspond closely to the positions marked on these maps.

The majority are brown 'egg-shaped' stones of various sizes and are clearly different from the other downland sarsens, having apparently been roughly shaped or at least chosen with deliberation and care. In some cases attempts had been made to break them (splitting wedge marks were recorded). As the sarsen breakers used splitting wedges from the mid-nineteenth century until the 1920s (King 1968), this suggests that these particular stones were broken before 1930 and, as was the norm, broken *in situ* (ibid, 90). They may even have been split in medieval times (Chapter 5). That they were broken at all suggests either a presumably illicit practice of removing boundary stones or that the stones were no longer recognised for what they had been.

The boundary between East and West Overton (Plates XIV and XV)

Two Saxon charters, S784 and S449, record the boundary between the estates of West and East Overton (Figure 4.1). Both mention *colta beorg* (S784) and *coltan beorh* (S449), 'Colta's Barrow', a now heavily denuded round barrow at the junction of the West Overton boundary and the *herpoth* (SMR 647; SU 12106937). To the east of Colta's Barrow the boundary was marked by a *crundel* (a quarry), *twegen dunne stanas* (two down or brown stones, mentioned only in the West Overton charter) and *scropes pyt* (possibly shrub pit). The *crundel* is clearly the quarry-pit just west of the modern reservoir (SU 12506928). In the Saxon period, the boundary cut the *crundel midde werdne*, in the middle, as it still does today (G Swanton, pers comm). The deliberate bisection of the quarry suggests that it was shared equally by the two estates, presumably for building material ('cob') and, possibly, for 'marl'.

The West Overton charter notes that the *twegen dunne stanas* are *estan colta beorg*, east of Colta's Barrow. Clearly close to *scropes pyt*, their exact position is uncertain, although the rectangular bend in the boundary immediately south east of the quarry (SU 12646918) suggests a location. Albeit in the wrong Anglo-Saxon position, two stones (SMR 108; SU 129695), now standing in the hedge 60m south of Down Barn, are probably the ones in question, but the West Overton charter cannot be convincingly reinterpreted to allow for their present position. They were not shown on the 'tithing' boundary on the 1794 map, which details many other standing stones, nor are they depicted on the 1819 map. If these are the 'charter stones' they must have been relocated. The move may have been much earlier than the 1794 map and then forgotten, or took place between 1820 and 1890, when they appear on OS maps in their present position.

Obscured by vegetation in the hedge-line, they are of apparently uncut sarsen and stand between 1m and 1.8m high. Both appear to have been set into an existing bank rather than being integral with its construction. In their present position the stones reinforce the line of the north–south through route, our 'Overton Ridgeway' (Figure 4.1), between the estates. Possibly something occurred to make it more important to emphasise the line of the track rather than the estate boundaries. Landmarks such as boundary stones are rarely moved without deliberation by agrarian societies, yet the ancestral virtues of the stones can powerfully underwrite a new arrangement. Something odd, but probably special, seems to be represented by these two stones set into a bank which has been in place for over 1,000 years.

In the West Overton charter (S784), *scropes pyt* lay east of the *crundel* and before the *west heafde*, the 'west headlands'. A likely candidate is a large round pit showing clearly as a cropmark and soilmark (SU 12826911) immediately east of the boundary hedge as it turns south towards 'Headlands'. The name is clearly reflected in a reference of 1312 to a *scrufeleput* in North Field (*see below*). Further south was the *furlanges west heafde*, the west headlands of the furlongs. This reference is from the East Overton charter; hence it was the west headlands of East Overton. This indicates that the land east of the boundary was under plough, just as the word *yrdland* suggests that land on the east of the East Overton estate was also under cultivation at the same time.

A little to the south of the headlands, both charters note two barrows, now re-identified from air photography (Figures 2.1 and 4.3). The boundary was forced to go *betweox tha twegen beorgas* (S784) here, so three almost right-angled bends were created. These bends are still mirrored in the modern hedge some 400m west of North Farm (SU 129685).

Pennings

Right in the middle of this small study area, at the junction between arable and pasture and Cow Down and Farm Down, were two 'Pennings', 'Higher' and 'Lower' (Figure 4.1). They were separated by a 'Drove' heading due east from The Ridgeway, past a barrow and two standing stones, to the eastern manor boundary just south of the Saxon 'headlands'. There was a Sheep House and Washing Pool at the north-east corner of Lower Penning. While it is possible these 'Pennings' are a faint etymological echo of the charter's *penne*, their significance is certainly functional. They show the continuing need for livestock enclosures, especially on the edge of the manorial pasture. These would be of use to both local and itinerant stock and, being the only downland available to the inhabitants of West Overton, would be under immense strain to satisfy their grazing needs. This is reflected by the remark on the 1784 map that the land had been 'injured by great quantities of sheep being drove over it'. Again, the importance of communications and movement is apparent. Major routes crossed West Overton's limited downland from three directions, Avebury, Bath and East Kennet and beyond, all being funnelled to a point some 300m west of the present Bell Inn (SU 125682). It was only from this point of convergence in 1784 that one road, now the A4, was hedged eastwards along both its north and south sides.

A park by The Ridgeway

Immediately north of the northern boundary of Anglo-Saxon West Overton is an area of enclosed fields, bounded on the west by The Ridgeway (Figure 4.1). Historically the area lay in Anglo-Saxon East Overton and was called Hackpen from the tenth century until it was divided and enclosed prior to the late eighteenth century. Enclosure created a block of fields between Parson's Penning and the northern tithing boundary of West Overton, 600m to the south. The fields present a markedly rectilinear plan and look very modern, though the northern edge forms a steady curve. This division of Hackpin Sheep Down into smaller fields was noted by Smith in 1885 who also recorded the name 'New Forest' for a narrow strip of land bounding the south side of Parson's Penning. There is no evidence for any woodland here over the last 5,000 years so the allusion may be to a 'forest' as a hunting ground, rather than a new plantation of trees. The two parallel fence-lines still marking the location of this odd field-name are unusual in this landscape of unenclosed vistas. Tenurially, it comprises the extreme north-west corner of East Overton tithing, labelled 'The Down Field' on the 1819 map. The same map calls the field immediately west of Down Barn 'Old Park Grounds' and its eastern edge contains the two standing stones described above.

Two decades before that map some of the land immediately west of the then New Barn (now Down Barn) was divided into five enclosures called 'Lower New Broke', 'Lower New Leaze', 'Upper New Leaze', 'Upper New Broke' and, right up against the far western boundary with Avebury, 'Park Grounds'. 'Lower' and 'Upper New Leaze' embraced the 'Old Park Grounds' of 1819. Such names are characteristic of new Enclosure and new arable and may indicate that these activities are taking place within a former emparkment.

No documentary or other evidence of medieval or later emparkment in this area has so far been found but that a park was attempted, or perhaps just intended, can hardly be doubted. The cluster of 'park' names strongly suggests the late eighteenth-century cartographers were aware that their farming clients were breaking up old parkland, however unlikely the downland location may seem. The long-term importance of the downs for sporting purposes must be remembered and many of the Wiltshire gentry enjoyed hunting rights outside nearby Savernake Forest. The Bishop of Salisbury's chase, for example, lay on the north side of the River Kennet (*VCH* IV, 423; *WAM* 48, 374, 376) and John Aubrey was hunting hereabouts in 1648 (Long 1862, 3). There is no evidence for the date of this putative park, although an early post-medieval origin might be suggested by analogy with the creation of nearby Clatford Park. This lay diagonally across our study area at the south-east corner of Fyfield parish (Chapter 11), was emparked in the early 1580s and disemparked about forty years later. Created by the post-Dissolution landlord, the First Earl of Pembroke, the brevity of its existence did not prevent it leaving a permanent imprint on the landscape and the name 'East Overton Old Park Grounds'.

A similar context and date would fit comfortably with the 'new' park on Overton Down, which, as part of East Overton manor, also belonged to the First Earl of Pembroke. If it was a manorial emparkment, the memory of it as 'Old Park Grounds' could still be strong in the late eighteenth century even if, like Clatford Park, it had been abandoned a century or more earlier. Such a park was not noted by Watts in his recent (1996) survey. Although further evidence is lacking, a park here seems probable and, *defaut de mieux*, is named (by us) 'Hackpen Park'. It was presumably created, or thought about, to provide protected grazing for deer, the better to hunt them across the downs as John Aubrey was pleased to do hereabouts not too long after 'Hackpen Park', on our interpretation, may have ceased to exist. There may be more to this than coincidence, but meanwhile the idea of a deer park on the Hackpen part of Overton Down sounds attractive in the context both of local history and the gentlemanly life-style of the Elizabethan and Jacobean period.

Chapter 5

The Northern Downland: Lockeridge Down to Manton Down

higher downland with ancient fields

This chapter examines three areas defined on Figure 3.4 in relative detail:

1 *Lockeridge Down*, including the northern area of Totterdown, and excavations OD I–III and TD VIII–IX
2 *Totterdown* Roman field system including excavations TD I–III
3 *Manton Down* long barrow area

LOCKERIDGE DOWN

The name 'Lockeridge Down' is a reinvention of an old one. A long, thin strip of land stretching west–east across the northern end of Overton Down and Totterdown (Figure 5.1) was historically called 'Lockeridge Down', a name which dropped out of use in the nineteenth century. Strictly it referred to land between the modern Ridgeway and the Overton/Fyfield parish boundary. East of this is, strictly speaking, the north end of Totterdown in Fyfield parish, but we include some of it in this section because the name 'Totterdown' is in print in relation to the next area of excavations.

More importantly, Lockeridge Down is archaeologically centred on linear ditch F4 – *tha dic with suthan Aethelferthes stane* (S449) – a relatively slight but significant archaeological feature (Chapter 2; Figure 5.1) whose course reflects the local geology and topography across on to Totterdown. At The Ridgeway, the ditch is at 250m above Ordnance Datum (aOD). The ground then falls quite steeply eastwards into the head of the sarsen-thick Valley of Stones from where it ascends a long south-west-facing slope to Totterdown Wood. The area is rough pasture dominated by sarsen fields (Plates VII, VIII and XVI; *SL*, figure 15) and apart from some disturbances connected with clay, chalk and sarsen extraction, it appears to have been used as grazing since the early centuries AD, an inference supported by documentary and cartographic sources.

During the Anglo-Saxon period this area of higher downland was referred to as *dun landes* and in the nineteenth century was called Lockeridge Down or Overton Sheep Down. The only evidence of ploughing, of mid-nineteenth-century date, is on the Clay-with-Flints north of ditch F4 and west of Totterdown. At the latter an enclosed field is still under cultivation and is strikingly pockmarked by mid/late nineteenth-century marl-pits (CUCAP AAU 89). The very openness of this landscape is also reflected in a preoccupation with defining boundaries. Such an urge was certainly there from the earliest documentation and by implication present in the need to dig linear ditch F4 in the Bronze Age. The linear ditch divides arable fields from unenclosed land (Figure 2.1), and may also have had a tenurial or proprietorial function.

Anglo-Saxon documentary evidence is concerned with *Aethelferth's* stone, no longer extant, but probably located *c* 100m north of linear ditch F4 where

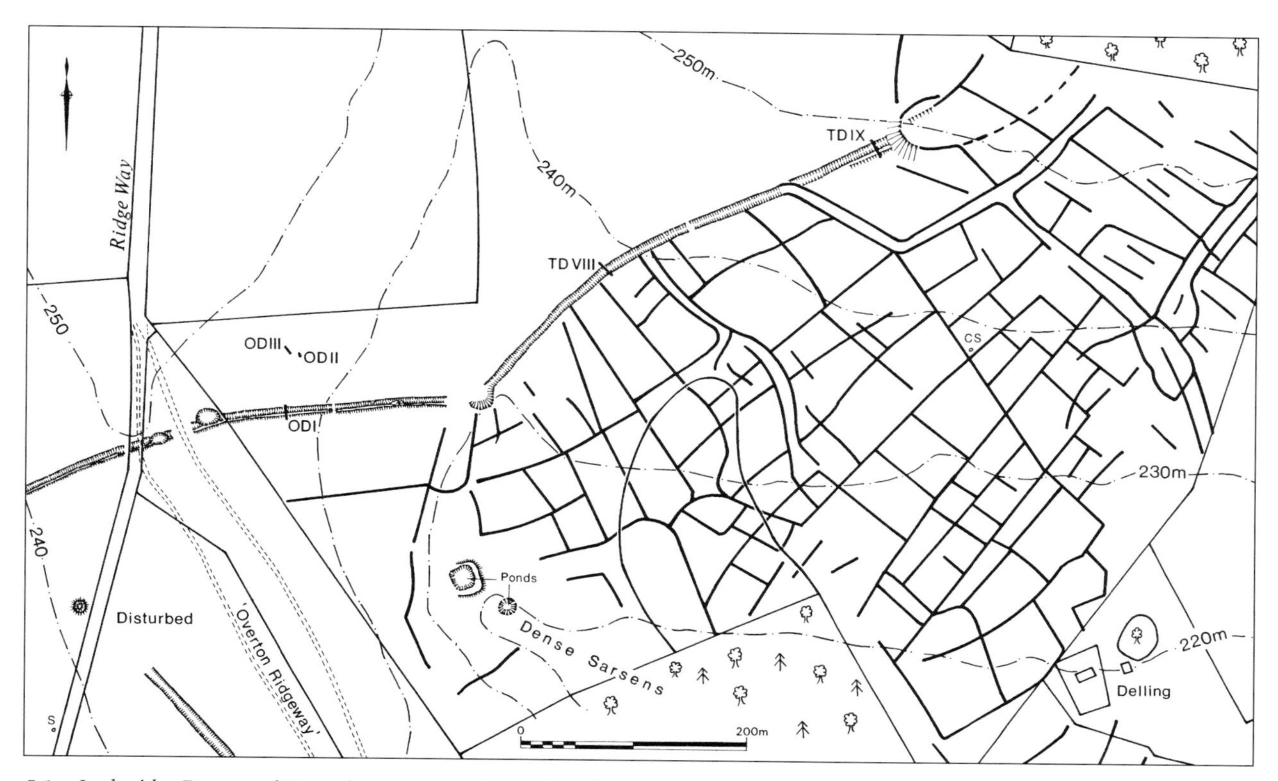

5.1 *Lockeridge Down and Totterdown: map outlining their field archaeology and excavations OD I–III and TD VIII–IX*

the parishes of West (formerly East) Overton, Avebury and Winterbourne Monkton met (an alternative interpretation is offered in *SL*, figure 17a). If the *egelferdeston* of the Winterbourne Monkton charter of AD 869 (S341) and *Aethelferthes stane* (East Overton charter S449) are the same, the estates, later the parishes in this area, may, on documentary evidence alone, have been established by the mid-ninth century (Chapter 16).

Lines of communication play a significant role over much of the downland. Lockeridge Down is no exception. Much of linear ditch F4 was incorporated into a network of lanes that characterise the early Roman landscape (Chapters 2 and 15). From the post-Roman period the evidence here, as we have already seen in Chapter 4, becomes increasingly concerned with The Ridgeway and the *herepath*. The Winterbourne Monkton charter of AD 869 (S341) goes from the *redeslo* to the *rigte weye*, 'the red slough to the straight way'. The red slough probably referred to the red clay covering this high downland, a resource used in the eighteenth and nineteenth centuries for brick-making (SU 125729). The 'straight way' may be the *herepath*, which is reflected in the present, markedly straight line of The Ridgeway from the boundary of West Overton and Fyfield parishes (SU 125724) to linear ditch F4. In the East Overton charter of AD 939, this ditch was approached from the south south west along the *herepath* (Chapter 2), at this point undoubtedly reflected in the line of the modern Ridgeway and not Green Street, which was the medieval and probably earlier downland route from Marlborough to Avebury (Plate VI; Figures 2.1 and 4.1; Fowler 1998, 32).

From a point immediately south of linear ditch F4 ran the *hric weges* (Figure 5.1), the ridge way. The line of this is probably continued today by a public footpath that runs not down the modern Ridgeway, but south east across Overton Down (*see below*). This point on the charter was marked by a stone in 1819, probably located *c* 25m south west along the fence on the east side of the existing Ridgeway. This marks the point where the boundaries between East Overton and Lockeridge tithings met. No longer recognisable, the stone was on the general line of the Romano-British hollow-way approaching from the south east along the spine of Overton Down (Chapters 2 and 16). The track veers to the north west, through an area disturbed by quarrying, to re-emerge by the present Ridgeway, close to the site of a round barrow (*SL*, figure 24) and *c* 130m south of the

Plate XVI Totterdown from the east (cf Figure 5.1) (AAU-94, © Cambridge University Collection of Air Photographs*)*

probable site of the boundary stone. It is suggested that this Roman downland track is to be identified with the 'ridge way' of the tenth-century charter.

Three small excavations (OD I–III) were undertaken on Lockeridge Down (Plates XVII and XVIII; Figure 5.1) and two sections were cut across linear ditch F4 on northern Totterdown (TD VIII, IX; Figures 5.1 and 5.2). Dating and context were the linking elements in this exercise, though each excavation had specifically different objectives. Detailed reports are in FWP 66, which includes appropriate plans and section drawings.

Investigations

The sites investigated (Figure 5.1) were: a split block of sarsen stone believed to be a Neolithic stone axe-sharpening bench or *polissoir* (cutting OD II); a short line of sarsen stones immediately west of the last (OD III); and the linear ditch F4 (cutting OD I).

A Neolithic polissoir *(OD II)*

> ... investigations carried out [around *polissoirs*] in the hope of discoveries have always proved unrewarding.
>
> Lacaille 1963, 193

The stone was discovered by Inigo Jones in 1962 and reported on, after 'repairing to the spot under his conduct', by Lacaille (ibid, 191) whose phrase so assuredly links the discovery to another Inigo Jones (Ucko *et al* 1991). Lacaille's (1963) comprehensive illustrated description and discussion need not be repeated here. The 2.1m-long stone had been split north–south and the western part removed. It is a remarkable stroke of good fortune that the polished and grooved patch at the stone's south-east corner (Plate XVII; *SL*, colour plate 11) has survived this destruction. Much of its upper surface has been repeatedly polished

Plate XVII Lockeridge Down, site OD II: excavation around the polissoir

and intercutting grooves cut previously polished areas. The polished area of the stone also makes it clear that use as a *polissoir* could only have occurred when it was in a recumbent position. It will never be known whether such stones were once common, although a further two polished, recumbent stones have been noted further south on Overton Down (G Swanton, pers comm).

The following is a summary of our excavation in 1963, undertaken after the discovery of the *polissoir*. Potentially one of the earliest visible features on the downs, information on its context was considered to be desirable, although the primary objective was to explore the possibility of Neolithic activity in its immediate vicinity.

Four small cuttings were excavated on three sides of the stone, but not on the west where part had been removed. No structures or significant features were found in plan and the stratigraphy was consistently straightforward (Plate XVII; FWP 66). Layers 1 and 2, a humic topsoil underlain by worm-sorted flints, were disturbed, probably by rabbits and the sarsen-breakers. The material appeared to be redeposited on top of an earlier ground surface, inferentially of medieval or earlier date (*see below*). At the north end of the sarsen bench the lip of a pit or trench was partly excavated. It showed clearly in plan as a feature dug into the top of an undated surface level with the disturbed top of the Clay-with-Flints; it was filled with flinty, clayey humus similar to that through which it was cut. In the top of that fill was a heavily weathered sarsen *c* 0.60m by 0.45m and a cluster of smaller, broken sarsen stones. The hole was at least 0.45m deep, its bottom as excavated marked by an increase in the density of flints. The evidence, though incomplete, suggested very strongly that the feature was part of a hole dug to take the *polissoir* as an upright stone. Excavation stopped at this point as enough had been done to demonstrate that, whatever the structural interest (which others may wish to explore), the immediate vicinity of the *polissoir* seemed unlikely to contribute significantly to our landscape objectives.

Perhaps the most interesting result of this small exercise was the establishment of the date when the stone had been split. A halfpenny of King John (1199–1216) and an iron wedge were found in layer 2 at a depth of *c* 0.15m. The wedge corresponded exactly with the wedge-marks along the split western edge of the recumbent stone. At a depth of 0.20m, half of an iron horseshoe, probably of late or post-medieval date, was recovered. This evidence seems to indicate active stone-breaking during the medieval period, the date when people were living at Wick (Chapter 10) and *Raddun* (Chapter 7).

Prehistoric activity was indicated by a small lithic assemblage (flint report in FWP 31a), including three micro-flakes, eight sarsen chips and a sarsen 'flake', but there was no debris that could be associated with stone-axe manufacture. More interesting is the possibility that the stone, once much larger, had originally stood upright. Further excavation around its north end would be necessary to settle the matter but, if it was once a standing stone, presumably that was before it was used as a *polissoir*. Such a sequence would contrast with polished sarsens reused in the West Kennet Avenue and in the West Kennet long barrow (Burl 1979; Piggott 1963).

Although excavation at OD II did not achieve its initial objectives, it did produce two unexpected results separated by a span of 4,000 years. A somewhat squat upright stone *c* 2.10m tall and 1.80m wide may have been erected in the early or mid-Neolithic period before being laid flat for grinding stone axes, presumably in the third millennium if not earlier. The same spot was then witness to sarsen stone-breaking, apparently as early as the beginning of the thirteenth century AD.

Stone structure (OD III)

Slightly uphill and a few metres north west of the *polissoir* is an irregular line of sarsens. Today it is more clearly visible than in 1963 and appears to be the fragmentary remnants of a line of upright and closely packed stones. Elsewhere on the downs such features delineate the boundaries of fields, but this line does not readily fit in with the adjacent field patterns to west or east. At what appeared to be its east north east end in 1963, a small, embanked depression was noted at a point where the line appeared to bend towards the *polissoir*. Had the point been as clear then as now, this small excavation would not have taken place. A single cut was excavated to check for any structure which might have been related to either or both the *polissoir* and the Beaker occupation which was known to exist at site OD I (*see below*).

The excavation proved the embanked depression to be a recent pit; a Home Guard or other military origin seems most likely. A spread of chalky material, which looked like a wall foundation with 'spill' to either side, proved to be lying top of a former but modern topsoil, interpreted as upcast from the pit.

A line of three stones was also exposed and showed the sarsen wall to conform to the description above: a line of single sarsens side by side and partly under the upcast from the pit. The stones were placed on, rather than in, a flinty layer between the top of the Clay-with-Flints and the bottom of the former topsoil, suggesting that their placement was relatively recent, ie, after the formation of the characteristic worm-sorted layer 2. A line of sarsens roughly placed at the edge of land clearance in the eighteenth or nineteenth century is a distinct possibility, though it remains undated archaeologically.

Nevertheless, twenty-four separate finds contexts were recorded in this small excavation, thirteen of them 'flint flakes' in layer 2 or the top of the Clay-with-Flints. A sarsen flake occurred in the last, and a leaf-shaped flint point, a beautiful implement, occurred in layer 2 right at the south end of the cutting. This material, in a similar context to that recovered in OD II, suggested activity in the area during the third (or fourth) millennium BC (*see also* OD I).

Linear ditch F4 (OD I; TD VIII and IX)

A ditch (Plate XVIII; Figures 5.1 and 5.2) with an accompanying bank to the south for much of its length has also been interpreted as a track for part of its course (F4 in Bowen and Fowler 1962; Lacaille 1963, 190, discussed above; Chapters 2, 3 and 15). Because it stretches west–east right across the northern part of the study area and is related *en passant* to a number of features, its composition, date and phasing were important. It provides a crucial horizontal landscape datum with a potential for both functional and chronological information.

All the questions were clearly not going to be answered by 4ft (1.2m) wide trenches. The chances of finding stratified and datable evidence were small but, in the light of detailed field examination, key points, which might provide some relative dating and structural evidence, were identified. In a conscious pattern of controlled variation, OD I was placed close to the

Plate XVIII Lockeridge Down, site OD I: west section of cutting through ditch F4

highest point of the ditch's course on Upper Chalk, immediately east and slightly down-slope of The Ridgeway, not far from the *polissoir*. TD VIII (*see below*) was meant to provide a marked contrast, testing for morphological variation within the topographic and geological situation. This was cut *c* 0.5km east of OD I, on Clay-with-Flints and a south-west-facing slope, where clearly defined, stonewalled fields were laid off to the south (Chapter 2, Plate XVI; *SL*, figure 24). TD IX (*see below*) was higher up the slope of Totterdown, again on Clay-with-Flints, but sited to test the observation that the ditch continued up-slope and under a 'Celtic' field lynchet after the track along it had turned off to the south east (Figures 2.1 and 5.1).

The description of the excavations through the bank and ditch begins with OD I and then, after a brief discussion of the north end of Overton Down, moves on to the two trenches on Totterdown (TD VIII and IX).

OD I. OD I (Figures 5.1 and 5.2; FWP 66; *SL*, colour plate 13) was excavated across the full width of the ditch and bank; the base of the latter was marked by a sarsen stone at its southern edge and remnants of the pre-bank old ground surface. From the bank and its erosion products, three flint flakes and five, probably Beaker/Early Bronze Age, sherds were recovered. The ditch was 1.34m deep below the old ground surface, cut entirely into chalk though presumably it had originally cut through a thin layer of Clay-with-Flints.

The ditch fill was characterised by its predominantly humic, rather than chalk, content and the near-horizontal layer of chalk across the upper part of that deposit (Figure 5.2). The humic material here is interpreted as the product of natural erosion and wind-blown deposition whilst the chalk layer is best seen as trackway metalling laid in the top of what would have been, at the time, a linear depression. The chalk layer contained a Beaker sherd, two flint flakes and an iron nail. The sherd and flakes are probably residual, having eroded from the ditch sides. The nail is, admittedly, a slender piece of evidence on which to hang a landscape, but its discovery may support a Romano-British date for the trackway. This period certainly witnessed a major reorganisation over the whole of the study area (*see* both Chapters 2 and 15).

The landscape and structural sequence suggested by this cutting is of Beaker activity cut through by a linear ditch and sealed by its bank. To the west, at the foot of Avebury Down, air photographic evidence shows this ditch to have cut through 'Celtic' fields so a post-Beaker horizon is certainly not surprising. Conversely, other fields in the same area are laid off from it to the north east (if the ditch there is indeed the same) and overlie it further up the slope. After a long period of deposition, the line of the ditch at OD I was probably reused as a track after receiving a chalk surface, probably (though not so-dated independently here) *c* AD 100. The earthwork grassed over and has remained relatively undisturbed ever since.

This evidence can be merged with that from the OD II and III excavations to propose a local landscape sequence for Lockeridge Down at the northern end of Overton Down:

- i Standing stone: Early/mid-Neolithic
- ii Axe-grinding bench and some flint/stone-working: mid-/Late Neolithic
- iii Flint-working/?occupation with Beaker pottery: Late Neolithic/Early Bronze Age
- iv Boundary bank and ditch (plus? field wall): Middle Bronze Age/Late Bronze Age
- v Trackway along Bronze Age boundary ditch: *c* AD 100
- vi Sarsen-breaking: post- *c* AD 1200
- vii Field-clearance and arable to north: eighteenth–nineteenth century
- viii Military activity: ?1940s

This area has high potential and has hinted at activity from the fourth millennium BC to the present. Whilst the excavation results are suggestive rather than conclusive, they provide a potential outline sequence in an otherwise unpromising locale on the northern limits of the study area.

Totterdown: TD VIII and IX The linear ditch and bank were next examined on the south-west-facing slope of Totterdown (Plate XIX; Figures 5.2 and 5.3) on the assumption that this was the same ditch, F4, as that on Lockeridge Down (Plate XVI). It certainly appears to be a continuation of it, but a slight doubt exists as to whether it is actually the same or an addition to it.

Cutting TD VIII was placed where stonewalled fields are laid off from the ditch's southern side. Remains of the bank were slight, being represented by a thickening of a layer of small flints and a single sarsen stone in the underlying Clay-with-Flints but probably marking the bank front. The inner edge of the ditch cut this layer (49) just in front of the stone. The ditch dimensions were similar to those in cutting OD I (Figure 5.2). The fill was not complex, indicating a long process of deposition (layers 5, 14, 43, 23a), with a progression from a primary fill (7) of coarser soil with large flints to a fine brown (wind-blown?) soil beneath the topsoil (14).

Twelve stratified artefacts were recovered from the ditch fill. Two were small flint flakes (archive, TD VIII, nos 4, 14), probably early/middle Neolithic, one in the primary fill, the other high on a tip-line; both were interpreted as representing pre-ditch activity on the 'natural' Clay-with-Flints, whence they moved into the ditch. Neither is weathered. The other ten objects formed an homogeneous group of Late Bronze Age potsherds, all of a sparsely flint-gritted fabric; they were similarly interpreted, not least because five of them were

Plate XIX Totterdown from the north: early fields laid off southwards from ditch F4 across foreground, with late enclosure (Allen 914, © Ashmolean Museum, University of Oxford*)*

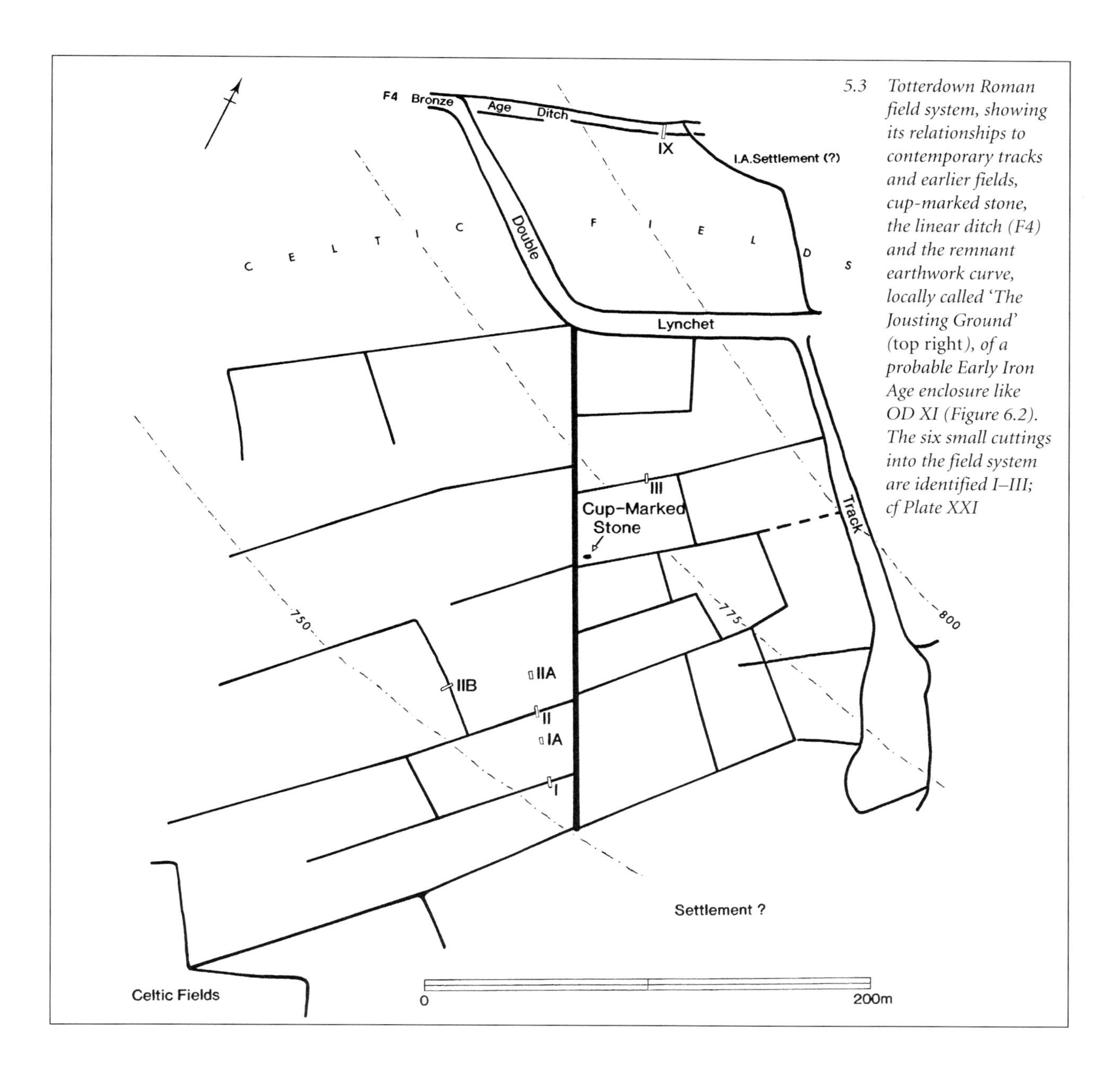

5.3 *Totterdown Roman field system, showing its relationships to contemporary tracks and earlier fields, cup-marked stone, the linear ditch (F4) and the remnant earthwork curve, locally called 'The Jousting Ground'* (top right), *of a probable Early Iron Age enclosure like OD XI (Figure 6.2). The six small cuttings into the field system are identified I–III; cf Plate XXI*

on the disturbed Clay-with-Flints beneath the slight bank. This material should therefore pre-date the ditch at this point, providing a *terminus post quem*. Given the absence of any later material, they suggest a construction date in the first half of the first millennium BC. This is somewhat later than a Middle Bronze Age horizon suggested for similar ditches in other places in Wessex but not in any way significantly different from the dating suggested for analogues on Salisbury Plan (Bradley *et al* 1994; cf Gingell 1992).

Of singular importance is the recognition of Late Bronze Age activity pre-dating – though perhaps only just – a major phase of landscape organisation represented by the construction of the ditch and the laying out of fields southwards from it, though it is possible that, despite appearances (Plates VIII and XIX; *SL*, colour plate 25), the ditch physically defined the northern edge of a block of fields already in existence. Whether that be true or not, this ceramic material is most likely to have been scattered in this area through manuring during a phase of activity preceding the digging of the ditch.

The third excavation across the bank and ditch was further up Totterdown (TD IX), approximately 140m north east of TD VIII. As with the previous trench, the geology is Clay-with-Flints over Upper Chalk (TD IX, Figures 5.1 and 5.2). The excavation was planned to demonstrate that the feature existed at this point where it had not previously been noted. This was because its course as a trackway swings off to the south east and becomes a low, double-lynchet trackway (Plate XIX; Figure 5.3). The ditch itself continues as a slight and overploughed earthwork, first beneath a 'Celtic' field lynchet and then beneath a larger, curving scarp, locally called 'The Jousting Ground'.

The bank, still on the south side of the ditch, was relatively well preserved close to the large scarp, having been respected by cultivation associated with the 'Celtic' field. Its southern edge was defined by a sarsen revetment, two courses of which remained. A spread of stones, probably collapse from the sarsen revetment, overlay a very thin layer of flinty soil. In this protected context, four grooves had survived, each scratched into the surface of the chalk for a depth of *c* 10mm–15mm. Slightly asymmetrical in profile, the southernmost ran obliquely across the cutting; the other three were fragmentary but parallel to the first and approximately parallel to the rear of the bank. The fragility of such evidence was demonstrated by its non-survival a mere one metre to the west, where a small cutting (IXa), with no collapsed revetment, contained no such grooves.

The grooves were interpreted as ard-marks. Their location suggests that they had been created when extra pressure was applied to the ard during the ploughing-up of the headland alongside the field boundary. Here the sarsen revetment of the bank was the field edge and a slight negative lynchet was created at its foot. In contrast, a few metres to the south west, cultivation had been carried over the bank and ditch and the northern edge of the ditch marked the limit of the field. A single grog-tempered (early?) Romano-British sherd on top of layer 2 flints (ie, at the bottom of the topsoil) was one of only two artefacts from TD IX.

The bank lay directly on top of Clay-with-Flints and was composed of flinty soil mixed with sarsens. A drystone revetment, its position marked by a ledge cut into the subsoil, had retained the front (north) of the bank. Flints lay on the ledge, trailing into the ditch to a small but cohering spread of drystone masonry. This had come to rest on the top of the main ditch fill, a brown, stone-free, probably wind-blown humus.

The stratigraphy suggests that both drystone revetments had stood for some considerable time and the collapse of the front was certainly late in the depositional sequence. The evidence from FL 1 (Chapter 7), coupled with other general considerations, suggests that the stone revetments relate to the early Romano-British period. Two of the four artefacts from this small excavation were Romano-British sherds from the sondage IXa, their context the equivalent of Figure 5.2, TD IX, southern end, layer 23a. Three sarsen stones and associated chippings high in the ditch fill along its outer edge almost certainly represent post-medieval sarsen-breaking associated with the fourth artefact, an externally glazed potsherd probably of seventeenth/eighteenth-century date.

The evidence from the three excavations across the linear ditch suggests that it is later than the Neolithic and Early/Middle Bronze Age (cuttings OD I, TD VIII) and earlier than Romano-British and post-medieval (TD IX). The complete absence of any sort of Early Iron Age pottery from all cuttings may be significant (the same was true of cuttings TD I–III, *see below*). The ditch, perhaps an extension of the original (F4), is associated with a block of stonewalled fields at the foot of Totterdown, independently suggested as 'early' on morphological grounds (Chapter 2). Furthermore, the ditch and its bank on Totterdown have unambiguously been overploughed by cultivation within a 'Celtic' field and are overlain by a curving lynchet, which, as at OD XI on Overton Down, encapsulates the boundary of a probable Late Bronze Age/Early Iron Age enclosed settlement. All the pointers suggest a date in the first half of the first millennium BC, perhaps somewhere in the eighth–sixth century, at least for the ditch on Totterdown.

But the ditch itself, a long landscape feature, may not actually be of only one build. It may well be that the structure sectioned in OD I is not the same as that in TD VIII and IX: the natural end for the former is at the head of the Valley of Stones below to its east (Figure 5.1, 'Dense sarsens'), where indeed the ditch is both discontinuous, not necessarily only from the disturbance there, and kinked to the north to continue up Totterdown. Its general relationship westwards on to Avebury Down seems to be with barrow groups as much as fields so it could well be that it actually comprises a western part running up to span The Ridgeway in the Early/Middle Bronze Age, to which an eastern length was added in the Late Bronze Age as cultivation spread on to new lands lying on Clay-with-Flints instead of the rendzina soils of Avebury and Overton Downs.

Whatever the precise chronology, its functions, first as a boundary feature and, more circumspectly, as a Romano-British track, seem certain.

TOTTERDOWN AND MANTON DOWN

We now turn eastwards to the old grassland of Totterdown and Manton Down (Figure 3.3). The higher reaches of these downs present their south-facing slopes towards the sun at an angle of characteristically 3°–5°. This gently undulating landscape, characterised by shallow re-entrants climbing through it from the Valley of Stones, is exposed to winds from the south west and icy blasts from the north east. Its elevation, some 200m–250m aOD, is relatively high for southern England and affords fine views (Plate II), including the hillforts of Martinsell to the south east and Oldbury to the south west.

Parts of Totterdown still retain a flavour of the pre-agrarian landscape of the region with spreads of sarsen stones that have escaped the extensive clearance witnessed elsewhere in the study area (Plate XIX; *SL*, figure 15). Due to its value as grazing land over the past seven centuries Totterdown also has a relict landscape of well-preserved earthworks (Plate XX). This contrasts with Manton Down, which has reverted to arable in recent years (Plate XXII). Nevertheless, most of this high downland persists as old grassland and we address the question of its antiquity. Our path follows a topographical pattern (*see* Figure 3.4) but, as it happens, each down proves different.

Plate XX Totterdown from the south, showing the Roman field system (AAU-84, © Cambridge University Collection of Air Photographs*)*

Totterdown: a field system and cup-marked stone (TD I–III)

'Totterdown' is probably a modern name, first appearing on nineteenth-century OS maps with no obvious antecedents. The area was probably part of the *pastura vocata Dyllinge* of the 1567 Pembroke Survey, a name preserved as Dillon Down in 1811. Of demonstrable pastoral use for at least four centuries, Totterdown displays earthworks of former arable use spanning at least four phases: respectively medieval, early Roman (our main concern here), probably Late Bronze Age associated with linear ditch F4 (*see above*) and an even earlier (?Bronze Age) phase. All are visible on Plates XIX and XX. Details of the excavations may be found in FWP 66.

A field system incorporating a cup-marked stone was identified and published early in the project (Fowler 1966, fig 9; Fowler and Evans 1967, fig 3, J and K; Lacaille 1962; *SL*, colour plate 12). Initially it was thought that these were contemporaneous, but fieldwork soon dismissed that idea, showing that the stone belonged to an earlier phase of land-use and had been unintentionally incorporated within a later field system. Nevertheless, the stone is of considerable interest in its own right – being a rare survival in southern England. Quite why it should be here at all is unclear. There are no parallels in the study area, nor any obvious immediately local context. It is, however, likely that blocks of cultivated fields were developing in the vicinity during in the earlier second millennium BC (Chapter 2) when the stone may well have been marked and, of course, a general context of 'land-marking' is provided on these downs by the round barrow cemeteries (Figure 2.4).

The field plan (Figure 5.3) is probably wrong in suggesting that the long straight, north-west/south-east bank is part of the original layout as marked by long rectilinear fields on a north-east/south-west axis. Their boundaries appear now on air photographs (eg, Plate VIII) to underlie the bank which, even if structurally later, still seems to respect the overall arrangement. Six small cuttings were excavated within this field system. Three were across the boundaries of adjacent fields and a fourth examined a nearby field boundary for comparative purposes (Plate XXI; Figure 5.3). The boundaries were slight low banks rather than simple lynchets; they did not, nor do they, show as clearly on the ground as in Professor St Joseph's superb air

Plate XXI *Excavation of Totterdown Roman fields. The landscape with figures marking cuttings TD I, II and III, and the cup-marked stone (on the right), looking north west towards ditch F4*

photograph (Plate XX). The aims of the excavation were to date the field system and to see if its boundaries contained any structural components.

Cutting IA (Figure 5.3), in the middle of a field, provided the baseline against which to compare sections through the boundaries. It showed a straightforward three-layer stratigraphy above the Clay-with-Flints, with the top of that subsoil disturbed in layer 3. All of the field-edge cuttings showed similar evidence with the addition of an extra layer between 2 and 3, taken to be the remains of a bank or the slight accumulation of ploughsoil against it. It may even have been nothing more than an unploughed baulk between arable plots. In cutting I the 'bank' effect looked as if it was largely created by such a baulk, accentuated by a furrow cut through the then topsoil to either side of it. Other than this the field boundaries contained no structure.

The dating evidence is reasonably clear. A few prehistoric items (but again no Early Iron Age sherds) underlay the area, hinting that Bronze Age cultivation may have occurred here, possibly providing a context for the decorated stone. The field system whose boundaries we excavated was dated by a small number of early Roman sherds, some from particularly significant contexts (detailed in FWP 66). There was no later material. At the time of the original investigation it was thought that this morphologically distinct field system was an outlying isolated group on high, marginal land. It can now be seen that it has a context in a general rearrangement of land allotment and use early in the Roman period (Chapters 2, 15 and 16).

Manton Down

Manton Down lies immediately outside the study area, but its potential contribution to the understanding of the downland landscape requires its inclusion here (Figures 3.3 and 3.4). Figure 5.4 embraces most of Manton Down in order to encompass the whole of field Block 11 (Figure 2.2) but the area of especial interest within is its northern half. This was old grassland until the 1950s when it was largely returned to arable. Consequently it produces crop and soilmarks not seen elsewhere on the high downs (shown in the top right-hand corner of Figure 2.1). Our prime interest is a prehistoric field system, here discerned for the first time, with a particular focus on its relationships to a linear

ditch, a probable settlement enclosure and the Manton Down long barrow. Indeed, the barrow lies within this field system, as strikingly revealed by aerial photography (eg, Plate XXII). The chronological depth is completed by ridge-and-furrow (Figure 2.3) and thirteenth-century activity at the nearby Beeches (Meyrick 1950).

The long barrow was aligned south south east–north north west and lay on a north-east-facing slope above the head of a coombe between Manton House and The Beeches (Figure 5.4). Of megalithic form, its former appearance, nature and structure are reasonably well documented in antiquarian literature, conveniently summarised by Barker (1985) who also describes the unpublished excavation which followed severe damage to the barrow in the 1950s. The long barrow effectively no longer exists as a field monument (*see below*) but its alignment is reflected in the historic tenurial units of this part of the study area. The area around it was called 'Manton Liberty and Field' in the late eighteenth century, its western edge abutting Clatford Down and stretching north west to the edge of Totterdown Wood.

In landscape terms, the barrow's interest is threefold. Topographically, it lies just above the floor of a dry coombe. In this relatively low-lying position it is in marked contrast to the prominent hilltop situations of the nearby long barrows of West Kennet and Adam's Grave. Secondly, as is argued elsewhere (Chapter 16), its position, akin to that of nearby Devil's Den long barrow (Chapter 15; *SL*, figure 18), was deliberately chosen so that the monument could act as a marker, even perform a boundary function, within a landscape of 'long-barrow territories' (Figure 16.3). Thirdly, the position became a fixed point in the evolution of a humanised landscape and was used to mark the corner of a field within a clearly defined prehistoric field system. Air photography indicates that the fields are later than the long barrow although theoretically they could still be Neolithic. The question cannot be answered, however, by morphology alone.

The crop and soilmarks of the field system (Figure 5.4) suggest a minimum extent of 160ha. The block extends from The Beeches in the north to barrow group M, close to the Kennet valley on the south. Over much of the area the fields are of long, rectilinear form with the long axis sharing the south-south-east/north-north-west alignment of the long barrow. In the immediate vicinity of the long barrow the fields are of a smaller and more compact nature, with the western edge partially defined by a 200m length of linear ditch. Traces of a lynchet on the air photographs may extend the line of the ditch for a further 400m to the south south east. Approximately 30m east of the linear ditch are the clear marks of a

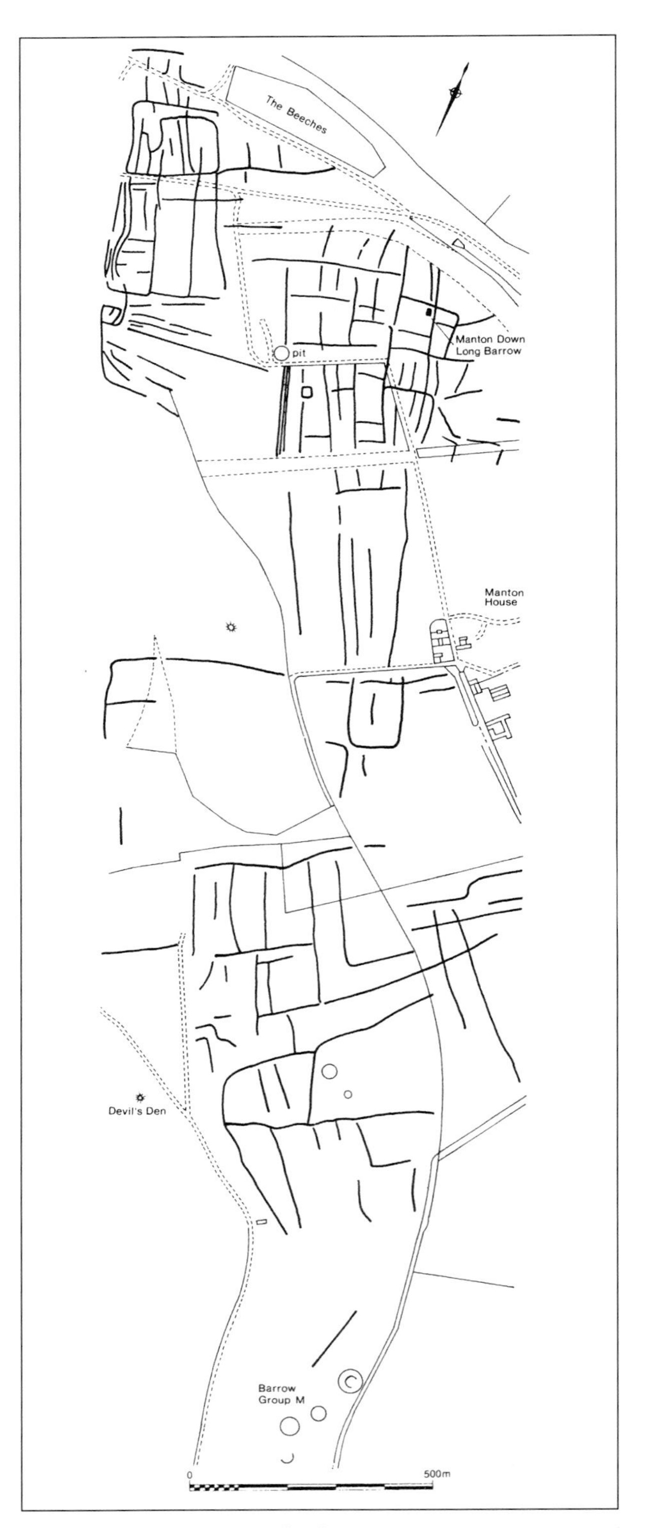

5.4 Map of the prehistoric landscape on Manton Down, showing the now-destroyed long barrow in the context of a field system with linear ditch and embanked enclosure

small square enclosure, its west side coinciding with a lynchet parallel to the linear ditch.

The integral patterning of fields, linear ditch and enclosure suggests contemporaneity and, although there is no direct dating evidence here, there are close parallels with Middle–Late Bronze Age landscapes elsewhere in Wessex. These are most striking at Rockley Down, 2.5km to the north (Gingell 1992), Boscombe Down East (Stone 1937), Thorny Down near Salisbury (Stone 1941; Ellison 1987) and in Cranborne Chase (Barrett *et al* 1991a).

The long barrow has a later historical niche too, for the discovery of its illegal destruction in the mid-1950s promoted a public debate on the treatment of Scheduled Ancient Monuments. The site of the barrow was subsequently incorrectly placed in the archaeological record (Barker 1984, 12–13), so, having gone to some trouble to replace it, we confirm that its proper location (SU 15137140) is where it was shown to exist on the OS 25in map in the 1880s and large-scale maps until recently. The remains of the barrow itself, however, have recently (1995–6?) been moved to the 'Old Chalk Pit' of earlier twentieth-century OS maps, the 'Pit (dis)' of current maps, in 1996 an irregular overgrown hole used as a dump for megalithic stones (*SL*, figure 20).

Plate XXII Manton Down long barrow and early fields: near-vertical view from the south (NMR 2115/0081, © Crown copyright. NMR*)*

Chapter 6

The Northern Downland: Overton Down and Down Barn in Pickledean

four thousand years of land-use

The domed centre of Overton Down is defined roughly by the 220m contour (Figure 3.3; *see* Figure 6.11). The generally smoother appearance of the topography when compared to the higher lands discussed in Chapter 5 reflects the underlying geology and more intense land-use. With the exception of two knolls of Clay-with-Flints, Overton Down lies entirely on Upper Chalk. The down in general slopes slightly from north west to south east, with a steeper drop into the Valley of Stones and a gentle incline into the head of Pickledean. Many of its sarsen stones have been buried, repositioned or cleared during cultivation 600 and more years ago (Plates VIII and XXIII). The area is old grassland and bears earthwork evidence of former land-use. This ranges from an unrecorded kerbed round barrow (SU 131705) to the experimental earthwork of 1960 (Bell *et al* 1996). The area appears to have grassed over about 2,500 years ago, witnessing only one major phase of cultivation and settlement in the early centuries AD followed by a brief period of cultivation in the thirteenth century. Otherwise, sheep, skylarks and racehorses have enjoyed their ideal habitat.

The sparsity of documentary evidence appears to confirm this picture of static land-use in more recent times. Long-term pastoral use is reflected in such names as 'Lockeridge Tenants Down', 'Hackpin Cow Down', 'Hackpin Sheep Down' (late eighteenth century) or 'East Overton Farm Down' in 1819. The last, today's Overton Down, was 'Ray Down' in 1773 and 1885. The only structure noted on the maps is a hut at the junction of Hackpin Sheep Down and Lockeridge Tenants Down, shown in this area in 1819. Crawford drew attention to two 'enclosures' in this area (Crawford and Keiller 1928, 125; here Plate XIX). The locations were ground checked under ideal winter conditions in 1996 and Crawford's three-sided 'enclosure E, F' found to be spurious, the image on the air photograph being a misleading coincidence of unrelated features. However, 'enclosure GHIJ' was confirmed. It consists of a small, rectilinear feature defined by narrow grooves marking, presumably, a wooden hut or sheep-pen overlying a 'Celtic' field lynchet. This could relate to the structure depicted on the 1819 map.

The parish boundary between the settlements of Fyfield and Overton crosses this open space and is clearly marked on late eighteenth-/nineteenth-century maps. One oddity is a portion of land, called 'Five Acres' in 1819, also shown as an enclosed area in the late eighteenth century, which juts into Lockeridge Tenants Down. Perhaps it is not entirely coincidental that this is also the one part of the north-east-facing slope of Overton Down where there is good surface evidence of former, probably post-Roman cultivation (*see* Plate VIII). The flora here was once claimed – erroneously

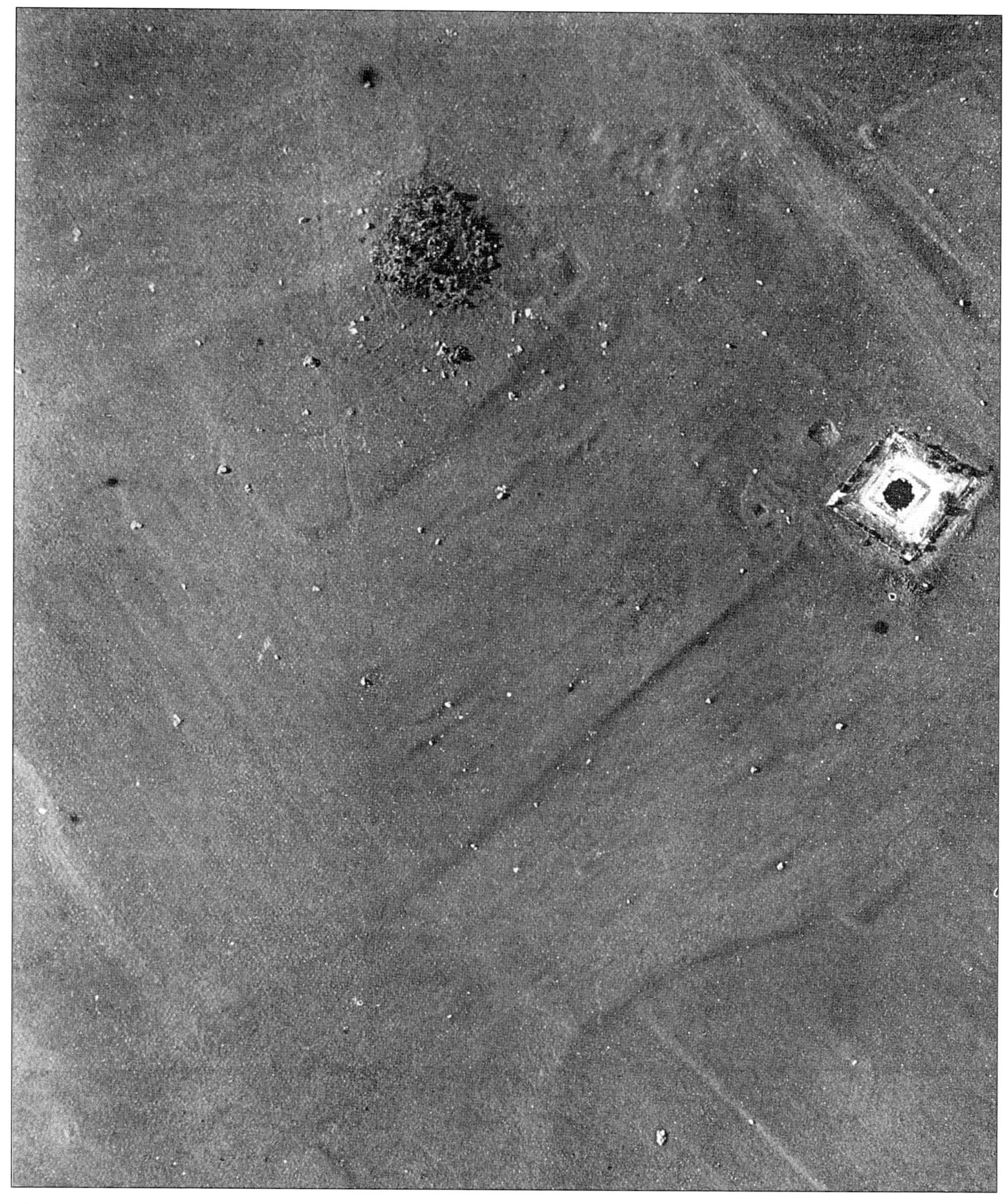

Plate XXIII *Overton Down: Crawford's near-vertical air photograph, originally published in* Wessex from the Air *(1928, plate xix, of which this print is a copy), which directed attention to the anomalous curved scarp* (lower right) *from which developed the excavation of site OD XI*

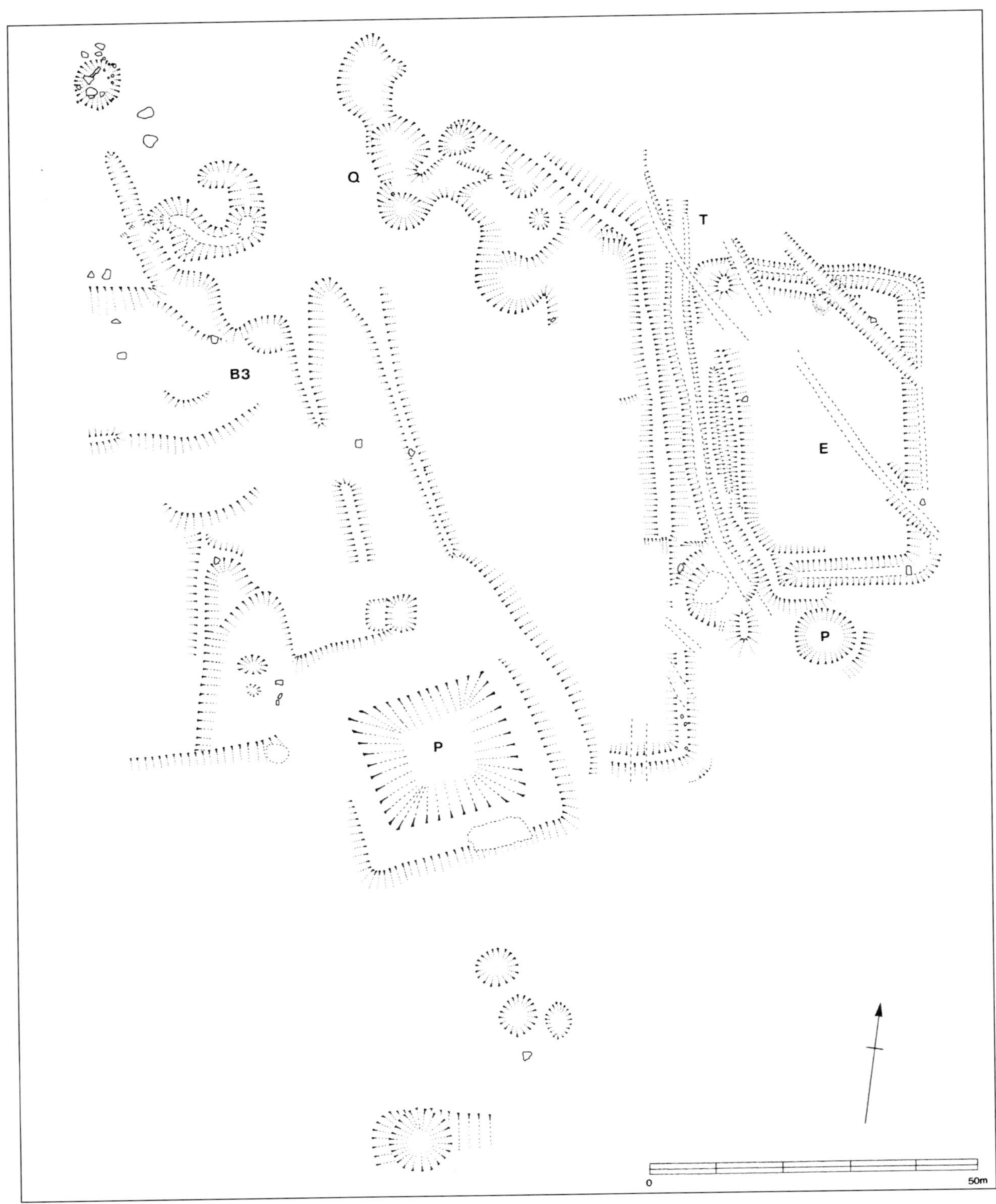

6.1 *Plan of central Overton Down showing earthworks of a Romano-British settlement (B3), partly overlain by a modern pond Q = quarrying; T = trackways; P = pond; E = enclosure*

(Thomas 1960, 61) – to be of particular interest because this area had never been ploughed.

THE FIELD ARCHAEOLOGY OF CENTRAL OVERTON DOWN

The archaeology of Overton Down is illustrated, at increasing levels of detail, on Figures 2.1, 2.2, 3.3, 6.1 and 6.11 (*see also SL*, figure 37, which shows, schematically, central and southern Overton Down). Focusing in on the central part of the down, from north to south, the main features are as follows (Figure 6.1).

The site of the experimental earthwork

Not discussed in detail here (see *Jewell 1963b; Jewell and Dimbleby 1966; Bell* et al *1996; and Plate LXVII).*
The archaeological results of the excavation conducted as part of the experiment can be summarised thus: the area was retained as a patch of pasture in the increasingly organised landscape from the second millennium BC onwards (cf Fowler 1963b; FWP 66).

Round barrow and 'gardens'

Some 250m south west of the experimental earthwork is a local high point and area of sarsens (Figures 6.1 and 6.11). The latter are conspicuous by their isolation, the former is delineated by ancient fields on three sides and settlement to the east (below), and was scrub-covered until the 1980s. Now clear, the knoll is occupied by a small, stone-kerbed and apparently undisturbed round barrow. This now prominent feature is significant, as few barrows are known on this part of the downs (Chapter 2, Figure 2.4). Photographed by Crawford in 1924 (Crawford and Keiller 1928, fig 24, pl xix, here Plate XXIII), the barrow is not visible on this photograph, but an area of narrow rig shows clearly. Interpreted by Crawford as 'gardens' (ibid), this evidence has never been visible to us on the ground.

Linear ditch/trackway

East of the experimental earthwork is a slight linear ditch heading south east along the spine of Overton Down. It is more a hollow-way than dug ditch, being the southern continuation of the trackway that branches off linear ditch F4, further to the north (Chapters 2, 5 and 16; Figure 16.6). In between, parts of its course have for long been marked by the OS on large-scale maps. Here it runs through a settlement and skirts the excavation site of OD XI before continuing to Pickledean, linking Romano-British settlements (Figures 6.1 and 6.11).

An earlier route on this line may have influenced the location of Early Iron Age settlement OD XI. A positive lynchet, overlying the ditch previously marking the southern boundary of the settlement enclosure, formed against the north side of the track. This suggests that cultivation was contemporary with the track and post-dates abandonment of OD XI (Figure 6.3). This could have occurred soon after abandonment, allowing this length of track to be of Early Iron Age date, cutting through Late Bronze Age fields as part of a new layout dividing the former enclosed settlement area into fenced fields (*see below*).

The linear ditch/track was apparently still recognised in the tenth century AD, for its line is almost certainly that of the *hric weges*, marking the boundary between East Overton and Lockeridge (Chapter 2). If correct, this is useful in supporting the air photographic evidence that the main route was originally along the spine of the down to the Romano-British settlement of Overton Down South (ODS, *below*). The tenth-century evidence also reinforces a point made elsewhere (Chapter 2) that linear features, once engraved on the landscape, tended to continue to function and could, over time, be both a line of communication and a boundary (cf Chapter 16).

The settlement area south of the experimental earthwork and on either side of the ditch/track contains three main features: an area of 'open' settlement; a rectilinear enclosure; and a large, rectangular pond (Crawford and Keiller 1928, 124–5, pl xix, here Plate XXIII, Figure 6.1).

Open settlement

The settlement area lies on the east/south-east side of the local high point. Slight irregular hollows mark it; some are undoubtedly the result of shallow quarrying and, perhaps, stone removal. Among these earthworks there appear to be building platforms, generally round rather than rectangular. The area has produced Romano-British pottery in molehills and other soil exposures. Overall, the evidence indicates a Romano-British settlement of *c* 2.5ha, related to an earlier track very probably still in use.

The earthwork enclosure

A small rectilinear earthwork enclosure with a small round pond on the east side of this track and within the settlement area (Figures 6.1 and 6.11, Plate XXIII; *SL*, figures 5, 37) was discovered and commented on by Crawford (Crawford and Keiller 1928, 125). He regarded it as part of the Romano-British settlement. It partly overlies the east side of the trackway, however, and is more probably post-Roman, morphologically similar to the slightly larger Down Barn enclosure, also cut into a Romano-British settlement, related to an adjacent, small round pond, and partly overlying the physical remains of a documented Anglo-Saxon trackway (*see below*). It is *possible* that this Overton Down enclosure is one of the 'missing' Anglo-Saxon sheepcotes, perhaps even Hackpen, a former place-name on Overton Down.

Square pond

The large rectangular is the most obvious of the earthworks here (Figures 6.1 and 6.11, Plate XXIII; *SL*, figure 37). Of early twentieth-century date and overlying Romano-British earthworks, it relates to the Meux estate's use of the area for racehorse training and stock-farming. The pond, like others in the study area, is a classic example of the 'dew-pond' advocated by contemporary agricultural improvers (Martin nd; Pugsley 1939). Ponds of this form are clay-lined rainwater reservoirs with a made-up 'apron' on one side to allow animals access without damaging the lining. The 'apron', in this case on the west side, was refurbished as a sloping area of sarsen stones during a well-intentioned but abortive youth-training scheme in the 1970s. The pond, now redundant and usually dry, is very much a monument to twentieth-century social, as much as agricultural, history.

OVERTON DOWN X/XI

A conventional report on this relatively major settlement excavation is available electronically (FWP 63; much more detail is available in the archive in FWPs 1, 2, 3, 4, 13, 14, 16, 17, 20, 27, 28, 34, 38, 42, 57, 61, 62 and 69; the archive includes all the finds, Boxes nos 3000–3026, 3094–5 and 3103). The description here is limited to the surface evidence and a summary of the principal findings of the excavation. The selection of illustrations is designed to present key evidence for an interpretation of the excavation in terms of landscape history.

A curved scarp or lynchet within a rectilinear field system on a 3° south-westerly slope just below the brow of Overton Down was first illustrated, but not remarked upon, by Crawford (Crawford and Keiller 1928, 124–5, pl xix, fig 24, here Plate XXIII; cf also Plates VII and VIII). Crawford was principally concerned to demonstrate that this area of 'old grassland' had once been cultivated, arguably in two phases. He proposed that earthworks markedly slighter than the lynchets, called by him 'parallel ribs', represented 'the "lands" of the ancient ploughing', anticipating by half a century the recognition in northern Britain of 'cord rig' (Topping 1989). Crawford interpreted the 'ribs' or ridges of cultivation as the marks of (at latest) Romano-British cultivation within prehistoric fields.

A detailed re-examination of this area led to the hypothesis that, whatever the date of the 'ribs', the curved lynchet which they respected might be a key to a longer local sequence than Crawford had imagined. This anomalous lynchet could plausibly have accumulated on the outside of the curving perimeter of a pre-existing enclosure. If an enclosure had existed, and could be dated, then the fields over its interior and perimeter must post-date it. The research attraction therefore was the possibility of establishing a *terminus post quem* for at least some of the early fields within the study area. Furthermore, fieldwork on Overton Down seemed to provide a rare opportunity to relate fields 'sandwiched' between two settlements of differing date and to examine horizontal stratigraphy across a landscape. Excavation where fields overlay a settlement (OD XI) was followed by excavation of a settlement (OD XII) overlying fields (*see below*).

It was also important to test Crawford's interpretation of the 'lands' as Romano-British. By the early 1960s the recognition of medieval activity on the downs had raised doubts over this interpretation (Chapters 2 and 7).

The two 'sites', OD X and OD XI, were part of the same archaeological complex. OD X was the excavation code for cuttings on and outside the perimeter of an enclosed settlement, and OD XI, by far the larger undertaking, was the code for all excavations within the settlement (Figures 3.3, 6.2 and 6.11).

Summary of the principal phases suggested by excavation

Period 1: c 2000 BC: Beaker/Early Bronze Age burials

Three Beaker graves lay in an open but not 'empty' landscape (Figures 6.3a and 6.4). In the wider local context (Chapter 2), they are in a landscape already containing earlier funereal monuments (megalithic and earthen long barrows) and near-contemporary and contemporary round barrows. The small, stone-kerbed round barrow only 130m to the north, described above, could well be contemporary. Other flat cemeteries may also have existed but are yet to be located. Settlements existed but are still poorly represented in the record except for the near-ubiquitous downland spread of worked flints and flakes (FWP 31).

The graves contained the remains of three inhumation burials. A child of about seven years of age

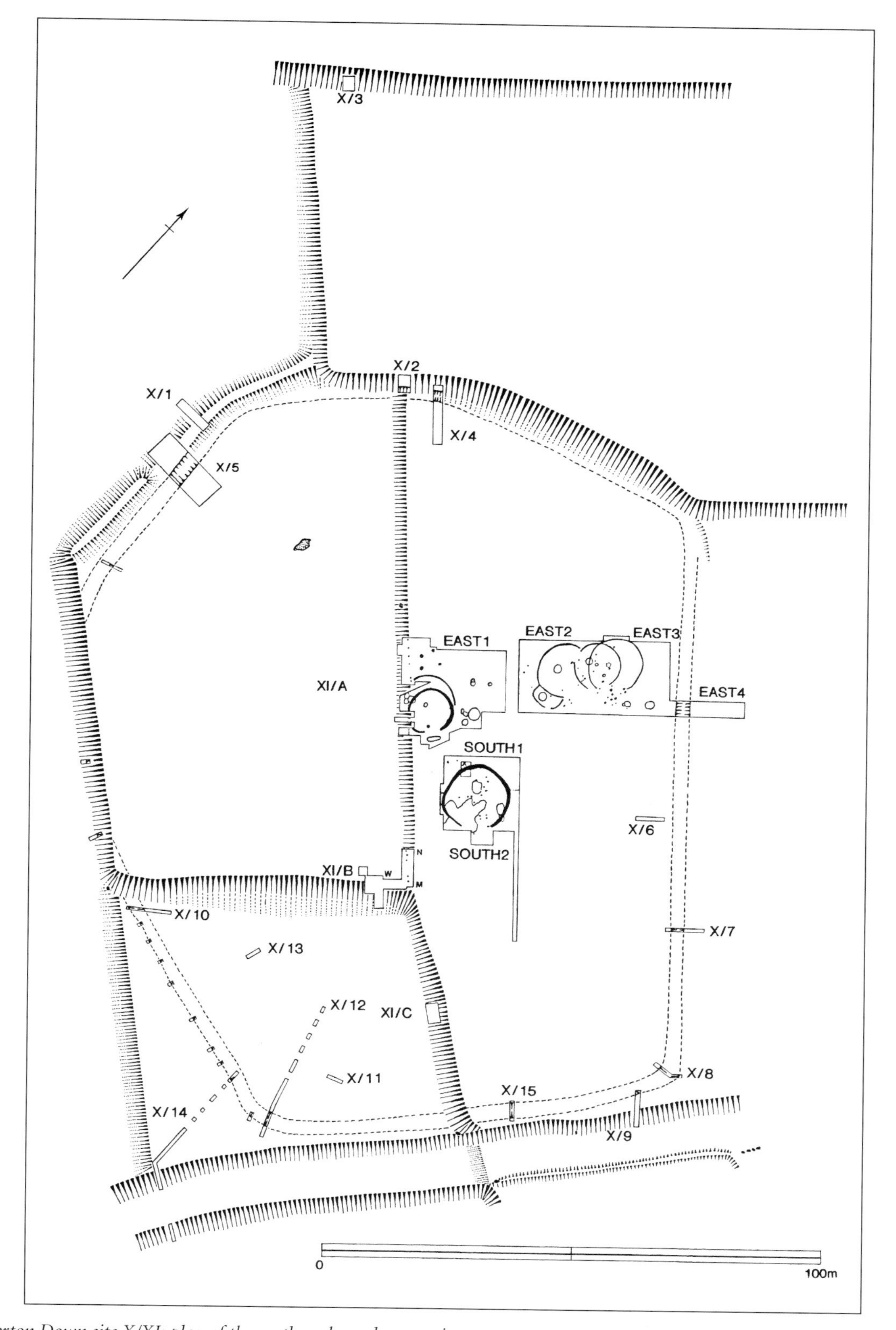

6.2 *Overton Down site X/XI: plan of the earthworks and excavations*

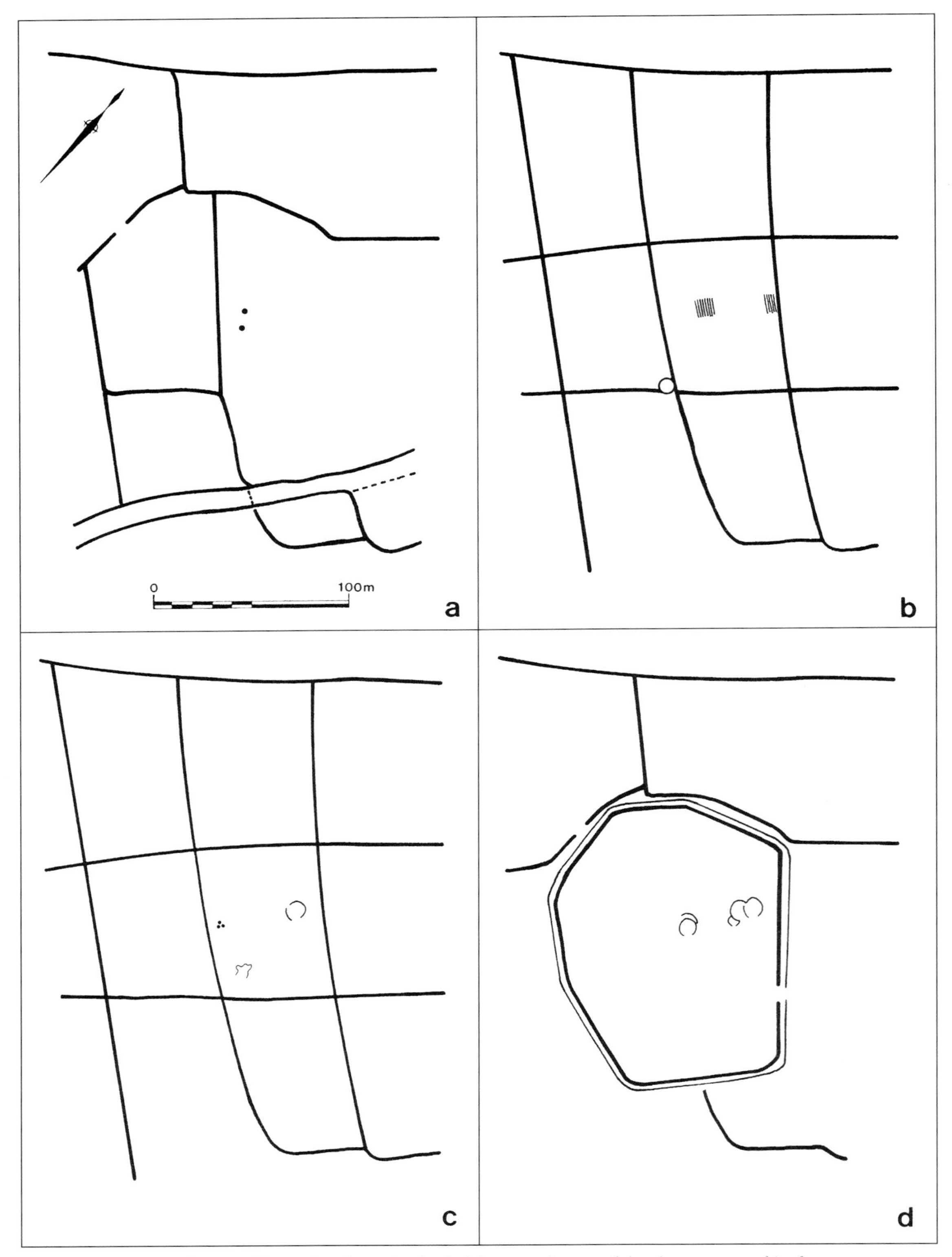

6.3 (a) to (g) *Site OD XI: diagram illustrating the main physical features of seven of the phases proposed in the preferred interpretation (see pages 82–92)*

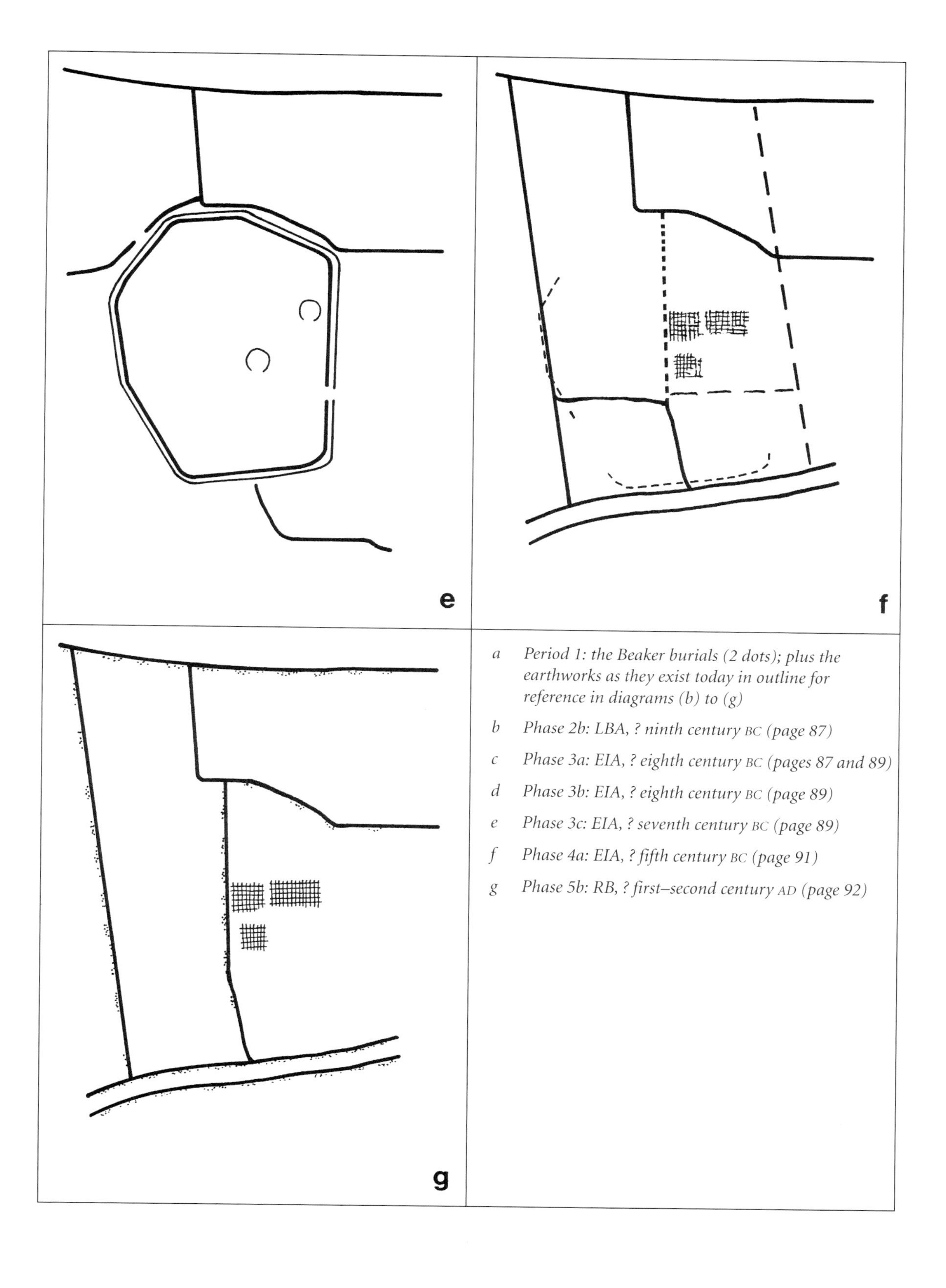

a Period 1: the Beaker burials (2 dots); plus the earthworks as they exist today in outline for reference in diagrams (b) to (g)

b Phase 2b: LBA, ? ninth century BC (page 87)

c Phase 3a: EIA, ? eighth century BC (pages 87 and 89)

d Phase 3b: EIA, ? eighth century BC (page 89)

e Phase 3c: EIA, ? seventh century BC (page 89)

f Phase 4a: EIA, ? fifth century BC (page 91)

g Phase 5b: RB, ? first–second century AD (page 92)

(Burial IA) was accompanied by an unusual Beaker vessel with all-over decoration in the form of paired finger- and thumb-nail impressions. It is fully described by Smith (1967) and discussed in more detail in FWP 63 (and FWP 63, fig 14; *see also SL*, colour plates 16, 31, figure 66). A large male in his twenties (Burial IB) and the lower part of a skeleton of a ?female adult (Burial II) were unaccompanied by any grave-goods but are clearly part of the same group.

Possibly fitting in at the end of this phase was a further pit, 23 (in cutting East 3, Figure 6.6), a small pit containing fragments of a possible cremation burial and other material which indicated an Early Bronze Age date.

Period 2: Middle Bronze Age field system
Phase 2a: the area later occupied by Early Iron Age settlement was part of an extensive, co-axial field system consisting of enclosed fields *c* 60m x 50m in size

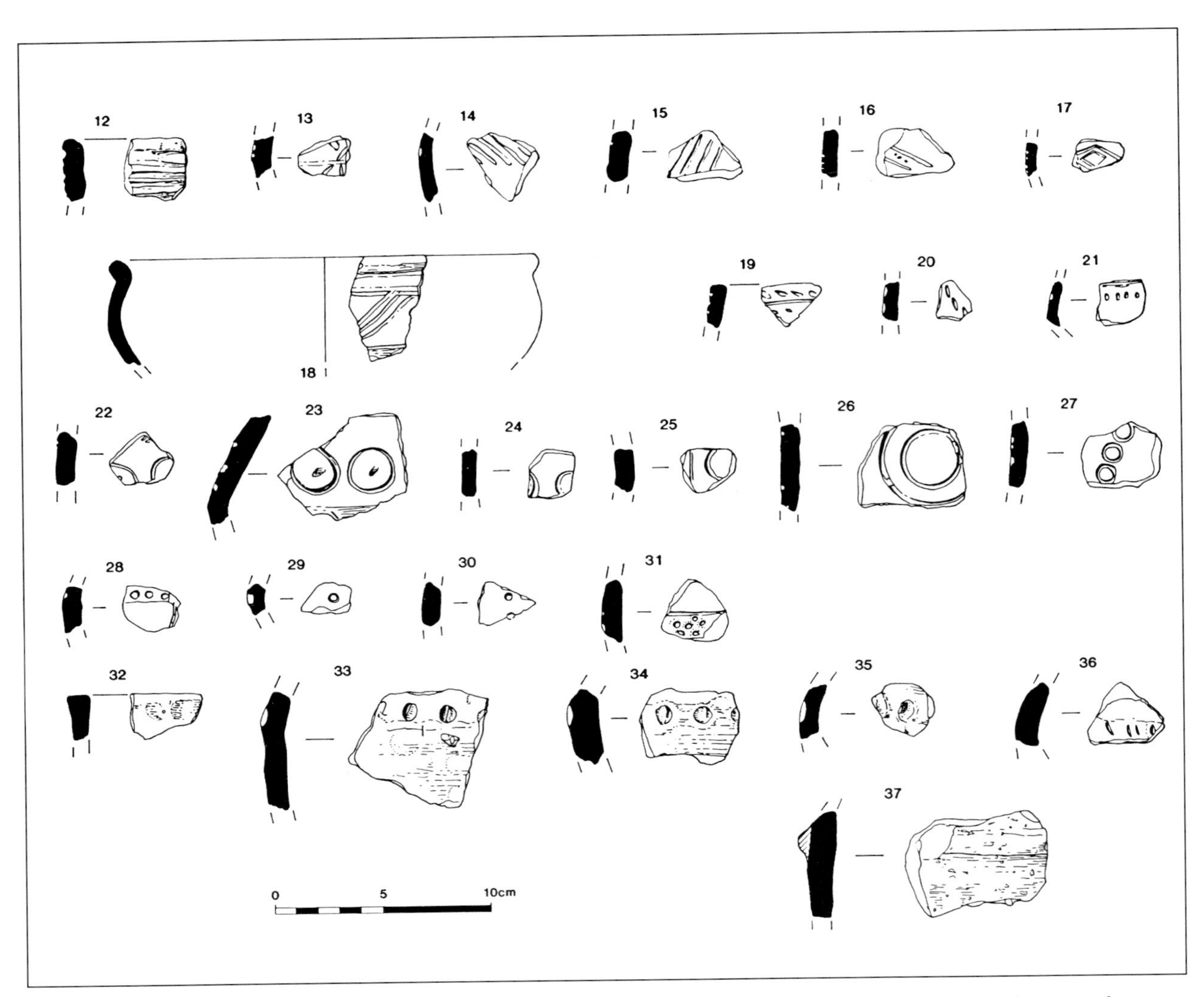

6.10 *Site OD XI, Phases 2b/3a: pottery: decorated sherds. All are incised except for number 37; all are body sherds except where otherwise stated (for explanation of fabrics,* see *FWP 63, and of GF numbers, see page xvi)*

12: rim, fabric C27, GF233; 13: fabric Q8, GF326; 14: fabric V28, GF209; 15: fabric Q8, GF471; 16: fabric M31, GF376; 17: fabric Q7, GF267; 18: rim/shoulder, fabric F11, GF344; 19: fabric Q8, GF208; 20: fabric Q6, GF330; 21: fabric M31, GF364; 22: fabric Q3, GF224; 23: fabric Q2, GF471; 24: fabric Q7, GF246; 25: fabric Q3, GF230; 26: fabric Q7, GF471; 27: fabric Q7, GF376; 28: fabric Q2, GF230; 29: fabric Q2, GF233; 30: fabric F10, GF233; 31: fabric Q3, GF224; 32: rim sherd, fabric Q3, GF237; 33: fabric Q3, GF397; 34: fabric Q2, GF232; 35: fabric Q2, GF219; 36: fabric Q3, GF385; 37: fabric S34, GF471

(Chapter 2; Figures 2.1, 2.2, 6.2 and 6.3a). This phase of cultivation established the open downland later reflected in the Early Iron Age enclosure ditch microfauna (Phase 3c/4); and may have triggered soil erosion (Chapters 14–16).

Phase 2b: ninth century BC*. Late Bronze Age occupation in a field within a field system:* a Late Bronze Age phase of activity was evidenced by a small but significant amount of pottery and metalwork, which tended to cluster around the south-western area of cutting South 1 and cutting B at a lynchet junction (Figures 6.2 and 6.3b). At the latter, one interpretation of numerous post-holes allowed the suggestion that a circular structure had stood at the south side of a field. The field, or part of it, was presumably taken out of cultivation while the rest of the system continued to be farmed. Cultivation may have scored some ard-marks in the bedrock surface. The pottery represented an ovoid jar and an applied cordoned vessel of Deverel-Rimbury type (Machling in FWP 63, fig 35).

Period 3: Early Iron Age occupation

This period embraces the main phases of Early Iron Age occupation, though it may well have developed out of Phase 2b. Its absolute date and the length of occupation are uncertain. The ceramics (Figure 6.10) suggest a date in the eighth/seventh centuries BC for its earliest phase, and occupation may have been short. Outside limits of ninth–sixth centuries BC are not unreasonable. The possibility of a 'three generation occupancy over a century' either side of 700 BC is one interpretation discussed elsewhere (Chapter 16). The various phases of the settlement produced a range of domestic artefacts, including worked bone/antler pins and needles, iron knives and awls, quern-stones, whetstones and spindle-whorls, and more personal items such as two iron and one copper alloy brooches. The animal bone assemblage included a range of domestic species with sheep/goat predominating over cattle and pig (Chapter 14). The settlement area included a number of pits containing what appear to be placed deposits of animal bone (eg, ox and horse skulls, articulated remains of cattle and pig) and other items, including sarsen blocks, non-local stone and decorated pottery.

No occupation of the site between the fifth century BC and first century AD was apparent.

Phase 3a: unenclosed Early Iron Age occupation within one specific (hypothetical) field: the phase is proposed on the basis of an initial Early Iron Age occupation in the

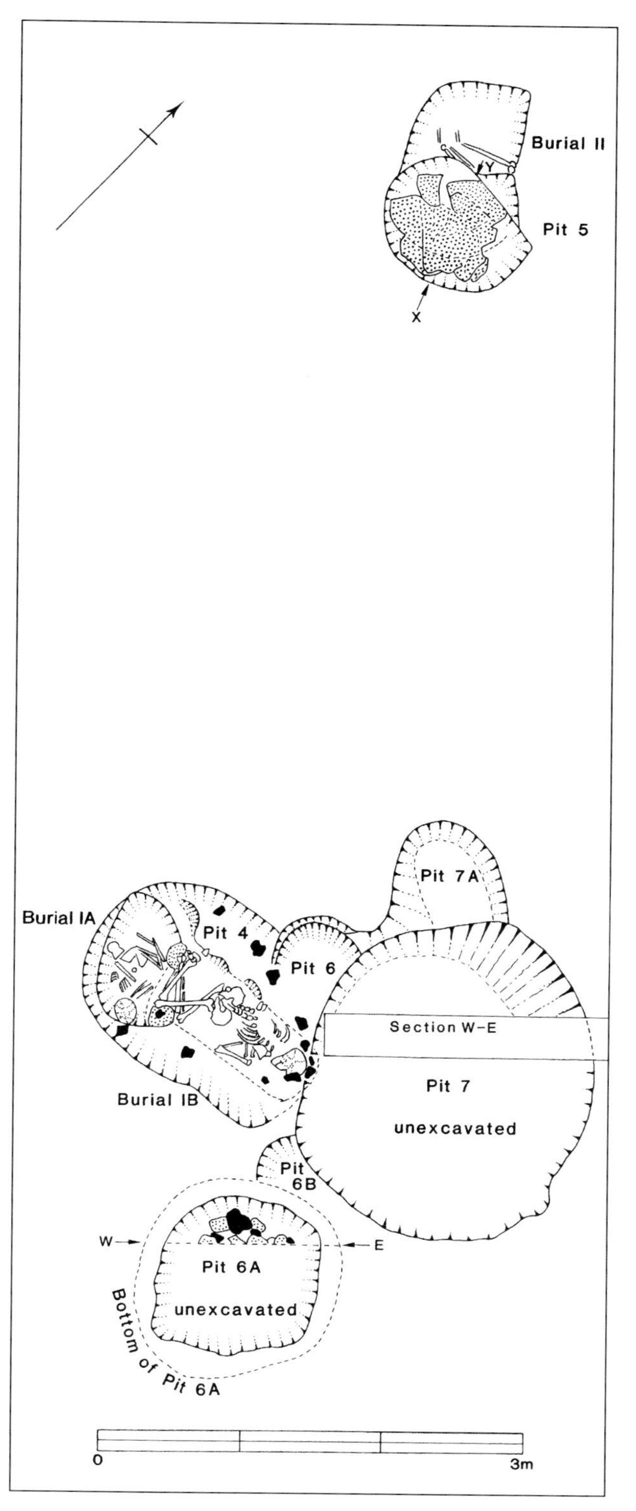

6.4 *Site OD XI/A, East 1: plan of Beaker graves (Burials IA, IB and II) and Pits 4, 5, 6, 6A, 6B, 7 and 7A. For location of this and all cuttings, see Figure 6.2 and FWP 63*

Plate XXIV (a) and (b): site OD XI/A, East 1. Sections through lynchet over Gully 1 (Building 1), east (top) and west (above) ends

Plate XXV *Site OD XI/A: intersection arcs of foundation trenches of Buildings 2 (G6) and 3 (G8) looking at cutting East 3 from the north*

form of an open settlement within an already existing field similar to that on which Phase 2b is based (Figures 6.2 and 6.3c). None of the early pottery associated with this occupation occurred in the enclosure ditch. Structural remains could include the earliest of three, possibly successive, circular buildings (B2 in Cutting East 2, comprising a penannular gully (G6) with a double-leaf, south-east entrance and possibly a hearth and central post-hole (Figures 6.6 and 6.8 (B2)). 'Working hollows' and the three earliest pits (1–3) could originate in this phase too.

Phase 3b: eighth century BC. *Early Iron Age settlement, complex building(s), in an enclosure cut out of and within a co-existing field system:* the main occupation phase was immediately preceded by major physical disruption when the size, nature and perhaps status of the settlement changed to take local precedence over arable fields. Digging some 400m of ditch (Plate XXVII) presumably associated with a bank created an enclosed settlement, occupying three times the area of the original settlement of Phase 3a. The arc of the enclosure ditch left a permanent mark on the landscape. Structures within the enclosure could have included two building complexes, probably of 'round-houses' with porches, hearths and annexes (Figures 6.2, 6.3d, 6.6 and 6.8 (B2, B1); Plate XXV). The ceramics fall within the eighth–sixth centuries BC (Machling in FWP 63, figs 35–36).

Many excavated but undated features, including pits and post-holes, may also belong to this phase (Figures 6.5 and 6.6). In particular, Pits 1, 2 and 3, under the lynchet later bisecting the settlement area (Figures 6.5 and 6.8), should all belong in this phase or earlier since they are cut by the gully of a Phase 3b or 3c structure, B1.

Phase 3c: seventh century BC. *Late Early Iron Age settlement, single round house, in an enclosure within a field system:* allowance is made for this possible 'late' phase of Early Iron Age enclosed settlement, which could have contained one or two buildings: either B4 alone or B4 and B3 together (Figures 6.2, 6.3e, 6.7 and 6.8). Ceramically, they appear to have been very close in time. Building 4 cut through stratified deposits in 'working hollows' containing material of Phases 2b, 3a and 3b (Plate XXVI). Phase 3b pottery in the enclosure ditch (cutting X/15) probably represents later deposition of residual occupation material, perhaps at this time rather than 'dating' a phase of ditch infilling.

Period 4: arable, followed by a long period of pasture

This phase accommodates a second major disruption: the ending of the settlement and, after a short interval, the return of the area it occupied to arable (Figure 6.3f). The settlement was abandoned by the mid-sixth century BC and possibly dismantled. Its area was incorporated into four new fields delineated by new boundaries on slightly different lines from the earlier system. One of the new north–south boundaries divided the former settlement enclosure roughly into two halves; it was fenced (Plate XXIV; FWP 63). The new fields, linked into the field system associated with the former

Plate XXVI Site OD XI/A, South 1: area of working hollows under Building 4, cut by its wall-trench (G1, Figure 6.7)

Plate XXVII Site OD XI/A, East 4: south face, ditch section

settlement, were cultivated, perhaps only for a short period and not apparently into the Middle Iron Age. Ard-marks cut across settlement remains (*see* Figures 6.5–6.7). The double-lynchet track past the south end of the settlement enclosure was probably inserted at this stage (Figure 6.2), respecting the Phase 3b enclosure ditch, the line of which had become a field edge and possibly an even more important boundary. The whole area then reverted to grass for several centuries.

Phase 4a: the site produced no 'Middle Iron Age' material, most marked being the absence of 'Wessex saucepan pots'. In land-use terms, there was sufficient time for a thin land surface to develop over the Period 3 occupation before the settlement area was returned to arable. Most of the ard-marks probably belong to this phase since they relate to the straight north–south fence against which a lynchet was beginning to accumulate (Plates XXIV and XXVIII; Figure 6.9). The cultivation was, however, short-lived.

Phase 4b: throughout the rest of the Iron Age, from around the sixth to, at the earliest, the first century BC, the downland became exactly that: a tract of grassland which, with two intermissions, it has remained ever since. There was no archaeological evidence of activity in the last centuries BC but, rather than a desert, it was much more likely to have been in regular use for stock-farming, perhaps including horses.

Period 5

A long period of some eight centuries when the site and its surrounds were basically grassland.

Phase 5a: first century BC. *Renewed but non-intensive settlement activity:* though not well attested, activity on

Plate XXVIII Site OD XI/A, East 2: ard-marks scored into the crumbly surface of the bedrock (Upper Chalk)

the site is once more attested archaeologically and may indicate occupation there or in its vicinity in the last century BC.

Phase 5b: first–second century AD. Renewed and intensified agrarian activity: as part of what is now recognised to be a general phenomenon in the study area, OD X/XI was recultivated from early in the Roman period (and indeed Phase 5a may be at the start of this phase rather than a separate, earlier one). What seems to have been intensive but relatively short-lived cultivation was preceded by the remarking of old, and the creation of new, field boundaries, with sarsen stones being placed along the existing prehistoric lynchets (Figure 6.3g). The phase is dated by early Roman material, presumably derived from manuring, mixed with earlier material ploughed up from the occupation deposits of Period 3. It was over before AD 200.

Phase 5c: third–fourth centuries AD. Some activity continuing: this phase was not attested structurally but is inserted to allow for some agricultural activity, pastoral rather than arable and certainly not occupation. It is represented by a small amount of late Romano-British material and was really a reversion to the long-term grassland regime interrupted by Phase 5b.

Phase 5d: fifth–twelfth century AD. Permanent grassland: crossed by the double-lynchet trackway to Pickledean coming off the 'Overton Ridgeway' down the spine of Overton Down (*see above* and Figure 16.6).

Period 6
Medieval strip cultivation followed by permanent pasture.

Phase 6a: medieval, probably thirteenth century, cultivation in strips: partly fitting into earlier land arrangements. The blocks were subdivided into ridge-and-furrow or, justifiably, 'broad rig', here *c* 8.5m (27ft) wide. The thirteenth-century date is taken from the similar evidence on Fyfield Down and its association with the *Raddun* settlement (Chapter 7). Virtually the whole of the Late Bronze Age/Early Iron Age settlement area was over-ploughed, though unploughed 'gores' exist either side of the straight north–south lynchet across the settlement (Plate XXIII). As represented by existing ridge-and-furrow, all the medieval cultivation over the settlement was north–south; butting furlongs, however, were oriented differently.

Phase 6b: fourteenth–nineteenth century. Permanent pasture: for sheep on the East Overton manor of the Bishop of Winchester and, later, the Earl of Pembroke. The present 'old grassland' developed.

Period 7: the twentieth century
Permanent pasture continued variously under owner and tenant management, but its use began to diversify from sheep-pasture to embrace non-agrarian functions including racehorse training, military training and scientific and recreational purposes (Chapter 17).

DOWN BARN AND PICKLEDEAN

Over some 500m south from OD XI the south-east slopes of Overton Down fall into a dry valley called Pickledean (Figures 1.3, 2.1, 2.2, 2.4 and 3.3). There, Down Barn is surrounded by a well-preserved field archaeology despite intensive agricultural use on all sides apart from the north east. It is a 'nodal point' in the landscape (Chapter 16). Two excavations were undertaken, OD XII and Down Barn Enclosure (details in FWPs 64 and 66). Unpublished vertical air photographs taken by O G S Crawford on 22 June 1924 have significantly assisted interpretation of the area (Plate XXIX).

The Down Barn area

The area studied here (Figure 6.11, Plate XXIX; *SL*, colour plate 19) is defined on the south west by a tenurial boundary at least one thousand years old: a trackway still in use, Romano-British if not earlier in origin, which bounds a post-medieval park (Chapter 4). Down Barn, on the south side of the coombe (Figures 3.3 and 6.11), is a common name in the Wiltshire agrarian landscape; Pickledean is an unusual one. It seems to derive from OE *pytteldene*, perhaps 'hawk-valley' (*PNWilts*, 306). The field immediately north of the Barn was *unum clausum vocatum Pikkeldean* in 1562. Pickledean curves north west up from the Kennet valley and contains two other barns (Pickledean Barn, mid-nineteenth century, and New Shed, mid-twentieth century). Down Barn is located at a slight widening of the coombe, at 180m aOD, about 2km from its mouth. It continues to climb gradually to the north west for a further 1.5km and merges with the smooth contours of Overton Down at 220m aOD. Generally the north side is steeper than the south, the opposite of the profile of the Valley of Stones (Figure 1.3). The flat bottom strongly suggests an accumulation of colluvial material, as indeed proved to be the case (*see below*). The dene is pasture throughout, though bordered on the south for most of its course by arable. In 1819 the two fields immediately

Plate XXIX *Down Barn (*bottom right*) with Down Barn Enclosure and pond to the left and the southern part of Overton Down above both, showing prehistoric fields, trackways, sites OD XII and XIII (*lower centre*), different sorts of ridge-and-furrow, the 'stone row' and settlement Overton Down South (ODS,* right centre*), the last as earthworks between the track and the fence and as cropmarks in the arable. Much of the detail in Figure 6.11 is based on this previously unpublished air photograph of 1924 (NMR ALK 7420 ORACLEEI, ©* Keiller Collection*)*

south east of the Barn were respectively under turnips and in 'fallows'. Such land-use reflects the historical fact that the Dean marks the boundary between old arable and pasture in East Overton tithing, an arrangement that may have existed in the late Anglo-Saxon period, though not earlier.

This area around Down Barn provides a key to the working of different landscapes. The visible archaeology on and in the ground can add to this sense of downland dynamics and, at the same time, can be interpreted in its light (Chapter 16). The following is a selective list with some commentary on the main archaeological features around Down Barn, working generally north west to south east but dealing with the four deserted settlements together at the end (Figure 6.11). Though distinguished as four, all may be components of one Romano-British settlement.

1 Early fields and integrated trackways

Early fields and integrated trackways, appearing both as hollow-ways and lynchetted tracks, cover much of the north-west half of the area (Figures 2.1 and 6.11). One of these, the *hric wege* of the Anglo-Saxon charter (S449), continues from the ridge-top of Overton Down and formed the boundary between East Overton and Lockeridge. It 'went south to the lynchets' (SU 136703), its line possibly corresponding to the trackway shown by Andrews and Dury in 1773. That continued through Fyfield Field to cross the Bath–Marlborough road just west of Fyfield village.

Two tracks branch off this ridgeway. One may be an alternative route down the slope and into Pickledean. The other is part of a through route from the north. These two parallel tracks are linked at their south-west ends by a terrace-way running along the north-east side of Pickledean. This is roughly parallel to the ridge-top trackway, as topography requires, and is cut into by the north-east side of the Down Barn Enclosure (*see below*). It becomes lost in the much disturbed area at the south-west end of ODS (Figure 6.13) but if it continued south east it must have run along the coombe floor, more or less on the line of the existing farm track. Archaeologically and in general, its line is close to or actually on the *lamba paeth*, which came down the east side of Pickledean and formed the east boundary of tenth-century East Overton (S449; *SL*, figure 65). Functionally, it seems plausible to think of sheep from the downs being collected on to the 'lamb's path' and shepherded down the dene along the trackway, *andlang weges*, into the *slaed* or bottomland where Pickledean gave on to the Kennet meadowland (around SU 143684, Figure 8.1; Smith 1970 ii, 127).

The line of the Romano-British trackway from the north east into the dene continues past the west side of Down Barn and is marked by a bridle path. This follows – or is followed by – the tenth-century boundary between West and East Overton. The track crosses the permanent arable of the Overtons to cross the Kennet by a ford. Its route is then continued by the road between the Overton villages (Chapter 9) and then south through the 'open' fields towards West Woods and the Vale of Pewsey. It is probable that this now little-used path, once of some local significance, formed an important north–south route through the area (Chapter 16).

2 Ridge-and-furrow

A zone of ridge-and-furrow can be subdivided into four different localised types (Figure 6.11 and Plate XXIX):

(a) 'Broad rig' north of the double-lynchet track, *c* 8m–9m (27ft) broad and laid out in furlongs, as demonstrated by Crawford (Crawford and Keiller 1928, pl xix).

(b) Similar rig to (a), east of the double-lynchet track and north of OD XII, probably part of the former open fields of Lockeridge tithing.

(c) Narrower rig south east of OD XII and east of the 'stone circle' (Figure 6.12).

(d) Narrow rig in a fan-like pattern running up the slope towards OD XII from the present gate into the National Nature Reserve north east of Down Barn. This is possibly temporary cultivation intake made during the Crimean War, a tradition recorded in 1963 as oral evidence from the late Mr Frank Swanton, then in his seventies (FWP 84).

3 'Stone circle'

So described on earlier OS 6-inch maps ('Mound' on the 1998 1:25,000 *Explorer* 167; Figure 6.12) and located east of OD XII, the feature is most probably, in descending order of likelihood, either an accidental configuration of sarsens resulting from disturbance in a much ploughed area; the very disturbed remains of a round barrow containing some form of sarsen structure; some other sort of mound; or the disturbed remains of a, possibly round, stone-based building, ie, a 'hut-circle' (if so it would be the only one known on the downs). Preparing a detailed plan of the remains (Figure 6.12) has not removed the scope for speculation, though it has significantly reduced the likelihood that the feature is a 'stone circle' and emphasised that it occurs on a patch of unploughed land.

4 Lines of sarsen stones

Three lines of sarsen stones occur on the downland of Overton Down:

i along the double-lynchet track described above where it turns to the north (on the OS 25-inch map): probably Romano-British and part of the first-century AD land reorganisation;

ii north of OD XII: medieval or later, associated with ridge-and-furrow of either 2b or c, above;

iii east of OD XII (on the OS 25-inch map): marking the line of a racehorse training gallop, probably the late nineteenth-century 'Derby Gallop' (Chapter 17).

5 A square pond

Marked on the 1819 map but filled in in the early 1970s (*see above*), the pond may be associated with the enclosure of the field, partly now defined by a bank, on its

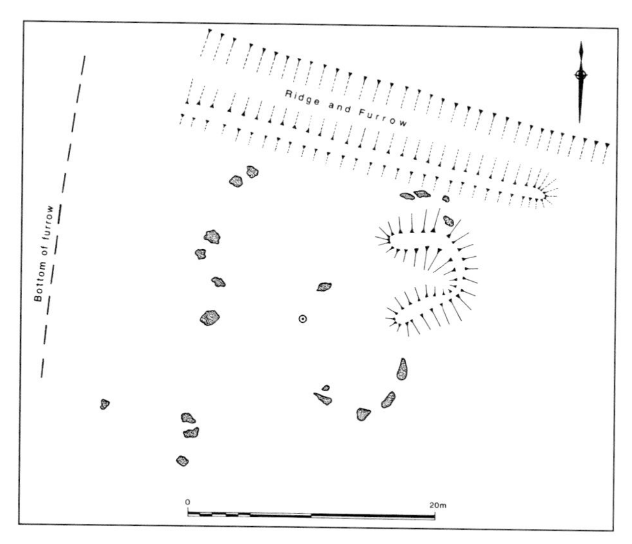

6.12 *Overton Down: plan of 'stone circle'*

south east. It lies just beyond the point where a double-lynchet trackway coming off the downs into the coombe turns south east. The pond, lying astride the coombe floor, may have incorporated an earlier, smaller one.

6 Down Barn

A cluster of buildings consists of Dutch barns of corrugated iron, replacing a thatched barn standing into the early 1960s on the site of the eighteenth-century 'New Barn' with its enclosed yard, which was shown together with its accompanying cottage by Andrews and Dury (1773). When first built it was on the northern edge of the permanent arable where it abutted the downland; its use was to overwinter stock on the downs. Immediately outside its northern yard wall is a small 'round pond' associated with at least one underground drain of sarsen construction; cf the square pond shown in 1773.

Down Barn cottage to the east, a Grade II listed building first depicted in 1773, with thatched roof, was home to permanent residents into the 1970s and, after considerable refurbishment, became available for 'holiday lets' in 1998.

Immediately to its east, now overgrown, is the concrete foundation of a former Nissen hut erected to service a military searchlight during World War II, used for farm purposes thereafter and subsequently revived in the 1960s and early 1970s as a makeshift dormitory and kitchen for this and the Experimental Earthwork projects (Jewell 1963b, appendix D). The superstructure finally collapsed during a storm in the 1970s and was cleared away.

7 Two standing stones

Two standing stones (SMR 108) on the eastern boundary of 'Hackpin Park' immediately south of Down Barn are discussed above (Chapter 4).

Overton Down South (ODS): Romano-British settlement

The settlement is impressive in extent even though only part of it survives as earthworks. Its plan, here reproduced with very minor amendments (Figure 6.13), has long been published and the site has been used in discussion of Romano-British settlement types in Wessex (Bowen and Fowler 1962, site B2; Fowler 1966). It lies up-slope from Pickledean, extending to the top of the ridge of Overton Down as a series of earthwork enclosures, some containing probable rectangular building sites. Sarsen stones obtrude in places from enclosure banks and interiors. It is quite possible that the rectilinear pattern of the settlement reflects its superimposition on an already existing pattern of enclosed fields.

Some enclosures within the settlement are divided by ledged trackways running south east off the main north-east/south-west track (*discussed above*). On the north-west side of that main track ploughing has reduced the earthworks. A fence on the east side of the earthworks marks the boundary of permanent arable, probably in existence by the tenth century when it is clearly referred to in the East Overton charter. This boundary between pasture and arable has sliced into the earlier settlement. It did not respect, nor does it now mark, the settlement's eastern boundary.

The full extent of the settlement, at least as represented by earthworks and potsherds, is as indicated on Figure 6.13. The pottery is homogeneously Romano-British and almost exclusively early *sensu* first–second century AD. The site is interpreted as a settlement of rectilinear form *c* 500m by 200m on a south-west/north-east–north-west/south-east orientation. Its full extent on the north west is still undetermined. It conveys an impression of having been laid out in an already much-used landscape beside a locally important point in the downland/valley communications network, alongside (or both sides of) an important trackway in that network.

Overton Down XIII: probable Romano-British settlement

A potential Romano-British settlement is marked by five possible buildings immediately uphill and north of the Down Barn Enclosure and slightly downhill and west of OD XII (Figures 3.3 and 6.11). It looks remarkably like

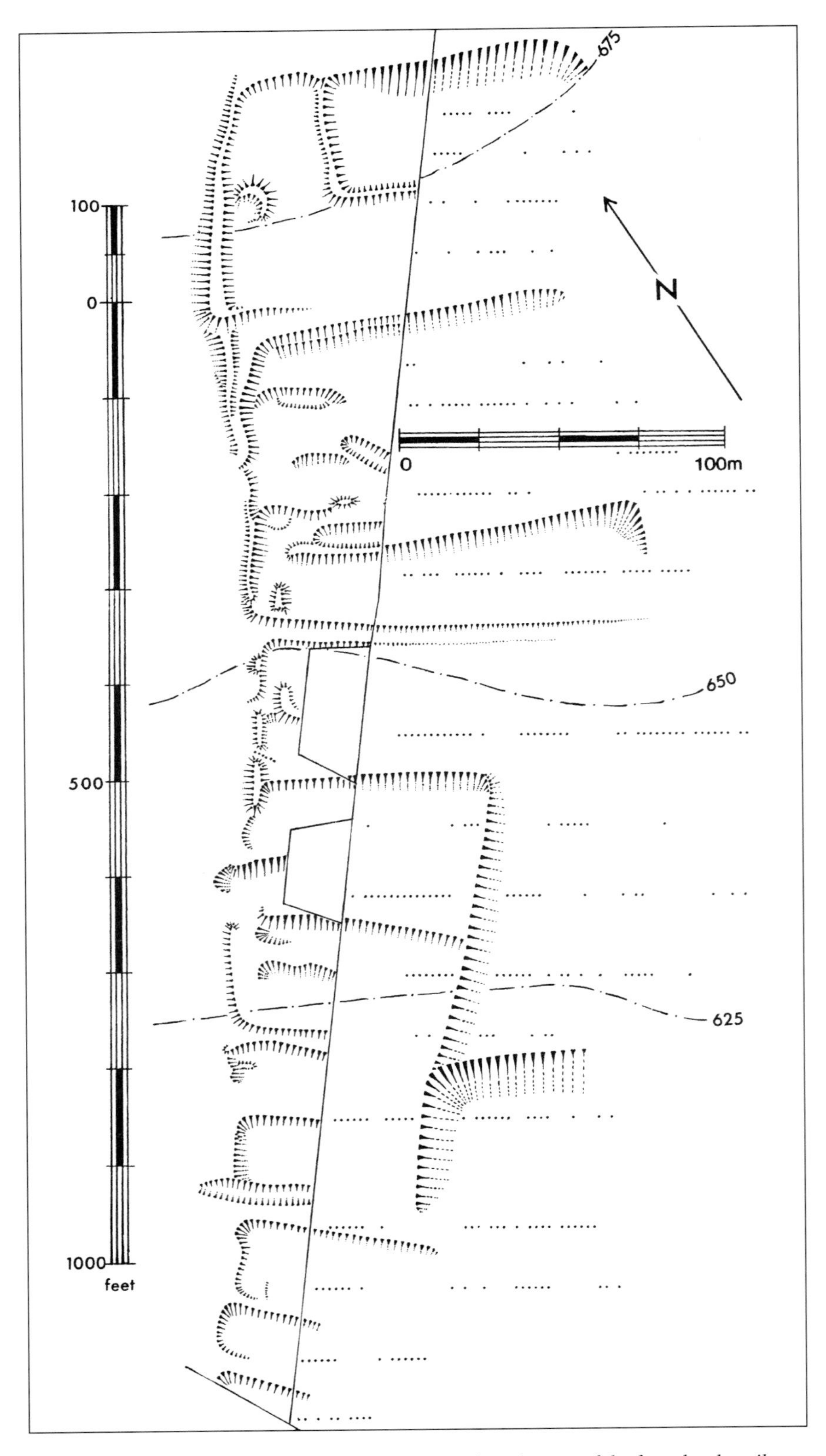

6.13 Overton Down South: plan of Roman settlement, preserved as earthworks west of the fence but heavily cultivated on its east

the latter before excavation, and was indeed recognised as a potential extension to it at the time (Fowler 1967, 26). It was only in the exceptional short-grassed conditions of August 1996, however, that it was possible to plan it and revise its status to 'probable'.

Down Barn enclosure: prehistoric stratigraphy, Roman occupation and a post-Roman earthwork

Two excavations of this site have been conducted by others as part of the project but neither has yet been fully published. The first, by J Scantlebury with boys from Marlborough College Archaeological Society, took place in 1962 and resulted in an interim report (Scantlebury in Fowler 1963a, 349–50), revised by this author here and in FWP 66. A second excavation, in 1996, is summarised below; a report by G Swanton is likely to appear in the *Wiltshire Archaeological and Natural History Magazine.*

The site was discovered in November 1961 (Plates XXIX–XXXI, Figure 6.14; FWP 66; *SL*, figures 31, 32). Trapezoidal in plan, the earthwork enclosure lies across the bottom of the narrow dry valley *c* 250m north and uphill of Down Barn (Figures 2.1, 3.3, 6.11 and 6.14), with old pasture to its immediate north on Overton Down and permanent arable to its south (*SL*, figure 31). The northern ditch of the enclosure cuts along a narrow terrace, or double-lynchet track, on the north side of the coombe, apparently a continuation of the Romano-British track running across the Overton Down landscape from OD XI (Figures 2.1 and 6.11). A pond lay beyond the south side of the enclosure, near an entrance. Inside, a low platform lay against the bank on each of the long sides. The whole site has been smoothed over by light cultivation in the early 1970s, reducing the sharpness of the earthworks and removing some of the critical detail that existed when it was surveyed in the early 1960s. The position, shape and size and relationships of this enclosure suggested it was 'late' in the local landscape sequence and likely to be of considerable significance. This has proved to be the case.

The 1962 excavation

The following is a summary of the published interim report (Fowler 1963a, 349–50), with additional interpretation by the present author.

Plate XXXI Down Barn Enclosure, 1996: section through the interior from post-Roman topsoil to Neolithic post-hole

Plate XXX Down Barn Enclosure from the north, with figures on the line of the main cutting across it in 1962 and 1996

Structurally, on the south west the ditch was *c* 1.2m deep, V-shaped and cut through a humic layer into the chalk. The bank was merely a low spread of soil and occupation material. 'Traces of what may be a small hut were found, defined by two parallel lines of small broken sarsens with a floor of packed chalk between.' This is the only record of this 'structure'; while its exact location and stratigraphic context are unknown, it probably lay in the central area of the western, post-Roman 'platform'. A structure here is of immense interest. The excavation produced 'large quantities of pottery and a considerable amount of animal bone, of which a high proportion appears to be sheep, iron nails, three very eroded bronze coins and the pin of a bronze brooch ... the whole assemblage would fit happily into a late Romano-British context towards the close of the fourth century or possibly rather later'. 'Two or three stray medieval sherds' were also found.

The context of this late-Roman material was on top of 'a sterile layer of fine, dark brown earthy clay, some 3ft [0.90m] thick at its deepest point and thinning out towards the sides of the valley'. This layer (clearly evidenced again in 1996, Plate XXXI) might 'represent a flood deposit in the valley bottom'; or be 'the result of accelerated soil creep and rain wash from arable fields on or immediately above its sides' (*see below* and Chapters 14 and 16).

Beneath this was another rubble layer lying on the Chalk. It apparently contained 'a grouping of large sarsen boulders suggesting some form of rectangular structure'. From the layer came a small number of 'sherds of undecorated, coarse pottery, rich red-brown in colour and containing a large amount of crushed chalk'. [PJF saw some of this in 1962 and thought that one or two might have been of 'Beaker' type, provisionally indicating an Early Bronze Age phase.]

Unfortunately, the excavation was not completed and the records and excavator have disappeared. In 1996, Bristol University re-examined the 1962 trenches as part of a small-scale training excavation.

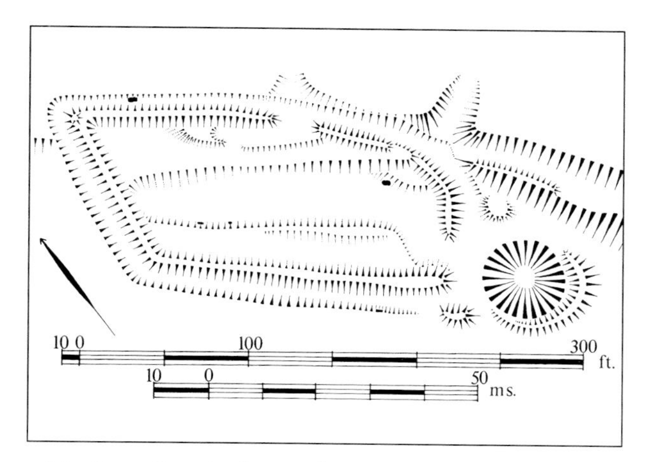

6.14 *Down Barn Enclosure: plan, 1964. The northern side overlies the terrace-way along the north side of the Dene; the pond has been largely infilled since the survey. The axis of the long trench excavated north east–south west across the site ran immediately east of the more westerly of the two sarsen stones along the front of the southern platform. An area excavation also took place in the centre of that platform*

The 1996 excavations

Under the direction of Gill Swanton, the trenches of the main north-east/south-west 1962 cutting were reopened mechanically. So much data and material were recovered that a full excavation report is now being prepared for separate publication by the director. Full access to the record has been generously provided by the director and forms the basis of this summary.

The 1996 excavation provided further stratigraphic detail and chronological depth (Mesolithic and early Romano-British). The chalk bedrock floor of the coombe lay some 1.5m below the present surface (Plate XXXI). It was overlain by a thick, humic old land surface with a Neolithic/Early Bronze Age horizon on or in its surface. Below this was evidence of structures (post-holes) and activity (flints, pottery) including Mesolithic material (rare in the study area). In broader terms the most important result was the dating to the Neolithic/Early Bronze Age of the old ground surface, broadly contemporary with the clear evidence elsewhere on these downs of 'Beaker' activity, for instance, at excavation sites OD I and OD XI (*see above*). This surface was overlain by a virtually sterile and thick layer of chocolate-brown humus stretching across the width of the enclosure from ditch to ditch and beyond.

It remains unclear whether this layer results from long slow accumulation or a sudden deposition (*see above*; further discussed *below*, Chapter 14). It is interpreted as the product of either alluvium or colluvium or both deriving from cultivation, particularly from the north (Overton Down) side. The layer is not securely dated, though the latest material in it was a few Early Bronze Age sherds. Overlying an Early Bronze Age phase, completely devoid of Early Iron Age material, and sealed by early Romano-British material, it seems likely that the layer represents a process, perhaps an event, in the second millennium BC after a 'Beaker horizon' (Chapters 2 and 14).

The bulk of the archaeological material came from an occupation layer above the thick humus deposit and apparently stretching across the coombe. Most of it was under or in the bank of the enclosure, or from the make-up of the platform on the south-west side of the enclosure. Characteristically late Romano-British material was rare; the material was predominantly of the first–second centuries AD. There is no doubt, therefore, that the enclosure itself is of late- or post-Roman date.

The total absence of material associated with it may, of course, be a reflection of a function as an animal fold, and it could therefore be of any date later than, say, *c* AD 400. The relative abundance of medieval and post-medieval artefacts, especially pottery, in Wroughton Mead and the Delling Enclosure (Chapter 7) contrasts sharply with their near-absence from the Down Barn Enclosure. This inclines interpretation towards its use in a post-Roman/pre-medieval phase, possibly aceramic. A cattle-pen or sheep-fold seems a likely function, beside a pond, on marginal land between arable and pasture and close to an intersection in local tracks and regional routes.

The enclosure, which could be the only visible part of a wider complex, may be one of the missing medieval sheep-cotes (Chapter 9). Those 'two or three stray medieval sherds', only evidenced in that published phrase, may be the slight but significant evidence indicating that here is the Overton equivalent of *Raddun* (especially triangular enclosure C; Chapter 7). An Anglo-Saxon origin, between the seventh and ninth centuries, when pottery was scarce, and before the tenth-century charters (which do not mention it) is another possibility. An implication of the lack of post-Roman material is that, whatever the date of its use, it was disused and forgotten before the thirteenth century when pottery became common locally. This makes the virtual absence of medieval pottery on Overton Down, and from the Down Barn Enclosure in particular, striking. Indeed, the two or three medieval sherds, perhaps indeed strays, draw attention to rather than dispute this absence on that line of argument.

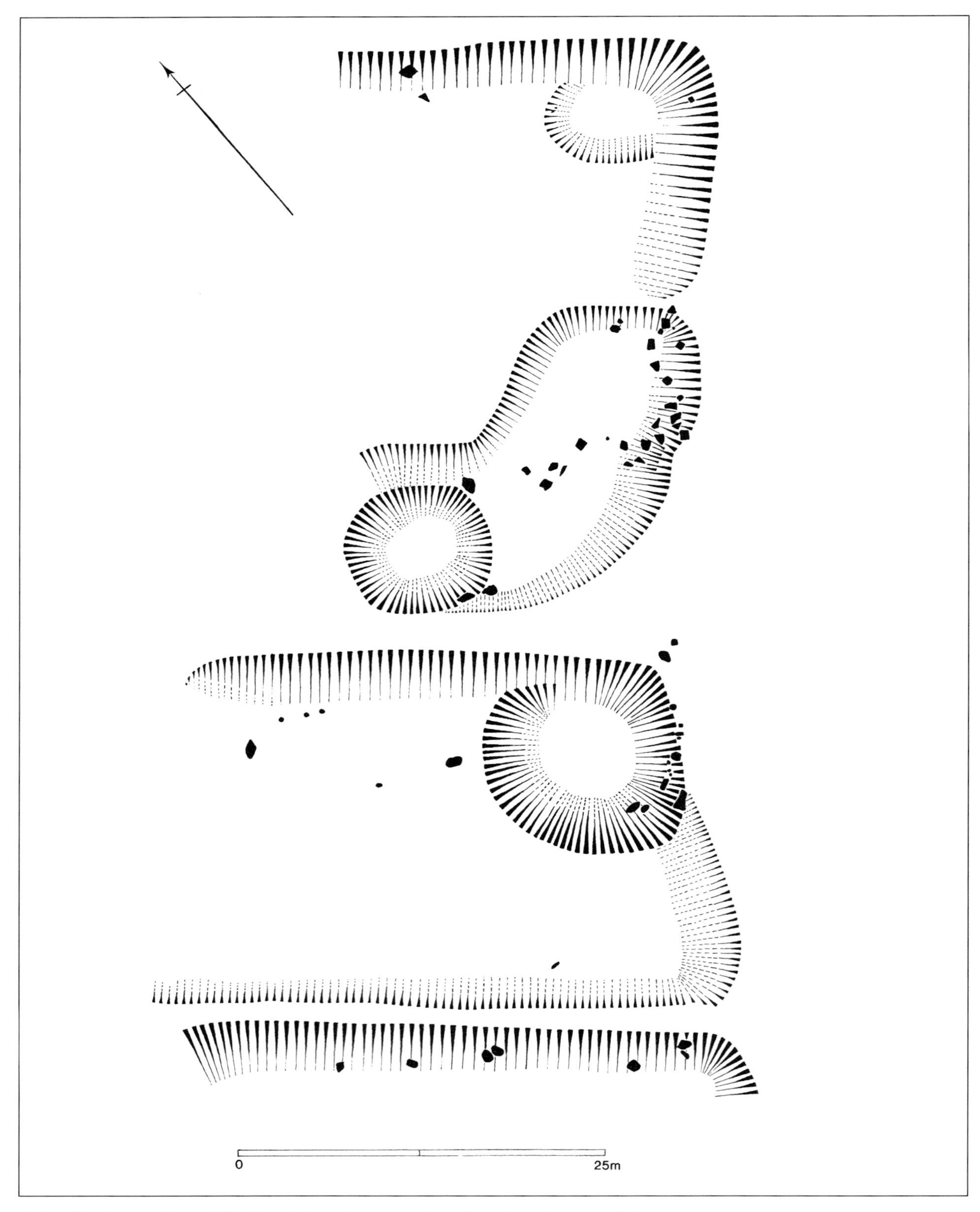

6.15 Site Overton Down XII: plan of the settlement earthworks and surface stones before excavation

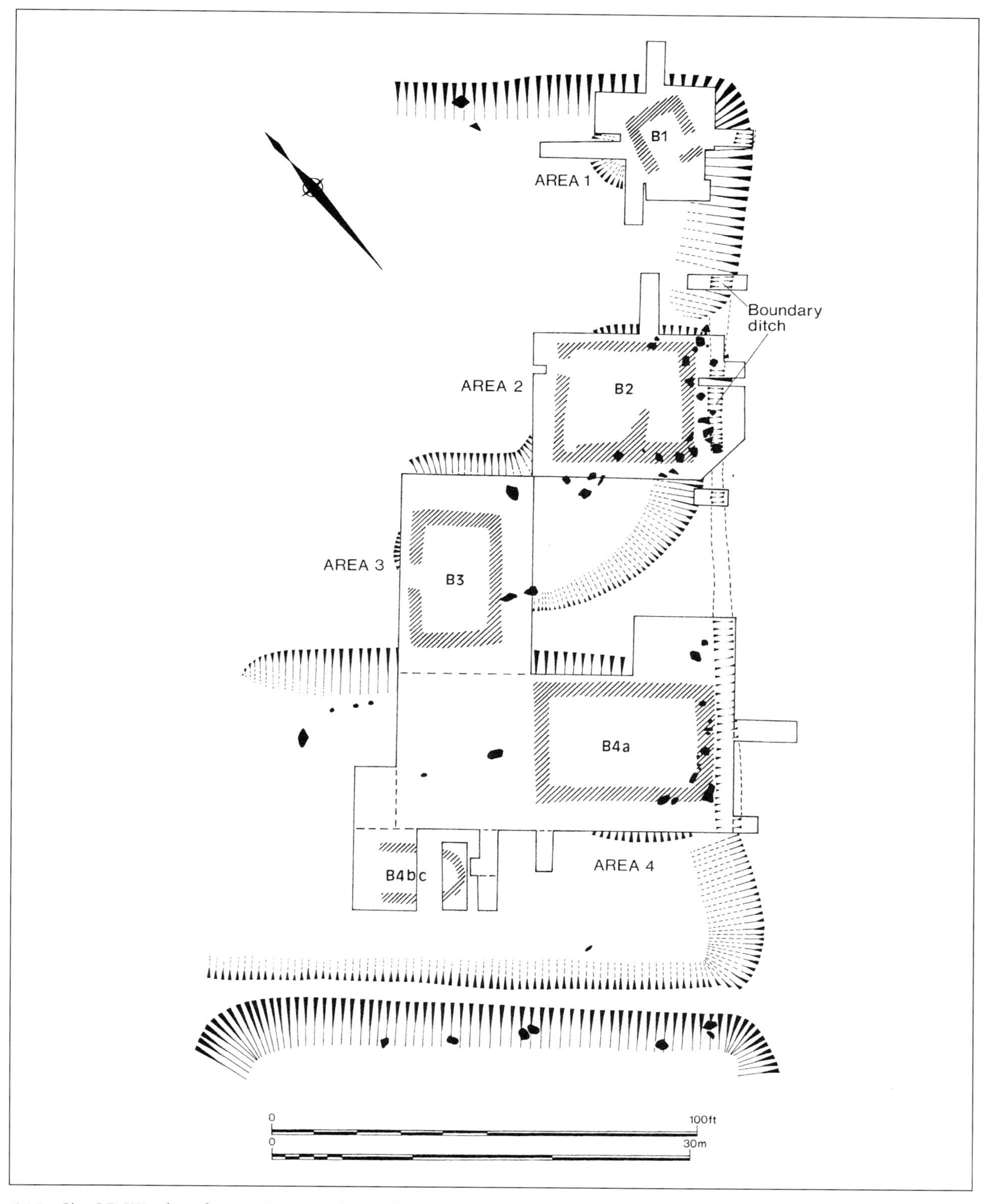

6.16 Site OD XII: plan of excavations superimposed on Figure 6.15, with the eastern boundary ditch and hatched outlines of the five main buildings excavated (B1–3, B4a, B4b/c)

Interesting though the enclosure is in its own right, particularly in hinting at post-Roman elements in the landscape, the site is even more significant because of the underlying prehistoric stratigraphy. The early phases point to the similarities with the evidence examined by Evans (*et al* 1993) along the Kennet valley. The subsequent sequence is interpreted as illustrating large-scale and probably widespread erosion on the downs in the second millennium BC (Chapters 14 and 15), a key factor, it is argued, in understanding this landscape. A monument-led approach can, apparently, produce bonuses.

Overton Down XII (OD XII): late Roman settlement

A full conventional excavation report is in FWP 64 with more detail in FWPs 5, 10, 19, 21, 32, 39, 52, 54, 58, 61 and 71. All the excavated material is in Devizes Museum, Boxes 3023 and 3027–3070.

The site lies uphill of and 150m north east of the Down Barn Enclosure and 150m north west of ODS (Figure 6.11 and 6.15, Plates XXXII–XXXVII; *SL*, colour plates 20–22). It seems improbable that it is unrelated in some way to either or both of those sites. In particular, it has to be seriously considered as a north-western part of the latter, rather than a separate settlement, though there are considerable chronological rather than spatial difficulties with such an interpretation. Long interpreted as a small, late Romano-British settlement (Fowler 1966), this view is no longer held for the reasons summarised below. The site lies on what was thought to be undisturbed old grassland and, having been carefully back-filled in 1966–8, remains visible very much as it was in 1965 when first surveyed (Figure 6.15).

Recognised as a discrete group of four, perhaps five, 'building platforms' during fieldwork, the site was initially interpreted as a 'complete' small settlement. As early as 1966, however, it was reinterpreted, 'heavy grazing of the area [then] having revealed several more probable sites of buildings indicating that there might be as many as a dozen structures scattered mainly in the corner of "Celtic" fields over an area about a 100 yards square' (Fowler 1967, 26). Doubts about its size and isolation have been reinforced on two grounds. As

Plate XXXII Site OD XII, Building 1, with quern in situ, *from the north*

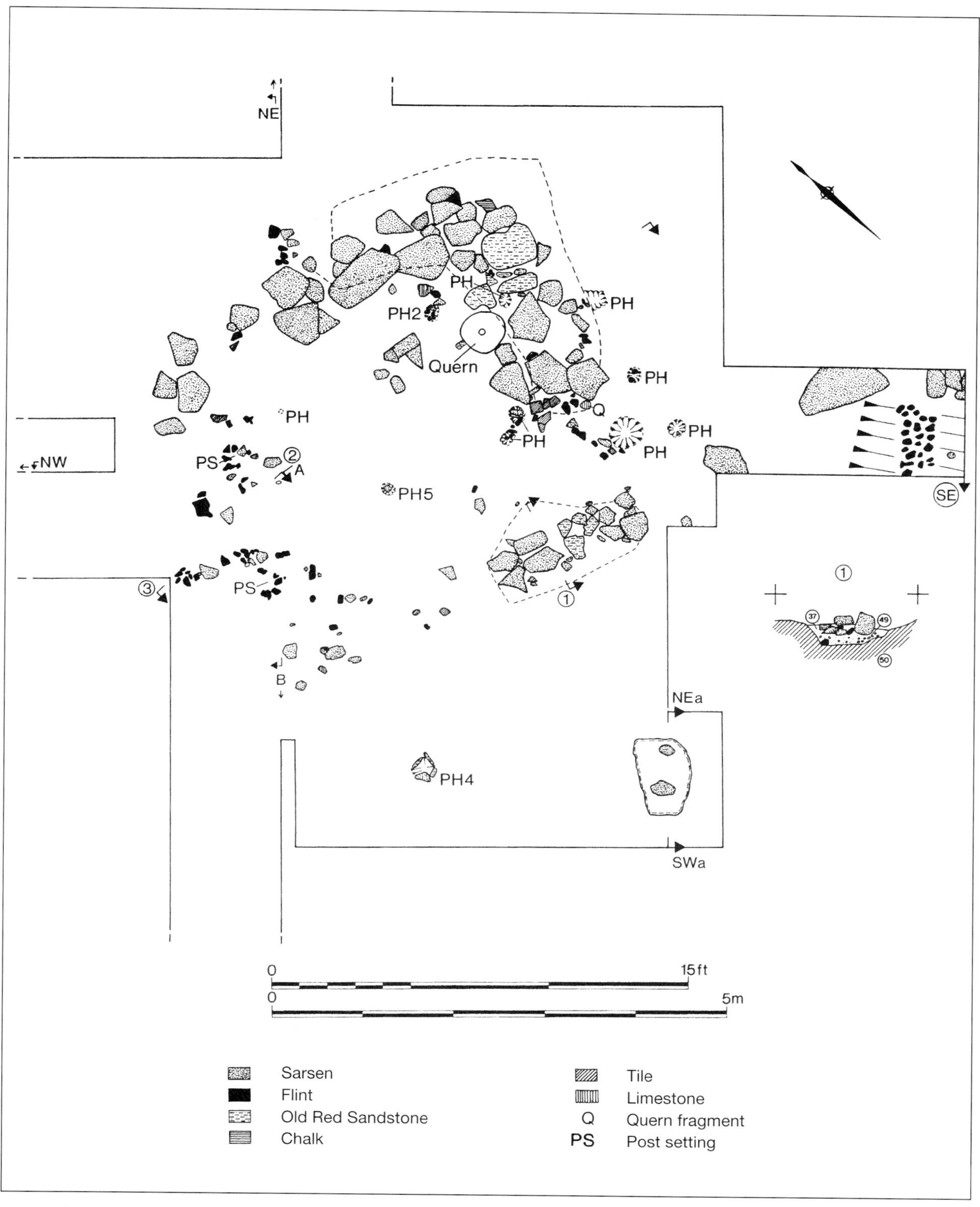

6.17 *Site OD XII: Area 1: plan of the stone-footed Building 1 with* in-situ *lower quern stone and related features, including the boundary ditch to the east. For section, see FWP 64, figure FWP 64.20*

mentioned above, it may be a north-western part of the settlement separately identified as ODS, the physical link between them having been flattened by cultivation and a training gallop. It may also have extended north west to a similar small settlement *c* 200m to the west (OD XIII, *see above*, Figure 6.11).

The extent of the settlement is an important point in both morphological terms and in a local context, but the key point for landscape history is that the original four 'building sites' were observed to be lying on top of 'Celtic' fields (Figure 6.15). Two structures were tucked into the north-east corner of these fields, and a third, larger structure, was sited above a lynchet *c* 1m high. The decision to excavate was inspired primarily by the continuing need to date such fields, and in this case, in contrast to the situation at site OD XI, in order to establish a *terminus ante quem* for the fields by dating the first phase of the settlement. Excavated in 1966–8, four areas were examined using a hybrid open plan quadrant system (Figure 6.16). The fields were satisfactorily dated to the later first/second centuries AD: cultivation did not apparently continue later than *c* AD 200 at the latest, so the area had for long been pasture when the grass-covered lynchets were chosen as building sites. The settlement was of considerable interest in its own right, both for its nature, still uncertain, and for the mid/late fourth–fifth century AD date of its main phases. Some of the material excavated, especially the glass (fourth-century, including an abundance of drinking vessels, very few closed vessel forms and some of the more unusual mould-blown vessels; Cottam *et al* in FWP 64; FWP 96) and some of the ironwork, was also of interest in its own right (Figures 6.22 and 6.23; FWP 64).

Excavation examined four buildings (1, 2, 3, 4A) cut into the lynchets, and a possible fifth structure was identified (4B/C). After activity involving timber structures from *c* AD 300 onwards, substantial occupation with stone-footed buildings began around AD 335, but, on pottery, glass and coin evidence, was primarily in the second half of the fourth century, continuing into the fifth, at least in Building 4A. The buildings were robbed for stone at an unknown date. No material later than the fifth century was found in the excavations. The site has since lain undisturbed, perhaps for 1,500 years, but only since its robbing, *not* since its desertion.

It is very tempting, and plausible in the light of all the dating evidence, to think of a dated succession of buildings in Area 4A, with Phase 1, a timber-framed hall or barn, in mid-century, Phase 2, a smaller, somewhat irregular timber-framed building in the later decades and around the turn of the century, and Phase 3, a stone-founded structure, coinless and with residual glass and pottery, standing in the early decades of the fifth.

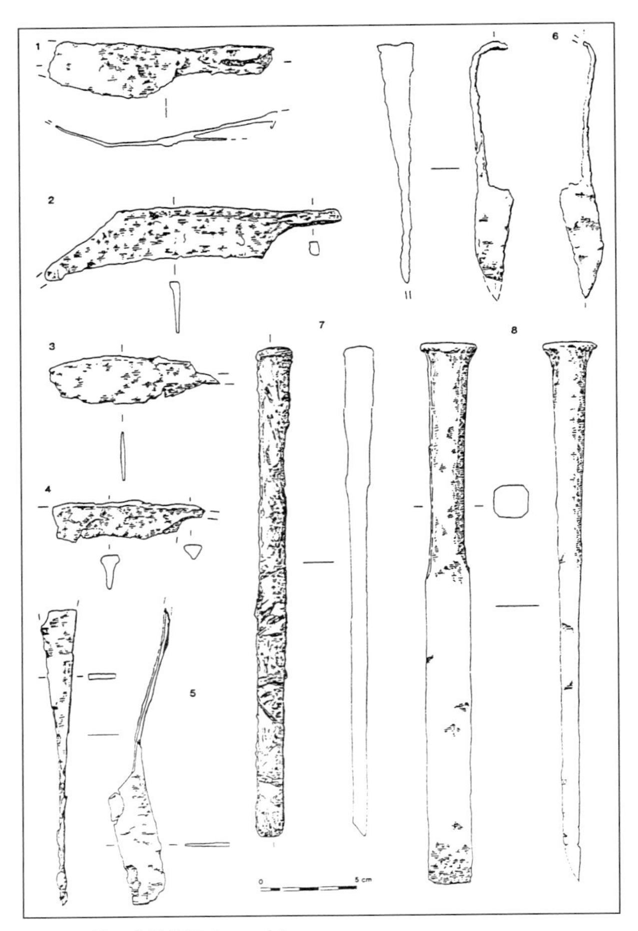

6.22 *Site OD XII: iron objects*

1: cleaver, SF310; 2: knife, SF260; 3: knife, SF208; 4: fragment of shears, SF297; 5–6: shears, SF322/SF333; 7: chisel, SF291; 8: chisel, SF316

In the light of this interpretation, given that in general Area 4 produced much of the later material, it is possible to suggest five stages in the settlement's history, each, as it happens, about thirty-five years long (cf Chapter 15; FWPs 64 and 95).

Stage 1: *c* AD 300: timber structures on Areas 1, 2 and perhaps 3.

Stage 2: *c* AD 335–70: first main occupation with stone-footed Buildings 1, 2 and 3, plus Phase 1 of Building 4A.

Stage 3: *c* AD 370–405: second phase of main occupation with Buildings 1 and 2 abandoned, 3 perhaps still in use and 4A, Phase 2, built.

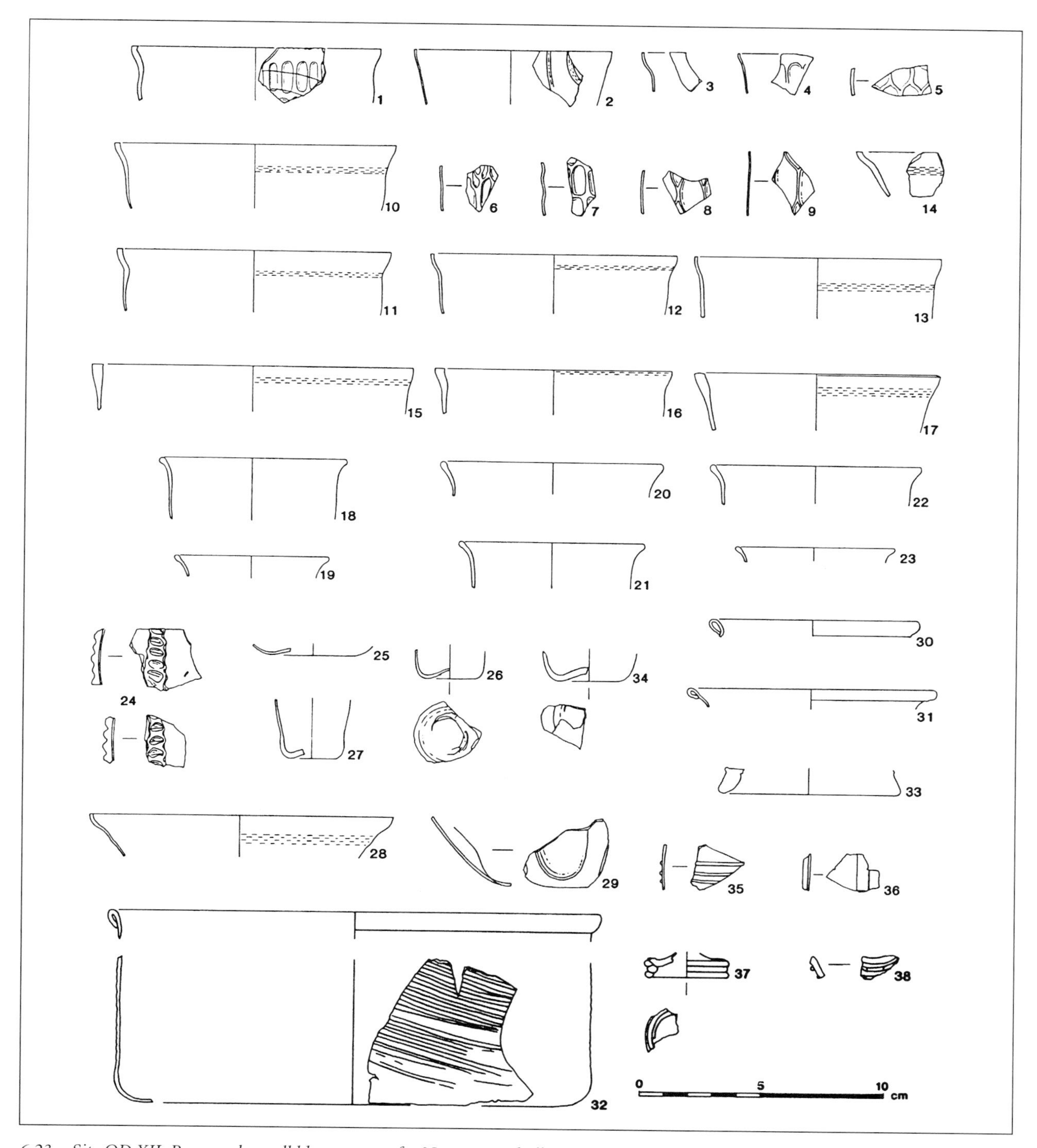

6.23 *Site OD XII: Roman glass, all blown except for Nos 1–9, and all pale green or yellow/green except for Nos 25, 26 and 33, which are green*

1–9: mould-blown cups and beakers; 10–16: beakers/bowls; 17–18: conical beakers; 19–21: thick green rim fragments from beakers or small bowls; 22–28: beakers with out-turned fire-rounded rims; 29: fragment with thick, applied vertical trail from cup or beaker; 30: hemispherical cup; 31: fragment with pontil mark from beaker base; 32–38: lower body and basal fragments of conical beakers

Stage 4: *c* AD 405–40: Building 4A, Phase 3, and perhaps Building 4B/C.

Stage 5: mid-fifth century AD : following desertion, the whole site was extensively robbed of its stone and other materials, fittings and contents for a building or buildings nearby.

There was evidence of two phases in Buildings 1 and 2 (Figures 6.17 and 6.18; FWP 64), two or possibly three phases in Building 3 (Figure 6.19), and certainly three phases in Building 4A (Figures 6.20 and 6.21). Construction took place in both timber and stone, but only in Building 4A is there clear evidence of an entire building constructed in timber. The small numbers of post-holes found beneath the stone phases of the other buildings may represent earlier fences or boundary markers. The walls, composed of unmortared, irregularly shaped sarsens with some flints, could not have stood to any great height and probably supported a timber superstructure. In several places the walls rest in a trench or step cut into the chalk.

The most striking feature of this group of buildings was the very regular layout, respecting the position of an earlier boundary and, in most cases, its alignment. The boundary, running north east–south west, had originally been marked by a ditch, although this was filled in by the AD 330s and certainly before the buildings were constructed (Plate XXXIV; FWP 64, fig FWP 64.4).

The function of the buildings was suggested by their morphology. Building 1 (Plate XXXII, Figure 6.17) was much smaller than the others and square rather than rectangular or sub-rectangular. It contained a quern *in situ* (*SL*, colour plate 22, figure 38) and a limited number and range of objects. Although over 11kg of pottery was recovered, this was by far the smallest quantity from any of the excavated areas and consisted almost entirely of coarsewares. The comparative lack of occupation material, together with the quern, suggests that it was a work-shed devoted to milling. Another quern-stone and many other quern fragments were reused in other buildings. Building 1 stood beside an earlier, 'depositional pit' whose contents included late third–fourth-century pottery and coins, animal bones, metal objects and glass fragments (FWP 64, fig FWP 64.10).

Plate XXXVI Site OD XII, Building 3: oven

Building 2 (Plate XXXIII, Figure 6.18) was divided into one large north-west room and one smaller south-east room. In the former was a hearth, in the latter was another 'depositional feature', this time a small cairn of broken sarsen stones (Plate XXXV). Building 2 produced a range of domestic objects, including spoons, needles, a handle, several brooches and other personal items and structural fittings such as metal ties, loops and staples. The largest surviving quantity of ceramic building material from the excavation came from this area and included at least one fragment each of *pedalis*, box-flue tile and *tegula*. The greatest amount of pottery (*c* 43kg) came from in and around Building 2 and included samian, New Forest and Oxfordshire finewares, Black Burnished ware and a relatively high percentage of large jar fragments. Parallels for the house-form in south-west England occur at Catsgore (Leech 1982), Bradley Hill (Leech 1981) and Gatcombe (Branigan 1977), and further afield at Hibaldstow, Lincolnshire (Frere 1977, 389). At all of these, two- and three-roomed buildings of this type have been identified as houses.

Building 3 (Plate XXXVI, Figure 6.19; *SL*, colour plate 20), like Building 1, appears to have been intended for working rather than occupation. A cluster of metal tools

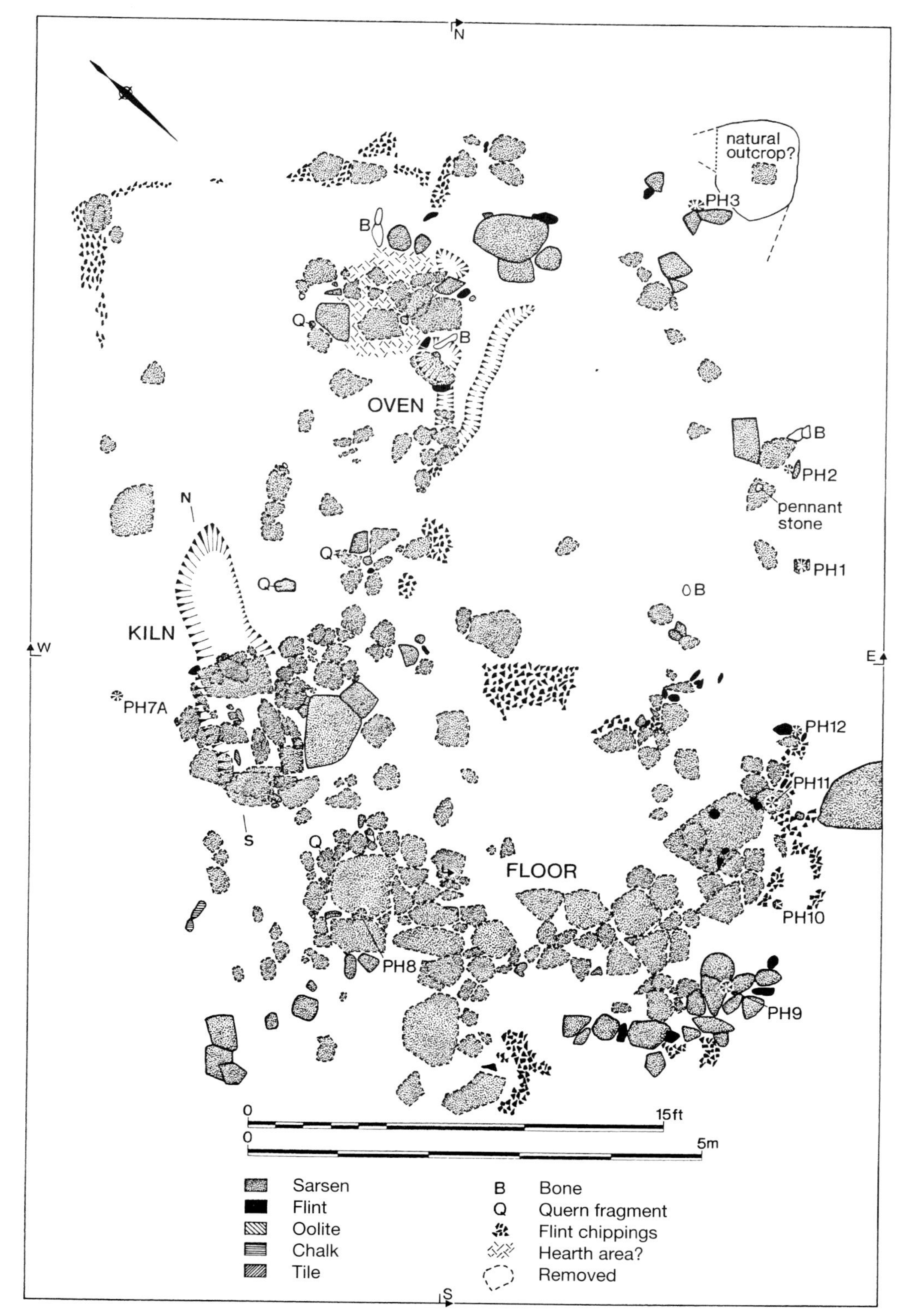

6.19 *Site OD XII: Area 3: plan of the fragmented remains of the stone-footed Building 3 with those parts of its sarsen-flagged floor still* in situ, *and both associated and earlier features. For section, see FWP 64, figure FWP 64.20*

Plate XXXIII Site OD XII, Building 2 under excavation

Plate XXXIV Site OD XII, Area 2: section through boundary ditch

Plate XXXV Site OD XII, Building 2, east end: deposit of fractured sarsen stones

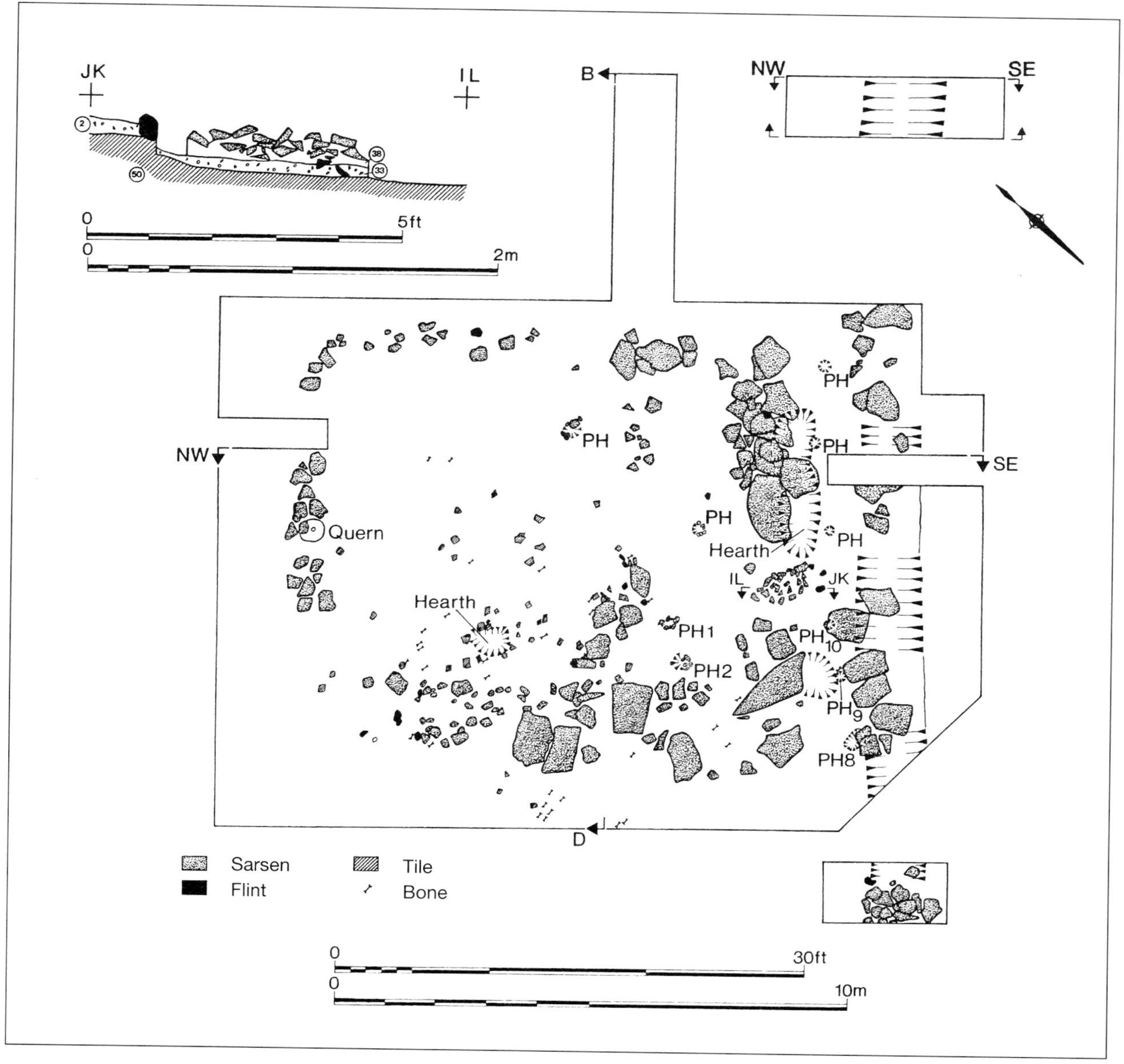

6.18 *Site OD XII: Area 2: plan of stone-footed Building 2 adjacent to the eastern boundary ditch;* see *FWP 64, figure FWP 64.20, for north-west/south-east sections*

indicative of metal-, wood- or leatherworking were recovered, including shears, chisels, a cleaver, a gouge and various knives (Figure 6.22; Hutchison in FWP 64, figs FWP 64.30–2). The heavily robbed nature of the remains makes it impossible to define the structural sequence with absolute certainty, but the interpretation offered here is of two stone-built phases, each involving some heating process. The walls of the Phase 1 building were fragmentary, but this structure incorporated a rectangular stone-lined hearth. The hearth of Phase 1 was overlain by the walls of the Phase 2 structure. This featured a stone floor and possibly an apsidal northern end. An oven with a clay superstructure probably belonged to the second phase (Plate XXXVI). Almost all of the finds from Building 3 were found in or beside the wall stones, indicating that the floor had been regularly swept (FWP 64, fig FWP 64.19). As excavated, the floor was incomplete because many of its flagstones had been removed.

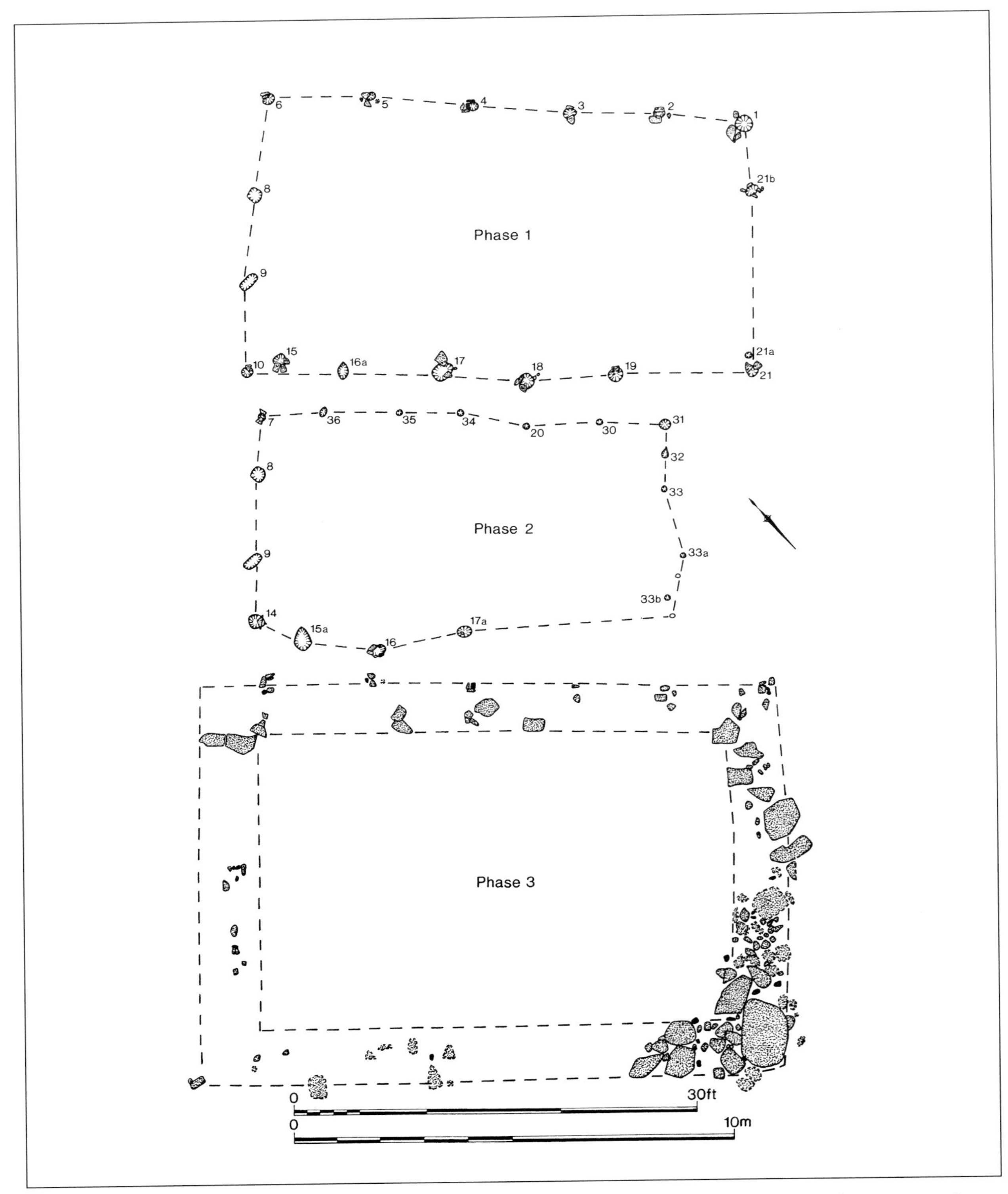

6.21 *Site OD XII: Area 4: interpretative plans of Building 4 showing three main phases of large timber-framed structure, small and irregular timber-framed structure and large stone-footed structure, dated respectively mid- and late-fourth century and fifth century* AD

Plate XXXVII Site OD XII, Building 4A

Building 4A appears to be of three phases, two of timber followed by one of stone (Plate XXXVII, Figures 6.20 and 6.21; *SL*, colour plate 21). The regular spacing of post-holes in the timber phases demonstrates that they were carefully planned and laid out. No trace of internal divisions or hearths survived and the structure may have been a barn or other agricultural building. Nevertheless, a quantity of finds was recorded, especially of late date in the site sequence (FWP 64, fig FWP 64.26).

Area 4B/C contained an oven with a clay superstructure adjacent to a floor of rammed chalk (Figure 6.20). It is possible that this area was bordered on three sides by sarsens, running roughly east–west, but in the absence of clear evidence, this should probably be regarded as an open-air working area. The evidence is stronger for a structure south of this. Sarsen walls shared a long axis parallel to that of Building 4A, each resting in a construction trench. It is possible that this structure had an apsidal east end. Similar structures have been recognised in agricultural settings at Catsgore (Buildings 2.10 and 3.13, Leech 1982, 17–21) and Hibaldstow (Frere 1977, 389).

This group of buildings may represent a single unit, comprising house, barn, work-shed and work areas, possibly related to grain-processing, storage, drying or parching, threshing and baking. The group as a whole can be paralleled at the Roman villa at Gatcombe, where the south-western part of the site was interpreted as a complex for the storage and processing of grain (Branigan 1977, fig 33). An alternative model of three paired units each of a house and outbuilding is offered elsewhere (*SL*, figure 40).

The dating and phasing of the buildings is discussed in greater detail in FWP 64. If OD XII was a specialised unit within a larger complex rather than a small settlement in its own right, then a specialist grain-treatment facility could explain its location within a small area of contemporary arable (*see above*). On the other hand, such an interpretation must also be examined in relation to the settlement's time depth, spatial patterning and broader, regional considerations (Chapter 15).

Chapter 7

The Northern Downland: Fyfield Down and Wroughton Mead

A place so full of a grey pebble stone of great bigness as is not usually seen, where they lie so thick as you may go upon them all the way.

A TRAVELLER IN 1644: *VCH* III, 186; *Diary of the Marches of the Royal Army*, CAMDEN SOCIETY 1ST SERIES, LXXIV, 151

In many instances I believe the dun name to be an English place-name given to a pre-English settlement in recognition of its characteristic situation … [it] sometimes denotes a flat shelf, often on the side of a high hill.

GELLING 1993, 142

Virtually every square metre of Fyfield Down shows evidence on the ground of former land-use. The down is densely packed with a more or less continuous extent of earthworks (Figures 2.1 and 7.1), and two excavations, one large, have demonstrated the time-depth. Here, however, in contrast to other downland areas locally, with their emphasis on the later millennia BC, our own era and in particular the most recent thousand years are well evidenced on Fyfield Down, both archaeologically and through documentary sources. Mainly for the benefit of anyone visiting the Wroughton Copse area now and seeing it, as indeed it appears on air photographs, as distant, remote and empty, we would stress that this has been a very busy place (frontispiece, Plates VII, VIII, IX and XXXVIII).

People have lived here, in small numbers, as at *Raddun* (Figure 7.5), and from time to time, as in Roman (Chapter 15) and post-medieval centuries (Figure 7.15), but the traditional use of the area is for grazing, of cattle as well as sheep (FWP 43). The earliest place-name is *Raddun* in 1248, arguably deriving from the OE 'red hill' or 'down' (*PNWilts*, 28; Gelling 1993, 141–9; *see below*). The area had previously been cultivated, periodically in the third and second millennia BC, and in the first/second century AD; it was probably under plough, for the last time, early in the fourteenth century (frontispiece, Figure 2.3 and *below*).

FYFIELD DOWN

The boundaries of Fyfield Down

One part of the East Overton charter of AD 939 (S449) is an attachment describing the bounds of a dairy farm and the downland. The nature and features of the charter, many of which were mirrored in the Pembroke Survey of 1567, clearly indicate that this attachment is describing land north and east of East Overton in part of what is now the north of the parish of Fyfield (hence its inclusion here). This identification has not previously been made.

Plate XXXVIII Fyfield Down from the south in 1960 (AAU-76, © Cambridge University Collection of Air Photographs*)*

This attachment starts at *mappeldrelen west weardre*, the 'western edge of the maple trees', then proceeds north up *anlang stan ræwe*, north 'along the stone row', then *on tha byrgelsas*, 'to the burial places'. From there the boundary went *suth andlang weges*, 'south along the way', *andlang hlinces on thæt suth heafod*, 'along the lynchets to the south headlands', and *adune on thæt slæd*, 'down to the slade'. Finally, it returned *up andlang weges eft to mappeldre lea*, 'back along the track to the lea of the maple trees'. Over 600 years later, a description of North Down, the Sheep Down on Hackpen, went as follows: 'beginning at Monckton's Down beside Balmere Peke, proceeding along Lollingthorne, thence to Rudge Banck, and then descending Bury Way via Mapple Dryley, and from there as far as by established paths between Clatford and Fyfield' (FWPs 11, 46; Straton 1909, 262).

The similarities are striking, and though these points across long-cultivated downland are difficult to locate precisely, the occurrence of Knoll Thorn (Lollingthorne?) on the 1811 map and the possible location of Balmere Pond near Glory Ann (SU 12837265; Crawford 1922, 54), clearly indicate we are dealing with an area which incorporated most of Fyfield Down. Stancheslade, which is surely the *slæd* of six centuries earlier, is believed to be the Valley of Stones (also called Stony Valley); indeed, Kempson (1953, 71) thought that the name derived from 'stan-chest-slade, the valley of the stone burial-chamber [Devil's Den]'. Furthermore, it is clear that in the tenth century, as in the sixteenth and nineteenth, a substantial area of downland now in Fyfield civil parish was precisely defined and managed as part of the East Overton estate.

Long-term land-use and the *Raddun* name are

underscored by the names 'Roddon Cowleaze' and 'Roddon Cowdown' in the 1567 Survey (Straton 1909, 259, 262). The latter was located over the southern part of Fyfield Down, east of Wroughton Mead (Plate XXXVIII). Its western boundary, a slight bank and ditch, cuts across early fields and the Romano-British trackway in its length from the north corner of Wroughton Copse north eastwards (Figure 7.1). It goes as far as the Fyfield/Clatford boundary where it turns towards a previously unrecorded megalithic barrow *c* 100m to the south east, passing its northern edge and then on down the parish boundary towards Long Tom (*SL*, figures 19, 62, 64).

Nor was this a boundary defining marginal land in the sense of being of little use and low value. The pasture on Fyfield Down was *valde bona* ('truly good', Straton 1909, 259). If such was the case 300 years earlier, such a resource may have been one reason for the appearance of a farm in this landscape. Indeed, two strands of evidence suggest that the charter's Anglo-Saxon cattle farm was in the Mead at *Raddun*: the locational and etymological link between Wroughton Mead, Roddon Cowlease and Overton Cow Down (Figure 7.1), and early to mid-Saxon (fifth–eighth centuries AD) sherds recovered from excavated Building 4 (*see below*). This interpretation denies easy assumptions that such use arose solely from the apparently marginal nature of the area. And even if it was marginal spatially, such an area can become valuable for specific uses, for example, pasturage and rabbit-farming, precisely because of its isolation and relatively low market value. The Priory of St Swithun's was granted the right to free warrening within its demesne land in Fyfield in 1300 (*VCH* IX, 192), probably, though not certainly, on Fyfield Down; by, or perhaps still in, 1880 a large warren existed on Fyfield Down. In 1910, Alexander Taylor killed *c* 14,000 rabbits to make downland gallops safer. And that has become the almost priceless facility now afforded by this (and Overton) down, precisely because of its remoteness. The down is part of one of the great racehorse-training stables of southern England, its landscape now laced with gallops, as has been the case for over a century and perhaps very much longer. Some gallops are in use (Plate LXVIII) but others are now redundant and quite as much part of its landscape archaeology as are prehistoric, Roman and medieval boundaries, trackways and fields (frontispiece).

Boundaries and trackways

Fyfield Down embraces several boundaries and three major lines of communication. All the latter were through-routes, not just local paths or tracks (Plates VI and XXXVIII; Figure 7.1). Of the boundaries, the most important in the East Overton charter of AD 939 (S449) has already been noted in its downland context (*see above*). This also became the boundary between the now defunct tithing of Lockeridge and Fyfield, and is still the boundary between the latter and the civil parish of West Overton. Its line is virtually impossible to follow on the ground, mostly through a dry valley bottom littered with sarsen stones, many quite large and close together (Plate III). Nevertheless, in the tenth century it followed a track 'down to the lynchets of the south headlands' where the steep southern side of the Valley of Stones rises towards 'Watkin's Gizzard'. In the coombe bottom itself, this track would have elided with the Romano-British track coming south south west off Fyfield Down past the west side of Wroughton Copse (Figures 2.1 and 16.6 and *see below*; FWP 52).

A local boundary at the south-east corner of Wroughton Mead is etymologically of some interest, though its implied archaeology has not been found. The boundary divided 'Cow Down' from 'Fyfield Tenants Down' along the bottom of the Valley of Stones. It is described as consisting of 'hills called Dillions' (1811). Dillions, or dillons, are 'earth-heaps to mark boundaries on the Downs' (*English Dialect Dictionary* 1900), the word possibly deriving from OE *daelan* or *dol*, meaning to divide or share (Smith 1956, i, 126). Dillions are also recorded in East Anglia, where they were known locally as *doles* or *dools* (Lawson *et al* 1981, 27). Totterdown's former name of 'Dillion Down' presumably reflects a similar origin (*see above*), and here gives its name to Delling (or *Dyllinge*; Straton 1909, 258; *see below*). Kempson (1953, 71) argued that the old name for Stony Valley was probably 'Delling Dene' or 'Den', which in turn was altered to 'Devils' Den'.

Of the trackways, the other important one here, perhaps even dominating this particular landscape, is the main east–west route from Marlborough to the east entrance into Avebury. In terms of landscape stratigraphy, this route is 'late', for its tracks and hollow-ways cut through everything else. Its nature as an unconfined way across open downland is particularly well illustrated on the east side of Overton Down (Plate VI; *see also* Plates VII and VIII and Figure 2.1). There, a whole bundle of hollow-ways spread across *c* 250m up the slope from the Valley of Stones, clearly marking the passage of heavy wheeled vehicles. This was before the early nineteenth century at latest, for by Enclosure (1815/16) this passage was formally demoted to 'Public Bridle Way and Private Carriage Road', limited to a

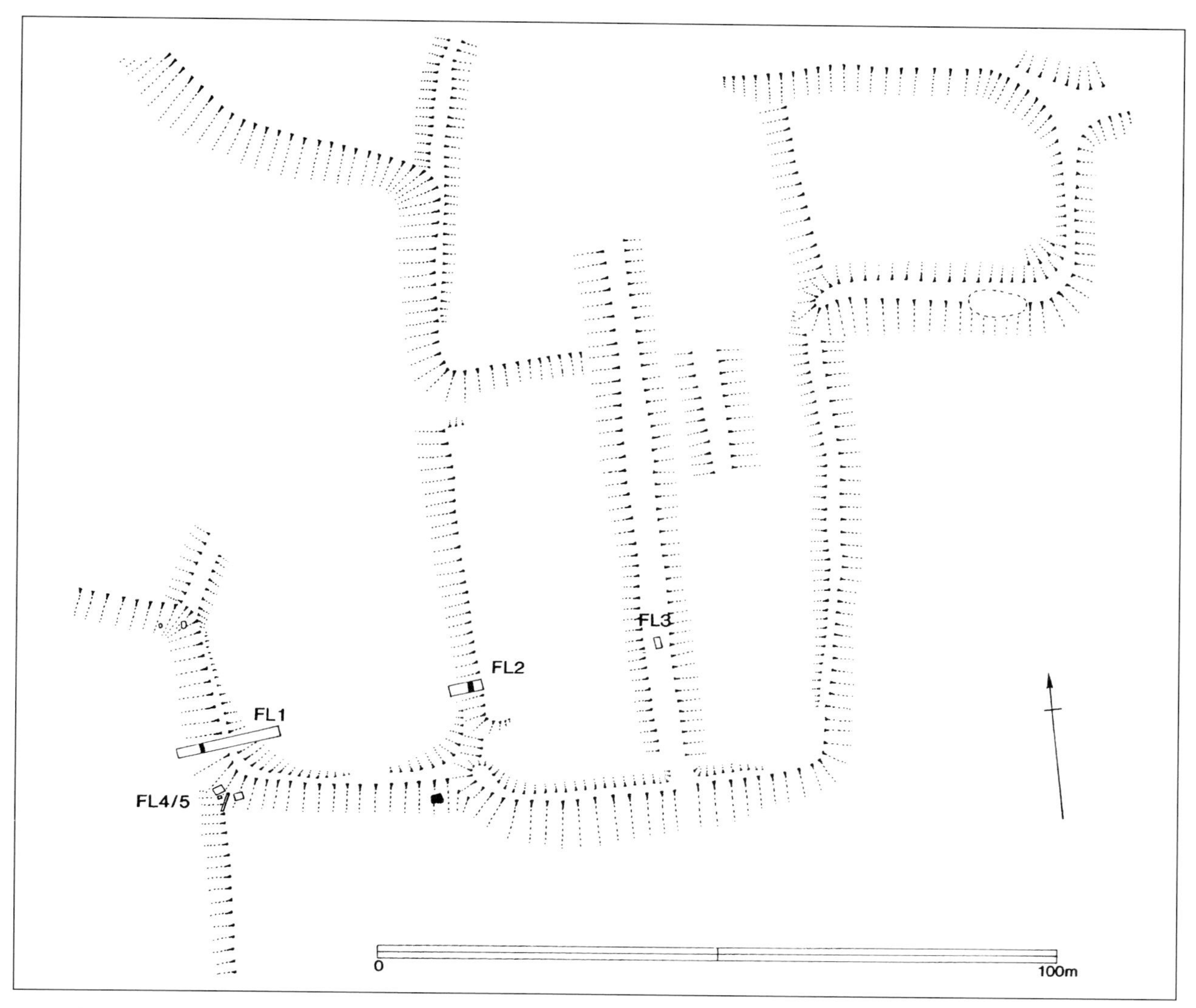

7.2 *Plan of prehistoric fields in the centre of Fyfield Down showing the positions of cuttings FL 1–5. The low bank marked by FL 3, and that to its east, are of overlying ridge-and-furrow. All the other scarps are prehistoric lynchets, modified in the first century* AD *by the addition of dry-stone walls*

width of 25ft (*c* 8 m). Then its line east of the northern point of Wroughton Copse was shown as bending south east towards Manton and not north east towards Rockley, suggesting that the latter way, now tarmacadamed, is a nineteenth-century development following Enclosure. Indeed, no such road past the Delling Cottage is shown on Dymock's 1819 map, while the line of what is probably the much older route to the north east is shown on the late eighteenth-century map taking off 'The Old Bath Road (Disused)' from the north corner of Wroughton Copse. This north-east branch appears to pass, at least in part, along the line of the Romano-British track through the earlier fields outside the north boundary of 'Rodden Cow Down' (*see above*; Figure 7.1).

The main Marlborough–Avebury track, 'the ancient trackway called Old London Way' in 1815/16 and, possibly, 'the London Wey' in 1567, is likely to have been in use as the principal downland route in medieval times and possibly earlier, though it seems to be later than the Romano-British arrangements and nowhere is it referred to in Anglo-Saxon sources. Versions of its authentic post-medieval name have already been quoted; nowhere is it referred to as 'Green Street' until the twentieth century (Fowler 1998). Given that the name of the Marlborough–Avebury track is, then,

Plate XXXIX *Fyfield Down lynchet excavations in 1961, looking from the west at the northern face of the main cutting FL 1, showing the dry-stone wall as excavated at the front of the ploughsoil accumulations visible in section*

historically neither *Green Street* nor the *Herepath* (Chapter 4), as shown on current OS maps, both the main downland roads hereabouts, this one and The Ridgeway, have spuriously antique names authorised only by the Ordnance Survey.

FIELDS

Prehistoric and Roman lynchets, Fyfield Down (excavations FL 1–3)

The initial reason for excavating the enclosures beside Wroughton Copse (*see below*) (Plate XXXIX; Figures 7.2 and 7.3) was to date the underlying fields, but by 1961 it was clear that the matter was not going to be resolved there (FWP 65). So it was decided to tackle the matter head on by excavating one or more large lynchets at the sides of demonstrably pre-medieval fields. After considerable inspection, one was chosen more or less in the middle of Fyfield Down (and of Major Allen's famous air photograph: *see* frontispiece). The choice was made because no ridge-and-furrow was visible in or over the field it bounded.

It was intended that this excavation would illuminate both chronology and questions of why and how such large lynchets had accumulated on a slope of only 3°. Even though the largest lynchet on the down was deliberately avoided, logistically, the excavation was akin to sectioning a hillfort rampart. That chosen was nearly 3m high, lying north–south along the west side of a field and just north of its south-west corner. The trench through it was 15.3m long (FL 1), with an addition through the lynchet uphill on the east side of the 'Celtic' field (FL 2; Figure 7.2). The line of examination was extended 25.8m further east to a test-pit (FL 3) in the top of the nearest ridge of ridge-and-furrow lying north–south in the 'Celtic' field adjacent to that sectioned. Four small cuttings (FL 4, 5) were also excavated right on the corner of the 'Celtic' field itself to elucidate the main structural question arising from FL 1 (Bowen and Fowler 1962, 105, pl iia; Fowler and Evans 1967). Differences here and in FWP 66 from those accounts and interpretations represent deliberate revision in the light of a re-examination of the primary evidence and a better understanding now of its landscape context (cf also *SL*, colour plate 14, figure 26).

The excavation

FL 1 was excavated by hand, layer by layer. Essentially, the soils consisted of small (<50mm) granules of soil and chalk and had clearly been pulverised to varying degrees (Figure 7.3). Layer 1/top of 23 contained a scatter of shrapnel fragments, presumably of late 1940s' vintage (*see above*); layer 23, the worm-sorted flinty residue from layer 1, contained a scattering of Early Iron Age and Romano-British sherds, mainly the latter. The bulk of the cultivation may well, then, have taken place by soon after, if not before, AD 100, by which time the top of the lynchet, essentially the present ground surface, had reached its existing height above the old ground surface. The question of dating is discussed further below.

Below layer 23 was as much as 1.20m of accumulated deposits (*see* caption to Figure 7.3 for layer descriptions). At their base, cut into chalk, was a shallow depression filled with light brown soil, flints and chalk lumps, probably a tree-hole (Evans 1972, fig 120; similar to one carefully excavated and similarly interpreted at the Overton Down experimental earthwork in 1992; Bell *et al* 1996, 76–7, 140, figs 7.12, 7.13). Lying directly on solid Upper Chalk was a light brown soil with flints overlain by a dark ginger soil with flints, small chalk lumps and flecks of charcoal. The latter was a disturbed, probably cultivated, old ground surface.

Well down the slope of the scarp forming the front of the lynchet, below but very near the present grass surface, was a small drystone wall (cf TD IX, Chapter 5). All the rest of the stratification was related to it. The wall itself stood on a ledge only 0.15–0.25m wide at the west end of layer 14. It consisted entirely of smallish sarsen stones, characteristically 0.30m across, all broken and packed around with large flints making up the body of the structure. A sarsen saddle quern was built into the bottom course (Figure 7.3). The wall had tipped forward a little; yet it had never been a large structure, for no tumble or collapse lay to its front nor was there any sign of robbing. Two or three courses at most probably constituted its original form. It would not therefore have kept animals in or out so its most likely function, if not just decorative, was perhaps tenurial, marking the edge of a property as well as a field.

Slightly more than one hundred sherds were retrieved from FL 1, all small and many abraded. Their presence can in general be regarded as the accidental by-product of manuring. Even those explicable in the lynchet as derived from the old land surface may have arrived there originally with manure in fields earlier than those of the 'drystone-wall' phase. The sherds range in date from possibly Neolithic to second century AD, with nothing later. In general, the sherds became earlier the deeper their provenance.

Originally, our interpretation envisaged the visible field system of the 'drystone-wall' phase being laid out in the mid-first millennium BC (Bowen and Fowler 1962, 105). Re-examination of the stratification, contexts and all the pottery indicates, however, that the wall itself was inserted in the later first century AD and not during the pre-Roman Iron Age. This 'drystone-wall' phase of fields on Fyfield Down is taken as fitting in with the locally widely attested period of rapid and substantial landscape reorganisation towards the end of the first century AD (Chapters 2 and 15).

A further cutting (FL 2) on the eastern side of the field from FL 1 (Figure 7.2) sought to establish whether a wall also existed there. The remains of a wall were indeed found, much more disturbed (probably by rabbits) than in FL 1 but of the same size and form. There was no good dating evidence in this case.

Since it now appeared likely that the whole field was enclosed by a wall, two small and rapidly excavated cuttings (FL 4 and 5, Figure 7.2) checked the presence or otherwise of a wall or walls at the south-west corner of the same field. Only one course of a former wall existed in FL 4; it did not bend round the field corner to the east and, although the evidence was inconclusive, if it continued at all it went straight on southwards. There was just the possibility of a gap, perhaps a gateway, in a southern continuation (FL 5), though the point excavated is shown as damaged by traffic ruts in Allen's 1934 air photograph (frontispiece); but then perhaps the downland track went for that point because the obstacle of a lynchet was absent.

A small test-pit (FL 3, Figure 7.2) was also dug to see if the soil was a greater depth at the centre of a rig in a pattern of ridge-and-furrow east of the 'Celtic' field already examined. It was not, and nor was there a flinty layer 2. Two implications were that the latest, presumably medieval, ploughsoil had been flint-free, and that the undulations of the ground surface reflected, or were reflected by, similar undulations in the surface of the chalk subsoil.

Conclusion

This little exercise on Fyfield Down succeeded in dating the lynchets and the fields they bounded to a beginning and periodic use from *c* 2000 BC onwards, ending with a terminal phase associated with drystone walling of the late first century AD. The earlier phases of activity, perhaps initially occupation but thereafter cultivation, involved ground disturbance and the accumulation of a lynchet along a line which seems to have remained a constant feature in a changing landscape throughout the second and first millennia BC, even though cultivation was not apparently continuous. The archaeology was successfully correlated with an environmental signal identified thirty years later (Chapter 14).

These famous Fyfield Down 'Celtic' field lynchets are in their existing form of early Roman date and were, at least in part, built. At an early stage of their last use, with drystone walling just showing among arable fields, the landscape would have looked totally different from the grass-covered downland sheep-runs and horse-gallops of today.

WROUGHTON MEAD AND WROUGHTON COPSE

Overlying and incorporating the sides of earlier fields, Mead and Copse together seem to have been enclosed in the later thirteenth century. The medieval Wroughton Mead stands out on all maps, and on the ground, as a diamond-shaped area of enclosed land with a wood, Wroughton Copse, at its northern corner (Plates VII, VIII and XXXVIII; Figures 2.1, 7.1, 7.4 and 7.11). It lies on a south-facing slope between the Copse and the floor of the Valley of Stones, a north–south distance of *c* 550m between the 215m and 180m contours. West to east, it is *c* 200m wide.

Its western boundary, including Wroughton Copse, is a trackway, almost certainly on the line of a Romano-British predecessor which may well have helped condition the Mead's exact position and shape. On the east, the boundary bank and ditch was originally above a shallow re-entrant off the Valley of Stones containing a large rectangular pond. This had been inserted by 1819, apparently designed to give stock access to water from both inside and outside the Mead. Its southern boundary now embraces the edge of the valley's floor but originally lay just above it; its northern boundary seems always to have included Wroughton Copse, the boundary bank and ditch originally heading south east towards the re-entrant but expanded to a new fence line by 1819.

The whole Mead is under grass, though parts of it have been cultivated on at least two occasions, and it is defined by a bank, sometimes accompanied by a ditch, now with either a fence or hedge on top, or both, and again of several phases. Wroughton Copse is of deciduous woodland, its understorey now rather open and straggly but with a canopy including large oaks.

The name 'Wroughton' Mead is derived from *Raddun*, which appears in a Winchester custumal of 1248 (*see above*). Arguably it developed into 'Roddons Close' (1567), 'Rowden Mead' (1773, 1821, 1885) and 'Roddon' (1819), with 'Wroughton Copse' appearing as such in 1885 after being 'Roydon Copse' in 1811 and 'Roddon Copice' in 1567.

7.4 Plan of the field archaeology of Wroughton Mead showing the fragmentary pattern of prehistoric field systems, clearance mounds, the local contexts of excavated sites WC and 10, and the successive enclosures of the Mead itself

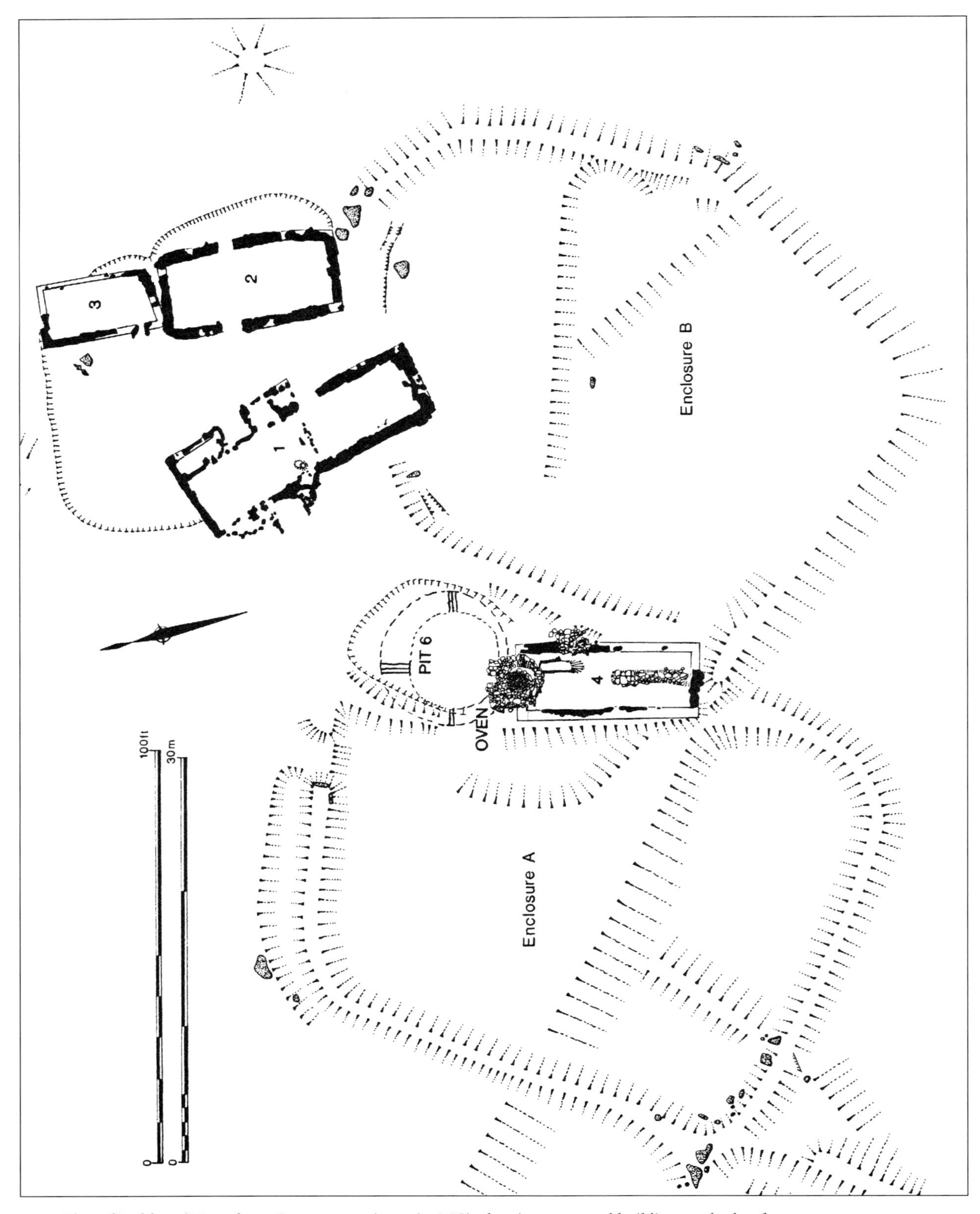

7.5 Plan of Raddun *(Wroughton Copse excavations, site WC), showing excavated buildings and other features*

Raddun, a medieval farm in Wroughton Mead, Fyfield Down

The site (Plates XL–XLVI; Figures 7.5–7.14; FWPs 65, 43, 59; *SL*, colour plate 23, figures 6–8, 42–45) acquired its name, not from the village of Wroughton situated about 10km to the north, nor because a Colonel Wroughton used to shoot over it at the turn of the nineteenth century but, as we have noted in discussing Wroughton Mead, because of medieval nomenclature. In the Winchester MSS for the period 1267 to 1318, the reference is to 'Raddon', and it is from the earliest documented evidence of 1247–8 that *Raddun* emerges. (Hereafter we use the name *Raddun* unless citing a specific text or map which uses an alternative.) *Raddun* is likely to mean a 'red hill', possibly offering a suitable site for settlement (Gelling 1993, 140–9). The superficially surprising 'red' element in a chalk landscape of green and white can be explained by supposing cultivation was undertaken and included at least some of the area of Clay-with-Flints overlying the Chalk around Wroughton Copse. Etymologically speaking, *Raddun* is certainly Old English in origin, perhaps very early English, suggesting cultivation hereabouts before AD 800 (ibid, 140–1). In fact, evidence for activity here in the early to mid-Saxon period was suggested by four organic-tempered pottery sherds (GF 413) as well as the documentary evidence of cultivation on Fyfield Down (*suth heafod*) and a cow farm hereabouts in the mid-ninth century (S449). Whatever its origins, a place called *Raddun* or *Raddon* certainly existed before 1248, as in that year there is a reference to land held by a Richard of Raddun.

'... I ascended the hill to a piece of down ... by the name of Rowden Mead, upon which we again meet the undoubted vestiges of an extensive British settlement.' So wrote Colt Hoare (1821, 45), as we discovered after our own independent search for a small enclosed site recorded from the air by St Joseph (Plate XL). Our initial fieldwork in April/May 1959 identified a complex of well-preserved earthworks in Wroughton Mead (Figure 7.4). The complex overlay 'Celtic' fields – the initial reason for starting an excavation of the two small enclosures on the air photograph that clearly looked as if they might comprise a settlement. Field survey confirmed that likelihood. Eventually, excavation (Figure 7.4, sites WC and 10) and documents (in the library of Winchester Cathedral) showed the site to span the four centuries between *c* AD 1200 and 1600, with non-occupational activity both before and after. Included in our interpretation is the identification of the settlement we excavated with the *Raddun* or *Raddon* of the Winchester documentation of the second half of the thirteenth and early years of the fourteenth centuries.

Documentarily, *Raddun* was a farm firmly tied into the administration of Fyfield as a subsidiary of East Overton Manor within the Wiltshire estate of St Swithun's Priory, Winchester. The farm, primarily for sheep, also served other functions, notably in arable cultivation, while archaeologically it passed through at least three main phases of structural change. Timber-based buildings were succeeded by buildings with stone-based walls; an all-purpose long-house was succeeded by a farm unit with specialist buildings, though only one house was occupied at any one time.

Excavation of the two enclosures (WC, enclosures A and B) and a third one to the north (enclosure C containing site 10), produced an outline of the site's history in structural terms, with a chronology comfortably within a range of ±10 years if we follow key pieces of evidence in the Winchester custumals blending them with relevant archaeological evidence (FWP 65; Figures 7.5, 7.9 and 7.10):

AD 1200–20: the first stock enclosure (B; Figures 7.5, 7.9a, 7.10a) probably with, outside it on the north, the early timber phase of Building 2 (Figure 7.7) as an animal shelter and a large pond (Figure 7.5, Pit 6).

AD 1220–60: the first long-house (Building 4, Plate XLI, Figures 7.5, 7.6) in the south-west corner of enclosure B, with an additional enclosure C being constructed against a prehistoric lynchet to the north around Site 10 (Figure 7.9b, 7.10b).

AD 1260–1300: three buildings (B1, 2 and 3, Plates XLIII, XLIV, XLV; Figures 7.5, 7.7, 7.8) replaced the long-house, forming a functional unit of house, later doubled in length, stable and animal shed. All had stone footings, those of B2 on top of the earlier timber building. Laid out across the northern part of enclosure B, which had probably become a garden, they necessitated the construction of a new enclosure (A) to the west (Figure 7.9c). Smithing occurred in the ruins of the long-house, and into its northern end was inserted the base of an oven, almost certainly for making bread (Figure 7.6). The whole farmstead lay within an extensive, embanked enclosure (D + E, Figure 7.10c), partly cleared of stones and presumably to provide some security for both animals and crops. This is the original 'Wroughton Mead', all subsequent versions of it reflecting the original shape of 700 years ago.

AD 1300–18: the farmstead disappeared from the record in 1318 and on archaeological grounds has clearly

Plate XL *Enclosures near Wroughton Copse, Fyfield Down, viewed from the east on the 1954 oblique air photograph by St Joseph that records the rediscovery of the site, later identified as* Raddun *(cf Plate X) (NX-76,* © Cambridge University Collection of Air Photographs*)*

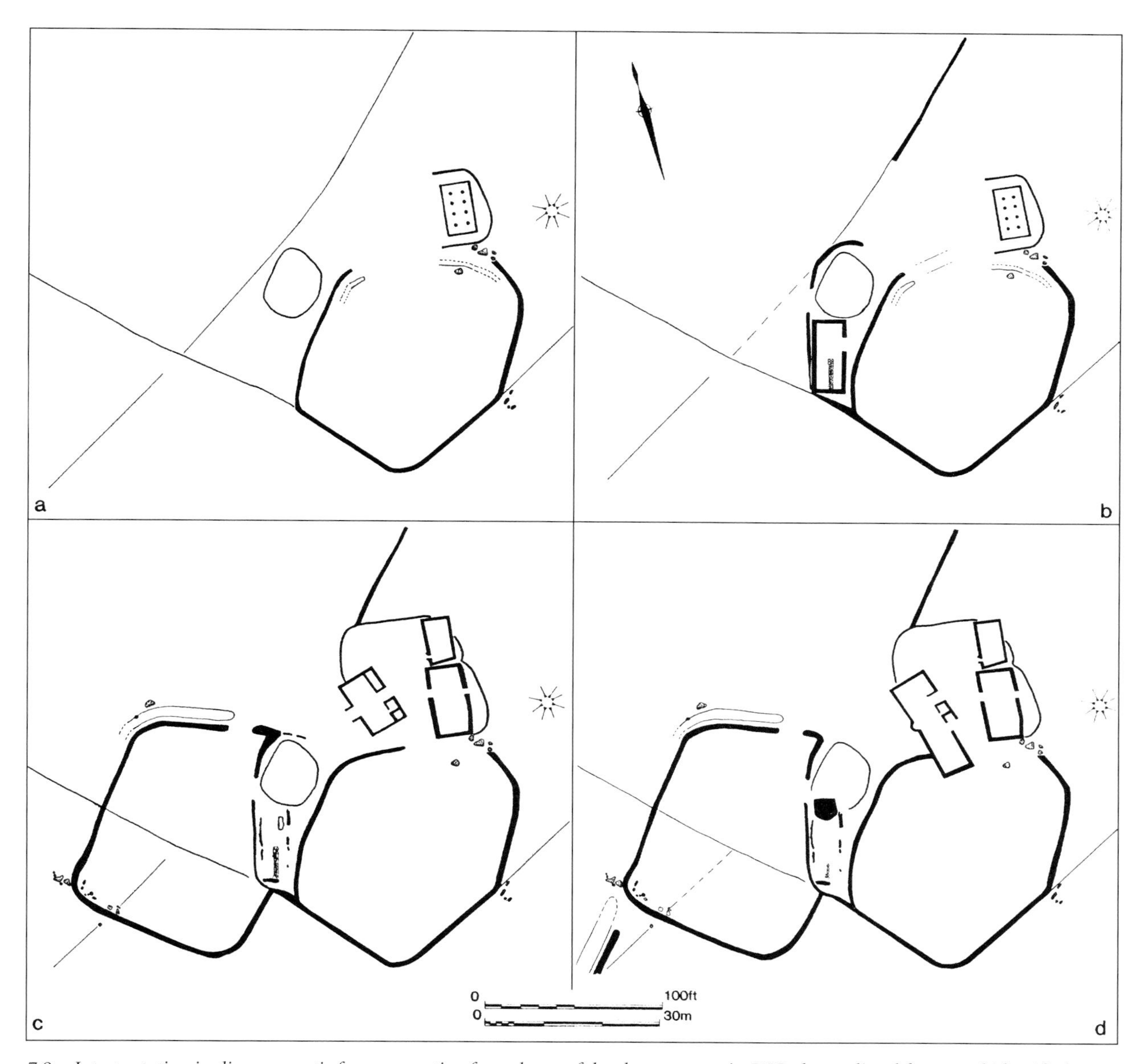

7.9 Interpretation in diagrammatic form suggesting four phases of development on site WC, the medieval farmstead identified as Raddun

a = in the early thirteenth century; b = mid-thirteenth century; c = late thirteenth century; d = early fourteenth century

been abandoned before the mid-fourteenth century; but site 10 in enclosure C continued to be used, probably as a sheep-house, perhaps occasionally occupied, during the fourteenth–fifteenth centuries (Figures 7.9d, 7.10d).

AD 1490–1650: a long, probably open-sided building stood on a low platform in enclosure C and could well be 'the grange' referred to in a document of 1493 (Plate XLVI).

Raddun *and its landscape: an interpretation*

The picture which emerges of *Raddun* in the early to mid-thirteenth century is one of a simple stock enclosure (enclosure B) for over-nighting animals, predominantly sheep, though goats and horses were also kept, with a watering-hole (Pit 6) outside it to the north west (frontispiece, Plates VII–IX, Figures 7.10a and 7.11). A timber-framed building (B2) was erected at the entrance to enclosure B (Figures 7.5 and 7.9a), a

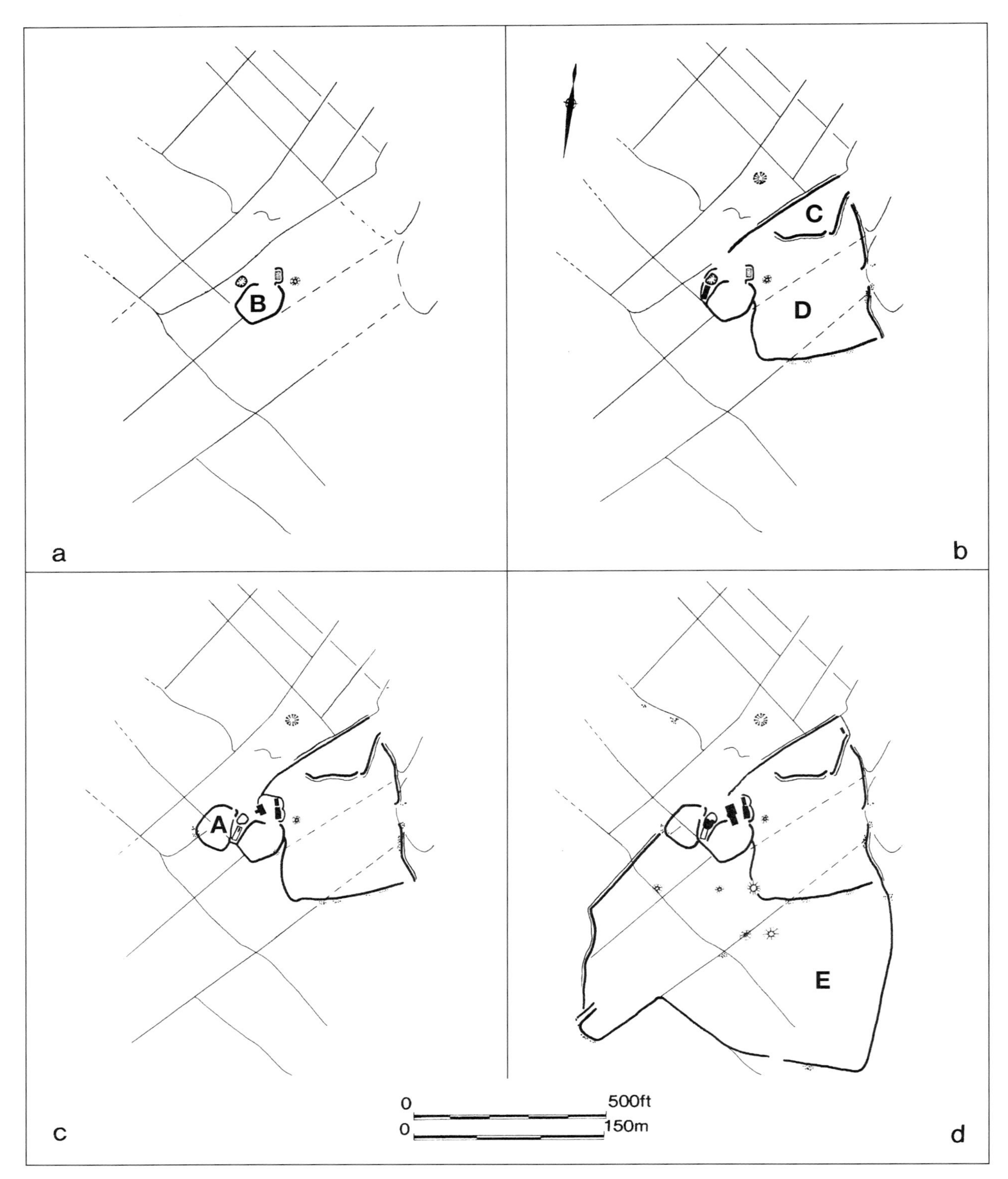

7.10 Interpretation in diagrammatic form of Wroughton Mead, suggesting four phases of development for the whole medieval complex superimposed on the physically existing framework of prehistoric/Romano-British earthworks

a = in the early thirteenth century; b = mid-thirteenth century; c = late thirteenth century; d = early fourteenth century

decision which was later to determine the positioning of the subsequent farm buildings in that area. This building, probably a shelter for milking and lambing, sat on a levelled ground surface. Further east, a larger enclosed area (enclosure D) was cleared of stones, many of which were placed along the banks of the enclosure or into some of the pits and ponds within the Mead. A further enclosure (enclosure C), with a ditch at its entrance and a hedge along its banks, was constructed in the northernmost corner of enclosure D.

With the construction of a permanent timber structure (B2, Phase 1), along with three enclosures, sheep farming had clearly became viable enough to necessitate a permanent dwelling at *Raddun.* Thus enclosure B was extended in the west and a long-house (B4), with a well-drained byre end, was constructed up against the enclosure bank (Figure 7.5). A family was now living at *Raddun* in a traditional dwelling, shared with animals. The archaeology and documentary evidence combine to demonstrate beyond all reasonable doubt that around the middle of the thirteenth century downland cultivation was occurring in strips on Fyfield Down (Plate XXXVIII, Figure 7.11). The significance of this association is far more than parochial, for in the first place it provides a context in which to place other, undated ridge-and-furrow in the neighbouring estates crossing these downs (Figure 2.3), and in the second it suggests at least one possible phase in considering the date of downland cultivation elsewhere in Wessex.

The animal bones discarded on the site (Chapter 14) indicate that the farmstead's economy was based on sheep husbandry. This is to be expected but it is useful methodologically to see two different sorts of evidence marching together. Corn, exchange/barter and a marked element of self-sufficiency were also important. *Raddun* housed the lord's oxen but presumably much of the corn harvested from the fields under cultivation to the east was eventually taken down to the manor farm and mill at Fyfield; yet the bread-baking oven built in the ruins of the long-house (Figure 7.6; *SL*, figure 45) indicates that *c* AD 1300 some was probably retained and ground at *Raddun.* The quantity of thirteenth-century pottery on the site (over 86kg) and the unusual presence of such items as curfews and some highly decorated jugs and pitchers (Figure 7.14) imply access to a wider trading network and a degree of wealth (Thomson and Brown in FWP 65; Figures 7.12–7.14). This evidence, coupled with the occurrence of armour-piercing arrowheads, the number of first-class meat joints consumed and considerable resource investment by the Winchester estate, all imply an above-average standard of living at *Raddun* from the mid-thirteenth to the early fourteenth centuries.

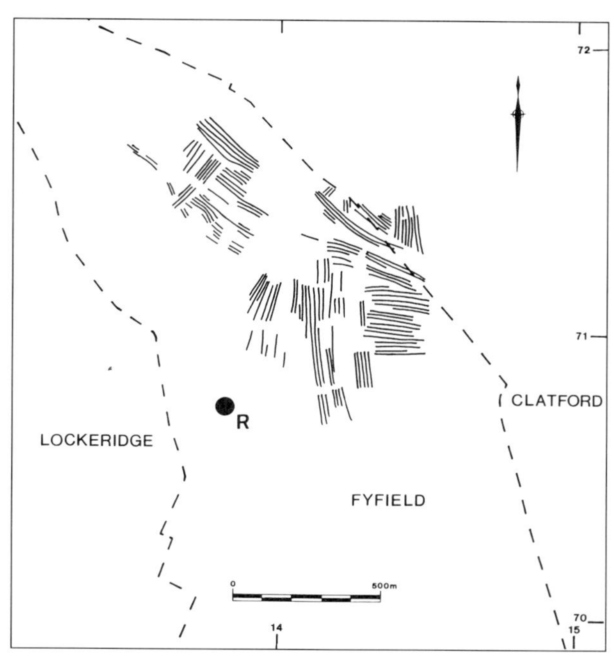

7.11 *Interpretation in diagrammatic form to illustrate* Raddun *(R) in its likely local landscape context in the mid-thirteenth century when part of its function was for one Richard to attend to the cultivation of downland arable, here taken to be represented by the blocks of ridge-and-furrow. The data used here come from air photographic, field archaeological, excavated and documentary sources (cf, frontispiece, Plates VII–IX and XXXVIII; Figures 1.2, 2.3, 7.1, 7.9 and 7.10)*

The finds from the later thirteenth-century house (B1, Figures 7.12 and 7.13, Plates XLIV and XLV) illustrate well the types of activities of the occupants of the farmstead. They were carpentering: their buildings had shutters and doors, both requiring catches, hinges, locks and bolts. They may well have been making storage chests and even furniture on the basis of the number of metal objects associated with binding wood that were recovered. In addition, knives and wood-working tools point, *inter alia*, to the making and repair of hurdles, predominantly for the penning of sheep. Leather-working tools, such as the awls, show they were working cattle hides – and, probably, keeping their own beasts on the 'Cow Down' over to the east beyond the arable fields.

Building 2, in all its phases, appears to have been an animal shelter (Plate XLIII, Figure 7.7). The first construction was a timber structure, possibly with aisles.

(a)

(b)

Plate XLI

Raddun, *Building 4: (a) from the south, showing the remains of the heavily robbed walls and the undisturbed central drain; (b) west side of intrusive 'firepit', showing the neat knapped flint and stone lining of a feature interpreted as part of a smithy, cut into a floor worn down below the north wall, surviving fragmentarily* top left *but otherwise indicated by a ledge*

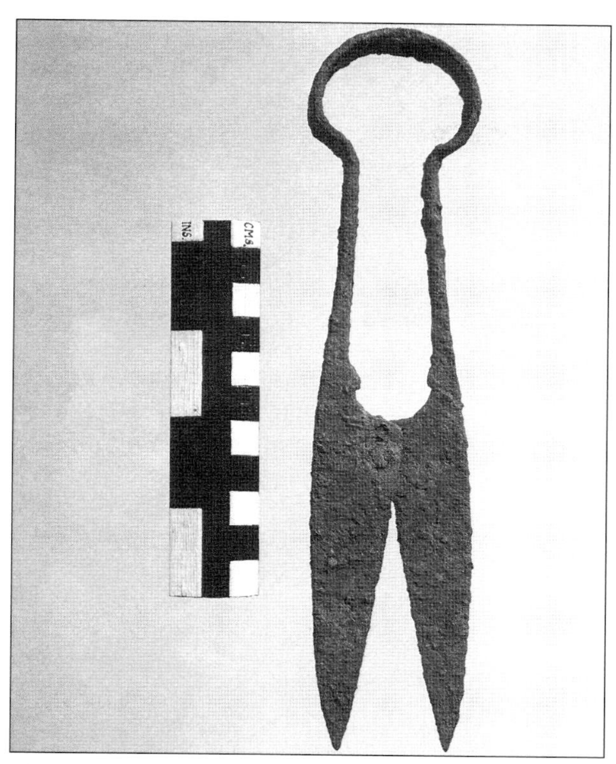

Plate XLII Raddun: *sheep shears from Building 1 (GF 89; Figure 7.12; FWP 65.32, 17)* (© Crown copyright reserved)

The skeletal remains and inference of stalls strongly suggest that B2 was where sheep, cows and goats were tethered to be housed, milked and tupped, and where they could give birth. This was later replaced by a building with sarsen stone walls, though, as with all the buildings at *Raddun*, these were only a few courses high. The superstructure would also have been timber, probably with a wattle and daub wall. This second building was probably for horses, producing numerous horseshoe fragments and nails, a prick spur, an armour-piercing arrowhead and a copper alloy object, possibly a harness mount, with Building 3 (Plate XLIII) for storage or possibly for pigs or hens.

On the whole sheep were of a type kept primarily for wool and though some were consumed at *Raddun*, the main source of meat was cattle and pig, with some chickens and ducks. It is therefore likely that the majority of the sheep, their fleeces and skins, were taken off site to markets or central stores. Yet, once again, a pair of sheep shears (Figure 7.12; Plate XLII; *SL*, figure 78) and other items associated with the preparation and spinning of wool indicate use on a domestic scale. Presumably, however, much of the craft and domestic equipment was taken off the site when its occupants departed. It appears that they took some, but not all, of the perhaps special things hidden, probably in a chest, beneath the left-hand doorpost of Building 1 (Figure 7.8, pit 2).

Plate XLIII Raddun: *Buildings 2 and 3 from the south*

Though *Raddun* was, above all, an important centre for sheep, the documentary references to cows, oxen and chickens, supported by the archaeo-environmental evidence for pigs, horses and dogs, demonstrate a varied and flexible husbandry. Nearly 70 per cent of the cattle and 60 per cent of the sheep bones can be classified as first-class joints; pig is represented by almost equal amounts of first- and second-class joints. Such figures strongly point to both cattle and sheep having been kept for meat, and indicate that the inhabitants maintained a diet of some quality (Chapter 14; FWP 40). Whether they were supposed to is another matter: both we and the lord might have expected such joints to have been due to the manor.

Goats were common, and no doubt kept for their milk and meat, and pigs were also kept and consumed at the farm. The age range for horses is fairly constant throughout the medieval and post-medieval periods; the suggestion of increased numbers of young animals in the thirteenth century may indicate horse breeding as distinct from horse-keeping and is possibly an echo of an earlier tradition (*see above*). The dogs represented could well have been sheep dogs. Three hunting arrowheads from enclosure B (FWP 65, figs FWP 65.36, 63, 64) are not particularly rare finds on rural medieval farms, and the evidence of partridge and fallow, roe and red deer points to hunting on the surrounding downland and probably in Savernake Forest too.

The three armour-piercing arrowheads from the site, on the other hand, are of particular interest (FWP 65, figure FWP 65.36, 60–2). They could have been used for hunting (cf Bond 1994, 127), but they may also suggest, however improbably, some sort of military activity in the vicinity. The proximity of the Knights Templar at Rockley and their tenanted estate in adjacent Lockeridge only some 500m south west of *Raddun* offers, however, a possible explanation. The arrowheads do not necessarily imply direct contact, for they could have been picked up by *Raddun* inhabitants while out shepherding.

The Delling enclosure and Delling, Fyfield Down

The Delling earthwork enclosure (Plates IX and XL, Figures 3.3, 7.1 and 7.15; FWP 66; *SL*, figure 46a) was so named by us but it lies on what was Lockeridge Down in East Overton, so it is now technically (just) in West Overton civil parish; Delling is firmly in Fyfield. The enclosure was discovered independently during field reconnaissance, though in fact it was published (upside down) as an air photograph at about the same time (Hill 1961) with the suggestion that it was a medieval or Roman farmstead. Detailed field survey showed that, like its neighbour in Wroughton Mead (Figure 7.4), it overlay early, probably prehistoric fields: the scarp dividing off its northern third is the lower edge of one such field. The enclosure was also shown to have a southern annexe. The whole looked post-medieval, perhaps associated with the pillow-mound across the coombe to its south (Figures 7.1 and 7.15). Despite its lack of a name, the enclosure was also considered as a candidate for the *Dyllinge* of the 1567 Pembroke Survey and possible precursor to the extant Delling Cottage, 300m to the north.

Dating the relict earthwork enclosure by archaeological means was hardly likely to add to precision in providing a *terminus ante quem* for early fields but it nevertheless seemed, at what was then still an early stage in the development of the project, that a useful purpose would be served by dating the enclosure itself. With the unexpected medieval date of the Wroughton Mead enclosures in mind, the idea of dating

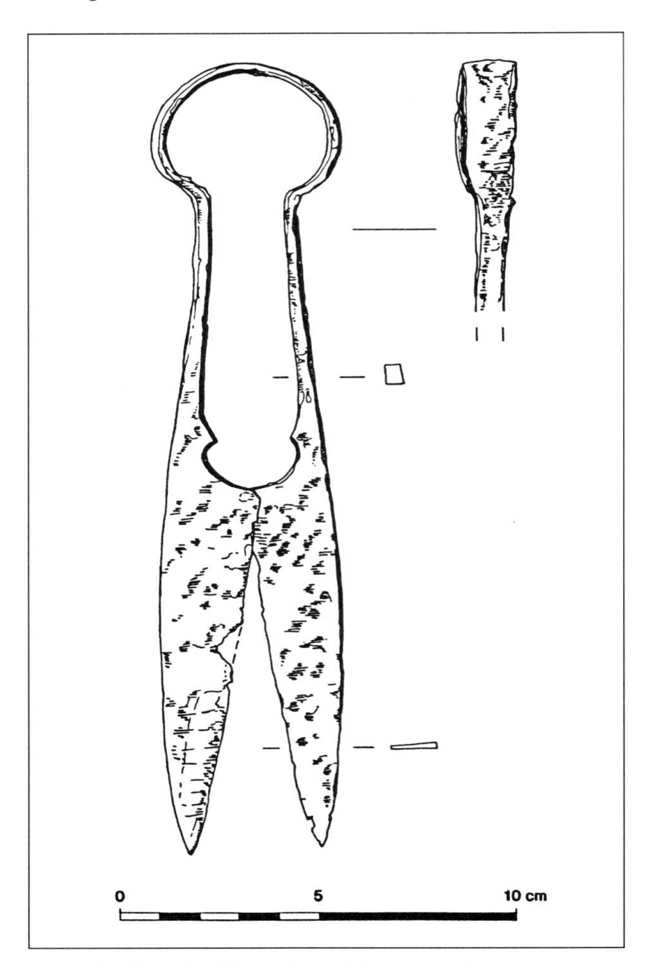

7.12 Raddun, *Building 1: iron object: sheep-shears*

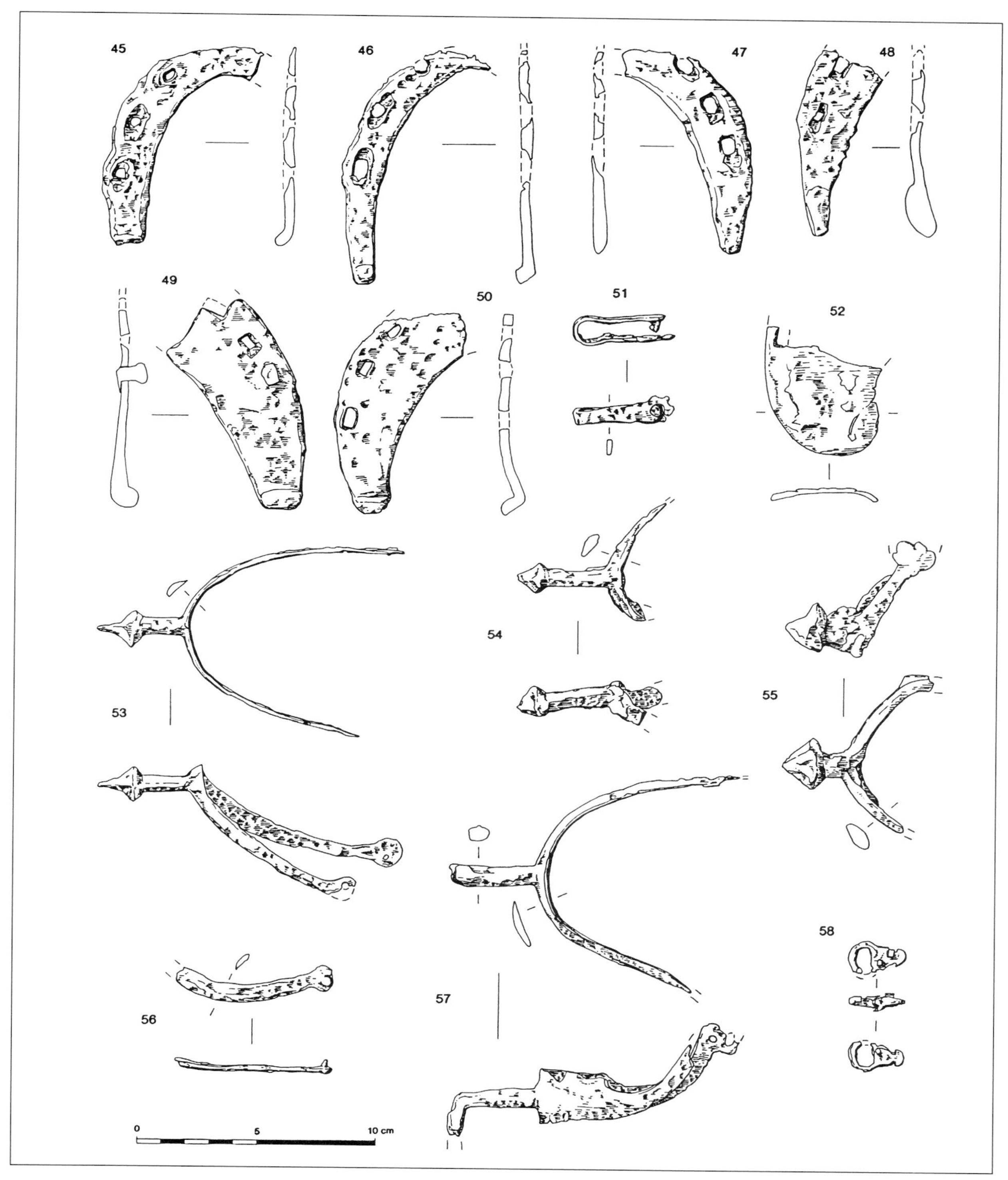

7.13 Raddun: *iron objects*

Horseshoes*: 45: GF187; 46: GF356 (B); 47: GF276; 48: GF510 (A); 49: GF122; 50: GF216; 51: strip, GF317; 52: oxshoe, GF234*

Prick spurs*: 53: GF44; 54: GF422; 55: GF559; 56: spur arm, GF345; 57: rowel spur, GF229; 58: small fitting, GF835 (A)*

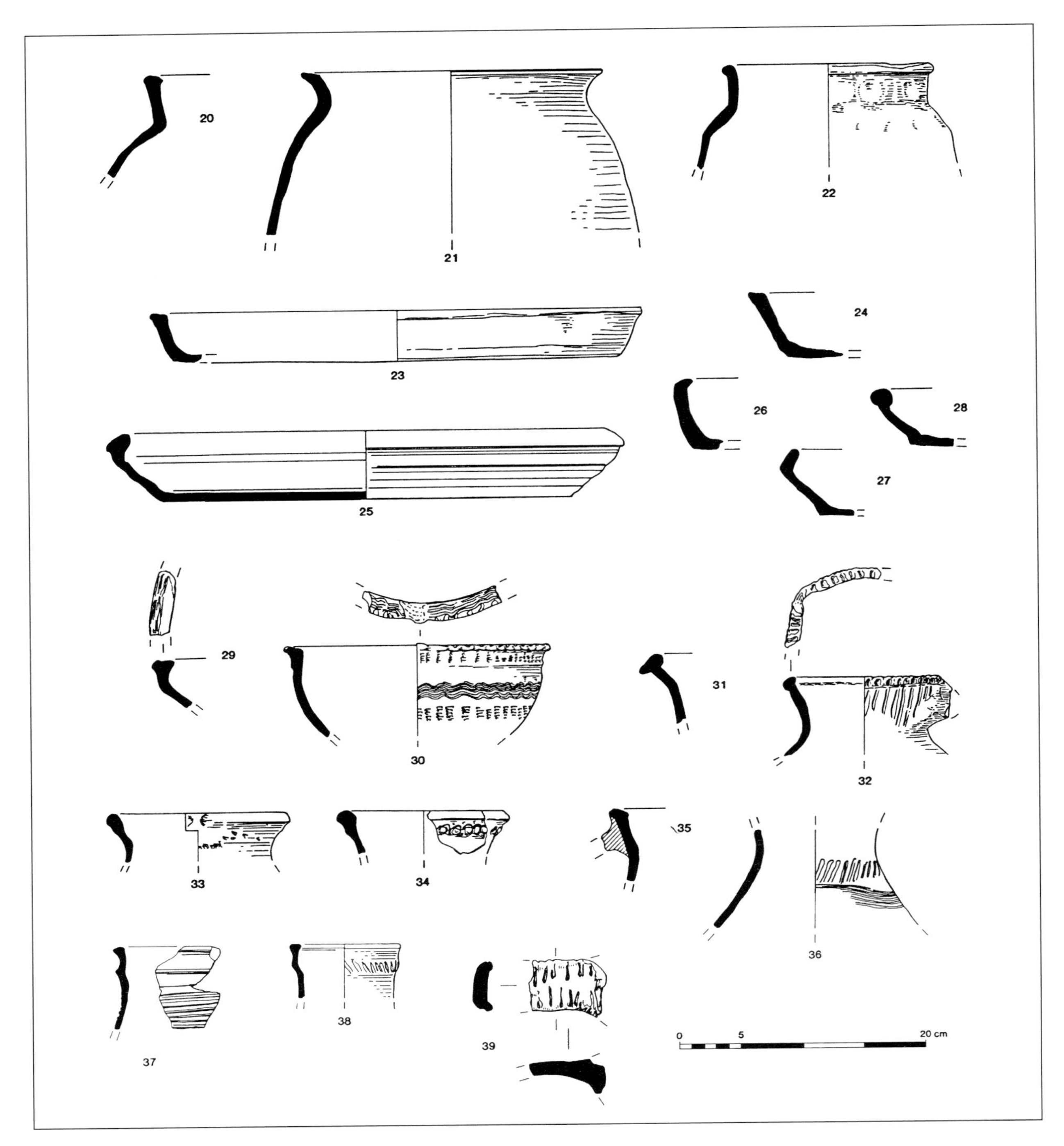

7.14 Raddun: *pottery*

Cooking pots*: 20: GF657; 21: GF657; 22: GF616*

Shallow dishes*: 23: GF293/318/376/512; 24: GF639; 25: GF167/200/156/193; 26: GF841; 27: GF317; 28: GF856; 29: dish, GF614; 30: ?pipkin, GF341/504; 31: deep bowl or pan, GF543*

Jugs*: 32: GF324/322/346/353/726; 33: GF100; 34: GF158; 35: GF548; 36: GF696/276/386; 37: GF546/544/459; 38: GF109/184; 39: strap handle, GF837*

Plate XLIV Raddun: *Building 1 from the south, showing how a platform has been made on the southerly slope to take the two-phase house, the newer end of which is nearer the camera*

another earthwork enclosure to post-medieval times was attractive, especially as that period was then archaeologically unrepresented on these downs by a settlement. Both *c* 1960, and again in February 1996, brick fragments were observed on the surface, suggesting the nature of these foundations and the post-medieval date of the structure. It was guessed that, if a building, possibly a house, had stood there, then its rubbish would have been thrown downhill into the depression. It was.

In fact, unknown to this writer, the enclosure had already been the subject of a small excavation in the mid-1950s by (Colonel) A Witheridge, then a schoolboy at Marlborough College. He thought the site might be 'Iron Age', cut two small trenches through its bank and ditch, and did no more when the three sherds he found looked to be medieval or later. This account was obtained orally from the Colonel in 1996.

A further small excavation was carried out by boys from Marlborough College under the supervision of J Scantlebury at the suggestion of the writer in 1961. Records and master have disappeared, and all attempts to locate both have failed. A small cutting was made into the depression below the potential house site; it was clearly into the top of a midden or rubbish tip (Figure 7.15) whose contents included pieces of yellow, internally glazed pottery with sgraffito brown decoration. Clearly the assemblage was post-medieval, probably of sixteenth- and seventeenth-century date (by analogy with material then being excavated from site 10, Wroughton Mead), but apparently with nothing later. The date of the enclosure seemed to have been established, inviting the suggestion that it could be the site of the documented mid-sixteenth-century *Dyllinge*. It is tempting to interpret the site more generally as the one farm in the area between, chronologically and spatially, *Raddun* of the thirteenth–fifteenth century and Delling of the early nineteenth century.

Delling, the existing cottage, is now the only roofed house on the downs. Clearly shown and named 'Keepers House' on Dymock's 1819 map, it was built between 1811 and 1819. It lay inside a fenced or hedged enclosure, roughly rectangular in plan and shown as

(a) (b) (c) (d)

Plate XLV *(opposite)* Raddun, *Building 1. (a): north-west corner from the west; (b): southern half of west wall from north; (c) and (d): pit with 'foundation' deposit under eastern wall*

A pit had been neatly dug into the chalk (d) and a padlocked chest positioned in the bottom along with an iron axe, a glazed jug and a cooking pot. Large stones were then carefully placed on the chest (c) and the pit filled with earth. The pit was next carefully filled, as indicated by the cross-fitting sherds from the top and bottom of its fill. Afterwards, the east wall of the building was built over the buried chest. Later, perhaps as the site was abandoned, the chest was broken open to remove the contents, the pots also being broken in the process

Plate XLVI Raddun*: detail of cutting 10 from the south, showing the east wall along the 'front' of the platform and long, late medieval building*

more or less square by Smith (1885) who called it 'Overton Delling'. This enclosure still exists as a slight bank and ditch on the ground either side of the now surfaced part of the trackway to Rockley which has been inserted since 1819. Part of the enclosure on Totterdown behind the house was recorded from the air by Major Allen (Plate XIX). The house was surrounded by an enclosed garden; another garden lay in the south-west corner of the larger enclosure which appears to have been aligned on the north side of the 'old London Road', enclosed in 1815/16 (*see above*). This road remains a pedestrian right of way between Avebury and Marlborough (*see* Plate VI and Figure 1.1.2).

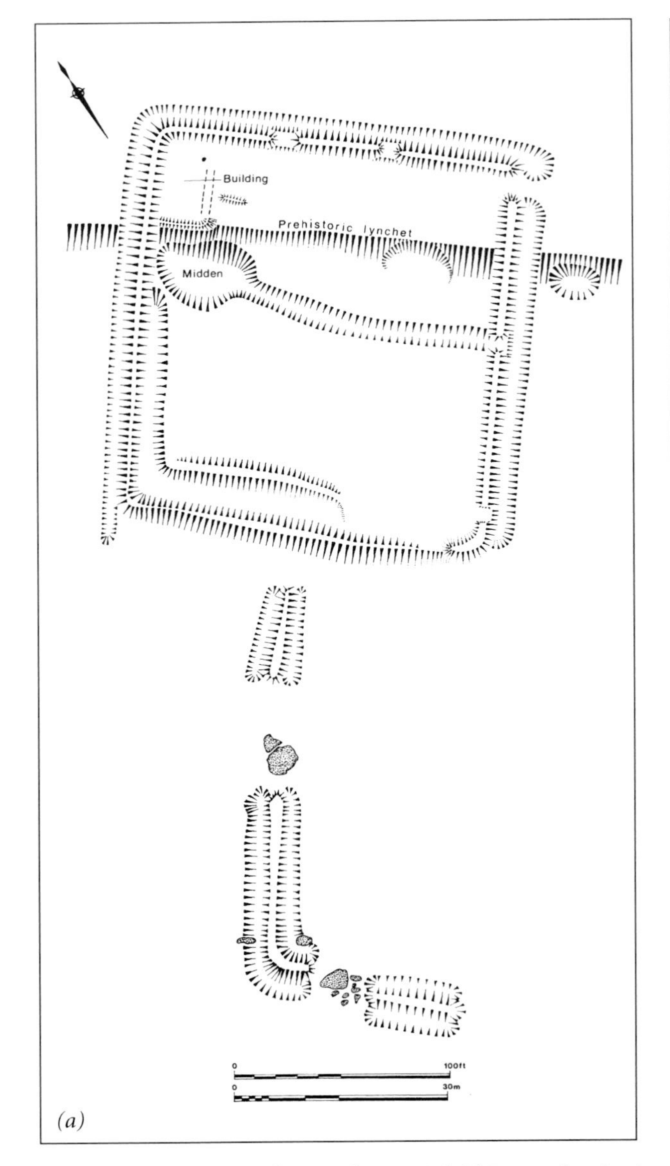

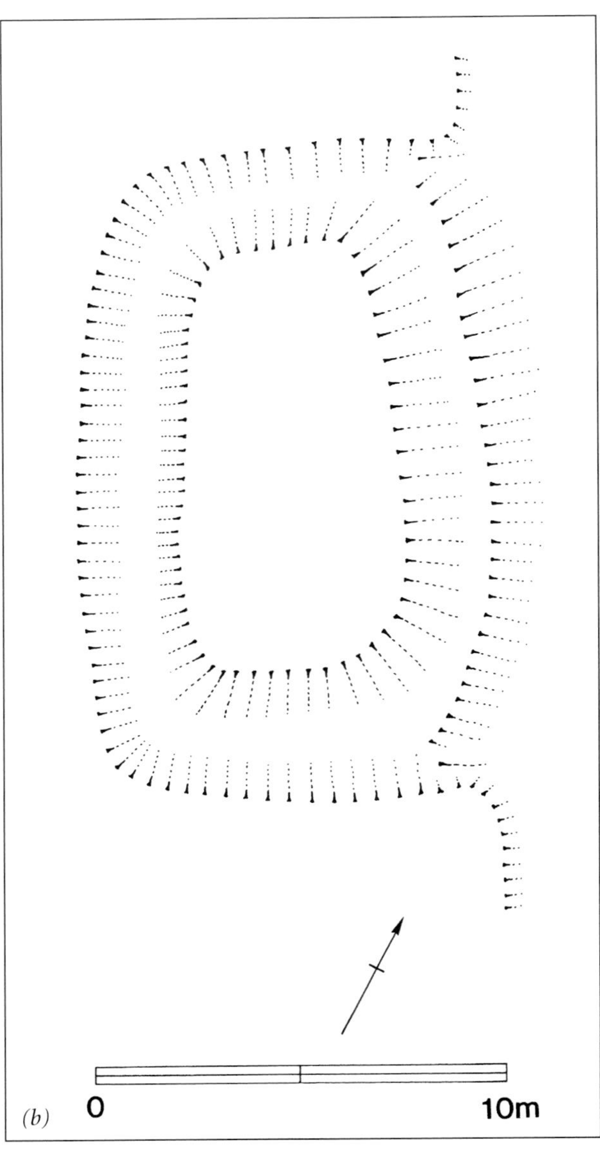

7.15 *(a) plan of the Delling Enclosure, Fyfield Down, showing its superimposition on a prehistoric lynchet, the location of a probable building in its north-west corner, the pit immediately on its south where a small excavation was conducted and the two-sided earth and sarsen stone annexe on its south west (cf Plate X and Figure 7.1)*

(b) plan of pillow mound near Delling Enclosure

OVERVIEW

Fyfield Down has been used since Neolithic times. Overall, its long-term use has been as pasture, but it bears witness to periodical cultivation. This probably began around 2000 BC but its impressive field systems developed during the second millennium BC. A major if short phase of cultivation occurred early in the Roman period and again in the thirteenth century AD. The last is of particular significance because it can be associated with a documented and dated thirteenth-century settlement called *Raddun*. Prehistoric settlements are unknown here, but a probable small one exists at a junction of Roman trackways and an enclosed farmstead almost certainly marks the site of a farm called *Dyllinge* in 1567. The trackways themselves range from through-routes definitely of Roman date and probably earlier to a main east–west route across the downs from Marlborough to Avebury closed to through traffic in 1815/16. Now, the main uses of the area are as continuing pasture, a National Nature Reserve, for racehorse training, scientific research and recreational walking.

Chapter 8

The Valley and its Settlements: West Overton

INTRODUCTION TO THE UPPER KENNET VALLEY

So far, we have looked only at the northern downland of our study area (Figure 3.3). We have moved as far south as the Roman road that roughly bisects the two parishes respectively along the northern and southern sides of the Kennet valley (Chapters 4–7). We now look at the whole area south of that road (Chapters 8–13; Figure 3.4). The same division was made by Colt Hoare early in the nineteenth century (1821, II, Station X, Marlborough, map facing p 3).

Along the northern fringe of this southern land block is the river and its narrow floodplain (here called 'bottomlands'). This is edged by river terraces, with locally prominent bluffs, two marked respectively by the churches of Fyfield and West Overton. The valley, and indeed the whole study area, contains three extant villages called West Overton, Lockeridge and Fyfield.

Throughout this southern part of the study area, our air photographic cartography has been comprehensive. This has now been independently supplemented by that of the RCHME (Figure 15.3; FWPs 85 and 86). Our documentary research was reasonably thorough; our fieldwork has been almost entirely at the reconnaissance level (Figure 3.2). We did not excavate south of the river, though a few others have (FWP 66). Nevertheless, for an area where the relatively little work done has tended very much to be either archaeological or historical, ours is a first step towards a more integrated understanding of it. Very much part of that is the realisation that the valley and southern areas are as important to an informed interpretation of the workings of the landscape within the study area as is the downland. Though to say so is a cliché by the 1990s, such was not the case in the 1960s when archaeology was very much oriented on the downs (eg, Grinsell 1958; Fowler 1967) while history dealt with the valleys and their villages. But, as we found, much about the northern downland can only be understood with reference to the valley and to the complementary range of resources both there and further south in the two parishes. It was, after all, in the valleys that power lay or, more correctly, that the local centres for the application and administration of distant power lay, certainly from Roman times onwards. The generalisation is demonstrated by medieval and later documentation, the only caveat being the late appearance on the scene of a few locally resident landowners.

That power was exercised from the villages, more specifically from the manor farms in the villages of West and East Overton and Fyfield. The manor farms operated through the mechanism of the manorial estates which, Lockeridge excepted, by and large came to coincide with the tenurial framework of the tithings (Figure 1.2; *VCH* XI, 104). These are the subdivisions of the historic parishes, long, narrow strips of land arranged roughly north west to south east across the grain of the landscape in territorial and proprietorial units. There are exceptions to this pattern, notably with Shaw, and changes occurred within it during medieval and later times, such as in Lockeridge, but essentially that pattern of land organisation has endured for at least a thousand years and, arguably, for much longer (Figure 1.2; Bonney 1976).

Much local proprietorial history and land-use has followed from this long-term stability. The tenurial framework presumably reflects basic economic realities, not least because to a marked extent it still exists in the divisions and operations of the contemporary agricultural landscape, even though the tithings are no longer fiscal units nor the civil parishes the same as the ecclesiastical ones. The size and shape of the historic parishes and their tithings together, then, became a major factor in the history of this countryside and in the creation of the landscape we see today. We have not attempted to write conventional parish or village histories here. Our concern, as always, is to select such evidence as we judge bears significantly on the landscape and its development.

Though this account continues to be led by topography, covering first the valley (Chapters 8–11) and then the higher land to the south (Chapters 12–13), its cross-grain is the human settlement and tenure imposed on that landscape. We therefore deal with the four villages along the valley from west to east (Figure 8.1), and then other settlements and abandoned archaeological features towards and in the woods to the south (Figure 12.1). Our treatment is not, however, even-handed. With so much material to choose from, we have had to select and have biased our selection towards different emphases in each place, hoping thereby to represent many of the features of the landscape history of the valley and southern uplands overall. The complexities of their treatment reflect those of the area, its settlement pattern, the morphology of the existing and deserted villages, the archaeology of earthworks and buildings, its tenure, its owners and its tenants, and their interactions through time (discussed with different emphases in *SL*, chapters 7–10).

The River Kennet and its flood-plain

The history of the valley-bottom itself (Plates XLVII–LIII; Figure 8.1) is undoubtedly complex. Some evidence of this below the present surface, and its scientific potential especially in geomorphological and palaeo-environmental terms, were provided in Evans' research

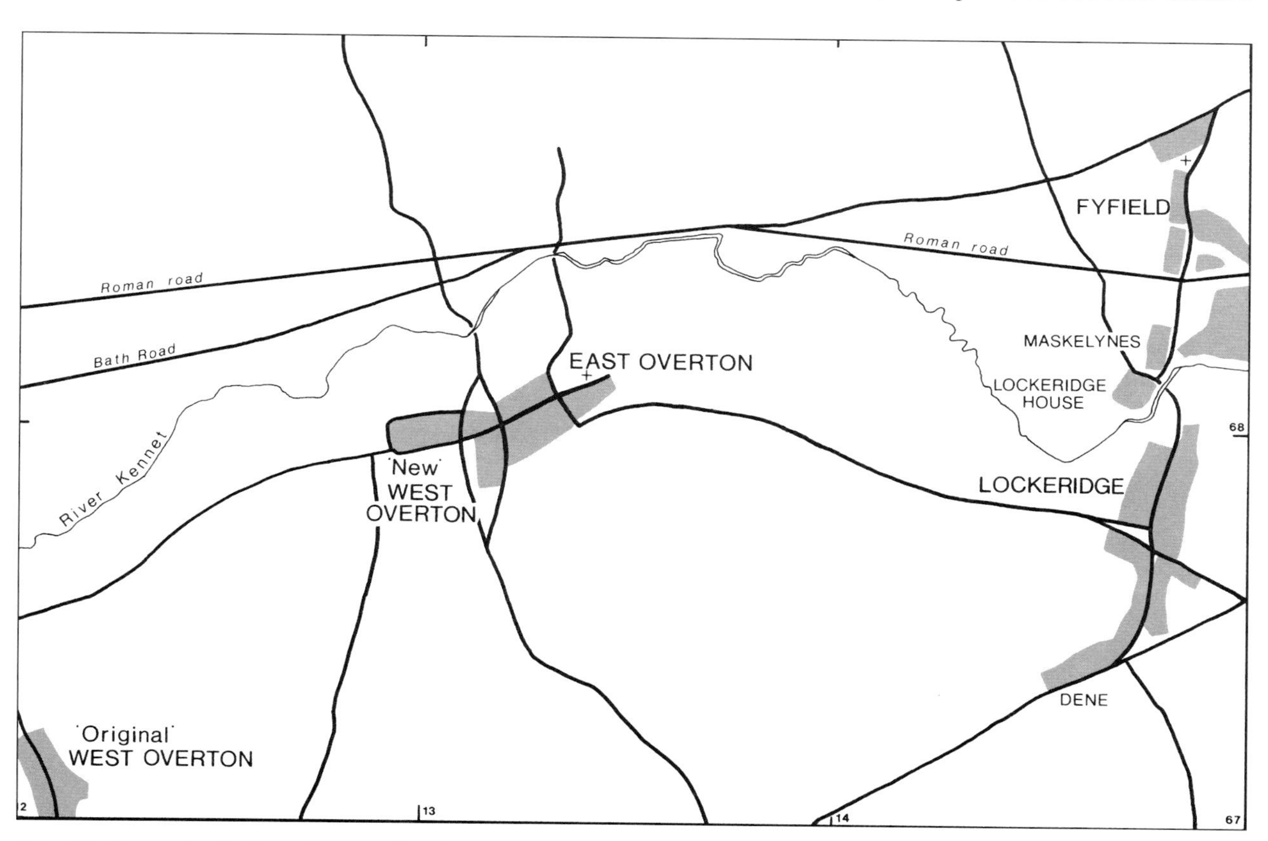

8.1 The 'bottomlands' of the Kennet valley showing (tone) the locations of the extant villages and former settlement areas, namely the three Overtons and four of the five Lockeridges (cf Figure 10.1) and parts of Fyfield at various times (cf Figure 11.1)

Plate XLVII *(a: above) and (b: opposite): the village of West Overton in 1924 portrayed on two previously unpublished overlapping vertical air photographs by O G S Crawford. Essentially, they show the plan of the historic village in its early nineteenth-century form, with the earthworks of the 'improved' water-meadows of that time still in operation. The lineaments of the two arguably late Anglo-Saxon 'planned' Overtons are also visible, as are the earthworks of the deserted part of East Overton in Ring Close. The apparent cropmarks of a 'planned medieval village' east of the church (and over the site of the round barrow discovered in 1993) were created by the strip arrangements of allotment gardens, which are no longer cultivated, though their remains have already archaeologically been misread as 'ridge-and-furrow'* (© Crown copyright reserved)

and more recent pipe-line work (Evans *et al* 1993; Powell *et al* 1996), summarised in Chapter 14. The essential point is that, while many local variations can be expected, the floodplain as presently defined had largely infilled by the end of the Bronze Age, with very little further deposition to raise the ground surface significantly over the most recent two millennia. The earthworks which lay on that surface early in the twentieth century (Plate XLVII) had been formed, then, on only 2m–4m of deposit, and could have been of any date from late prehistoric onwards. None is known to have been Roman or earlier and most are or were post-medieval.

Four main uses of the valley-floor, all involving the control of water, are documented and, to an extent, represented archaeologically (Plates LXII and LXIIIb). They are: fishing and fish-storage; withy-production; power generation for mills; and grass-production for both pasture and meadow, increasingly through the management of water meadows.

Immediately on the north side of the ford (later bridge) across the Kennet on the common boundary between the Overtons stood a salt-house in the mid-tenth century (S449; Plate XLVII). Though nothing is known about the fabric of the building itself, it was almost certainly used to store the manorial fish and other foodstuffs; probably, too, it stored imported, salted fish. It lay at a critical point on a busy, north–south route through the chalklands, a route identified here as

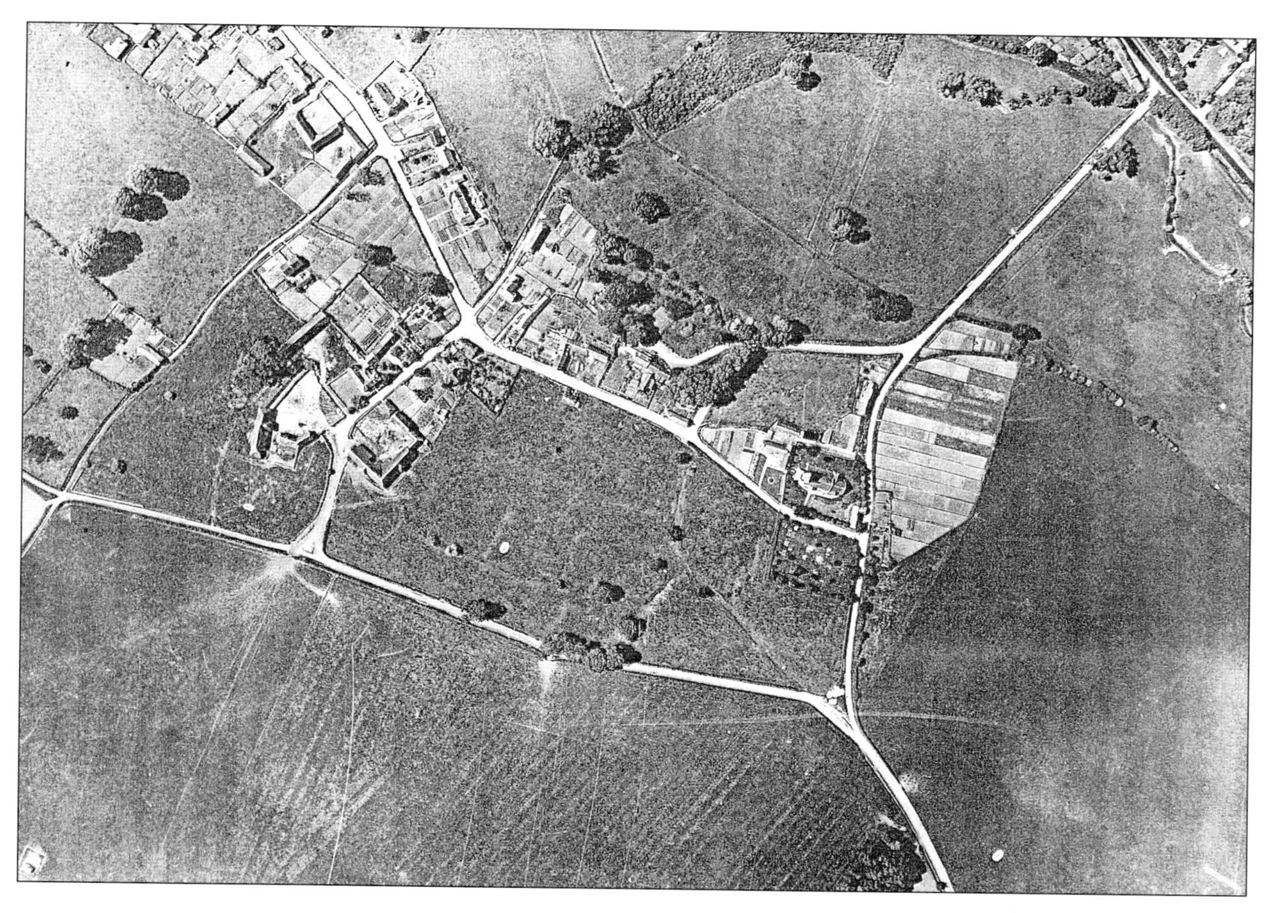

the 'Overton Ridgeway' (Chapter 2; *see below*) coming off the Marlborough Downs, crossing the Vale of Pewsey and following either tracks across Salisbury Plain or along the Avon valley down to the sea.

Another use of the valley bottom was to grow willows. An osier bed belonging to the Duke of Marlborough lay on the other, southern side of the site of the salt-house, still on the East/West Overton boundary and beside the through-way but, probably more relevantly, at the end of what used to be called 'Watery Lane' (now 'Frog Street' or 'Lane') coming north out of the villages (Plate XLVII; Figure 8.2). Another osier bed, belonging to the Priory of St Swithun's (*VCH* IV, 419), lay to the east, possibly at Stanley Mead (SU 141681). Both areas, according to Andrews and Dury (1773), were provisioned with ditches dug to them from the Kennet to divert water to feed the water-hungry willows. The trees were presumably managed to produce a steady supply of withies to meet the never-ending demand for sheep-hurdles, in particular, but also for use in building work, as in lath and plaster, and basket-work. Other managed willows lay along the banks of the river. Trees still grow thickly among a cluster of sarsens by the Kennet at 'Slandly Copse' (SU 142682) where a Roman coin was found (*WAM* 43, 140) at a time when a light railway line existed to take sarsen stones away (King 1968, 92).

The River Kennet was also a power-source for water mills. The West Overton mill, for example, was part of the landscape here from the eleventh until the mid-nineteenth century (*VCH* XI, 198); its site is now marked by a large tree just upstream from the road-bridge behind the Bell Inn (*SL*, back dust jacket). Most of the visible archaeology on the floodplain, mainly in the form of slight earthworks, represents attempts to regulate the flow of the River Kennet to produce grass (Plate XLVII). Though doubtless fragments of earlier arrangements survive, the remains generally date from Inclosure in the early nineteenth century. A phrase in the East Overton charter (S449), nevertheless hints that some artificial arrangements for controlling the river flow may have existed as early as the tenth century: *juxta dirivativus fluentium successibus* ÆT CYNETAN *in illo loco ubi ruricoli antiquo usu nomen indiderunt* UFERAN TUN ('by a series of off-takes from the Kennet in the place known of old to the local inhabitants as *Bank Far*') (Brentnall 1938a and b, 119; Kerridge 1953). Perhaps such 'off-takes' were early canalisation work to drain the valley bottom to increase the land available for pasture and to reduce the area of marshland around the early settlements (Chapter 14).

Further schemes to manage the river were undertaken along the Kennet valley in the medieval and early modern periods (Brentnall 1950, 304; Kerridge 1953, 55, 111–12; Andrews and Dury 1773). The purpose of these ventures, apart from supplying power to the West Overton mill, was, presumably, to continue the work begun in the Saxon period and drain land for pasture and hay. A system to 'float' the water-meadows appears to have already been in existence by June 1814 when the Inclosure Commissioners ordered 'that the several brooks, streams, ditches, watercourses, funnels and bridges ... shall at all times from henceforth be sufficiently deepened, widened, cleansed, scoured and kept in repair by and at the expense of the respective owners or occupiers' (WRO 79a/1).

The creation of relatively well-engineered, substantial and extensive water-meadows which were such a feature of the landscape and its economy in the nineteenth century is due to this 'Act for Inclosing Canals in the Tythings of East Overton and Lockeridge'. The document is as vivid now in its evocation of a valley-bottom landscape as it was then in its specification for the better management of a vital resource. The valley bottom was divided into seven stems, each a small area of meadow in its own right with its own sluices and channels which could be flooded separately and independently off the main float when necessary. The main float south of the Kennet was 8ft (*c* 2.4m) wide and dug to 3ft (*c* 0.9m) in depth. Starting at the dam 60yd (*c* 55m) above West Overton mill, the water was regulated by a series of culverts, an aqueduct and sluices as far as Lockeridge village (SU 14556794), before the water fell back into the Kennet.

The regulation of the water meadows was of the utmost importance to ensure the exact amount of water spilled from the culverts into the meadows at the right time, not only of the year, but the right time of each day and indeed night. Here in the Kennet valley, the first stem was floated for four days and four nights, with the third to seventh for two days and two nights from 1 December to 4 April and from 5 May to 1 July for each stem in regular succession. The significance of the dates was that they gave the owners control of, respectively, the first grass growth and 'early bite' so crucial to the welfare of over-wintered sheep and, secondly, the hay crop (Kerridge 1953, 105–18; Atwood 1964, 403–13). By 1819 much of this canalisation work had been completed, and at Parliamentary Inclosure in 1821 arrangements were made to pay the owner of West Overton mill, Edward Pumphrey, £27 at Michaelmas yearly for turning out of the mill dam to irrigate the water-meadows along the valley floor (WRO 79a/1).

Much of this work, presumably also including some surviving earlier elements, is recorded in some detail on later nineteenth-century large-scale maps. The system was apparently still being maintained then. It was not operating, however, on the earliest air photographs of the Overtons in the mid-1920s (Plate XLVII), though its earthworks are visible then, and on later air photographs up until the mid-twentieth century. Ditches were still coloured blue on the 1961 OS 1:25,000 map, but about then and in the following decade the remains suffered a great deal during agricultural improvement. Among the remains still are impressive earthworks on the floodplain immediately north of St Michael's Church (Plate LXIIIb) and stone structures including a leat and 'clapper-type' sarsen-stone bridge down 'Watery Lane' (Frog Street).

SETTLEMENT IN THE BOTTOMLANDS

After the middle of the first millennium AD, villages *sensu* nucleated settlements, both extant and deserted, occur only in the southern area of the two parishes (Figure 8.1). Yet each tithing contains only one nucleated settlement at a time during the same fifteen hundred years, a position which persists today. Outlying settlements also existed, for example the small twelfth-century settlement of Upper Lockeridge (*see below*).

All three present-day villages, West Overton, Lockeridge and Fyfield, lie in the valley of the River Kennet. Overton and Lockeridge, like East Kennet to the west and Clatford to the east, are on the south bank, their buildings now just above and back from the floodplain. The newly recognised Upper Lockeridge is further south still. Fyfield, like West Kennet to the west, is on the northern side. This locational distinction is conceivably significant, perhaps hinting that their positions are remnants of a pre-Saxon settlement pattern, for both lie close to the main east–west Roman road and contemporary settlement (Fyfield, Plate LI, Figure 11.1; West Kennet, Powell *et al* 1996, fig 5).

The villages were previously called, respectively, East Overton, *Ovretone* (DB) and, earlier still, *Uferan tun* (S449), *Locherige* (DB) or *Lokeruga* (AD 1142), and *Fifhide* (DB). A fourth settlement lies in the modern parish of West Overton. It was called *Vuertune* by the *Cynetan* in AD 972 (S784), *Overtone Abbatisse* in 1332 and later 'West Overton' (*PNWilts*, 305–6). In other words, there were two Overtons where now there is one (*see below*). The former East Overton is particularly well documented and therefore appears to dominate both West Overton and, to its east, Lockeridge and Fyfield. This evidential impression reflects an historical reality (cf Chapters 9–11).

Though none was along the riverbanks, at least four other smaller medieval settlements formerly existed in the study area: Shaw, the pre-tenth-century settlement of West Overton, Lockeridge Dene and Upper Lockeridge. To them, to fill out the medieval settlement pattern, can be added the outlying farmsteads of *Aethelferthes setle*, Walter of Thanet's demesne, *Raddun*, *Attele* and *Hacan penne*, Heath and Park Farms, Fosbury and Spye Park Cottages, Delling and a settlement near Boreham Wood. Of these, only Dene is located beside the old 500ft (152m) contour which so accurately picks out the sites of the four certain village settlements of the Anglo-Saxon period – the three Overtons and Fyfield.

All the valley-bottom villages have 'shuffled' a little in their positioning (Figures 8.1, 10.1 and 11.1). Each is looked at in more detail below (Chapters 8–11), but to illustrate the generality here we note that the first settlement of West Overton moved across its estate, leaving a church and an apparently vacant site behind it. The 'new' West Overton, in existence by at least the tenth century, subsequently spread westwards to the manor and then the new manor farm of the early nineteenth century. Early East Overton village is likely to have been formed on the knoll on which the later church was built (Hase 1994, 58), and spread south as well as west, bumping up against the boundary of the West Overton estate. It then either shrank or shuffled north. In a location shielded from the north-east winds, Dene was superseded in the twelfth century by a new Templar planned settlement of Lockeridge. Much later, the new Lockeridge imploded to an extent with some large houses around its original core at the Dene crossroads, before again stretching northwards with a new estate village of the mid-nineteenth century. Fyfield has shifted around the most, including down on to the floodplain. Its complex *pas de deux* around its church is discussed in Chapter 11.

West Overton

The two Anglo-Saxon and medieval settlements of West Overton and East Overton were at the centres of separate manorial estates until fairly recently, so they are treated individually here. In this account, the name 'West Overton' means the manor and former Anglo-Saxon estate of that name, not the present civil parish nor the present village of that name. This West Overton estate, naturally enough, lay in the west of the study area, west of East Overton. 'East Overton' is the name correctly used for the settlement, which lay to the east around St Michael and All Angels Church, in the dominant manor. Its land, unlike that of West Overton, stretched far on to the northern downs (Figure 1.2). In the thirteenth century, East Overton and Fyfield (not East Overton and West Overton) formed a combined manor called Overton whose landlord was St Swithun's Priory, Winchester. 'Overton' in this study refers to this large medieval estate (cf FWPs 43, 44, 46 and 48).

Confusingly, both of the major settlements of West and East Overton came to lie side by side on either side of a common boundary certainly in existence in the tenth century, and they remained ardently distinct communities until the major changes of the early nineteenth century. The reallocation of land during the eighteenth and nineteenth centuries has also had the effect of removing such tenurial complexities, leaving the modern inhabitant in a single village that is today called West Overton. To have given the name 'West Overton' to the new civil parish and also to the main village in it is, of course, historical foolishness, topographically misleading and grossly unfair to the memory of the major manor hereabouts, East Overton. The present civil parish and the large village within it both result from a combination of West and East Overton, yet the composite ends up with the name of the inferior element. It is a great pity that the name 'Overton' was not used for both the present civil parish and village.

Settlement in West Overton

The early medieval West Overton was a slightly crescent-shaped strip of land, running north to south east for about 5km, and only just over 1km across west to east at its widest. North of the Kennet, the land was divided into arable and pasture (Figure 4.1), then, moving from north to south, there were and are the 'bottomlands' and the river terrace, with the main post-Roman settlements. A good area of open downland suitable for arable then stretches south and uphill for 1km past Hill Barn (SU 12806665; Figure 8.3; cf *SL*, figure 58), with a broad dry valley beyond suitable for pasture and formerly cultivated in early fields (Figure 12.3). The main estate stretched a further 1km up into the woods, ending at two features with good Anglo-Saxon names: *scyt hangran* (Chichangles, later Pumphrey Wood) and *Eadgardes gete* (S784) on Wansdyke (*see below*, Figure

13.2). A separate area of land, 'Savernake Grounds', lay some 2km to the east of Pickrudge.

A charter of AD 972 is our first documented reference to the West Overton estate. In that year King Edgar granted ten hides of land *Æt Vuertune*, at *Cynetan*, to lady *Ælflæd* (S784; Grundy 1919, 240–7), possibly a nun at Wilton Abbey. This estate, which by the thirteenth century became *Westovertone* (*PNWilts*, 305) to distinguish it from the estate of the same name to the east, corresponds to the *Domesday* estate of *Overtone* (*VCH* II, 129). In AD 972 there was, or had been, a church at West Overton. The Saxon charter begins at the *chiricstede*, the site of a church, at a point between Lurkeley Hill (SU 121665) and the ford, now bridge, across the Kennet (SU 119676; Costen 1994, 98). This early church would have been on the east side of the boundary line with East Kennet and could have been near the sixteenth-century Orchard Farmhouse (SU 119674). It is perhaps more likely to have been in an area referred to on the 1794 map as 'Church Ditch' (SU 121670), where The Ridgeway and 'Double Hedge Way' fork (Figure 8.3). This reference may, however, be proprietorial rather than topographical.

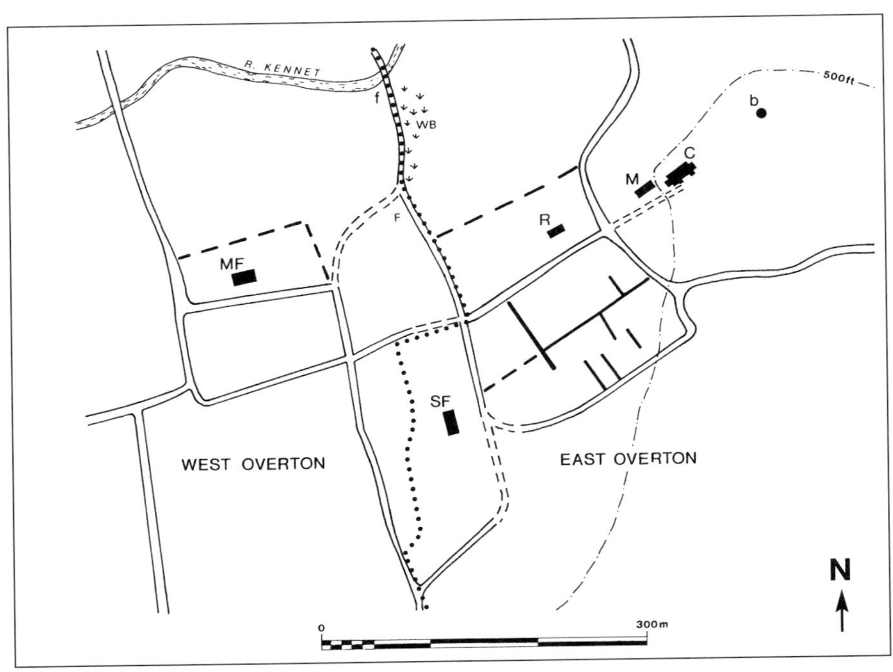

8.2 *Schematic map of West Overton village to show its possible origins as two regulated late Saxon villages, West and East Overton*

MF = Manor Farm; F = Frog Lane; WB = Withy Bed; f = ford; R = rectory; M = manor house; C = church; b = round barrow; SF = South Farm

The deacon at Wilton obtained over £2 out of the church at West Overton in 1291 (*VCH* III, 236), although this does not necessarily indicate a building; the money is likely to have been raised from tithes. If a church was standing in the late thirteenth century, however, it is likely to have been either on the tenth-century site in Anglo-Saxon West Overton or it may have been a relatively new church in the emerging nucleated village accruing to the east (*see below*). Wherever it was, the church has since disappeared and its position been lost. There is no evidence of another church in West Overton before that name was applied to both villages, a nineteenth-century development as a result of which St Michael's, the church of East Overton, found itself in West Overton without moving.

As the tenth-century *chiricstede* lay at the western edge of the Anglo-Saxon West Overton estate, it is likely that the early tenth-century village also lay here, abutting the east side of The Ridgeway and the road to West Woods (Figure 8.1). The land nearer the Kennet, where cottages stand at present, is referred to as 'Home Plot', 'Home Close' and 'Home Mead', indicative not only of a piece of land, probably cultivated, in a river-side position, but also the site of a settlement (Smith 1970, 226–31; Gelling 1993, 43). This rather scant evidence hints that *Vuertune* was a street village stretching from a ford at the northern end to a church at its southern one, with the church probably also acting as a place of worship for travellers along the *herepath* (*see above*).

For several recent centuries, however, perhaps for over a thousand years, the main occupation has been on the eastern edge of the estate, that is, around the demesne farm in the western part of what is now called West Overton village. The 1km-shift eastwards to this area may well have occurred before the charter of AD 972. Moreover, this charter, but not the earlier one of East Overton (*see below*), refers to a stone in front of the *burg gete* at the eastern edge of the bounds, firmly indicating that a settlement lay there then. The *burg gete* stone lay south of the ford, possibly towards or at the end of Frog Lane, perhaps a little like the large, erect sarsen at the corner of 69 Church Lane (SU 13196800) or near South Farm (Chapter 9). Either way, on balance it would seem most likely that the *burg* was West Overton. Furthermore, it is tempting to interpret the charters' difference as possibly indicating that a new settlement, planned and likely to have been rectilinear, or even fortified, was laid out against the west side of the West Overton estate's eastern boundary in the mid-tenth century (Figure 8.2). If the settlement required a *burh-geat*, then it is likely other buildings, such as a church, kitchen and hall – all essential possessions of a thegn (Yorke 1995, 251) – were also components of this postulated late Saxon village. The late pre-Conquest period saw the emergence of many regulated villages in the chalklands of Wiltshire (Smith 1970, I, 62; Biddle 1976, 128–9; Lewis 1994, 187–8) and of course not even Alfred or Athelstan's *burghs* were all either large or in any sense urban (Biddle and Hill 1971, 81–5). On the other hand, any difference may have become only slight by the later tenth century, with *burg* more or less synonymous with *tun*. *Burg* in any case may not have carried any special significance in this context. Furthermore, the rectangularity discernible in the western parts of present-day West Overton village could represent a new settlement created in, for example, the later thirteenth century when 'West Overton' first appears as *Westovertone* in 1275 (*PNWilts*, 305), implying a need to distinguish a West one from the East one. On the neighbouring estate, even an outlying farm called *Raddun* was being significantly reorganised at that time (Chapter 7).

The main features of West Overton village suggesting an element of planned layout were present when it first appears on maps from the late eighteenth century. This banishes any thought that the geometric characteristics are anything other than historic; they cannot be attributed to planning, and certainly not the planning of a whole village, during the age of agricultural improvement, though of course some individual elements, such as a new road and the rectangularity and symmetrical placing of house and farmyard at West Farm, can be ascribed precisely to such a phase. Yet there are few if any 'old', *sensu* pre-1700, houses in what was 'new' West Overton. The argument rests on whether or not the western part of the present village can be separated out in plan as a distinct unit with a certain rectilinear symmetry about it (Figure 8.2).

A central west–east street with equidistant and parallel back-lanes can be discerned, roughly at right angles to the known line of the estate boundary between West and East Overton estates. This intersected a 'back-lane', still extant and a public right of way, running west from the present cross-roads in the village centre. Anglo-Saxon West Overton as a distinct settlement probably lay north of that back-lane, its plan created on the western side of, but up against, the boundary with East Overton. Such an origin would have fashioned an early version of the cross-roads at the centre of the existing village, though perhaps a little further west, offering attractive explanations of the origin of two features still in the village-scape: the sharp bend at the south west of the village, where a path leads off the present road and between houses to turn east into the lane behind the properties, could be the south-west corner of the planned settlement, and similarly an equivalent south-east corner could explain the continuation of that back-lane to its junction with the cross-roads now at the centre of the village. More particularly, it could explain the neat dog-leg in the boundary itself at that cross-roads, as if the boundary was providing a niche into which something fitted or was actually going round something which was now there. This new village, as proposed here then, was aligned along two west–east roads, each end closed by north–south roads running respectively from a bridge and a ford southwards to the fields. Its plan was of a little rectangle of properties 150m west to east by 75m north to south (Plate XLVII).

The farm, presumably the predecessor of the medieval manor house, lay at the west end of the village, on the north side of the street, on a site now just a grass field (that the 'PO' of the modern 1:25,000 map has been placed there [SU 12956803] makes the point about the field's present emptiness). That West Overton has also lost its farm in addition to practically everything else makes the following 1631 record poignant as well as useful in landscape terms. The farm consisted of :

> … a dwelling house of four ground rooms lofted over, much ruined, a barn of 7 rooms, a cart house of 2 rooms, well repaired, a backside, garden and orchard (in all 1½ ac.), closes of meadow called Short Close adjoining the dwelling house (2 ac.), Long Close (3 ac.) and the Penning (1 ac.), a meadow called Custom Mead (12 ac.) ... and 274 ac. of arable in the common fields … North Field 62 ac., in the West or Little Field 60 ac. and in the South Field 152 ac.; and common pasture for 20 horses, 25 kine and 400 sheep. Reputed 5 yardlands. Worth 100l.

Kerridge 1953, 77

This land and its arrangement are reflected in the property attached to the farm on the pre-Inclosure maps for this estate. Land west of the farm was called 'Home Grounds', 'Home Mead' and 'Custom Mead' (which by 1783 had become 'Custard Mead'). The land to the north was also 'Home Mead', though Smith referred to it as 'Paddock' in 1885. By 1818, the former demesne farm, then called West Overton Farm, had consolidated its holdings by acquiring the vast majority of the agricultural land: managing 330 acres (*c* 135ha), with 10 acres (*c* 4ha) of water-meadows by the Kennet. The other successful farm at the time was Park Farm at Overton Heath, which had by the same date increased its acreage to 200 acres (*c* 81ha) (*VCH* XI, 196; *see below*).

West Overton Farm now lies at the west end of the village road. The new farm was built *c* 1825 (ibid, 189) with slate roof, Flemish brickwork, two storeys and double-pile plan. Almost the whole of the nineteenth-century model farmyard, as with West Overton's former South Farm (*see below*), has been destroyed since our project began, but the farmhouse still stands rather grandly looking out over the river towards the Bell Inn (*SL*, colour plate 32). Now listed Grade II, it is one of the several Georgian 'working farm/country seat-type' houses positioned, like North Farm across the river, to some visual effect occasionally in our study area. Such residences indicate a certain prosperity and a statement of social status during and just after the Napoleonic Wars. The means, indeed social need, to build such was presumably being provided by the emergent 'winners' in the increasingly formalised rural hierarchy resulting from Enclosure.

The farm still closes the village's west end. There, unlike its north side where its limits are defined by the river and floodplain, there is no topographical constraint, yet it seems to be an historic limit, perpetuated up to the present. A corrugated iron Wesleyan Methodist chapel (Plate XLIX) characteristically stood on this village edge

Plate XLVIII *Bridges in the bottomlands: the site of the former bridge (or bridges) crossing the Kennet at the position of a ford in the Overtons' Saxon charters actually on the estates' common boundary, here followed by the 'Overton Ridgeway'. The ruinous remains, just upstream of the modern footbridge, are of the northern bridge abutment(s); similar remains lie in the river's southern bank, immediately right of the camera position*

Plate XLIX West Overton Methodist church (destroyed), positioned at the historical west end of the former village of West Overton on a site occupied since the early 1960s by a bungalow

until it was replaced by a bungalow in the early 1960s. Curiously perhaps, residences once did lie on the north, but immediately across the river. Four cottages, one of which was a 'Parish House', stood east of Overton Bridge and the New Inn, first mentioned c 1815 and renamed the Bell Inn by 1823 (*VCH* XI, 185). These cottages were marked on the 1802 map and again in 1819 but by the 1880s they had gone. Earthworks now mark their site (SU 12856831).

Continual changes in detail become apparent within a village once cartographic evidence becomes available. We see, for example, that an apparently substantial house of 1794, the home of a woman called Lettice Sweetapple, had been replaced by a terrace of brick houses by 1885. One is now the Post Office. It seems likely that similar change was happening earlier. Contemporary observation suggests the same, yet notes elements of plan continuity. Houses were built, for example, where a house had stood. Modern development has now replaced much that existed thirty-five years ago, though its control through town and country planning legislation has reinforced the village morphology, even where new housing has filled in what were once spaces. West Overton has retained its historic shape.

People

The Kingmans farmed here in the seventeenth century, the Cooke family in the eighteenth. Edward Pumphrey became tenant in 1784 and his family held West Overton Farm, of 232 acres (*c* 94ha), into the nineteenth century. The Tax List of 1332 lists thirteen tax payers in *Overtone Abbatisse*, paying a total of 21s 3d (Crowley 1988, 58), exactly the same number of inhabitants noted by the *Domesday* recorders, though clearly the total number of inhabitants was many more (cf *VCH* IV, 310). In 1567 the estate contained, besides three freeholders and a

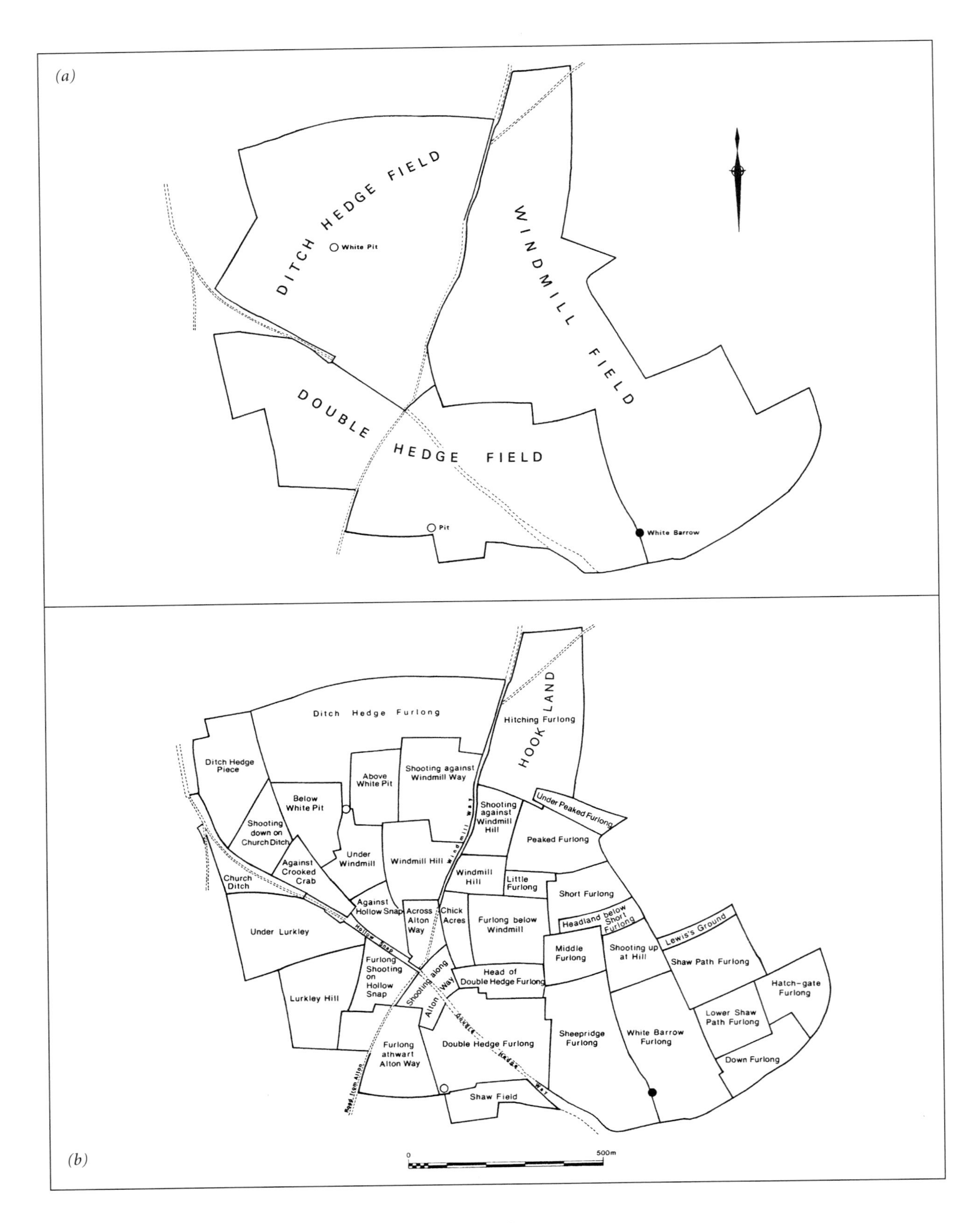
(a)
DITCH HEDGE FIELD
White Pit
WINDMILL FIELD
DOUBLE HEDGE FIELD
Pit
White Barrow
(b)
Ditch Hedge Furlong
Hitching Furlong
HOOK LAND
Ditch Hedge Piece
Above White Pit
Shooting against Windmill Way
Below White Pit
Shooting down on Church Ditch
Shooting against Windmill Hill
Under Peaked Furlong
Peaked Furlong
Against Crooked Crab
Under Windmill
Windmill Hill
Windmill Way
Windmill Hill
Little Furlong
Short Furlong
Church Ditch
Against Hollow Snap
Across Alton Way
Chick Acres
Furlong below Windmill
Headland below Short Furlong
Under Lurkley
Hollow Snap
Middle Furlong
Shooting up at Hill
Lewis's Ground
Shaw Path Furlong
Furlong Shooting on Hollow Snap
Shooting along Alton Way
Head of Double Hedge Furlong
Hatch-gate Furlong
Lurkley Hill
Lower Shaw Path Furlong
Sheepridge Furlong
White Barrow Furlong
Furlong athwart Alton Way
Double Hedge Furlong
Down Furlong
Road from Alton
Shaw Field
0
500m

8.3 *Cartographic analysis of the pre-Enclosure open fields of the manor of West Overton, based on a map of 1794. They lay, somewhat unusually, in one contiguous block, over Windmill Hill south and south west of the village and manor farm. The map shows,* opposite, top left, *(a), the three fields making up the whole of the manor's common arable fields, with two large pits and a round barrow (also shown in the next two maps as visual markers for the reader);* opposite, below, *(b), the furlongs, with their names, as blocks of land making up the three open fields;* above, *(c), all the individual strips within the furlongs, with those of three named individuals selected to show the number and distribution of their strips across the thirty-nine furlongs making up the three open fields (cf Figure 16.2)*

cottager, eleven customary tenants who paid annual rents totalling £7. There was about the same number in 1631 (cf Kerridge 1953, 77–81), though in 1706 there were twenty-four manorial tenants, the most substantial of whom held no more than 30 acres (*c* 12ha). In 1794, seventeen tenants held 560 acres (*c* 227ha), again most holding about 30 acres (*c* 12ha) each (1794 map; cf also discussion of population figures in Chapter 11). In 1802, 551 acres (*c* 223ha) in the open fields, common meadows and pastures of West Overton were enclosed at the expense of the Earl of Pembroke, with the earl's seventeen tenants receiving a total of 385 acres (*c* 156ha). Lettice Sweetapple, for example, received about 40 acres (16ha), nearly twice the average allotment (*SL*, figure 58). Rents immediately rose from £655 to £916 *per annum*, this income becoming available for improvements to the estate (*VCH* XI, 195–6).

The estate and its resources

In 1086, *Overtone* was worth 100s and paid geld for ten hides as part of the 231-hide Wilton Abbey estate, an estate with a gross income of £246 15s, the highest recorded for any nunnery in England (*VCH* III, 233). In *Domesday*, seven hides and half a virgate were in the West Overton demesne, worked by two serfs with two ploughs, whilst elsewhere on the estate three villeins and eight bordars had two ploughs to farm the remaining two hides and three and a half virgates.

The land contained a diversity, but not a particularly large quantity, of resources in *Domesday*. These are represented by 5 acres (*c* 2ha) of meadow, 20 acres (*c* 8ha) of pasture and 8 of woodland, and a mill that paid 10s. In the late eleventh century, therefore, West Overton had less pasture and woodland than Fyfield, though it had more meadow. In addition, whereas East

Overton had land for seven ploughs, West Overton had land for four, the same as Fyfield. Considering West Overton was a ten-hide estate and Fyfield a five-hide one, the apparent anomaly of both being worked by four ploughs suggests that West Overton and Fyfield, each with thirteen recorded workers, had similar amounts of land under cultivation in order to feed a similar population.

At the very western edge of the estate, the Kennet was crossed by a *straetford* (S784) in the tenth century. The *straet* prefix suggests that the ford was metalled (Costen 1994, 105), no doubt because the *herpoth*, the modern Ridgeway, was a busy thoroughfare for people and animals at the time. The river here was crossed by 'Pigeon House Bridge' in 1794. At the eastern extremity of the estate the river was again forded (SU 13116823), just as it had been in the tenth century (S784). By 1783, this crossing was over 'Four Bridges', shared between the two Overton estates, as it is today by a modern footbridge. Remains of where these bridges stood are still evident (Plate XLVIII). Further west by the mill, the road north from the village crossed a wide ford in 1794 where the present Overton Bridge stands (SU 12856821), with a bridged pathway across Mill Ham just to its west taking pedestrians.

West Overton's arable and pasture in the medieval period can be located with some precision. It is very likely that the arable lay west of the present village along the road to East Kennet and to the south in an area called 'The Common Fields' in the late eighteenth century. The land north of the Kennet and part of Boreham Down may also have been under cultivation, though both are more likely to have been rough grazing, with richer, if circumscribed, grazing also available along the 'Meads' of the Kennet.

Cultivation in 'The Common Fields' area is indicated in the tenth century from the charter's *langan hlinc eastewerdne*, 'the east side of the long lynchet', though as elsewhere such might have been a lynchet from earlier field arrangements, particularly as it was in fact located on downland – *scyfling dune* (S784) or simply *dune* (S449). The lynchet may well be under the prominent hedge, called 'Lewis's Ground' in 1783 (SU 134668), which formed part of the tithing boundary until relatively recently. This is the same place as the *riht gemaere*, or 'straight baulk', of the East Overton charter, a reference which is more convincingly to a contemporary field division. In the latter charter a reference to the boundary with West Overton being a *west heafdon*, or 'the west headland of the field' (SU 139665), also bespeaks tenth-century arable on Boreham Down. To argue for adjacent arable in West Overton on the evidence of what was happening on the other side of the hedge, however, is obviously risky. We met the same interpretative problem in the village (*see above*).

Similarly, Saxon West Overton's pasture is not so well illuminated by its own charter as is that of East Overton by its. A tentative guess would place West Overton's pasture in the tenth century across the whole of its area north of the Kennet, with further grazing north and south of the Alton to Lockeridge road on, respectively, *scfling dune* and 'Tenants (or 'Allyns') Downe' in 1567 (Straton 1909, 146), which in the late eighteenth century was tithe-free downland.

In 1544 the manor was granted, along with most of the Wilton estates, to Sir William Herbert, First Earl of Pembroke, and his wife Anne. West Overton, along with Overton Heath, then descended with the Pembroke title (*VCH* XI, 189). In 1567 it contained 168 acres (*c* 68ha) of arable and 7 acres (*c* 3ha) of meadow and supported a large flock. By the late medieval period, judging by the 1567 Survey, and possibly as early as the thirteenth century, part of the land north of the modern A4, which may have been pasture since the Saxon period, had been brought into cultivation. In the late 1700s the arable lay in the eastern half of that land unit, but in 1567 it had lain in the western half. The development of this new arable is likely to have begun lower down the slopes and have continued northwards, up the hill to the limits of the West Overton estate (Figure 4.2). The management of 'Allyns Downe' was also carefully organised so that it lay fallow during alternate years with restricted grazing rights for the Lord's farmer and tenants (Straton 1909, 146).

A change in the management of the common arable seems to have occurred between the late seventeenth and late eighteenth centuries. Whereas in 1567 many tenants farmed land 'in the part north of the stream', seemingly that western side of 'North Field' discussed above, by 1783 this land had been divided into larger furlongs called 'Upper', 'Middle' and 'Lower Field' and was farmed by the estate. The tenanted land then lay solely in the common fields south west of the village in 'Ditch Hedge', 'Double Hedge' and 'Windmill Hill' Fields (Figure 8.3).

At the time of Edward Pumphrey, around 1800, West Overton Farm consisted of Farm Down and the Fields, Pennings, droves and sheep houses, all north of the Bath road. The farm also held 'Home Ground', 'Home Mead' and 'Custard Mead' on both banks of the Kennet and a portion of 'West' or 'Little' Field, possibly the early name for 'Ditch Hedge Field'. The manorial pasture was reckoned at 177 acres (*c* 72ha) and called 'Cow' and 'Tenantry' Downs, 'Mill Ham' and 'Church Ditch' in

1794. The arable south of the river, which was farmed by the tenants in strips, was divided into 'Ditch Hedge', 'Double Hedge' and 'Windmill Field'. 'Hookland', 'White Barrow Furlong' and 'Hatch-gate Furlong' are among the smaller fields recorded. Two pits, presumably for marling, lay in this area, as did a barrow, all of which can be located accurately, and the site of the windmill, hence Windmill Hill, can be fairly accurately guessed (late eighteenth-century maps; Figure 8.3). The demesne farmer was allotted 131 acres (*c* 53ha) at Inclosure in 1802 (*VCH* XI, 196), at which time there were two downs within the manor: 'Cow Down', of 100 acres (*c* 100ha), simply called the 'Down' after 1802, and 'Allens Down', of 40 acres (*c* 16ha) south of Lockeridge Lane in Pumphrey Wood.

Over the centuries woodland clearance has reduced the acreage of Savernake Forest drastically, so that today it lies some 3km east of West Overton. In the eleventh century the manor of West Overton was reported as possessing 20 acres (*c* 8ha) of wood (*see above*), though this is unlikely to be an accurate record of the total wooded area at the time (cf Yorke 1995, 242–3). Up until the late fourteenth century, however, the Forest, or at least its laws, extended well into this study area. Its former extent is reflected in the extant names of Pumphrey and Pickrudge Woods, even though neither are of 'ancient woodland' as both have been replanted in the last fifty years.

The Anglo-Saxon charter suggests such management has been in train for at least a millennium: it refers to leas and a grove (Gelling 1993, 192–4, 198–207; Rackham 1996, 46); *ers lege*, now 'Hursley [?Horse lea] Bottom', or 'lez bottom de Hurseley' in 1567 (SU 148662); *lorta lea*, now Lurkeley Hill (SU 123663); and *mere grafe*, 'Pond Grove', now at the very western edge of Wools Grove (SU 143663). If the *lea* of *lorta lea* indeed indicates a woodland clearing, then, coupled with *langan sceagan*, the woodland of the late Saxon period may well have extended from Pumphrey Wood and Pickrudge on to the estate of Shaw and further west on to Lurkeley Hill (Figure 14.2; *SL*, figure 56). In the tenth century, Pumphrey and Pickrudge woods were called *scyt hangran* (S449), a name which suitably describes this wooded slope (Smith 1970, I, 233; Gelling 1993, 194–6), with the *scyte* element possibly referring to the wood being at the corner of the estate (*PNWilts*, 306). This was 'Checheangers Coppice' of 25 acres (*c* 10ha) in 1567 when the manor contained a total wooded area of 86 acres (*c* 35ha), all of which were considered dissafforested lands of Savernake (*VCH* XI, 198). In 1631 the central area of Pumphrey Wood was open, roughly wooded and coppiced for grazing purposes, within a more densely wooded area, still of 86 acres. The remnants of much of the woodland boundary bank between it and neighbouring plantations remain (*SL*, figure 14) while Mr Pumphrey of West Overton Farm left his mark on the landscape by renaming the wood after himself in the 1820s (*see above*). A derivation of the OE original had survived into the early nineteenth century with 'Upper' and 'Lower Chichangles' (1802 map; *SL*, figure 71) and 'Chick Changles' (OS map, first edn). In total, *c* 80ha at the southern edge of West Overton is today wooded, a similar area to that in the late nineteenth century.

In 1567 the tenants of North Newnton, Wilcot and West Overton had common rights on (West) Overton Heath. This also retained its earlier name of 'Abbess Wood', reflecting its previous owner and its previous use as land 'said to have been formerly part of Savernake Forest' (1783 map; Brentnall 1941). The link to Wilton Abbey survives in the name of a modern cottage, albeit now in the parish of Wilcot (SU 16966518).

The bounds of the heath, which are delineated in the Pembroke Survey, stretched for 1¼ miles (2km) and covered 64 acres (*c* 26ha), whereas the late eighteenth-century West Overton Heath covered an area almost twice as large. This increase in the heath's size came sometime between the late sixteenth and late eighteenth centuries during the Pembroke tenure, perhaps in compensation for the formation of Clatford Park to the north (*see below*). The insertion of the park, judging by the boundary features of the 1567 Survey, certainly reshaped the northern part of the heath, so it is unlikely the late eighteenth-century maps reflect the medieval unit to any great extent. However, they show the land-use to be predominantly arable by then. Today West Overton Heath contains a former Wesleyan Methodist chapel and one or two houses including the Dog House, site of the former Old Dog public house (*VCH* XI, 187; SU 16856493).

Chapter 9

The Valley and its Settlements: East Overton

South of the Roman road the tithing of East Overton stretches 4.5km from North Farm to Bayardo Farm. It is mostly 1km to 1.5km wide, narrowing to just over 100m in Hursley Bottom. Its successive zones of land-use to the south are similar to those in West Overton. They include at the extreme south-east end, beyond the woods at 210m aOD, 'Heath Grounds', marking a different habitat and the late enclosure of rough pasture (*see above*). The associated building was 'Heath Barn' in 1889, with 'Heath Cottages' some 400m to the west; its present name, 'Bayardo Farm', is after a Derby-winning racehorse (*PNWilts*, 307; SU 16006507).

Here we first look briefly at the landscape in terms of conventional documentary sources, then we turn to the village itself (*SL*, colour plate 1) before considering what we have selected as the most appropriate aspect of East Overton for detailed consideration, that is, the medieval manor and its workings in the landscape of the estate.

MEDIEVAL ESTATE AND VILLAGE: PRINCIPAL DOCUMENTARY SOURCES

In AD 939 King Æthelstan granted fifteen hides at *uferan tun* (East Overton) to Wulfswyth, a nun (S449). The bounds of this charter are, for the most part, identifiable on the ground with considerable precision. We have already followed them in part on the northern downs (*see above*, Chapter 4), along the wooded southern marches and in relationship to Wansdyke (*SL*, 106–110; Fowler forthcoming c) and through the village (Chapter 8), and we touch on them again in discussing the adjacent estates of Lockeridge and Fyfield (Chapters 10 and 11). Overall, coverage of the Saxon bounds and their inferential landscape in the study area is relatively full (Figure 14.2). The boundaries themselves remain influential in a distinctive factor of the East Overton landscape, that is, its manorial documentation and its scholarly study.

In 1086 the Bishop of Winchester held *Ovretone*, and it paid geld for fifteen hides. The land here was for the support of the monks at Winchester, so the land had been transferred from Wulfswyth, or her successors, to Winchester at some point in the preceding 150 years. There was land for seven ploughs, with eight and a half hides and two ploughs in demesne, thus leaving the villeins' five ploughs to farm the remaining six and a half hides. The demesne of East Overton was worth £8 in the time of Edward, but only £6 when the Bishop received it (*VCH* II, 120–1).

The estate contained 15 acres (*c* 6ha) of meadow. The pasture was 8 furlongs long by 4 broad and the woodland 5 furlongs long by 2 broad (ibid). If a furlong can be taken to be *c* 200m (Coleman and Wood 1988, 29), then the area of pasture would have been about 130 ha (320 acres), roughly three times that of the woodland and probably lying far to the north on Overton Down, which was pasture in the tenth century. The woodland, again a narrow, rectangular strip, perhaps covering an area of about 40 ha, reflects the size and shape of the woodland areas of 'Wools Grove', 'Wells' Copse' and 'Little Wood' as far as Wansdyke within the tithing of East Overton today (*see* Figure 12.4). So probably in the later eleventh century, as today, the woodland lay in a long, fairly narrow area at the very southern tip of the East Overton estate. Arable was most probably on the Fore Hill (north of the Bath Road) and south of the village on Bitham Barrow Hill.

VILLAGE MORPHOLOGY

The village now called West Overton is superficially a street village oriented west–east along the terrace bordering the south side of the flood plain of the River Kennet. As we have noted, it actually comprises two villages, historically West Overton and East Overton, so here we are concerned with the eastern part of the present village and the manor of which it was the centre (Figure 9.2, Plate XLVII).

About half-way along the street of the present village called West Overton, a geometrically awkward cross-roads gives off what were in the 1960s minor lanes to north ('Frog', or 'Watery', Lane) and south to South Farm (Figure 9.2, Plate XLVII). This was the boundary between West and East Overton in the tenth century as described in Chapter 8. The main road continues up towards the church on a slight, but locally prominent, eminence at its east end.

THE CHURCH AND SOME VILLAGE HOUSES

The eastern part of the village now called West Overton seems, even at the end of the twentieth century, to be the 'old' part of the village (*SL*, colour plate 1). The church is there, much older than its present neo-Gothic splendour (*SL*, figures 48, 50), and so too are some older-looking houses (Plate LIV; *SL*, colour plate 24); to their south is a grass field, Ring Close, full of the earthworks of abandonment (Figure 9.2).

The church of St Michael and All Angels visually dominates the village and, because of its impressive tower, is a prominent landmark both along the Kennet valley (*SL*, figures 48, 50) and looking south from Overton Down. Undoubtedly its site is ancient, though not the present building (Pevsner 1963, 504–5; Anon, *Church of St Michael*). Its axis is oriented well north of east and, significantly or otherwise, is aligned on a round barrow 100m to the north east (Powell *et al* 1996, fig 6). The tower is of 1883, an addition to an almost complete rebuild of 1878–9 to the design of C E Ponting (*WAM* 45, 615). This was occasioned by the dire state of the then standing church, apparently a fourteenth-century chancel with a fifteenth-century nave (Plate La). Though the Victorian nave followed the plan of that church, most of the building shown before rebuilding on Plate L was removed. Parts, however, remain: the chancel arch, for example, was moved to its present position in the side aisle next to the organ chamber, and the seventeenth-century bells, one dedicated to St Margaret, were retained (*WAM* 2, 342), but, so unstable is the tower, they do not now ring, their tintinnabulation replaced by a tape-recording. During the work, fragments of a still earlier church were found, and two early consecration crosses that were unearthed were built into the external east chancel wall where they can still be seen (Plate Lb). The present somewhat dramatic church is, therefore, the third known church on the site, taking stone construction certainly back to the thirteenth century, possibly the twelfth, and perhaps earlier still. If there had been such an earlier church it would be interesting to inspect its remains for Roman material.

More old buildings occur near the church than exist in the other villages, the two principal ones being along the one-sided 'Street' of 'old' East Overton. Coming up the slope from the 'Saxon cross-roads' eastwards towards the church (Plate XLVII), on the north side is a group of nineteenth-century buildings with older origins, followed by the sarsen-built West Overton House (SU 13276808), formerly the Old Vicarage, that almost certainly incorporates at least some of the 1496 Rectory (Greatrex 1978, 190; *VCH* XI, 200). In 1567 the vicar paid no rent to the Lord (Straton 1909, 261); twenty years later the *Vicaridge Howse* comprised 'one Dwelling House with an Orchard, a Garden, and Court or Barton, a Barne, a Stable and several Closes of Pasture or Mead', and in 1671 the *Vicridge house* had 'a Parlor Hall and Kitching, one Barne of fower Bayes and a Stable' (FWP 46; 1588 and 1671 Terriers). The Old Vicarage was rebuilt in the late eighteenth century and faced with rendered brickwork, its three bays and two storeys plus attics setting the standard for Mr Pumphrey's later West Farm House at the other end of the village.

In contrast, the range of low, thatched cottages lying next to the Old Vicarage, to the east, is acutely perched on a sharp corner as the road dives down to the north. In fact, this is a hollow-way still in use, it being as much as 3m deep below ground level where it curves around to the north east on to the floodplain. The cottages are of seventeenth to eighteenth-century date, and built of sarsen with brick dressings. Windows now replace their former end doors.

Across the top of the hollow-way and again up-slope is The Old Manor House (SU 13366812). Now a private residence, it appears to incorporate at least parts of the medieval manor house (*VCH* XI, 189). In the thirteenth century, the *curia* at Overton included a hall, chamber and kitchen, together with the agricultural buildings: the

Plate L *St Michael's Church, West Overton: (a) in 1877, as depicted on a sketch of 'the old Overton church' before restoration (above), photographed in the 1960s from the original then in the church but now disappeared; (b) consecration crosses built into the outside face of the Victorian eastern gable of the chancel (below)*

great gate, the great barn and tithe barn, the oxshed, stables, granary and a mill. In the late fifteenth century the manor had its own courtyard, which included stables and other farm buildings, some of which were tiled or slated (Greatrex 1978, 190). Indeed, sixty years later, a sketch of the manor house clearly shows a slated or tiled roof. By then, too, the manor house had undergone substantial changes: main east and west wings had been added, though the central part of the house then, and still today, remains visibly of the proportions of a medieval hall. The 1567 sketch, an 1877 sketch (Plate La) and external appraisal of the standing building (*SL*, colour plate 24) allow a metrical reconstruction of its Elizabethan form and appearance (Figure 9.1; location on Figure 9.2).

Despite the many later additions and alterations, the house retains sixteenth-century features. It is basically of sarsen build with a timber-framed and tile-hung upper floor and, south centre, a canopied porch which appears to be eighteenth century, though Pevsner (1963, 505) dated the door to the sixteenth century. Those late medieval wings are not evident on the early nineteenth-century maps, though a new, longer eastern one was built in the mid-nineteenth century. A little later part of the house became the village Reading Room as part of the welfare provision of the Meux estate. Its subsequent modifications, not least to give it its present rusticated Arts and Crafts appearance, may well have taken place just after World War I.

Two more modest secular buildings remain at this east end of the village. To the north, tucked into the slope downhill of the church, is Church Hill Cottage (SU 13366818), seventeenth-century or earlier, with colour-washed brick and thatch, remains of a probable cruck and a sarsen rear wall. Immediately east of the church are Nos 74/75 Church Hill, now two cottages but formerly one cottage and probably seventeenth-century at the latest, to which the Verger's Cottage was added in 1746 (date-stone). Both are thatched and seemingly crowd into the churchyard, adding their contribution to the antique effect at this east end of the village. On the other side, to their east, they seem to be roadside cottages, for the road they front was there in 1815.

This road led north across the river to the turn-pike and, on the west side of the junction, the George Inn (SU 13296844), first mentioned in 1736. It was one of five roadside buildings that stood here until *c* 1930 when, at the same time as those in Fyfield (Chapter 11), they were demolished during road widening. Opposite them was, and is, North Farm. This originated as a set of new barns, present in 1801 (*VCH* XI, 185). A house was later built fronting on to the turn-pike from above an impressive, sarsen revetment. This could have been built after a generation or so, judging by the architectural and cartographic evidence, making it one of the later of the seven modestly grandiose farmhouses in the study area (cf West Farm, West Overton; *SL*, colour plate 32). Its

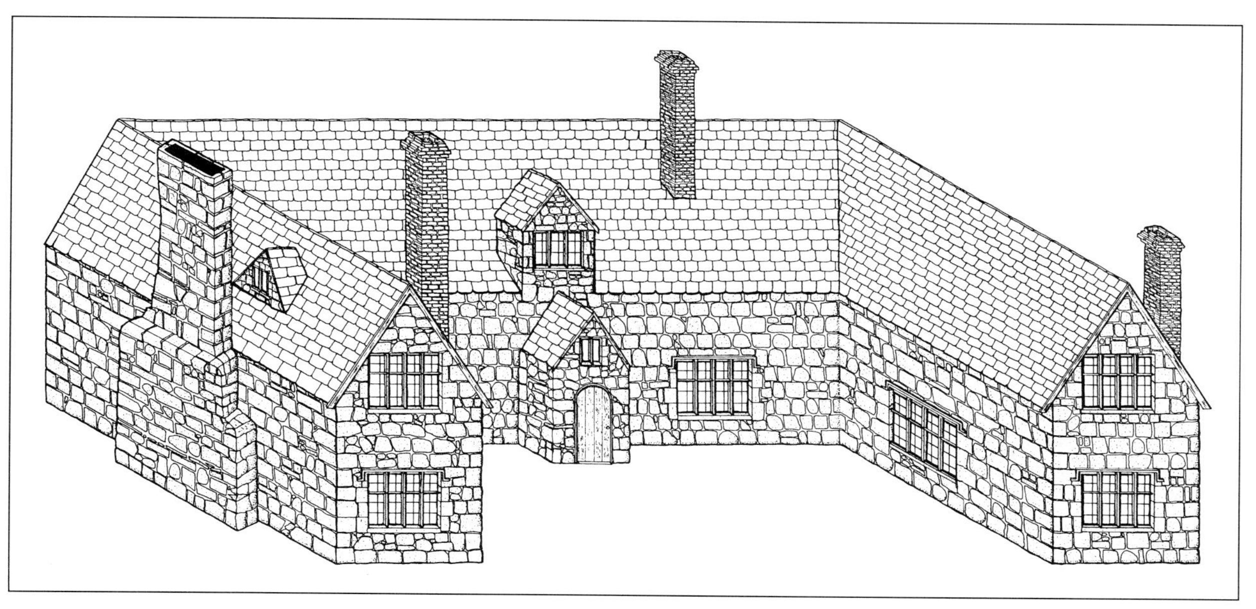

9.1 Axionometric reconstitution of East Overton manor house as it appeared in 1567, based on a contemporary illustration in the Pembroke Survey of that year adjusted to 3D-metrical approximation

significance is, however, sociological rather than architectural, for it well represents the availability of new money and the out-reaching of habitation beyond the confines of the medieval village as a consequence of Enclosure. Another generality suggested by the buildings of East Overton is that a significant rebuilding may also have occurred around 1600 when, perhaps, sarsen replaced timber-framing on some ordinary houses as part of a modernisation on a newly secular estate.

Lanes, lines and earthworks: a village archaeology

The sweep of the road eastwards through West Overton and south to South Farm masks the former existence of a cross-roads on the Anglo-Saxon boundary between the two Overtons. From the north along the boundary comes Frog Lane, still a footpath (right of way) and bridleway and part of the line of the tenth-century 'Overton Ridgeway' to and from the northern downs (Figure 16.7). Now it crosses the Kennet by a modern footbridge, beside at least one earlier bridge of which the stone abutments are visible in both banks of the river (Plate XLVIII) where, presumably, there had earlier been a ford. *En route* southwards to the village it passes The Wilderness or Withy Bed (Plate XLVII; *SL*, figure 48), a splendid stone 'clapper' bridge, a main water leat of the 1821 'floating' (*see* Chapter 8; Plate LXIIIb), and the earthworks of the briefly new but long-discarded early nineteenth-century road to the west.

The lane continues south from the cross-roads along what has now become the major road leading to a new housing estate on what used to be the space occupied by South Farm (Figure 9.2). The boundary itself diverts westwards before the farm, apparently to include it in East Overton. Nevertheless, this line, largely a lane, from river almost to South Farm, provides the main north–south spine of the village, roughly at right angles to its west–east axis. South Farm itself, destroyed in the 1960s, had a south range dating from the seventeenth century, with major extensions of *c* 1800. A splendid range of timber buildings enclosing a farmyard opposite it was already seriously dilapidated in 1960 and has since been destroyed. Its lands were sold in 1995, including the field, Ring Close (1819), across the lane on its east.

Ring Close

This grass field slopes gently uphill from South Farm to the church of St Michael (Figure 9.2). It is full of earthworks (Figure 9.2; *SL*, figure 50). These represent a totally deserted part of East Overton. The earthworks contain hollow-ways of former roads or tracks, the sites of former buildings and the banks around closes or gardens. Some deeper, larger hollows and other superficial features almost certainly represent disturbance after desertion, probably quarrying and robber pits for the sarsen stones of former walls and foundations.

A possibly more significant pattern appears to be present: a distinction between the earthworks to east and west respectively of the north–south hollow-way which, although it now turns sharply west with the main road into the village from the north east, continues southwards as a hollow-way into Ring Close (Figure 9.2). East of it is another hollow-way. In the area between those two and the church the earthworks are slight and mainly consist of a cluster of ten or so sites apparently of former buildings, bounded on the east by a low bank. To the south, across the east–west hollow-way, is a 'blank' area. The inference is that the east–west hollow-way formed the southern boundary of this part of the settlement.

In contrast, west of the main north–south hollow-way, the earthworks are more upstanding, more rectilinear and apparently enclose somewhat larger spaces. Their almost planned appearance is emphasised by the presence on their south side of 25m of a non-symmetrical hollow-way, almost certainly a relic from an earlier phase. The earthworks themselves, including perhaps six potential sites of buildings, relate to another east–west hollow-way, debouching into the area of the barns of the former South Farm where it would have intercepted the southern continuation of Frog Lane. They also relate to a parallel bank 60m to the north. That bank, and its parallel hollow-way, form the axes of the western part of this earthwork complex. Neither continues east into the area south of the church; but both are parallel to the eastern end of the present village street that now leads up to it (Figure 9.2).

This complex of well-preserved settlement earthworks presents a particularly good Wiltshire example of a shrunken or shifted habitation site, existing not in isolation out on the high ground, as at Shaw (Chapter 13), or alone along a river valley, but actually in a village. Whether they represent a whole medieval village which then shuffled sideways and northwards to its present alignment along the 'Street', or whether they represent just part of a village which became partly deserted, that is, which shrank as distinct from moving sideways, is uncertain. The earthworks may also be the remains of a late addition to an existing village, an expansion perhaps in the twelfth/thirteenth centuries when population growth may have triggered an intake beyond the village's 'Anglo-Saxon' shape and size. So far, no

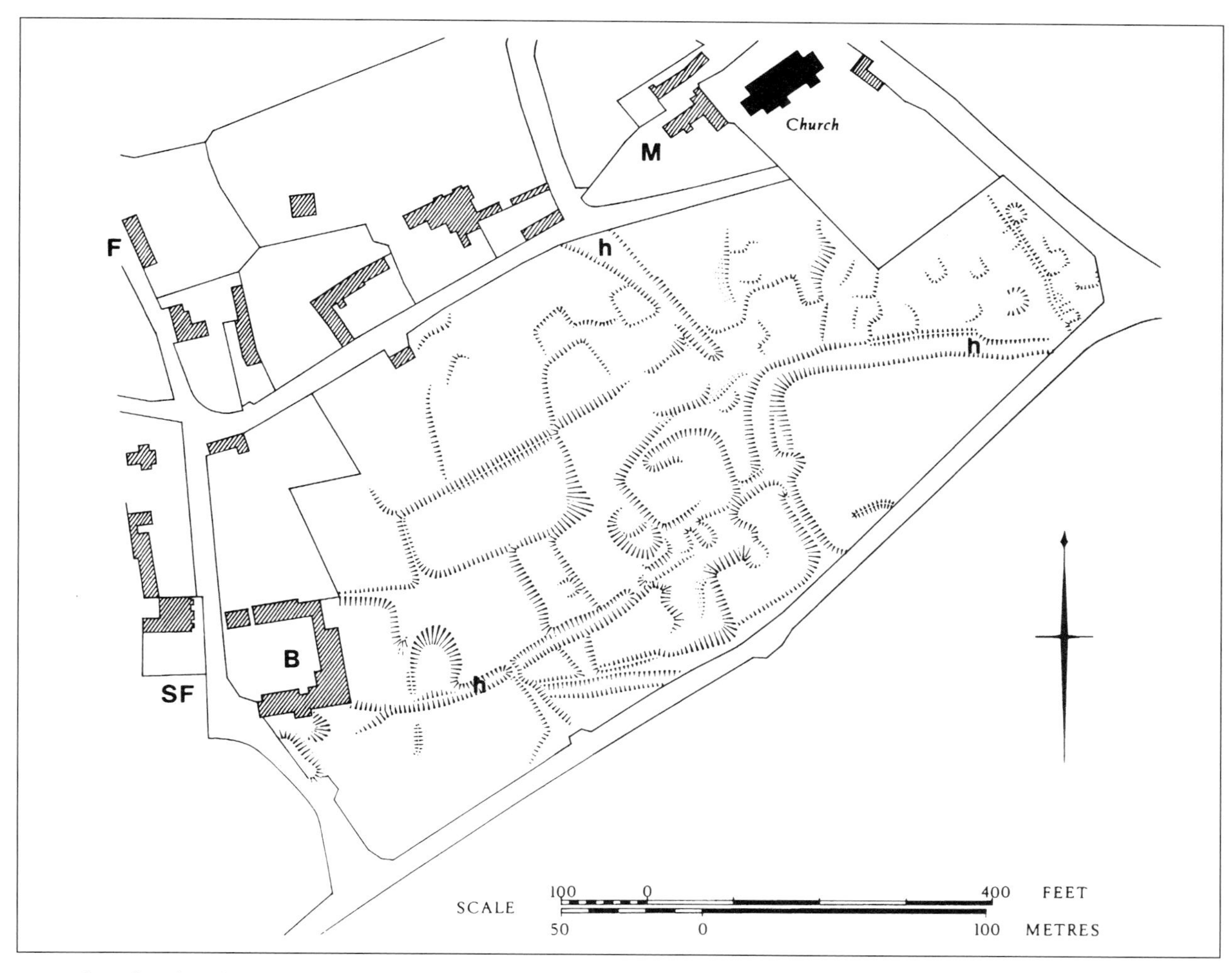

9.2 *Plan of earthworks in Ring Close, opposite East Overton manor house (M) and St Michael's Church in the village now called West Overton. Historically, the earthworks represent a deserted part of the former East Overton village*

SF = South Farm; B = barns (now destroyed); h = hollow-way; F = Frog Lane

documentary evidence has come to light bearing on this desertion, whether it be the result of shuffle, shrinkage or expansion.

A more complex model potentially takes the village story back to its beginnings as a village (in the absence of any pre-Anglo-Saxon evidence from the site; cf Fyfield, Chapter 11). We suggest that the hollow-way coming off the floodplain round the north-west side of the 'ancient' church site continued southwards as the north–south hollow-way dividing the earthwork remains as distinguished above. Presumably it then climbed southwards to the local resources of arable land, pasture and woodland beyond, as the 1793 map shows, while its east branch south of the church led south east to Lockeridge. These north–south and west–east tracks, we suggest, bounded at least the south part of an early settlement, perhaps a nucleated village, centred on the church. Potentially this little knoll and its slight slopes above the floodplain is the site of the Anglo-Saxon settlement of *Uferan tune*, 'Bank Farm' (*PNWilts*, 305, but for an alternative interpretation, *see* Chapter 16).

The rest of the earthworks, west of that north–south hollow-way, can be interpreted in conjunction with the eastern part of the present village street as forming a characteristic rectilinear planned village added to an existing nucleated one (as well demonstrated in south Somerset: Ellison 1973; Lewis 1994, 188–9, a recent discussion of medieval planned settlements in Wiltshire, does not assemble similar evidence). Planned or otherwise, the village seems to have expanded in a reasonably logical way to the west of its nucleus as far as – but no further than – South Farm and the western boundary of the manor of *Overton Abbatisse* (1291; *PNWilts*, 305). The boundary – and perhaps with it the village and also the farm? – was still basically where it had been in the tenth century.

A possible planned village could therefore have existed some 200m² in size, that is 200m from the 'old' north–south hollow-way on the east to the probably equally 'old' north–south lane along the east side of (or through?) South Farm to the west; and about 200m from the back of the properties along the north side of the northern street to a similar position at the back of the properties on the south side of the southern street (the present road along the south side of the earthworks). This rectangular plan is divided exactly in half, at 100m from both property boundaries as envisaged above, by the west–east bank running through the centre of the earthworks (Figure 9.2).

At some later date the southern half, and eventually the southern strip of the northern half, of this rectangular village was deserted or possibly even cleared. Six buildings appear to be depicted in these fields on the late eighteenth-century manorial map; their positions reinforce the 'early' pattern of the north–south hollow-way and the settlement core to its east. Only one building is shown in the area of the 'planned' earthworks and that had disappeared some twenty-five years later. Then, only one building, perhaps a shed or barn set up in the hollow-way, is shown in the two fields on the Enclosure Award Map (1815). Only some hachures mark its site on nineteenth-century OS maps. The houses on the north side of the approach to the church have, then, enjoyed an effectively clear view southwards across a green and pleasant field for at least 200 years.

Whether we are looking at the earthworks of abandonment, slow desertion, 'village shuffle' or manorial clearance, a fairly safe inference is that the change from habitation to grass took place long before 1800. Specific times could include the late sixteenth/early seventeenth centuries when we can see other indicators of change on a post-Dissolution estate, and the fourteenth–fifteenth centuries when two local downland settlements, *Raddun* (Chapter 7) and Shaw (Chapter 13), were deserted. A glimpse of contemporary village poverty, real or more probably pretended, may be afforded in the *Nona Inquisitions* of 1342. Then an attempt by the King to extract help from each parish to finance the war with France found a none-too-enthusiastic group of East Overtonians claiming they could give nothing 'because they all live by agriculture'. More generally, following Hare (1994, 167–8) and trends of demographic and economic decline balanced by local stability (Lewis 1994, 177–83), recent discussion points, albeit for different reasons, to the same period, but the best evidence, short of excavation, is most likely to come from an as yet unnoted documentary source.

Agriculture and land-use on the manor

The following passage is entirely based on the research of, and largely on the text of, Dr J N Hare. His original text, with the tables referenced here, is available in the archive (FWP 43). This is a shortened, and edited, version of it.

AD *1248–1400*

Overton manor initially comprised two separate units, (East) Overton and Fyfield, both visible in *Domesday Book*. They were physically divided by a sub-manor, which later became the manor(s) of Lockeridge (Chapter 10). For about 200 years after *Domesday* the manors of East Overton and Fyfield remained economically independent, being valued in 1210 at £16 and £8 respectively, with both possessing their own manorial *curia*. The merger of the two manors had begun to take place early in the thirteenth century, when the prior of St Swithun's withdrew his manors of East Overton and Fyfield from the Selkley Hundred and included them in his own hundred of Elstub; by 1248, the two had been combined and were called Overton. By 1309 the process was certainly complete and the manors were interdependent. Overton, though, remained the larger; by 1280 the *curia* at Fyfield had become a vacant croft, leased out to rent.

The combined manor of Overton formed part of one of the richest estates in medieval Wessex. It is illumined in particular by a group of twelve manorial accounts from 1248 to 1318 (WCL). Despite their limitations, these documents nevertheless allow us a picture of the manor of Overton, much of which would be familiar elsewhere in the chalklands, amidst rural economies dominated by grain production and sheep farming (Scott 1959; Hare 1981a, 1981b, 1994). On the other hand, Overton manor also shows its own peculiarities particularly in the exceptionally large scale of its pastoral farming. In this respect, Overton was one of the pre-eminent manors within a major pastoral estate.

Most of the demesne agriculture was carried out in the open fields which lay around the settlements and which was shared with the peasantry. Here, as was usual in the chalklands, the fields were divided into two with half the land remaining fallow each year (Hare 1976, 1981a, 1981b; Harrison 1995). Such apparently wasteful use of fallow made sense given the importance of large-scale sheep farming in the chalklands and the use that could be made of the fallow by the large sheep flocks, both for feed and as a recipient of dung. It was extensive rather than intensive land-use, but it also provided an agriculture appropriate to the soils and agricultural needs.

In its cropping regimes, Overton was typical of the chalklands around. All the crops (wheat, barley, bere, drage, oats and a small amount of vetch) were sown in alternate years in a single field, although this was carried out with flexibility.

Other fields also appeared. In 1267 the entire demesne sowing had been described as in the *northfield*. From 1311 a *southfield* appears, producing a maximum sowing of 39 acres (*c* 16ha) of demesne, and a small *northfield* (at *scrufeleput*) also appears in 1312. *Southfield*, *eastfield* and *westfield* all included land at *Schflyn(g)don* (1312, 1316), presumably on Bitham Barrow Hill (the *scyfling dune* of the AD 939 charter; *see above*), and perhaps representing further expansion of arable to the south of the river.

At Overton the demesne, whilst being a large-scale producer of wheat and barley, placed less emphasis on these than did the rest of the parish. In 1311, for example, the rest of the parish produced about two and a half times the grain of the demesne. By contrast, oats production was relatively insignificant amongst the tenantry; in the extreme year of 1307 all the oats produced came from the demesne, perhaps reflecting the cultivation of large areas of downland by the lord.

The thirteenth century witnessed demographic growth here as elsewhere (Scott 1959, Hare 1994), though the accounts give us little indication of how this pressure of population could be coped with. Essentially, there seem to have been three possible areas for further expansion: on *Shuflyndon* and White Hill, around and in the woodlands to the south and on the thin chalkland soils to the north. The widespread production of oats on the demesne may have reflected its expansion on poor-quality chalk downland, perhaps reflected in its turn in the blocks of downland ridge-and-furrow (Figure 2.3).

Overton was one of the most important sheep-farming manors of the cathedral priory estate. In a tax assessment of 1210, Fyfield was stated as having 100 and Overton 300 sheep, the two townships still being treated as distinct manors (*VCH* XI, 194). This places the two flocks among the smaller ones in Wiltshire on this estate. A transformation in the scale of sheep farming soon occurred, however, which far exceeded growth on any of the other priory properties. By 1248, the manor had 2,256 sheep, a five-fold increase from 1210, but a figure that it may never have reached again. As such, Overton, along with Enford, were to remain the most important sheep-farming manors on the estate, exploiting the extensive thin downlands for this lucrative cash crop.

The 1248 account shows the demesne flock at its greatest, whereas the accounts from 1267 show decline that continued into the 1280s when it reached the lowest levels in our accounting period, perhaps reflecting the rising rents and the increase in the acreage of downland under plough. By the end of the thirteenth century through to 1318, there was a new stability at a higher level of about 1,650 to 1,700 sheep. By 1318 it had reached 1,882. The wool was sent to a central wool store, initially at Barton Priors outside Winchester, where all the estate wool would be sold in one large contract. The sheep also produced meat, some of the stock for slaughter being sent to the cathedral priory, with others being fattened and sold in the area (and yet others being eaten at *Raddun*; *see* Chapter 7). Cheese was made from sheep's milk and the sheep themselves not only produced manure but spread it too.

During the winter months sheep were normally kept in enclosed shelters from which they could be let out as appropriate (Ochinsky 1971, 337–9; Hare 1994, 161–2; Dyer 1986). At Overton, the sheep were kept in three different sheep-houses from 1248. Subsequently these were described at Attley (Audley's Cottage/Hillside Farm; *see below*), at *Raddun* (Wroughton Mead, Chapter 7) and at Hackpen (Overton Down, possibly at Down Barn, Chapter 6).

The sheep-house consisted of an enclosed yard, the sheep-cote itself and appropriate accommodation or shelter for the shepherd as at *Raddun*, Phase 1 (Chapter 7). Inside the compound were thatched buildings, whose roofs were regularly being repaired, with the sheep-house itself a substantial timber-framed structure. At Attley apparently major works in 1282 involved three crucks being used to repair or extend one building, followed by repairs to the ditching around the sheep-house and hedge-planting. The sheep-house at Hackpen had three doors and was completely rebuilt in 1318, at a cost of £3 4s 3½d, with a mason being paid 3s 8d to construct the stone foundations. The timber frame of eight couples of forks or crucks, which cost 26s 8d, suggests a seven-bay structure, perhaps between 21m to 34m in length, a similar length to the late fifteenth-century grange excavated within Wroughton Mead (Chapter 7, site 10; FWP 65). At the same date, rafters and laths were purchased and 36 acres (*c* 14.5ha) of straw were reaped to provide the thatch, an investment that demonstrates the importance of such buildings in the profitability of the estate. Such buildings would have been subdivided with internal partitions, as rebuilt at Hackpen and *Raddun* in 1282, and used to store hay and winter feed. Various features in this description can be recognised in the archaeological evidence at *Raddun* (Chapter 7) to the extent that it seems legitimate to ponder whether the rebuilding of Building 2 is not actually dated to 1282 (*see above*).

Though sheep were the most common form of demesne livestock, they were only one of the many types of animal kept (FWP 43, table v). Cows and pigs were of course essential but so were the oxen, and to a lesser extent horses, needed to work the estate. As with sheep numbers, the first account for 1248 shows some contrasts with later ones, although a single account should be used with caution. It shows a herd of cows over twice the size of any known later example here and a herd of swine half the size of any later average. The twenty-two cows provided a large herd by comparison with other farms in the area during the Middle Ages. Oxen showed a fairly constant number of about thirty though, as with numbers of other livestock, it fell in the 1260s and 1280s. Pigs increased in number in the late thirteenth and early fourteenth centuries, with large herds of over fifty comparable in size to those at the priory manor of Enford (WCL 1311, 1316). The presence of pigs in such numbers would have had a marked effect on the landscape if they were daily herded on the manorial feeding grounds, presumably mainly in and around the edges of the southern woodlands; and that same presence reinforces our view of those woodlands as a resource important for the local economy.

After 1318 the account rolls cease to survive and the evidence is fragmentary. The Black Death of 1348–9 probably killed over 40 per cent of the population nationally and locally, and in Wiltshire, as elsewhere, the plague and other factors resulted in a low population until about the end of the fifteenth century (*VCH* IV, 38–42, 295; Hare 1994, 163–4). In the late fourteenth and fifteenth centuries, individual tenants were accumulating holdings, opening up gaps in the fabric of existing villages and occasionally, as at Shaw, producing village desertion itself (Crowley 1988, 10–11; Hare 1994, 165–8). What evidence we have for Overton is largely about the development of demesne farming rather than about the peasantry.

Some indications as to the scale of the demesne agriculture can be found from the stockbook of St Swithun's Priory from 1389 to 1392 (WCL Stockbook). It provides us with a snapshot of the pastoral farming of a great estate at the peak of its activity, reflecting the continued demand for wool and food both from the priory itself and from the increased prosperity of much of the chalklands based upon the growth of the cloth industry in south and west Wiltshire (Hare 1976). Arable production seems to have continued on a similar scale as before since there was no significant fall in the number of horses and oxen compared with the earlier figures. The cattle herds also remained comparable to those of the early fourteenth century, but both pigs and sheep showed signs of substantial growth of demesne production. Thus by 1390, the number of pigs had more than doubled from an average of 48 to 133, and Overton then possessed the second largest herd on the cathedral estates.

While sheep flocks did not reach their exceptional levels of 1248, they nevertheless showed a substantial growth (12 per cent) from the high levels of the early fourteenth century. The flock grew to 1,882 in 1318 and sheep farming continued on a very large scale. In 1334/5, wool sales were responsible for 83 per cent of the receipts of Overton manor, while in 1390 Overton possessed the largest sheep flock on the priory's estates with the exception of Barton Priors (which also exceeded it in pigs). Such large-scale sheep farming in the middle and late fourteenth century would have continued to ensure an important role for *Raddun*.

AD 1400 to the early twentieth century

The stockbook suggests that at the end of the fourteenth century demesne agriculture continued with little change. But at some point the lord leased out the demesne of Overton, a likely date being in the second quarter of the fifteenth century. Here sheep flocks continued to be run and financed by the lord (Hare 1985, 85–6), but the process of retreat from direct farming occurred at some time after 1453. By 1496, it was complete (Greatrex 1978, 189–90). A lease of that date illustrates two aspects of the landscape: the combination of arable and pastoral economy typical of the sheep/corn husbandry of this area, and the continuation of large-scale demesne arable farming. A sharp fall in oats production may reflect a longer-term trend, a decline of population and shrinkage of arable. The lease also included 819 sheep and ten cartloads of hay in the lord's sheep-house in the grange at *Raddun* (Building 10), showing that the site of the thirteenth-century farmstead remained an important centre for sheep production. The terms of the lease also suggest that some of the manorial buildings had followed a trend found elsewhere in the area and shifted from thatch roofs to slate, or, more likely here, tiles (Hare 1991). Again, the *Raddun* evidence can be interpreted as reflecting that exactly (Chapter 7; FWP 65, site 10).

The next lease in 1512 was to Thomas Goddard (WCL Register D&C. f). The Goddard family were wealthy farmers who could have continued the large-scale farming of Overton, despite the retreat of the cathedral priory, the old ecclesiastical landlords, from direct involvement in agriculture (Hare 1981b, 9–13). In 1541, Winchester chapter received a royal grant of the manor of East Overton, which it reconveyed to the Crown in 1547. Subsequently, the Crown granted it to Sir William Herbert, First Earl of Pembroke (*VCH* XI, 188), with the manor itself coming to Richard Kingsmill by the mid-sixteenth century (Straton 1909, 257–60). The 1567 Survey illustrated that the medieval large-scale, capitalist estate economy established by the priors of St Swithun in the thirteenth century had emerged seemingly unaffected by the Dissolution. It also named twenty-one tenants (ibid, 251–62), indicating a normal adult population of about sixty. This is approximately of the same order of numbers as that discussed below for Fyfield and Lockeridge (Chapters 10 and 11), which, in this instance, could be interpreted as also indicating a long-term demographic recovery by the local community from the mid-fourteenth-century crisis.

About 1800 the manorial estate was divided into North and South Farms on either side of the Bath road. At Inclosure in 1821, the lord of East Overton was allotted some 800 acres (*c* 324ha) in the East Overton tithing (*VCH* XI, 195). Fifty years later, much of the land of the historic manor was acquired by the Meux estate, which also included Lockeridge, Fyfield, Glebe and Clatford Park Farms and West Woods. Into the early twentieth century both Overton and Fyfield Downs were exploited, as part of the Meux estates in north Wiltshire, for sporting purposes (ibid, 188, 195).

Chapter 10

The Valley and its Settlements: Lockeridge

INTRODUCTION

Lockeridge is a thin sliver of land squeezed between East Overton and Fyfield tithings (Figure 1.2). Its northern reaches included Pickledean and downland on its north, stretching as far as the 'lost' Lockeridge Down lately subsumed as part of the northern area of Overton Down (Figure 5.1). Lockeridge tithing is divided by, and simultaneously closely linked to, the River Kennet (Figure 8.1) The river runs west–east across its middle; the lands on its banks feature largely in the manorial history, though the former tithing was never concomitant with a single manor like its neighbours. The land that constituted that tithing is now in the civil parish of West Overton, reflecting an ancient arrangement in which much of it was closely linked to historical East Overton (Chapter 9). It shared a common western boundary with former East Overton and, of course, a common eastern boundary with Fyfield which has enjoyed – suffered from in some respects – an even closer historical relationship with the episcopal manor of East Overton (*see below*). These boundaries are essentially of late Saxon date at latest, and their detailed histories are complex, especially in the twelfth and thirteenth, sixteenth and nineteenth centuries.

The range of resources from north to south, from valley side to high woodland, is similar in the southern parts of the very narrow tithing (Figures 10.1 and 14.2) to those already noted for their western neighbours. Its sliver of woodland has to be considered archaeologically, however, as part of the relatively large amount of southern woodland which contains plentiful evidence of prehistoric and later activity (Figure 12.1; Chapters 12 and 13; Plate LII), though there are considerable complexities, tenurial and chronological as well as spatial, in recognising which manor/tithing/parish a place is in at a particular moment (cf *SL*, figure 55). In 1543, for example, Lockeridge manor contained 29 acres (*c* 12ha) of woodland (*VCH* IX, 197), now probably represented by Lockeridge Copse (33 acres, *c* 13ha, in 1900). In the late eighteenth century other woodland in Lockeridge included 'Rising Coppice' west of Fosbury Cottage, 'Fosbury Coppice' and 'Henley Wood'. All then, as today, were experiencing active woodland management. The outstanding feature of Lockeridge's history, however, still reflected in its landscape and notably by the site of the main village itself, was its association with the Knights Templar in the twelfth and thirteenth centuries. Rockley, a former tithing now in Preshute civil parish, is deeply and quite unavoidably involved with Lockeridge during this 'Templar phase' (*see below*) and, as a result, in later times too (Figure 10.3).

The landscape of Lockeridge is well documented. Much of the documentation is to do with property in one way or another, in particular a series of charters and surveys occasioned by a constantly shuffling proprietorial interest, much concerned with customs and services as well as delineation. A discernible strand of continuity runs through the changes and incremental complexity. The presence, individually, institutionally and documentarily, of the Knights, first the Templars, then the Hospitallers, enables us to observe in landscape terms the creation, organisation and fragmentation of a particular estate. Its twelfth- and thirteenth-century Knightly charters are at the core of a discussion of Lockeridge. They allow a different perspective, the latter in comparison with the contemporary Winchester documents (Chapters 9 and 10), the earlier ones from their somewhat unusual place in time between the conventional sources of the local historian in the later eleventh and thirteenth to fourteenth centuries. Such evidence was not available to such an extent for our studies of the Overtons (Chapters 8 and 9) or Shaw (Chapter 13), yet the landscape of Lockeridge and Fyfield does not appear distinctly different. Nevertheless, the impact of the Templars was considerable locally and, as we shall argue, is influential still in the late twentieth-century landscape. Present-day Lockeridge village, for example, is only 1km distant from its nearest neighbour, Fyfield, but for the Templars in the twelfth century, that distance could have been even less (Figures 8.1 and 11.1) if our location of the *Domesday* Lockeridge is correct (*see* Figure 10.1).

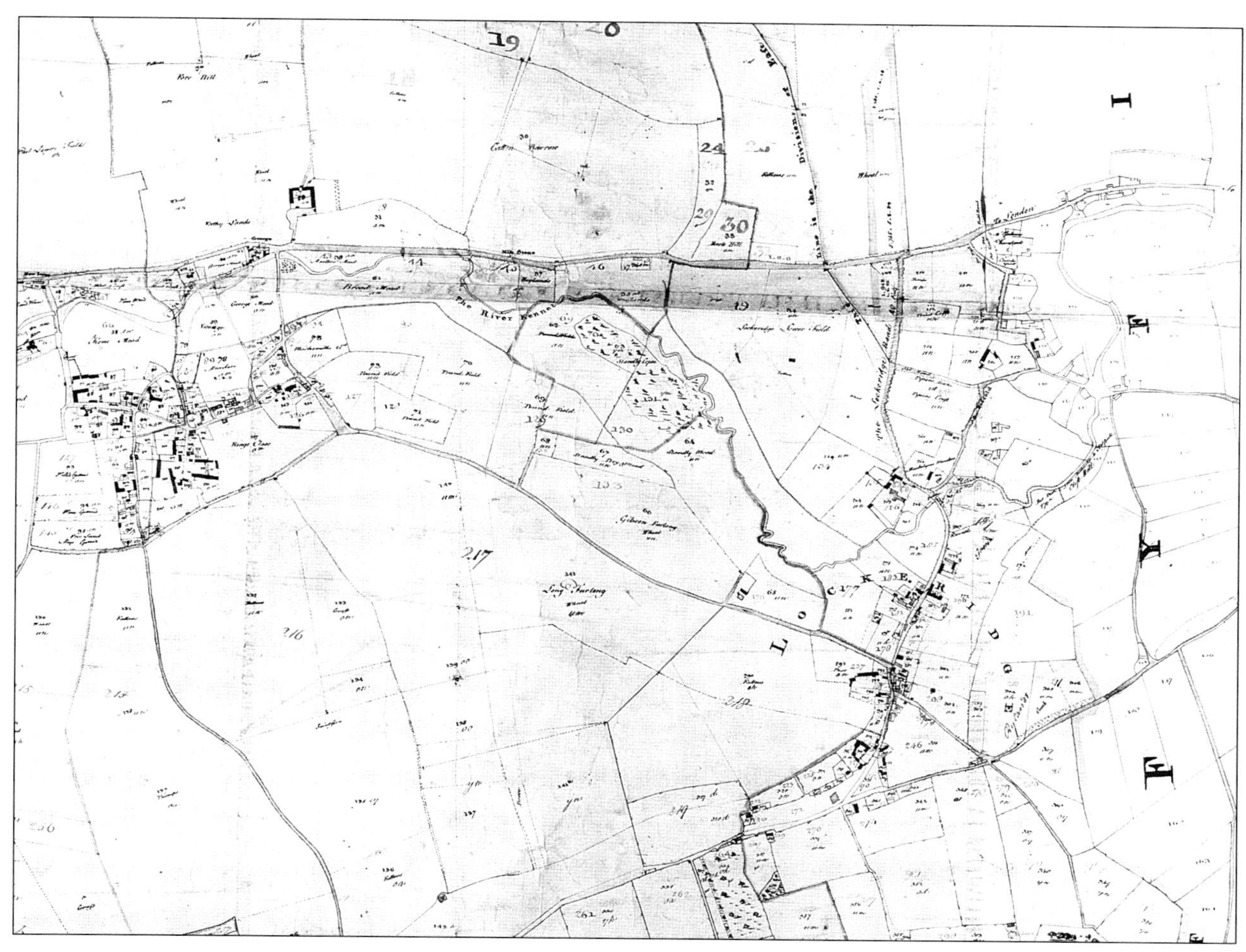

Plate LI *The valley landscape in the early nineteenth century as depicted, in colour on the original, on Abraham Dymock's map of 1819 (WRO/778/2) between the villages of West Overton* (on the left) *and a fragmented Fyfield 2km to the east. The River Kennet is shown with the subsequently 'lost' Overton mill* (extreme left)*; the major leat east from Overton is already in place, even though the date is just before the river's Parliamentary 'floating'. West Overton manor farm exists but West Overton Farm does not; North Farm exists as three sides of a courtyard of farm buildings but apparently with no farmhouse; only one building is shown in Ring Close. Three elements making up present-day Lockeridge village are clearly delineated; the fourth, the Meux estate village, was yet to come. Post-Enclosure, the pattern of fields and communications is very much as it is now; many of the field names are still in use, but the crops are of 1819 (cf Figure 8.1)*

PRE-MEDIEVAL LOCKERIDGE

A long barrow lies on the spur of downland forming the north side of Lockeridge Dene and overlooking Lockeridge village from its south west (*see SL*, colour plate 25; Featherstone *et al* 1995). It lay in 'White Barrow Field' (1815), close to the boundary between the estates of East Overton and Lockeridge (Figure 10.2). To its north east were round barrows and enclosures, all showing as cropmarks in very dry conditions. The remains of the mound are, however, still visible as a low rise on the ground, despite some two centuries of arable implied by the field name. The long barrow should be called 'White Barrow'.

The siting of the long barrow is locally explicable in topographical terms by reference to its position on high, but not the highest, ground above a dene, similar to Manton long barrow (Chapter 5). More generally, if theoretically, it accords with the proposal that the long barrows of this south-west corner of the Marlborough Downs were knowingly sited in relation to each other and to possible 'territories' (Chapter 16). It may not be just coincidence that the estate boundary, documented in medieval times, passes close by it. A long barrow claimed as existing on the high ground east of the village can almost certainly be discounted (SU 14936779, G24; cf Figure 16.3; Barker 1985, 23).

The village itself was the site of an Early Bronze Age burial. In a deposit comparable in some respects to the Beaker burial on Overton Down site XI (Chapter 6), a man of about fifty years of age was buried in a flat grave with a 'Bi beaker and flint dagger'. 'Parts of a second skeleton were also found' (*VCH* I.i, 120). Another beaker also comes from Lockeridge, but is unlocated. Whether or not that too was from the valley, the burial behind the school suggests that people were using not only the downs but also the low valley terrace where the village subsequently grew, 3,000 years before Lockeridge is likely to have been founded (*see below*).

A group of round barrows lies immediately south west of the village on the spur of high ground, one of a number of such groups and singletons on the high spots along the south side of the river valley (cf the round barrow in a similar position east of West Overton church; Chapter 9, Figure 12.1). Their presence seems to have affected the course of the Lockeridge/East Overton boundary, a relationship observed elsewhere in the study area (eg, Figure 4.2). Here, its line is marked by the thick hedge running towards the barrows from the south west and then going round their east side (*SL*, colour plate 25). Three large round barrows seem to have stood here, a typical sized group for this area. The darker, smaller circle may well be the site of the windmill documented in 1564 (*VCH* IX, 198).

The rectangular enclosure marked by thin ditches in the centre of the photograph (*SL*, colour plate 25) is currently unusual in typological terms for this area, though it is in general familiar on gravel subsoils (Whimster 1989). It is much larger than other enclosures examined in the study area and indeed than the characteristic Bronze Age types of the Marlborough Downs (cf Gingell 1992). It might be a ploughed-out medieval or post-medieval stock enclosure of the sort discussed on the Wiltshire Downs elsewhere (Fowler *et al* 1965) and exemplified here by that at SU 116693 on the west side of Figure 2.1; but, overall, it seems more likely to be a settlement enclosure, with an approach track from the south, of late prehistoric or Roman date. Indeed, were it not for its position and the absence of the familiar outline of a 'cropmark villa', it could well be interpreted as a villa enclosure, similar to that interpreted as such at 'Headlands' (Chapter 4, Plate XV). The authors of *PNWilts* observed over fifty years ago that one possible derivation of the place-name 'Lockeridge' was from OE *loc(a)-hrycg*, 'ridge marked by enclosure(s)' (*PNWilts*, 306).

The settlements

The origins of the settlements at Lockeridge are unknown; it is even uncertain how many existed. It has always been assumed that there was but one village where present-day Lockeridge stands, and we begin with that. Here, however, we argue that there were at least five main settlement sites, one of them coincident spatially with the present village, one to the north and two to the south (Figure 10.1, respectively Dene, Lockeridge House, 'Templar Lockeridge', Upper Lockeridge and Victorian Lockeridge). Though, of course, the first four settlements may have been successive 'Lockeridges' rather than different, contemporary settlements, that there were forty-one poll-tax payers in 1377 suggests the contemporaneous existence of several settlements, or at least one larger than West Overton (*VCH* IV, 316).

Lockeridge village

The estate of *Locherige* is assumed to have contained a main settlement, assumed to have been at the site of the present village; but it could as easily have been either around Lockeridge House, close by the river as internal evidence in the *Domesday Book* entry requires, or less probably, at Dene (Figure 8.1; *see below*).

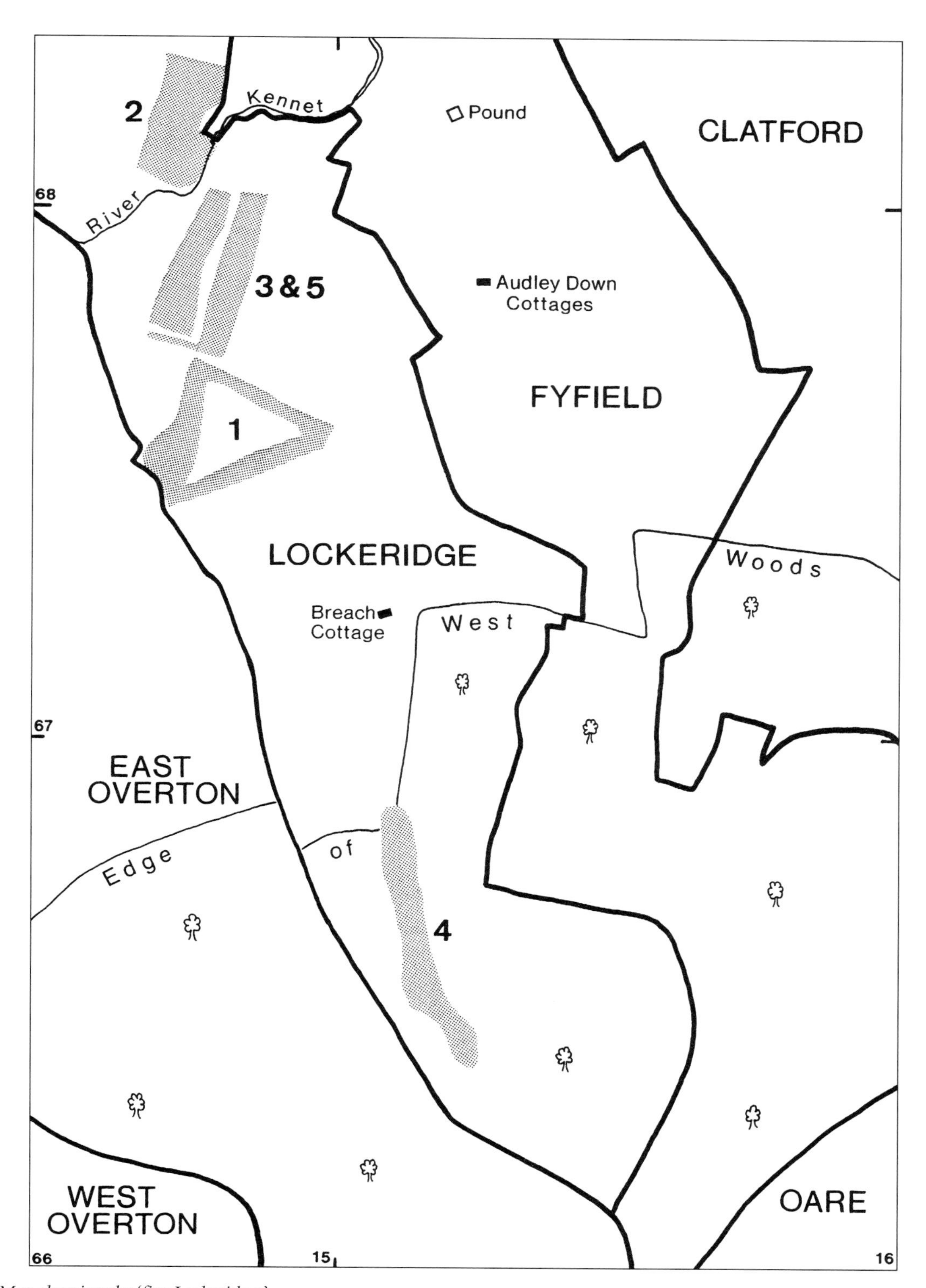

10.1 Map showing the 'five Lockeridges'

1 = Dene; 2 = 'Locheriga' (Lockeridge House); 3 and 5 = Lockeridge: Templar and Victorian; 4 = Upper Lockeridge, including Spye Park and Fosbury Cottages

Plate LII *Lockeridge village, Dean, Boreham Down and the northern edge of West Woods on 4 November 1946. Lockeridge House is* extreme top, centre. *Hursley is the clearing* centre bottom, *with convergent tracks; the tenth-century estate boundary between West and East Overton runs westward from it, along the curving hedge-line, on the north of which are the already plough-damaged remains of 'Celtic' fields. On the right-hand side of the plate, top, is Attely, a medieval sheepcote;* centre is *onomatopoetic Breach Cottage, and* lower, *the isolated field and house indicate the location of thirteenth-century Fosbury (CPE/UK 1821,* © Crown copyright/MoD*)*

Plate LIII *Lockeridge: cropmark of 'back street' or boundary ditch along the existing property boundaries on the east side of what might have been a planned village in the twelfth century, photographed in 1995 (NMR 15367/49,* © Crown copyright. NMR*)*

The present village is not picturesque and presents an image of a somewhat eclectic sprawl of domestic buildings (Plates LV–LVI), but it is of considerable historic and structural complexity (Figure 10.1, 3 and 5). Our suggestion is that, unlike the other valley villages (all pre-Norman), it is relatively recent, being a twelfth-century foundation. It is arranged mainly on a rectilinear plan based on a north–south street stretching from the edge of the floodplain, ending at an 'old' centre around a dog-leg cross-roads (SU 147677) and a now rather ragged southern end in Lockeridge Dene (*see* Figure 10.2 *in pocket*). The rectilinear village is now

Plate LIV

West Overton: a 'traditional cottage' constructed from sarsen stone, with a thatched roof, the front facing on to the village street and the gable end on to a medieval lane. Formerly (?eighteenth century) a row of cottages modified in the nineteenth century with brick quoins, jambs, lintels and chimneys, plus new fenestration, including upper 'eye-brow' windows

Plate LV

Lockeridge: a traditional cottage with thatched roof, timber-framing and sarsen stone walls, gable end on to the village street and fronting on to a side lane which may be medieval in origin

Plate LVI

Lockeridge: Victorian estate cottages, tiled, of brick with 'Tudorbethan' jettying, timber-framing and chimneys, facing the road

dominated by later nineteenth-century buildings and modern insertions, so is morphologically somewhat fragmented; but the remnants of a back-lane parallel to the village street on both west and east remain and, in 1995, an air photograph located a large straight, linear ditch (or possibly a road) on the east parallel to, and apparently integral with, the village's rectangular plan (Plate LIII). The village appears to have been laid out in a co-ordinated way at some stage, but whether this was during the development and tidying up of the nineteenth century or earlier, possibly much earlier, is unclear. The buildings today are almost entirely post-medieval. Mid/late twentieth-century construction apart, others represent the remains of a model estate village built during the later nineteenth century by the Meux estate (Plate LVI; *VCH* XI, 185).

The temptation to see this estate village as an unconscious revision of a planned twelfth-century Templar village is strong, but there is little firm evidence to encourage the thought. The late eighteenth-century map of East Overton manor shows about sixteen buildings in Lockeridge north of the cross-roads, arranged either side of the north–south road but with hints of both rectangularity and former larger size; it, and early OS maps, help give perspective to the scale of the Meux reshaping of Lockeridge. The strongest pointer to the planned, medieval origins of the village is now the new evidence of the large eastern ditch. This is hardly likely to be modern, yet is clearly integral with the village form. Whether it is twelfth century is a different matter.

Templar villages with indications of possible planning are recorded elsewhere, notably in the RCHME's investigation of Willoughton, Lincolnshire (Everson *et al* 1991, 22, 218–20, figs 19, 153). At Willoughton, however, the village plan appears to include two additions to a settlement core around the church, one of them being identified as 'the moated Preceptory of the Knights Templars [sic]' (Everson *et al* 1991, 22). At Lockeridge, in contrast, the village's purpose, wherever it lay, was to support a small community living in a preceptory 6km distant, rather than in the village; so a Lockeridge Templar village, planned or otherwise, would not have contained the equivalent of the layout over the south-west portion of Willoughton. Otherwise, the argument for a new 'Templar' Lockeridge at the moment rests only on two suppositions: that a former settlement at Lockeridge House was the *Domesday Locherige*; and that the name shifted across the river to a new settlement on a new site perpetuated by present-day Lockeridge village. This latter could have occurred when new landlords took over a newly created estate in the mid-twelfth century and planted a new settlement to attract tenants to work it – and pay rent.

The Anglo-Saxon Locherige?

Topography is the first factor to appreciate and that takes us, so we argue, a little away from the situation of the present village to the 'old' river crossing and a locally marked bend in the course of the Kennet. We therefore look at the site of the one modestly outstanding example of domestic architecture in the study area, Lockeridge House, as potentially the early site of Lockeridge.

The 'old' road south from Fyfield village forded, and from at least 1819 bridged, the River Kennet immediately east of Lockeridge House (Plate LI and Figures 8.1 and 11.1). The house's name indicates, correctly, a crossing into another land unit, from Fyfield into the former tithing of Lockeridge. The house itself is misleadingly located in Fyfield by Pevsner (1963, 225). In fact, as one would expect of land treated since the early fourteenth century as part of Winchester's East Overton estate, it lies in the civil parish of West Overton. A fine Queen Anne mansion, Lockeridge House stands alone in a locally prominent position, now approached through impressive gate-piers with, as Pevsner deliciously remarked, 'very big pineapples' (ibid; *SL*, figure 51). The front is ostentatiously turned away from present-day Lockeridge and, perhaps significantly, overlooks earthwork settlement remains, essentially of a toft-and-croft pattern, towards present-day Fyfield (Figure 11.1). The earthworks, the river crossing, the historic road pattern and the house itself comprise a small complex. This suggests the possibility that the house perpetuates an earlier settlement site between the present-day villages of Fyfield and Lockeridge, on an old boundary at a river crossing on an 'old' main road.

The site of former buildings on the north bank of the Kennet, directly north east of Lockeridge House, was originally noted as the 'shrunken village remains' of Fyfield (Bowen and Fowler 1962, 102, fig 1) because the majority of the former settlement remains are indeed in what is now Fyfield civil parish. Part of the spread of (now largely destroyed) settlement earthworks, previously described as the 'remains of Fyfield village ... south of the church' (ibid), were actually in the tithing of Lockeridge (Figure 11.1). This misunderstanding is because the boundary in this riverside area between Fyfield and Lockeridge has followed two courses. The later (1819) and current one ran along the line of 'Piper's Lane' or 'Old Road' (the Roman road, Figure 11.1) and then south, down the former main road, now footpath, to the Kennet, keeping along the eastern side of 'Pipers Croft' and 'Maskelynes Meadow' (1819; from the name of the mid-thirteenth-century owner 'Walkelinus' Swift?; FWPs 44 and 45). Previously, however, the boundary had gone directly east to meet the Kennet by the ford, now Back Fyfield Bridge, on the old (Roman) route to Clatford (SU 15086840). This places a roughly square area of land, including some of the settlement remains, in Lockeridge in the late eighteenth century. This particular plot was still referred to as Lockeridge Meadow a century later (Smith 1885, map section XV, K.VI).

Allowing for modern disturbance, there appears to be a gap between the 'Lockeridge' earthworks and similar medieval earthworks north of the Roman road. Such a gap is reflected in the early maps. This allows the interpretative possibility of two distinct settlements, one a southern part of Fyfield village and the other further to the south but on the north bank of the river, with no obvious name.

Through the settlements ran a road, now a footpath, from Fyfield church towards Lockeridge House and past a surviving cottage (SU 14786816) which, like the earthworks, seems to relate to it. It is suggested overall that such remains and Lockeridge House reflect the former existence of a previously unrecognised and therefore unlocated settlement. Lockeridge House and the present bridge seem to be but the latest excrescences on a site as old as any other bar Fyfield along the bottomlands. We suggest that in 1086 it was called *Locherige* and that it could be the location of late Anglo-Saxon Lockeridge. If this were so, then the area could be that identified a little later as the demesne of Sokemond or Walter of Thanet. Perhaps Lockeridge House stands on the earlier foundations of one of these manors just as Fyfield House stands on the site of Fyfield manor house (Chapter 11).

Dene

It can only be suggested, not proven, that the second of the settlements was around the triangular green at 'Dean' (1820 and 1889 maps). Here are clear signs of a formerly greater extent of settlement (Plate LII and Figure 10.1, 1), which may represent shrinkage on the southern fringes of Lockeridge 'main village' or the fragments of a once separate settlement. Several of the oldest standing buildings in the village are either side of the cross-roads, where the southern end of the village street, perhaps more significantly the long-distance road from the downs through Fyfield, across the Kennet and then on through West Woods, intersects a local road between Clatford and Overton (Figure 8.1). The latter itself branches off a genuinely old road, 'Market Lane' or 'Royal Lane' (Smith 1885, 207), providing an important access between Marlborough and the Vale of Pewsey. Perhaps significantly, this road passed through our postulated early settlement of 'Dean', but by-passed the present village of Lockeridge as if it were not there.

The conjunction of roads on the floor of the dry valley, now with plenty of space and a few houses round about it, as if around a triangle, perhaps marks the area of a separate, and possibly early, settlement (Plate LII). Smith (1885, map section XV, K.VI), following Andrews and Dury (1773), shows twenty-five buildings arranged around the triangle of roads at 'Dane'. So there is no doubt that it was an area of settlement; if originally a separate settlement rather than a southern extension of the parent village, then it may have been the settlement leased from the Bishop of Winchester in 1086 which subsequently formed part of Richard Quintyn's demesne (*see below*). As we have seen elsewhere (eg, Chapters 4, 8 and 9), such a settlement may be right on an ancient boundary: the tenth-century charter boundary of East Overton (S449) takes *wodnes dene* as one of its estate markers. Such a cramped site, and the road's dog-leg, can be explained topographically – they relate to a local 'pinch-point' created by the hills to the south west and north east of the cross-roads. Together they emphasise the morphological difference between the north and south parts of what today is generally perceived as the single village of Lockeridge, a difference pointing to the separate origins of two different settlements.

While the matter of origins may defy field and cartographic evidence, here documents lend weight to the idea of a separate place with its own identity. The Pembroke Survey (1567), for example, noted that the tenants of Richard Kyngesmyll had a toft, a messuage and a virgate at 'lez Deane' and a *clausum* at 'Deane close' (Straton 1909, 259). The late eighteenth-century map of East Overton manor clearly shows present-day Lockeridge split into two either side of the cross-roads. The southern part has some eighteen buildings arranged around the triangle of roads. The Pembroke Survey also mentions *Stonydeane* (ibid, 262) near *Old Berye* and *Connyfelde* as adjacent fields. The names incidentally attest to sixteenth-century rabbit farming and the uncleared state of the dene-bottom, preparing the way for its acquisition by the National Trust in the twentieth century to maintain its uncleared state.

Other settlements

Lockeridge tithing contains two probable medieval farmsteads, with no indication that either has earlier origins. Breach Cottage (Plate LII and Figure 10.1; *SL*,

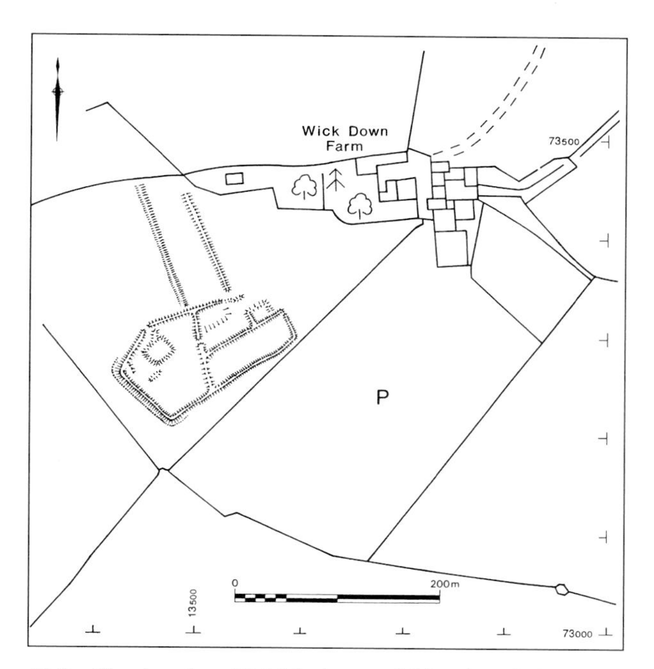

10.3 Plan based on OS 25-inch map, fieldwork and air photography of an enclosure forming part of the large extent of apparently medieval earthworks (P) around Wick Down Farm, Rockley, now in Preshute CP (top centre *on Figure 2.1). Clearly later than an underlying pattern of seemingly toft-and-croft earthworks, the enclosure is suggested as the site of the preceptory of the Knights Templar in the twelfth century; the chapel may well have been in its north-west quarter (cf the air photograph reproduced in* SL, *figure 46b)*

figure 57) represents a farmstead associated with assarting on the woodland edge. It was called 'Lockridge breache' in 1567 (Straton 1905, 263) and its presence then is more likely than not to reflect an earlier, presumably medieval, origin (*PNWilts*, 424). The tenants of Lockeridge were certainly making inroads into the forest in the late twelfth century (*VCH* IV, 417–18), today perhaps reflected in the clearances south and west of Breach Cottage and south and east of Spye Park Cottage (SU 15106685; Figure 10.1). That cottage ('Keeper's House', 1819) stands on or near the site of an earlier settlement called 'Fosbury', a name reflected in the nineteenth-century Fosbury Cottage a little to the south of Spye Park but actually attested in 1270 (*PNWilts*, 306; *VCH* XI, 190; Crowley 1988, 58).

Dene Farm (SU 146677), with a misleading name suggesting a closer link with Lockeridge Dene than was the case, lay on, or to the west side of, the East Overton/Lockeridge estate boundary, and historically lies in East Overton.

Outside the tithing and the present civil parish, but closely linked to the history and land-use of Lockeridge, is Wick Down Farm, now in Preshute parish, formerly the tithing of Rockley. Around (and presumably beneath) the present farm is a considerable extent of settlement remains (Figure 2.1), characteristically of deserted medieval appearance but unresearched for present purposes except in one respect. Superimposed on the earthworks is a prominent rectilinear enclosure on relatively low-lying ground near the farm (Figure 10.3; *SL*, figure 46b). It is a strong claimant to be the site of the Templars' preceptory; ground and air photographic inspection suggests the site of the chapel itself may well be in the north-west corner of the enclosure at a slight but sharp angle in the northern bank and ditch.

THE ESTATE

Lockeridge tithing contains no church, nor, it would seem, has it ever done so. This absence of a permanent fixed point, as St Nicholas's provides in Fyfield (Chapter 11), may well be partly reflected in a settlement pattern that has been both shifting and dispersed. At least three areas of settlement have existed, two of them the separate manors of Lockeridge and Upper Lockeridge, the third argued here to be Dene. The River Kennet divided the tithing into two roughly equal-sized land units (Figure 1.2). Both the northern and the southern unit were likewise subdivided, with the subsequent estates held by different people. Numerous grants of lands to outside landowners included, notably, the Order of the Temple and the Priory of St Margaret, Marlborough, followed much later by the Earl of Pembroke and the Duke of Marlborough.

North of the Kennet

The area which runs northwards from the banks of the Kennet to the high downland around Delling was, and is, a long, narrow parcel of land (Figure 1.2). Its northern part, that is to say more or less the whole of the area called Lockeridge Tenants Down in the eighteenth century (Plate VII), was in the tenth century surrounded by the East Overton estate (S449). By the mid-thirteeenth century it was held by a certain Sokemond. The southern half of this northern land unit is the area encompassing Pickledean, as far east as the boundary with Fyfield, and as far south as the Roman road and the Kennet. It may have formed a one-hide demesne estate leased by the Templars to Walter of Thanet in the 1250s (cf *VCH* XI, 192).

South of the Kennet

The southern portion of the tithing seems to have stretched at least from meadow land just north of the river to the southern edge of the Fosbury clearing; possibly it continued as far as the heathland area west of Barrow Copse. The area also consisted of two estates, both reckoned at two hides at *Domesday* when both were held by Durand of Gloucester (*VCH* II, 121, 148). One of these two-hide estates was granted by Humphrey de Bohun, a descendant by marriage of Durand, to the Templars; the other was granted to Humphrey's Knight, Richard de St Quintyn.

Quintyn's land probably lay in West Woods, centred around Fosbury Cottages (SU 152665), and may have included land at what was known as 'Heath Grounds' in the late eighteenth century as well as some meadow land by the Kennet. Much of this two-hide estate had, by the later thirteenth century, been acquired by the de Macy family, who in turn granted some of it to the Templars and some to the Priory of St Margaret, Marlborough. The Priory, with a grant of a further 40 acres (*c* 16ha) in 1294, was thus able to create a manor, called Upper Lockeridge, in and around West Woods. After the Dissolution (1539), it passed through various hands, including those of Anne of Cleeves, until 1759 when it was sold to the Duke of Marlborough. It then descended, with other lands at Lockeridge, as part of the East Overton manor (*VCH* XI, 190).

The rest of what had been Quintyn's estate similarly enjoyed various owners, including John of Fosbury, until, in the early sixteenth century, it belonged to Richard Benger. References to 'Quyntons lands … once property of Robert of Berewyke and before him Henry Attwood' (Straton 1909, 141) and a settlement near 'Fortesbury' known as 'Hardings [which was] once called Bengers' (ibid, 264), in and around the woods at the southernmost end of Lockeridge tithing, support the location of this thirteenth-century estate in the clearing south of Lockeridge Copse. It seems likely that Breach and Spye Park Cottages represent the vestiges of Upper Lockeridge, together with the nineteenth-century buildings called Fosbury Cottage and Forest Lodge. By 1768, this land had passed to the Duke of Marlborough.

The Templar estate (*see below*) included the present village of Lockeridge and the land south as far as Lockeridge Copse. On the Templars' suppression (1308), this two-hide estate passed to the Hospitallers, and in 1543 the manor passed from them to secular landlords. In 1719, it was considered a free tenancy of East Overton manor, then belonging to the Pembroke family. The second Viscount Chetwynd of Lockeridge House sold it, along with Clatford Park, to the Duke of Marlborough in 1756. Twelve years later the Duke had acquired the manors of Lockeridge, both north and south of the Kennet, and Upper Lockeridge, as well as the land deemed part of the East Overton estate (*VCH* XI, 189–90, 196; Figure 10.1).

Domesday, *the Templars and* Locherige

It is necessary to reconstruct the *Domesday Book* arrangements on the ground in order to envisage the Templar estate and land-use on it in the twelfth and thirteenth centuries, especially in the southern woods.

In 1086 Durand of Gloucester, the sheriff, held the two-hide estate of *Locherige*. One hide was in demesne and there was land for one plough, with one villein and two bordars with a serf on non-demesne land (*VCH* II, 148). Worth 30s, *Locherige* was worth relatively more at 15s a hide than West Overton's 10s a hide, though *Locherige* was undoubtedly the smaller, covering less than half of the area of West Overton and less than a third of East Overton.

Between 1141 and 1143, this two-hide *Domesday* estate was granted in totality by Miles to the recently formed Order of the Temple (Lees 1935, 207; FWP 18, folio 65v, 250). As such, the Templars obtained a settlement called *Locherige* with 6 acres (*c* 2.5ha) of woodland, 12 acres (*c* 5ha) of pasture and a meadow by the river. Miles, however, did not reside at *Locherige*, so the day-to-day management of the manor was left to *Ricardi de Sancto Quint'* (Lees 1935, 207, n 16). As the Templars were now to take over the manor of *Locherige*, Miles 'gave an exchange for that same land according to [St Quint(yn)'s] wish' (FWP 18, folio 65v, 250). The land exchanged was a second estate in Lockeridge, owned by Miles at the southern end of the tithing (*see below*).

Another two hides of land in Lockeridge granted by William de Beauchamp (d 1170) to the Templars at Rockley between 1155 and 1169 (FWP 18, folio 65v, 248; *VCH* XI, 190) were not, however, merged with the Templars' two-hide manor of *Locherige* to form a four-hide estate (*VCH* XI, 190). The grants by Beauchamp and Miles were almost certainly of the same two hides (FWP 18, folio 65, 247; Lees 1935, 208–9, n 10).

Another hide of land at Lockeridge was granted by Robert of Ewias to the Templars in the mid-twelfth century (FWP 18, folios 28–30v), so by the end of that century the Templars' preceptory at Rockley managed a three-hide estate called *Lokeruga* (cf Lees 1935, 209, n 3). It becomes clear that the Templars were, of course, keen to increase their rental income from their new estate, which was but a means to an end in their eyes, and that they attempted this in several ways. The most obvious were the encouragement of new settlement and the clearance of waste or wooded land for cultivation; in one case the rent in fact increased almost three-fold in just one or two decades (FWP 18).

The next set of grants of land at Lockeridge to the preceptory at Rockley appears to date from the 1240s and 1250s. 'Richard of Sokemond of Lockeridge', for example, granted freedom of entry to and from the Temple lands and his holdings at Lockeridge for the Templars' men and tenants at Rockley (FWP 18, folio 65v/66, 251). This freedom allowed

> free and quit entry to and exit ... with all their types of livestock for the grazing of their commons, as well in meadows as in pastures and in other places, ... everywhere and wherever my lands and theirs are shared and lie co-mingled without any contradiction and impounding and impediment by me or my heirs or my assignees.

It continues:

> so that moreover my men and tenants of Lockeridge shall not have nor be able to have any entry there with their cattle, except me and to my free men holding a tenement of the same their own common land.

This demonstrates, first, that Sokemond was a man of standing in Lockeridge, presumably holding a manor with tenants and freemen and a large area of grazing land in his own right. Indeed, his surname may derive from 'Sokeman', a term for a pre-Conquest freeholder (Coleman and Wood 1988, 53). Secondly, grazing land given to the Templars was regarded as shared, even 'co-mingled', with Sokemond's land. Thirdly, in arranging this agreement with the Templars, Sokemond was ensuring he retained rights to graze cattle on the land, while excluding those of his men at Lockeridge without a legitimate claim to such rights, just as Quintyn had done with his grant of two pastures (*see below*).

The charter continues:

> Besides I have conceded ... that the aforesaid brothers of the Temple and their men and tenants shall hold ... free and quit, well and in peace, their land and tenements at Lockeridge and whenever they should wish to cross my land and tenements and to plough, cultivate and improve their lands and tenements, according to their wish, to carry off their fruits [revenues] from there.

This second part in particular usefully conjures up a picture of the busy mid-thirteenth-century landscape, though of course its purpose was to regulate a crucial function, that is the means of communication within part of that landscape.

This function was repetitively evidenced archaeologically on the downs. Here, a document provides a clear indication that the tenants of the Templars need to cross Sokemond's land to carry out their daily business. In particular, as tenants on a fragmented estate, they needed to commute between the brothers at Rockley and the land holdings at Lockeridge. Each of these points implies that Sokemond's land lay between the preceptory at Rockley and the village of Lockeridge. This would locate it north of the river, with the focus on grazing rights and the notion of shared land pointing to an area of land called 'Lockeridge Tenants Down' in the late eighteenth century, Delling Copse in the twentieth (SU 131711; Plate VIII; Figure 2.1).

Some time in the early 1140s, Miles of Gloucester gave his knight St Quint' another plot of land in exchange for some the knight had previously held at Lockeridge, before, that is, Miles granted it to the Templars. This land in exchange, it appears, was also in Lockeridge, as in the mid-thirteenth century we learn that Richard Quyntyn, no doubt a descendant of St Quint', granted a dwelling-house and a meadow to the Templars (FWP 18, folio 66v, 254). That this land is also leased from the Bishop of Winchester, at least until 1243 (*VCH* XI, 189), indicates that it is very likely to be the same as the '2 hides all but ½ virgate' estate leased from the Bishop of Winchester in 1086, though held by Durand of Gloucester (*VCH* II, 121). In the mid-thirteenth century, as in the late eleventh, it seems this land 'could not be separated from the church' (ibid). If correct, then Miles gave his knight St Quint' an estate deemed part of *Overtone* (East Overton) in *Domesday* but which later was regarded as part of the tithing of Lockeridge. Its inclusion in an extended Lockeridge estate is possibly a result of the reorganisation of the estates of the Bishopric of Winchester and the Priory of St Margaret after the Dissolution.

We can place part of Quint(yn)'s estate as, from the grant to the Templars, we learn that it consisted of an 'entire meadow with messuage, yard and other appurtenances, that lies between the meadow of the said Templars and [that of] Alexander Man, indeed which meadow ... the former Eylwinus once held in the town of Lockerugg, and two pastures for the grazing of cattle in all places in which my cattle ... graze' (FWP 18, folio 66v, 254). This meadow, with its house and yard in the town of Lockeridge, would have lain close to the Kennet, and the pasture, if our previous interpretations are correct, could only really have been situated south of Lockeridge village, perhaps even in the clearing around Fosbury Cottages or, further south still, at Heath Grounds.

The de Macy family's holdings in Lockeridge were substantial enough by the end of the thirteenth century for William Macy to grant 60 acres (*c* 24ha) of land to the Priory of St Margaret, Marlborough, in 1281 (*VCH* XI, 190). This estate was increased in 1294 when Philip le Frauncey and Richard le Eyr granted a messuage and a further 40 acres (*c* 16ha; *VCH* III, 317), thus creating a 100-acre (*c* 40.5-ha) manor for the Priory, called 'Upper Lockeridge' (*VCH* XI, 190), on the higher ground to the south. This holding lay in woods, so would have bolstered the Prior's already considerable pasture and timber rights in the forest (*VCH* III, 316–17). Indeed, the Prior of St Margaret was allowed to take small tithes from his manor in return for a yearly payment of 3s to the vicar at East Overton, an arrangement which continued into the later eighteenth century (ibid, 200), though not always smoothly. In 1588, for example, a 'controversye and sute of lawe' arose between the vicar of Overton and Richard Browne of Lockeridge over the tithes of a farm in Lockeridge 'som tyme belonging to the Priore of Saint Margaret' (FWP 45).

Among others, two generalities emerge from such detail here. We see, for example, the importance of the very restricted meadow land to the actual working of a landscape visually dominated in general by its unenclosed downlands. Though local owners retain their share, the medieval grants and agreements show how the meadow lands close to the Kennet were embraced by the Templars to give the brothers' estate an agrarian viability, in stock management especially, and thus a new-found ability to draw on the resources of bottomland, as well as forest and downland. The grants also indicate a changing ownership pattern emerging at the same time. Certainly Quintyn's 'old' demesne estate, and seemingly others too, were being split up, but not amongst numerous off-spring nor because of modified feudal allegiance, but because of emerging families, such as the de Macys, with money to buy land.

The Templars' tenants Beauchamp's charter of 1155–69 gives no indication of size, though the land was 'rated' at two hides in the Inquest of 1185. Lees (1935, cxxx) believed, however, that this did not reflect two areal hides and indicated that some demesne land in Lockeridge remained unassessed. We do know from the Inquest, however, that the estate contained at least 48 acres (*c* 19.5ha) of arable and pasture, as well as some assarted ground and 6 acres (*c* 2.5ha) of common land, and was shared by nine *cottars* with holdings of between 1 and 10 acres (*c* 0.5–5ha) each, paying a total of £2 14s 10d (FWP 18, folios 28–30v; Lees 1935, 53).

Two of these *cottars* with holdings of 1 to 2 acres (*c* 0.5–1ha) also rented crofts, suggesting outlying homesteads within a landscape being worked severally. It is therefore particularly interesting to note that, between them, they owed 2s for an area of assarted land. We can infer from this assarting that in the mid- to late twelfth century clearance of woodland was proceeding, perhaps making 'leas' within the wooded areas and inroads around the forest edge. Indeed, in 1198–9 Lockeridge had to pay 10s for new and old waste within Savernake (*VCH* IV, 417–18). Perhaps the particular assarted land referred to here was around Lockeridge Breach Cottage (SU 150672). Whatever the date of this 'breach', the name illustrates the process (Gelling 1993, 233–4; *PNWilts*, 424), but we cannot know from this twelfth-century documentary evidence the scale or pace of the process. It is tempting to relate the activity then to the demands of new estate management (ie, the Templars), who needed good and active tenants to sustain their economically non-productive religious life-style (cf Hooke 1997).

This number of tenants suggests a fairly large village in twelfth-century terms. It is therefore tempting to see the medieval planned street village of Lockeridge as a Templar creation at this time, particularly as we know from the Inquest that the tenants worked meadow land which was almost certainly by the Kennet. Part of the tenants' arable may then have lain in the area called 'Rayland Hedge' (SU 148677; Straton 1909, 263) or 'Rylands Field' (1811), to the east of the village, as well as south of Lockeridge Lane, with the Templars' woodland centred at Lockeridge Copse.

A dispute about customs of service The Inquest of 1185 also detailed the customs of service that the tenants of Rockley, who held the majority of the land granted to the Order, were expected to deliver (cf Barber 1995, 251–7). In general, the terms were similar to those expected of Richard of *Raddun* working half a century later a little further south for another ecclesiastical landlord (Chapter 7). The list of services indicates a mix of arable and pastoral farming, suggesting the Rockley holding could not have been confined to the downland; indeed, its existence relied on its holdings in the richer valley bottoms. We have not elucidated the detail of the whole Rockley estate on the ground as we have done with other land units in our study area, but the landscape it occupied stretched from the high ground at some 200m aOD in the Man's Head/Wick Down Farm area (*c* SU 1474) towards Rockley Manor some 50m lower at the junction of three coombes just over 3km to the south east (the area lies eastwards from the Wick Farm enclosure at SU 134733 across the top of Figure 2.1). This area contains a perhaps surprising natural variety, for example of soils of different aspect, so the services could be reflecting both a local agrarian diversity and the widespread residences of the tenants, for example at Lockeridge, on a fragmented estate with access to a wide range of resources.

OVERVIEW

Overall, Lockeridge appears to be and actually is an insignificant place in historical terms; yet its complexity archaeologically and documentarily provides an insight into the detail of the working of a landscape which is probably typical of many places in southern England. Internally, its history and many of its landscape features today reflect the particular complexity of its proprietorial arrangements dominated by external institutions and individuals using the place for their own ends, notably to support a religious life. Such purposes did not necessarily serve the interests of Lockeridge.

In 1338, for example, receipts from the Templar estate totalled £20, compared with £26 8s in 1185 (Lees 1935, 58). As outgoings amounted to £7, the profit of £13 was sent to the preceptory at Sandford, Oxfordshire, from where the Templars' estate was being administered after their abolition (*VCH* XI, 196). Rome, Winchester, Wilton, Blenheim – at least Sandford is a change, though its addition to the list of recipients of the fruits of local labour reinforces the point that the landscape of our study area is very much a landscape resulting from proprietorial exploitation.

Yet at the same time the place cannot be understood either tenurially or economically without reference to its immediate neighbours. Clearly, in general, the Lockeridge landscape reflects common concerns with downland, bottomland and woodland; yet it is very much a place with a distinctive character, the odd one out between the Overtons and Fyfield.

Chapter 11

The Valley and its Settlements: Fyfield

INTRODUCTION: TOPOGRAPHY AND SETTLEMENT PATTERN

Fyfield tithing is another long, thin tithing divided by the River Kennet, more or less, except at the south-east end, coincidental with Fyfield civil parish (Figure 1.2). It is aligned north north west–south south east, embracing a wide expanse of downland to the north, *c* 1km of river bank and floodplain, and, historically, an attenuated strip of land probing south into West Woods. The Kennet, which generally flows west–east, here runs south–north up the middle of the parish before turning east again as it meets the relatively harder Lower Chalk at the foot of Fyfield Hill (*SL*, figure 52). The village sits on this slight bluff.

As the estate's name indicates, Fyfield paid geld for five hides. In the time of King Edward it belonged to the Sacrist, probably the treasurer, of the church (at Winchester) and was held by Alsi, a monk. Its not uncommon place-name might also imply that it contained five arable fields if it had emerged as a distinct unit between the mid-tenth century and 1086 (Gelling 1993, 236). Whether or not it was such a relatively late creation as a distinct estate, its origins apparently lay in a core of holdings, evident in *Domesday*, to which was attached an area of downland belonging to East Overton. The modern civil parish essentially reflects the resultant tithing's limits on the north side of the river, but to the south it is today larger and more cohesive than the medieval tithing, mainly through the inclusion in 1896 of the former Clatford Park and a detached block of heathland around 'Heath Barn' (now Bayardo Farm, SU 161651; *VCH* XI, 183, 186).

Fyfield shares a common western boundary with Lockeridge; both have enjoyed a close historical relationship with East Overton (*see below*). Most of its eastern boundary is with the historic tithing of Clatford. These boundaries are essentially of late Saxon date at latest, and their detailed histories are complex, especially in the twelfth, thirteenth, sixteenth and nineteenth centuries. Fyfield's eastern boundary is the formal edge of our study area. The study of this tithing and parish suffers, therefore, in that whereas we can try to appreciate Lockeridge in the context of similar studies on both of its long sides, we have to draw a line somewhere and have not examined in similar detail Fyfield's eastern neighbours, Clatford, Manton and Rockley, all now absorbed in Preshute civil parish (Brentnall 1950). Nevertheless, we have to look over the boundary: Manton has already been involved on the downs (Chapter 5) and Clatford becomes involved with Fyfield in the woodlands and is included within the 'Ridgeway zone'. Rockley was touched on in discussing Lockeridge (Chapter 10).

The modern parish embraces a relatively large amount of woodland to the south which is on the edge of an extensive spread of prehistoric material discussed, for convenience, in Chapter 12. It is likely that ground disturbance and intensive survey here would add to that distribution, especially in the area of the former Savernake Park on Overton Heath, in the woods on its west significantly called Broom Copse, and to the north in the former Clatford Park. On present evidence, however, the woods have scarcely been inhabited since before the Roman period. A medieval sheep-house lay at 'Audley's Cottages', now – in another prosaic name change which literally wipes a historic landmark off the map – just Hillside Farm (SU 152678). The particular pity in this case is that the place, on the northern fringe of the historic woodland, is one of the oldest known, continuously inhabited locations in the study area outside the villages. No other evidence of permanent settlement south of that site exists in the parish. The two Park Farms on the edges of modern Fyfield's woods are both post-medieval, each reflecting the former presence of one of the two post-medieval parks (*see below*).

Savernake Grounds (1794), formerly a detached part of the tithing of West Overton (Figure 1.2), is part of Fyfield civil parish. Originally the area formed part of an estate at North Newnton and Oare granted in the late ninth century by King Alfred to Aethelhelm (S348) and to Wilton in AD 934 by King Athelstan (S424; Grundy, 1919, 320–1; *VCH* X, 126, 128). As part of *Safernoc*, five *cassatos* or crofts stood here (S424; Brentnall 1941, 391–2), demonstrating that settlement existed in the woods as early as the ninth century. Perhaps these crofts housed huntsmen and woodcutters involved in woodland management.

CLATFORD PARK

It seems likely that Clatford Park, a long, rounded area sitting in the south-east corner of Fyfield parish, was carved out of the estates of Clatford, Fyfield and Oare. Archaeologically, the park is still defined on the ground for much of its circuit by a low bank and ditch, though its northern limits are uncertain. Cartographically, the fact that the whole of its southern boundary is perfectly clear as a continuous, curving boundary as far north as junctions with Wansdyke to west and east suggests that the park utilised the earlier earthwork as its northern boundary. But slight if discontinuous and ambiguous banks and ditches on the ground and air photographs hint that the park extended north of Wansdyke. Its western side may well have passed between Henley Wood, in Lockeridge, along the Fyfield boundary to include Fyfield Wood. The latter has a suggestively curving northern edge, from which the park's north-eastern side could well have easily passed down to the sharp angle in the course of Wansdyke.

The park itself is quite well documented as a holding, though its actual shape and boundary are not described. After the Wilton Abbey estates were acquired at the Dissolution by Sir William Herbert, the First Earl of Pembroke, it appears land in this area was apportioned between West Overton, which obtained Savernake Grounds in return for its previous heathland hereabouts (Chapter 8), and Lockeridge, which obtained or retained Heath Grounds. The other part of this estate was turned into a park by Sir Thomas Wroughton (*VCH* XI, 192), fairly clearly in the early 1580s. The 1588 Terrier from East Overton church refers to 'a certain Parke lately inclosed on the Est' (FWP 46) and the *Note on Parks in the County of Wilts* (1583) describes the park 'as being three miles in circuit' (Watts (1996) lists a short bibliography of Wiltshire deer parks, though does not refer to either of the two parks identified here). This last information is not as helpful as could be hoped, for a park confined to the south side of Wansdyke would have had a circuit of *c* 2½ miles (*c* 3.8km), while the larger park suggested above has a circuit of somewhat over 3½ miles (*c* 5.4km). Nevertheless, whatever the details of size, the fact of a later sixteenth-century park offers a possible context for what seems to be the post-medieval emparkment of Savernake Park directly south of Clatford Park and of land beside The Ridgeway in East Overton (Chapter 4; Figure 4.1). The former provides the occasion for 'Park Farm' (SU 166651) in the extreme south east of the modern parish, and the unusual conjunction, with 'Clatford Park Farm' (SU 164661), of two Park Farms only 1km apart but belonging to two different, post-medieval parks.

Clatford Park seems to have been disemparked about 1631 by the then owner Richard Goddard (*VCH* XI, 192), perhaps because of the need for more productive land (Watts 1996, 92–3). By 1717, the land had been acquired by Viscount Chetwynd (of Lockeridge House; *see* Chapter 10), who sold it to the Duke of Marlborough in 1756. It then remained part of the manor of East Overton until 1896, when it was transferred to Fyfield civil parish (to which was also added the extra-parochial area of Overton Heath).

THE VILLAGE

Fyfield village today is not much to look at (Figures 8.1 and 11.1; *SL*, figure 52). Historically, however, it has had an interestingly chequered career, surprisingly calamitous for what appears to be such a quiet place. In landscape terms, its significance is that its nucleus has apparently remained at the locally prominent and dry site of the Roman villa/church/manor farm for nearly two millennia.

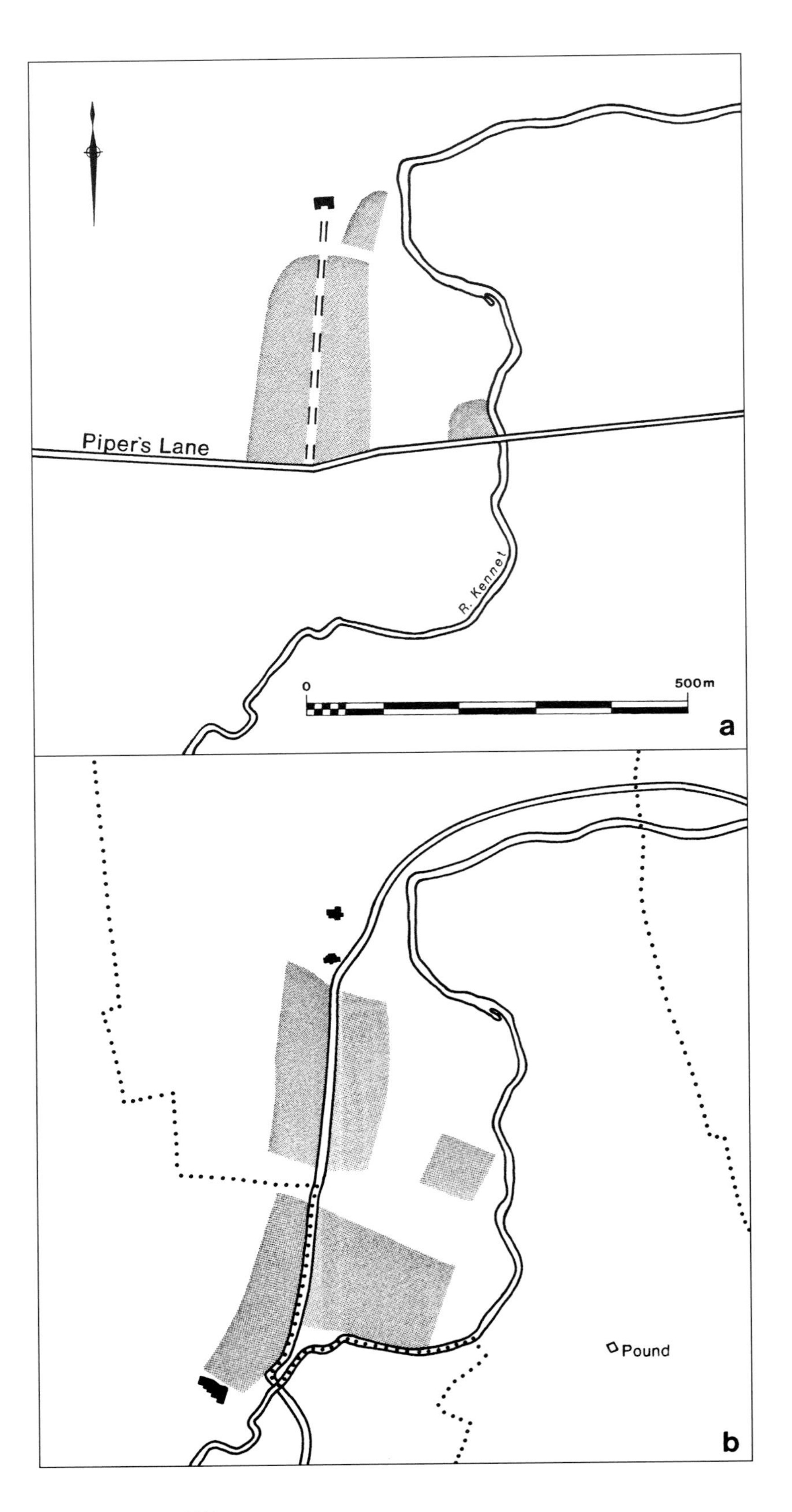

11.1 Maps interpreting the settlement archaeology between Fyfield House and Lockeridge House

(a) Roman: *a villa is shown where Fyfield House now stands, linked by a conjectural side-road (on the line of a medieval road), to the certain main Roman road later called Piper's Lane; we guess it may have crossed the Kennet on a bridge. Stipple, here as on all four maps of this figure, indicates habitation areas, known or probable.*

(b) Medieval: *Fyfield manor house stands on the site of the villa, with St Nicholas's Church immediately on the south, both at the head of a 'street village'. South again is an extensive settlement area on the flood plain, here interpreted as 'old' Lockeridge associated with a hypothetical predecessor of Lockeridge House (*lower left*).*

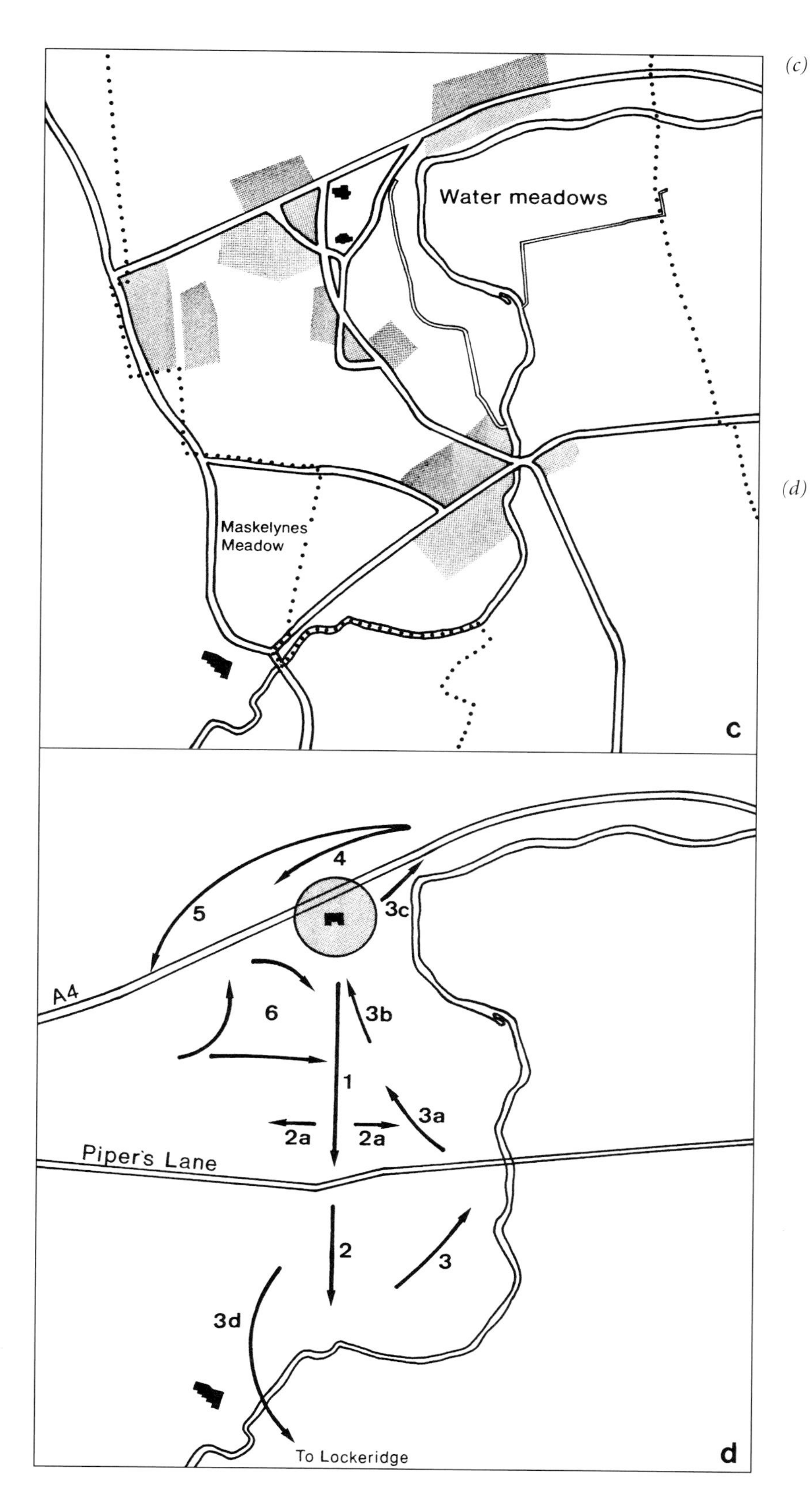

(*c*) Post-medieval*: largely based on historic maps and standing buildings, this diagrammatic amalgam of evidence over two centuries conveys an impression of fragmentation (but* see *(d)* below*), with multiple habitation areas (all only contemporary in the 1990s) and many roads/tracks (all documented). Fixtures remain Fyfield House/church and the two river crossings, on the Roman road to the east and to Lockeridge on the south, but the two main areas of medieval occupation (*see *(b)* above*) were largely deserted by the late eighteenth century at the latest.*

(*d*) 'Settlement shuffle'*: this diagram attempts to convey some impression of the dynamics which have created the pattern in (c) through a long-term process of sequential habitation movements around the core of Fyfield villa/manor house (black in stippled circle). Six main stages are suggested.* **1** *indicates the Roman link between villa and road with its likely associated settlement area, and* **2/2a** *the spread by medieval times across and along that road.* **3** *suggests a possible double movement, both back to fixed points at the core and Roman river crossing (***3**, **3a**, **3b***) and outwards across the river to present-day Lockeridge (***3d***) and eastwards along the north bank of the Kennet (***3c***). Settlement there and along the toll road is attested in the eighteenth century and later (***4***), drastically affected by road improvements c 1930 when inhabitants were moved to new housing to the west (***5***). The later twentieth century is evidenced by new housing, returning habitation close to the core area and in the north end of the 'Roman/medieval street village' (***6***).*

The church of St Nicholas developed beside or actually on a Roman villa, now the site of Fyfield House, formerly the 'Manor House' (Figure 11.1a, b). Together, these four buildings (*see below*) seem to have provided the all-important fixed point in Fyfield's landscape for one thousand, perhaps even two thousand, years. The village itself, never large and probably always somewhat scattered, seems to have comprised several different areas of habitation, some occupied at the same time as is the case now, but in general shuffling through time around the core area within the physical limits of the valley bottom and its immediate edges (Figure 11.1d). It is almost as if it could never quite make up its mind which route it was related to – one or other of the main west–east roads along the valley or the through traffic off the downs and across the valley. Perhaps it would be nearer the mark to see it as having been pulled this way and that by the disposition of the river and by changes in the relative importance of these routes resulting from proprietorial whim locally and economic fortune generally.

Fyfield and West (East) Overton are the only existing villages in the study area with living churches, though West Overton and Shaw, both now deserted, once had churches too (Chapters 8 and 13). Fyfield is the only village, and its church the only church, on the north bank of the river between Avebury and Marlborough – both of which, like Fyfield, were Saxon settlements with a church on the downland side of the River Kennet. No other settlements existed north of Fyfield village between Roman times and *Raddun* in the thirteenth century (Chapter 7); on the other hand, narrow though the tithing was along the river, another settlement area, perhaps part-Fyfield village and part-early Lockeridge, lay on the floodplain south of the Roman road and close to Lockeridge House (Chapter 10; Figure 11.1b).

Today Fyfield House, a neo-Gothic building just north east of the church (SU 14846875), lies on the site of 'Manor Farm', occupied by a Mr Tanner early in the nineteenth century. In 1815 John Goodman owned the manor; the house was called 'Mr Goodmans Homestead' in 1819. Presumably it was during its construction that 'a rude Roman pavement was found on the property of Mr. Tanner immediately on the right of the turnpike at Fyfield' (Colt Hoare 1821, 88–9). The discovery was probably either seen by, or reported first hand to, Colt Hoare because, curiously, he gives no authority for the information. It seems reasonable to accept, nevertheless, that a Roman building, probably a villa with at least one mosaic, lies underneath the 'Homestead', itself overlying the medieval manor house. Judging from subsequent finds, the Roman occupation area extends along the path above the river to the north east. It may well exist under the graveyard and even the church. Despite these finds, neither the structure nor the contents of the church of St Nicholas hint at a story beginning earlier than the Norman period. Its chancel was built in the early thirteenth century and its roof is probably late seventeenth century (*VCH* XI, 200; *SL*, figure 53).

The earliest maps (late eighteenth century) show a small village straddling two lanes from the church to a wide ford across the Kennet south east of the church. There, a minor node of settlement is always shown on historic maps right up to the few present buildings beside a modern bridge, Back Fyfield Bridge (Figure 11.1), possibly successor to a Roman one carrying the road across the Kennet close to this place. The southern branch of the nearest Fyfield had to a village street went towards Lockeridge House through what appears to have been a scatter of low-density occupation across the floodplain. Some at least of the earthworks could well represent events in the mid-nineteenth century when much of the village was abandoned after a series of floods and a disastrous fire (*VCH* XI, 187).

To the north, the street went up the slope past the west side of the church. This lane was realigned during the Inclosure years, as was the 'Old Field Road' which became the 10ft-wide (3m) 'New Road' up to Fyfield Hill or 'White Acre' (Figure 11.1). In 1743 the Bath to London road was turn-piked (*VCH* XI, 184), one of the most significant developments in the landscape and economic history of Fyfield since the Roman road was built some 1,700 years earlier. This capital investment – again largely inspired from outside – also tended, long term, to drag the village northwards and realign it west–east rather than leave it clustered around the church and orientated southwards.

Certainly from the opening of the turn-pike other habitation areas have developed, not only on the little knoll around the church but also along the Bath to London Road. That to the north east, on the parish boundary, may well be a pre-turn-pike hamlet; it was certainly there in 1773 (Andrews and Dury 1773). It included Ivy House Farm with, on its south, a row of seven or eight houses and the Fighting Cocks inn in the early nineteenth century (ibid, 186–7; Figure 11.1c). In the early 1930s, the inn, Congregational chapel and many cottages on the south side of the road were demolished when the (now) A4 was widened. The village was physically sliced in two, west–east. Its physical, and probably social, fragmentation was further ensured when Priest Acre Cottages, neatly enshrining Fyfield's ecclesiastical landlord five centuries previously but outside the historic settlement area, were built to rehouse the dispossessed on previously unoccupied land defined by a corner in the parish's tenth-century western boundary (*see below*). The village has continued to suffer, or prosper, depending on the viewpoint, in the second half of the twentieth century. In another major change in its form and fabric, it has experienced, without the benefit of archaeological surveillance, aesthetically questionable housing development immediately west of the church, which nevertheless has returned the focus of village living close to the historic core. In the core itself, the graveyard is full and major works were carried out at Fyfield House in 1998–9 without appropriate archaeological advice. No addition to Colt Hoare's information has been recorded. Fyfield's long-term drama is probably more typical of southern English villages than more cosy models.

THE ESTATE AND PARISH

Medieval Fyfield, in contrast to the Overtons, had relatively little woodland, though it contains a 1km-length of river bank out of all proportion to the tithing's width of *c* 400m (Figure 11.1). While it had come to acquire extensive areas to the north, its agricultural viability was enhanced by having an area of heathland at the southernmost end of the estate in the early nineteenth century, though, as we have seen elsewhere (Chapter 9), this was certainly not unusual in this area. The origin of this land is unclear (*see* Upper Lockeridge, Chapter 10; Savernake Park, *above*) and was made more so by the creation of Clatford Park in the 1580s. Prior to emparkment, however, the land was certainly heathland, though 'lez Heathe' of 1567 (Straton 1909, 264) was not described as being part of Fyfield. Heath Grounds, as the area was called in the late eighteenth century, is marked on one map of the period as belonging to the manor of Lockeridge, so prior to its acquisition by the Revd Fowle, the vicar of Fyfield in the 1810s, it is possible that this had always been the case (*SL*, figure 55).

Domesday

The medieval manorial associations between Fyfield and East Overton emerge as a result of both *Fifhide* and *Ovretone* being held in the late eleventh century by the Bishop of Winchester (*VCH* II, 120–1); but it had passed to a certain Edward in *Domesday* when there was land for three ploughs, with three hides in demesne along with two ploughs and a serf. Elsewhere on the estate were three villeins and nine bordars with two ploughs, farming the remaining two hides. On record in Fyfield, then, are four landholders and tenants, two by status, one by status and name, and one by name; and land for three ploughs (although four are recorded) and thirteen workers.

These figures might suggest a minimum adult male population of about 25 to 30, perhaps indicating a total population of 125 to 150 on the estate (cf 150 in 1841, 152 in 1891, 134 in 1971; *VCH* XI, 186). There were twenty-two poll-tax payers in 1377, again a figure of the same order, possibly representing a population rise to forty or fifty during the twelfth to early thirteenth centuries and its halving at and after the Black Death. This rough consistency hints that, even in the later fourteenth century, Fyfield's working male population was supporting between 100 and 150 people locally, never mind episcopal and monastic establishments some 60km away. Extrapolated, such an estimate points to the whole study area containing, and supporting, a working male population of between 100 and 200 and a total population in the 500 to 1,000 range round about AD 1100. This figure remained viable thereafter, with fluctuations, until external events boosted the population in the century either side of 1800 (cf Lewis 1994, 177–84).

THE MANOR

As discussed above, the two manors of Fyfield and East Overton were amalgamated and run as an entity in the later and post-medieval periods, giving the combined estate of East Overton and Fyfield, on the *Domesday* figures, twenty hides with at least 17 acres (*c* 7ha) of meadow, 346 acres (*c* 140ha) of pasture and over 123 acres (*c* 50ha) of woodland. This resource dwarfed neighbouring West Overton and Lockeridge, whose joint income was £116 when they were assessed together in 1309 (*VCH* IV, 299).

The combined manors consisted of eleven holdings of one virgate and thirteen holdings of a half virgate, held for the usual agricultural services and small money rents in the late thirteenth century. The *virgater*, who acted as *woodward*, was excused certain of the usual duties but was instead bound to carry the lot and crop of the manorial timber to the lord's court. The *half-virgator* at *Raddun* looked after two of the lord's plough-teams at the ox-yard there (Kempson 1962, 113; Chapter 7). In 1299 the combined manors of East Overton and Fyfield supported 717 ewes, 400 hoggasters and 322 lambs.

By 1697, Fyfield manor was administered separately from East Overton, although perhaps much reduced from its original size. The manor was then in the hands of Thomas Fowle and remained in the Fowle family until 1840 (*VCH* XI, 192). This family undoubtedly renamed 'Atlyes copice', Fowle's Copse (SU 154669) –

another indication of the sensitivity of place-names to proprietorial change.

Estate boundaries

The AD 939 eastern boundary of East Overton formed, in several places, the western boundaries of Fyfield and Lockeridge. Moving southwards across the downs and along *pyttel dene*, the boundary comes *on hole weg* then *eft on cynetan*, to the hollow-way and then back to the Kennet. The hollow-way is possibly the Roman road which leaves the course of the A4 (SU 13856851) and crosses the valley bottom heading towards a ford or bridge, near Spring Cottage (SU 15086840). In the fields between the A4 and the Kennet, however, surface evidence is more of an *agger* than a hollow (*SL*, figure 67), though excavation in 1997–8 showed its remains to consist of both elements (G Swanton, pers comm). A little further east, as the road descended the slope down to the river south of Fyfield church, it was indeed a hollow-way until *c* 1970 (*SL*, figure 52), a state in which it had existed since at least the eighteenth century. As the 'Old Road' or 'Piper's Lane' (Figure 11.1a) to the inhabitants (1819), it was a main thoroughfare through Fyfield before Inclosure.

Nearer the river, the boundary line between East Overton and its neighbour to the east led to the south headlands, *on thæt suth heafod*, quite probably a continuation of the *hlinc ufeweardne* (S449; Chapter 9). We suggest, therefore, that in this particular case we, like the Saxon boundary-walkers, can distinguish two sorts of agrarian landscape feature: a *hlinc*, a 'Celtic' field lynchet, and a *heafod*, the edge of contemporary arable. The latter indicates that some land on Fyfield Hill north of the A4, in what was called 'North Field' prior to Inclosure (*SL*, figure 52), was under cultivation in the late Saxon period.

The bounds of four dairy farms and downland are described in an attachment to the East Overton charter (S449). We locate the area on Fyfield Down, so the document is discussed in Chapter 7.

The boundary with Clatford

The early nineteenth-century line of Fyfield's eastern boundary followed markers on the hills north of the village. These were either dillions, stones or lynchets, or a combination of the three. The estate line on the higher downs certainly followed a line of stones, shown on the nineteenth-century OS 25-inch maps, as does the modern parish boundary. One of these stones is undoubtedly 'Long Tom' (SU 14387128; *SL*, figure 62), a remarkable sarsen which may have been erected then but which could have been in position for a long time, usefully marking a boundary in an otherwise topographically featureless area of downland. It stands on a long, straight stretch of boundary, which was part of the tidying up by the Parliamentary Inclosure Commissioners, at least on the map, of the boundary's irregular, historical line. Another prominent stone, at the eastern end of 'Temple Hedge' in 1811, stood at the most northerly end of Fyfield, where the parish boundaries of Winterbourne Monkton, Berwick Bassett and Preshute meet that of Fyfield by The Ridgeway (SU 12567297; Figure 1.2).

The southern boundary

At the far south east of the modern parish, land later divided into Clatford Park, Savernake Grounds and [West] Overton Heath was attached to Oare in the ninth century (*see above*). Fyfield abutted that land on its north, making it likely that the original southern boundary of the 'Five hide' estate was Wansdyke (Figure 12.1). If our argument that early boundaries ignored Wansdyke because they pre-date it is correct (Chapter 13), then it may follow that an estate which follows Wansdyke for a considerable length is a later, post-Wansdyke, creation. Hence, Fyfield estate would be later than *c* AD 500. This hypothesis is supported by the name Fyfield, which is late Saxon in creation (Gelling 1993, 236), and by the East Overton charter which clearly indicates Fyfield was not a distinct entity, as it is today, in the tenth century. It also suggests that Oare, as a separate estate, is a late Saxon development too.

SOME FEATURES IN THE MEDIEVAL LANDSCAPE

The references to fields in the manors of Overton and Fyfield in the records of the Bishop of Winchester (Chapter 9) clearly indicate the operation of an open-field system, with four fields named after the cardinal points and one, *Munkfield*, so called after the ecclesiastical overlord (1280; cf *Munkmede* 1248). 'La inlonde and above the church' (1312) seems to imply some recognition by implication that there might have been some sort of 'outfield', at the time when an anonymous successor to Richard of *Raddun* was ploughing up parts of Fyfield Down well outside the permanent, common fields. *Gravelesputte* (1312) indicates the exploitation of the mineral resource in the bottomland, perhaps at the supposed gravel pit close to the east end of 'Piper's Lane' and cutting the line of the Roman road as it made for the Kennet crossing at Back Fyfield Bridge (Free 1950, 11); at about the same time someone was breaking up sarsens high on Lockeridge Down (Chapter 5, site OD II; FWP 66).

In 1719 some lands in the open fields and common meadows of East Overton and Fyfield were inclosed. Yet the splendid map of 1811, made for the Revd R C Fowle a decade before Parliamentary Enclosure, represents a familiar landscape. One small but vital feature in practical terms was a pound (SU 14226817), now badly damaged but still recognisable just south of the sewage works at a kink in the lane to what used to be the *Attely* sheep-cote.

OVERVIEW

Overall, Fyfield seems to have been a place affected, rather than a place effectively in control of its own affairs. The Roman road and villa were obviously imposed from outside. Its shape and position hint that it may have originated as a distinct land unit only at a late stage in the fragmentation of a larger estate, perhaps in the tenth century. The estate was always the junior partner in its externally imposed partnership with the medieval, episcopal East Overton manor and was, it seems, fairly peripheral to the adjacent Templar estate and their higher concerns. Later, it was but a minor part of the estate of the Duke of Marlborough and could only reorient itself when others inserted their new turn-pike. It only moved towards parity with its neighbours in access to adequate woodland with its acquisition – courtesy of external decisions by others – of Clatford Park in the late nineteenth century. Similarly, while various recorded floodings and burnings may have been acts of God – also, in a sense, externally imposed – the external, destructive decision to widen and straighten the turn-pike, now the A4, followed by the external planning permission to allow residential development, became the actions which have largely made the village look the way it does as the twentieth century closes.

Chapter 12

The Southern Upland: Denes, Down and Woodland

South of the villages are the broad, north-facing sides of the Kennet valley, of Lower and Middle Chalk but tending to be covered with Clay-with-Flints to east as well as south (Figure 12.1). Here lay large expanses of open fields, especially in West and East Overton (eg, Figure 8.3); in general probably, and in places certainly, they themselves succeeded earlier arable arrangements. The southern edges of the arable throughout time seem largely to have been remarkably consistent, stopping more or less along the same line at the northern fringe of the woodland that has permanently occupied a large area of Clay-with-Flints (Figure 12.3).

Beyond, to the south east, reflected primarily in place- and field-names, lay a small area of heathland. The area overall contains a greater diversity of resources than the supposedly favoured downs, a fact probably reflected by its archaeology which, from the Mesolithic period to the twentieth century AD, is as consistent and varied as the better known material on Fyfield and Overton Downs.

A WOODLAND ARCHAEOLOGY

Monumentally, most impressive is a long barrow (G12; SU 157656) in West Woods just north of, though not mentioned in, the boundary of the tenth-century charter (S449). The 210m-contour crosses its eastern summit, its highest point at 4.3m. The barrow's principal features are clear (Figure 12.2; *contra* Barker 1984, 18) since it stands proud in managed woodland with well-spaced beech trees, though some trees stand on it. The oval-shaped mound is 40m long and 33m wide, with side ditches of crescentic plan along each side, respectively 43m long and 10m wide on the north side and 39m long and 12m wide on the south. A shallow extension 17m long has been dug north eastwards from the eastern end of the southern ditch, spoiling the overall symmetry in plan and leaving only a narrow eastern causeway. The original western causeway is 20m wide. The mound appears to have suffered only superficial disturbance on its top ('Ex' on Figure 12.2) but it was in fact surreptitiously trenched from the north about 1880. A report was never published but Passmore (1923) recorded the finding of a four-sided, rectangular stone chamber with capstone beneath a central cairn of small sarsen stones.

Morphologically quite unlike the classic rectilinear long barrows at nearby East and West Kennet, as a plan type it does not feature in Ashbee (1984, chapter 3) but it has close parallels in Hampshire (RCHME 1979b, fig 2, nos 14 and 15, fig 3, nos 19 and 20) and West Sussex. North Marden (Drewett 1986) in particular, though smaller, is almost an exact mirror image in plan. Though slightly wedge-shaped and lacking the distinctive curving side ditches, the newly discovered White Barrow 2km away on the hill above Lockeridge Dene (*see above*; *SL*, colour plate 25) is of similar size. It is tempting to envisage both long and oval barrows existing in forest clearings rather than on open hills in the fourth millennium BC (cf Barker 1984, no. 10), and we argue below (Chapter 16) that one way of looking at them in this landscape is as markers on the edges of community territories. Ashbee's map alone (1984, fig 8) suggests the thought.

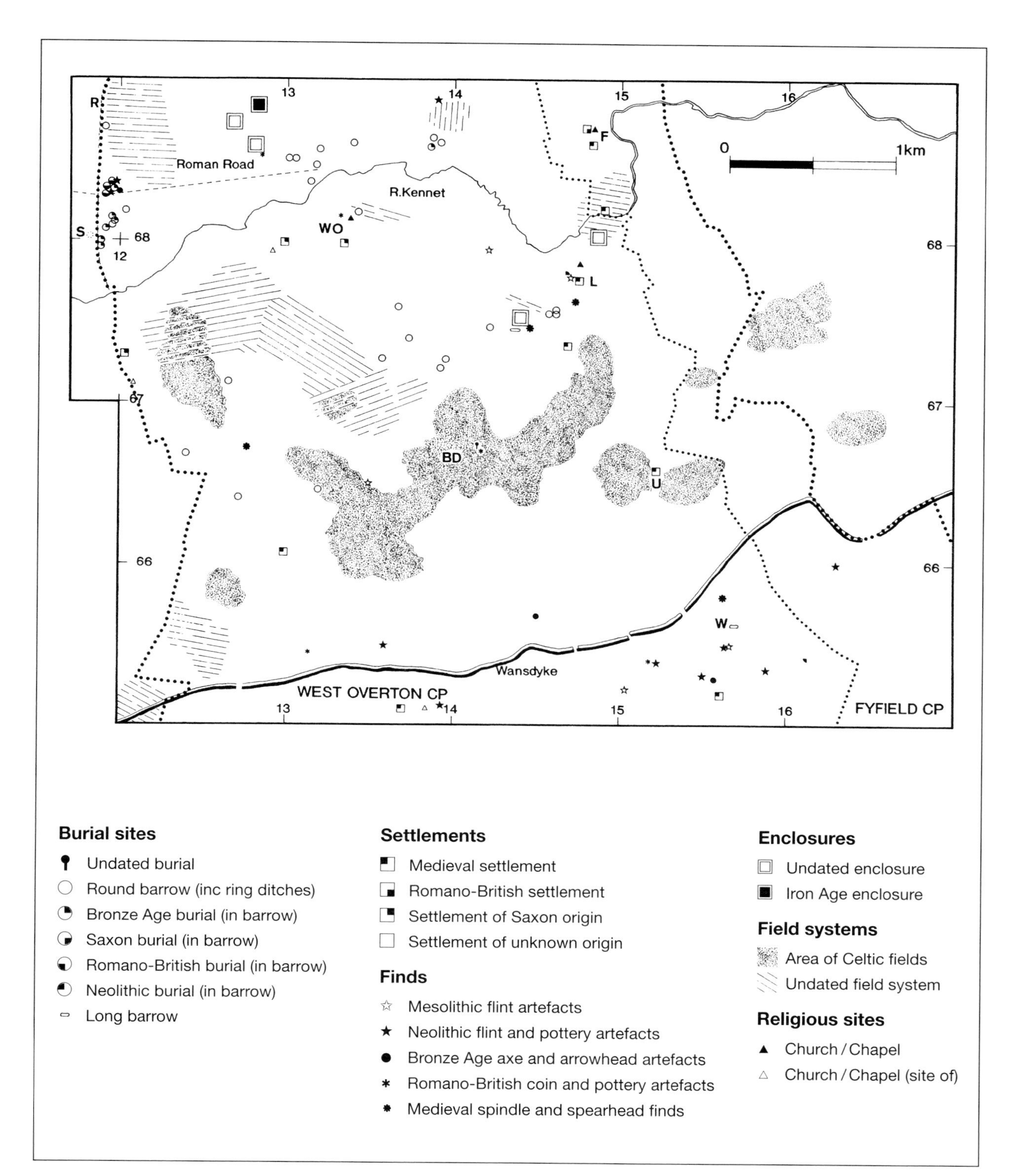

12.1 *Map of the southern part of the study area graphically displaying the archaeological data recorded in the SMR*

BD = Boreham Down; F = Fyfield; L = Lockeridge; R = The Ridgeway; S = The Sanctuary; U = Upper Lockeridge; W = West Woods long barrow; WO = West Overton

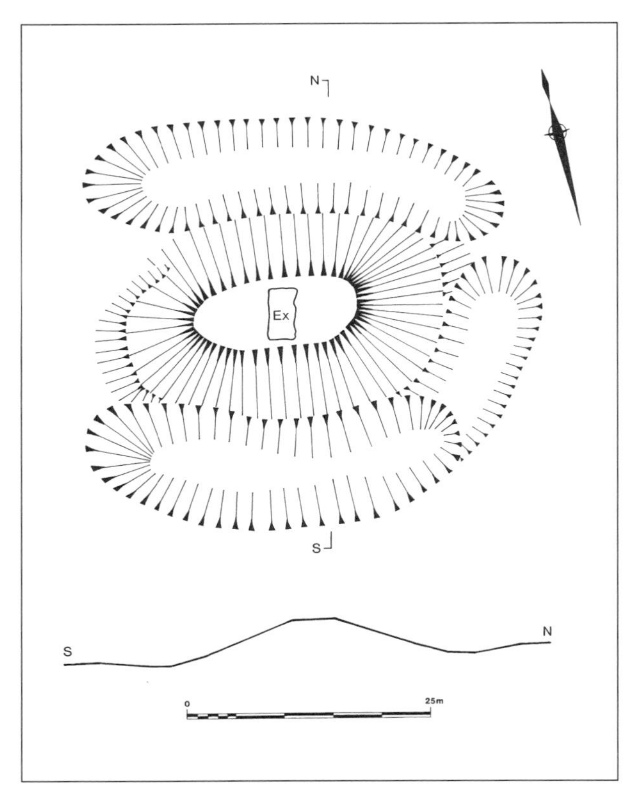

12.2 Plan and profile of West Woods long barrow

Ex = nineteenth-century excavation pit

Mesolithic flint debris has recently been reported south of the long barrow (*WAM* 89, 152), part of an extensive area of working long recorded by the Bull family of Bayardo Farm but not yet translated into a systematic study in the field. Nor has the considerable amount of Mesolithic debris and flint tools yet been comprehensively studied, though Brown (1997) has made a brave and single-handed start. The collection includes early Mesolithic material; so fresh is so much of it that somewhere in the vicinity quarries or mines in addition to knapping areas may be anticipated. The possibility of settlement too, along the edges of the woods and on the south-facing plateau overlooking the Pewsey Vale, is strengthened by evidence excavated in 1997 on Golden Ball Hill, immediately outside the south-west corner of the study area (Anon 1997).

Indeed, as noted below (Chapter 13) when discussing the equivalent area around Shaw 2km west on the south-western fringes of the woods, this plateau has emerged as a significant activity zone in its own right, especially for prehistoric times. In former Lockeridge tithing (West Overton civil parish), for example, Neolithic material has been recorded both to the south east of the long barrow (*VCH* I, i, 120) and west (Burchard 1966). The latter, only 1.5km east of the Shaw finds, 'hints that this area, just within the Clay-with-Flints belt, may well have been free of trees at least temporarily during this period'. A possibly Early Bronze Age rectangular flint knife and large barbed and tanged flint arrowhead were found in 1848 during grubbing-up where Pickrudge Wood now stands (*WAM* 44, cxlviii, 99). Collectively, the evidence suggests at the very least a phase or phases of Mesolithic, Neolithic and Early Bronze Age activity involving occupation, burial, probably hunting and collecting, and possibly even tree-clearance and farming, perhaps along the edges of clearings in permanent, deciduous woodland.

The recognition of Romano-British pottery, probably 'mainly first century AD in date', is locally of considerable interest. It was found in an area, freshly ploughed, of 'patches of soil containing burnt matter', some with chips of sarsen, 'markedly affected by heating, and the occasional fragment of bone'. The conclusion that these patches 'are the result of tree and scrub clearance' is uncontentious but that such work was 'in Roman or later times' seems unproven (Burchard 1966). Such clearance could have been earlier too, and indeed the evidence may hint at a repetitive practice. Nevertheless, whenever else clearance may have occurred, the hint that land clearance may have been carried out in the (by implication, second half of) the first century AD fits in very well with an interpretation, independently witnessed on the downs, arguing for a locally widespread phase of landscape reorganisation and exploitation *c* AD 100 (Chapters 2 and 5–7).

Air photographic inspection and new RCHME photography plus field reconnaissance have also added, especially during 1995–7, a number of ploughed-out barrows, including the new long barrow above Lockeridge Dene with its nearby group of circular cropmarks and rectilinear enclosure (Chapter 10; Figure 10.2; *SL*, colour plate 25). Another small rectilinear enclosure (SU 138672) is morphologically more akin to Middle Bronze Age ones elsewhere on the Marlborough Downs. Fragmentary air photographic traces of a larger, more circular enclosure on Lurkeley Hill suggest a site akin to 'Headlands' and OD XI, with traces of probably prehistoric fields to its north and east in the area subsequently cultivated in the furlongs identified in relation to 'Crooked Crab', 'Hollow Snap' and 'Alton Way' (Figure 8.3).

There were also other, smaller medieval settlements south of the valley villages, on higher ground and apparently secondary in settlement history. The largest, Shaw (Chapter 13), is an exception which, although up at around 229m (750ft) aOD, could well be primary to the Saxon settlement pattern and even a medieval repetition or continuation of an earlier place of habitation. The generality of these higher places is that they represent colonisation outwards as well as upwards from the valley settlement pattern from Roman times onwards. Some seem to originate in the late Saxon period, others appear in the twelfth and thirteenth centuries. All are now wholly or partially deserted.

Boreham and Boreham Wood

Place-name evidence suggests the former existence of an otherwise unknown early settlement of *Burham* or *Borham* in or on the sides of *Wodens dene* climbing south west from Lockeridge. A derivation from the Old English *(ge)būr ham*, or 'peasant's homestead', has been proposed (*PNWilts*, 305–6). On etymological and topographical grounds a location for this homestead at or near Boreham Barn (SU 128660), or further to the east on Boreham Down, is presumed. At the latter, the prefix 'Bore-' may well be a modern corruption of the fields named *Baresfeld* and *Old Berye* which occupied the same area in 1567. The etymological reference could have been to the, at that time, well-preserved earthworks of the pre-medieval fields on the down (Figure 12.3). It is conceivable, therefore, that a farmstead, the *ham*, lay in an apparently otherwise unoccupied Anglo-Saxon landscape of the upper reaches of Lockeridge Dene, perhaps in the vicinity of the 1970s' Greenlands Farm (SU 14216694) near the boundary between the Overtons. One of the reasons for suggesting that area is the demonstrable frequency with which settlements occur beside or astride boundaries in the study area (for example, Chapters 4, 6 and 8).

It is more likely, however, that Boreham derives its name from the barrow, called White Barrow in 1794, which lay across Lockeridge Lane in Windmill Field (SMR 740; SU 13206649; *PNWilts*, 423; Gelling 1993, 127–8). The reference to the *haethene byrgils*, a heathen burial-place, in the West Overton charter (S784) between Shaw and *lorta lea*, Lurkeley Hill, clearly suggests White Barrow was the *haethene* landmark (cf Bonney 1976, fig 7.7). Indeed, *Burh-ham* is regarded as 'part of the earliest place-name-forming vocabulary of the Anglo-Saxons' (Gelling 1993, 128). Such heathens were presumably sub- or post-Roman Britons or pagan Saxons, rather than prehistoric people. Pagan Saxon interments in prehistoric barrows are a notable feature of central and northern Wiltshire (Yorke 1995, 170), as is demonstrated on Overton Hill some 2km to the north west (Chapter 4). Indeed, if White Barrow did contain one or more pagan Saxon interments, it would add a

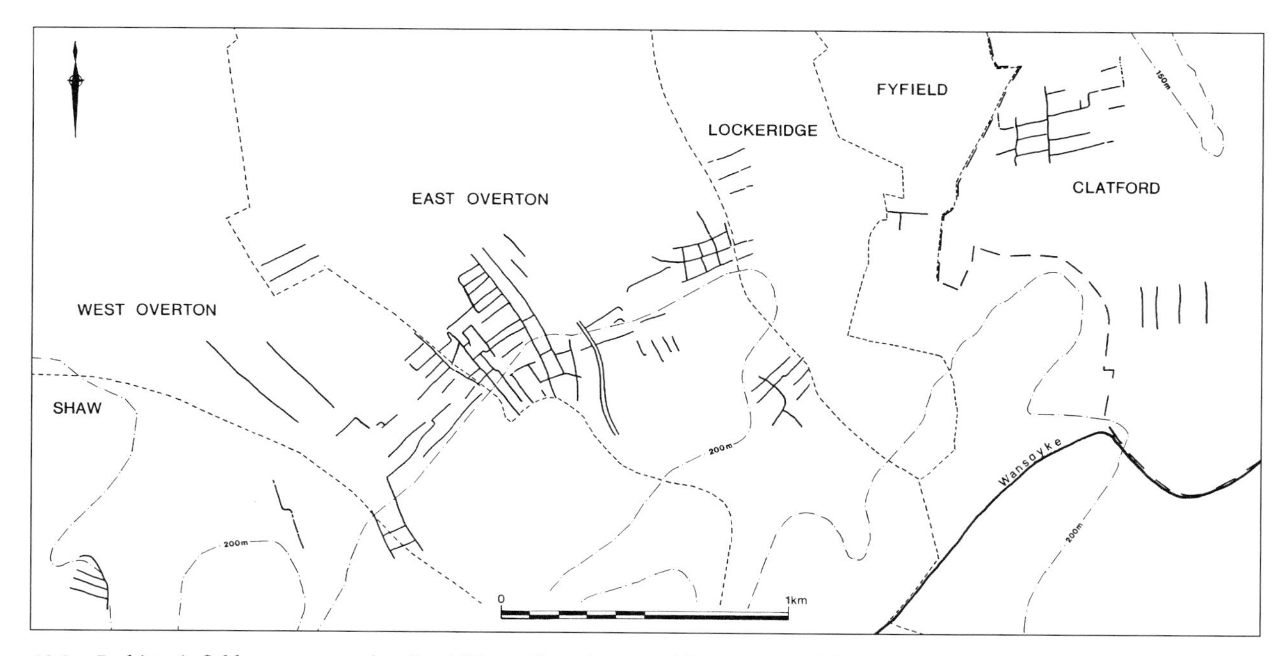

12.3 Prehistoric field systems, now hardly visible at all on the ground but as mapped from air photographs over the southern part of the study area, with tithing names and boundaries

further element to the early Saxon settlement in an area particularly rich in heathen place-names (ibid, 166; Fox and Fox 1958, 40–2; Gelling 1978, fig 11).

Saxon activity nearby is also indicated by the recovery of an iron spearhead after deep ploughing just north of Hill Barn (SU 12756675; *WAM* 68, 186; Devizes Museum, Acc No. 70.1973). The *haethene byrgils* may be recalling an event or recognition across at least four centuries though, more prosaically, such a reference could equally well be following the more recent disturbance of inhumations during Anglo-Saxon cultivation. The barrow was finally ploughed out soon after Inclosure in the early nineteenth century, hence its whiteness.

Abutting the Shaw–West Overton boundary is Boreham Wood. Like Chichangles and Wroughton Copse (Chapter 7), this wood has been forested for several centuries, possibly for more than a thousand years, and is notable for the boundary bank which surrounds it. In 1330 the bounds were predominantly demarcated by pits (Brentnall 1941, 424–5), many of which can still be seen, and the woodland covered a slightly larger area, so this bank is likely to demarcate the maximum extent of the wood in the post-medieval period, about 30ha. Documentary evidence clearly indicates woodland here in the tenth, eleventh and thirteenth centuries (S784; *PNWilts*, 305; *see* Shaw, Chapter 13). *Burham* was regarded as a wood in its own right by the mid-twelfth century and, though it pertained to the *Schaghe* estate, it came under the Forest laws as part of Savernake (Brentnall 1941, 424). In 1225, for example, Boreham Wood was 'put out' of Savernake Forest, meaning it was no longer regarded as part of the royal forest, thus allowing Henry de Luny and Thomas of Kennet to cut and sell the timber before Boreham reverted to 'ancient forest' law the following year (*VCH* IV, 418; Bond 1994, 124). Today Boreham Wood, although still confined to its banks, which have remained a constraint, covers only about a third of the embanked area.

THE ANGLO-SAXON BOUNDARIES: SOUTHERN EAST OVERTON

In AD 939 King Æthelstan granted fifteen hides at *uferan tun* (East Overton) to Wulfswyth, a nun. The boundaries of the land so granted can be identified on the ground with complete or near-certainty for most of their length, and depicted on a modern map (FWPs 11 and 68). Here we look at but a small part of the evidence as exemplar, namely a short stretch between East Overton and Lockeridge.

From the earliest cartographic evidence, the boundary between Lockeridge and East Overton ran south from the *hlinc ufeweardne*, the upper side of the lynchet at the southern edge of the arable at the end of Pickledean, to the Kennet, thus suggesting the charter's *hole weg* was the Roman road hereabouts (SU 141684) or a predecessor of the Bath–London Road, later the A4. The exact divide is complicated by the very nature of the resource: this is valuable meadow land, and from the twelfth century onwards it was to experience numerous changes in ownership (Chapter 9).

South of the *weg*, the boundary crossed the river *cynetan* either just east of Lockeridge House (SU 14786812) or just east of 'Stony Dean' (1819 map) at the bend in the river (SU 14556792). From the Kennet to a prominent *ellene*, an elder tree, the boundary then moved to *wodnes dene* (Lockeridge Dene) across the top of the spur of land just west of Lockeridge village. This locally prominent point was marked by a group of large-diameter round barrows (*see below*; *SL*, colour plate 25). After Dene the next point is *wuda on mær wege*, the wood on the boundary, or possibly pond way (Gelling 1993, 26). A bridleway climbs the hill from Lockeridge Dene into Wools Grove, the *wuda*, arguably as 'a way along the balk of a ploughland' (Grundy 1919, 241 n. 1; Figures 8.1 and 12.3).

The boundary then proceeded to *hyrs leage up to wodens dic on titferthes geat*, to Hursley Bottom then up to Wansdyke and then to Titferthe's Gate (Plates LIX and LXI, Figure 13.2). The likely route passed eastwards around West Woods, keeping 'Little Wood' in East Overton as described in 1567 (Straton 1909, 262) and later followed on nineteenth-century maps. The boundary would then meet Wansdyke and run along or beside it for *c* 300m to *titferthes geat* (cf Brentnall 1938a, 127; and below). The 'gate' was at the south corner of Wells' Copse and Little Wood and at the west corner of Barrow Copse where a track passes through Wansdyke (SU 15406570). Two of the major, long-distance 'through-ways' identified in the study area (Chapter 16) converge on this point. They, and the boundary, continue south east past the long barrow (SU 157656; Figure 16.3).

The next stretch of this boundary is described in *SL* (chapter 9, figure 63), taking it along the southern and up the western side of the estate. Here, we merely extract one point of interpretation that is of wider significance. The boundary reaches the northern edge of the woodland and, beyond, what is now pasture (SU 143666) which may well have been arable in the tenth century. A thin strip of hedge, baulk and trees remains today, possibly the remnants of the Saxon *hlince* along

Plate LVII West Woods long barrow from the south east

what became the northern edge of *Allen's Higher Ground* on the 1794 map. A prehistoric field system formerly existed on Boreham Down (Figure 12.3) immediately to the north west, however, and the *hlince* is more likely to have been an earthwork of that than evidence of tenth-century cultivation (cf *below* on Lockeridge Down).

The estate boundary from here followed the existing hedge line *to west heafdon*, to west headlands, marked by two erect stones in 1784 (again echoing arrangements to the north on the downs; *see* Chapter 4). Being *western* headlands, they demarcated the western limits of the ploughland of East Overton in the tenth century. The boundary line then moved across downland, *north ofer dune*, and north west up the hill to a still very distinctive hedge in the modern landscape, *thaet riht gemaere*, the straight balk or boundary (SU 132668). It marks the tithing boundary and the northern edge of the strip called 'Lewis's Ground' on the late eighteenth- and early nineteenth-century maps. This is the *scyfling dune* of the West Overton charter, echoed by the name *Schufly(g)don* in 1312 (WCL; DM, Kempson Notes). The *scyfl* element refers to the gently sloping nature of this hill (Gelling 1993, 186–7). Its further course down to and through the Overton villages is discussed in Chapter 9.

TRACKS, FIELDS, DOWNLAND AND FOREST

Such is the close relationship between the topographic and man-managed resources south of the river that it is difficult, and to an extent misleading, to attempt to separate them. Yet downland and forest are separate spatially, with the fields largely confined to the former. The tracks link all three elements (Figure 16.6), here and right through the two parishes, and are discussed in general for the whole of the study area in Chapter 16.

South of the river, the earliest fields are prehistoric. Their incidence, pattern and form, as recoverable largely from air photographs, are displayed on Figure 12.3. This map, in fact, also shows a striking relationship between the tracks and the woodland; and indeed between them and the early fields fringing the northern side of the woods. They all respect one another. The ways, all without exception elements in the great north–south 'Ridgeway route' (Chapter 16), feed into the same entrances to the woodland and the nodal points within it. They pass through the woodland along lines in part still in use today and suggestively already old when they were in part picked up in the tenth-century charters. An implication of this is, of course, that the woods were themselves already spatially structured as this relationship developed, by late Saxon times certainly, in Roman times probably. We may use 'probably' because these ways and tracks also respect the ancient fields: with one exception, they go round the areas of enclosed fields.

The exception rather reinforces the observation, for the line through the broadest part of the central area of fields is clearly a hollow-way on 1946 air photographs on the line of the common tenth-century boundary between West and East Overton. This particular relationship both indicates that these fields were no longer in use, for the track just cuts straight across them, and suggests that in general the lines of the tracks and the field areas are broadly contemporary components of a cohesive landscape. This latter point is further emphasised by the line of another track-cum-hollow-way that formerly existed as earthworks and soilmarks on 1946 air photographs. It loops around the north-east edge of the same field group (Figure 12.3), joining the through-track at both its ends. This appears to be the original track, physically and archaeologically an integral part of the system of ancient fields; but subsequently made redundant by a more direct way across the, by then, abandoned fields.

This original track, and its fields, are likely to have been in use in the landscape of the early centuries AD and may indeed have originated then as part of the specifically 'Roman' landscape (Chapters 2 and 15); though their origins may well lie much earlier in the later second millennium BC. In either case, they carry with them a strong suggestion that the pattern of lines on the landscape probably is an old pattern, with elements in it certainly going back to early medieval and Roman times and, perhaps, even earlier.

After the establishment of the early Saxon settlements an expansion (or reclamation?) of arable is likely to have occurred just up from the valley floodplains in the seventh or eighth centuries. Initially, for West Overton, this would probably have been just north of the river probably at the site of 'Headlands' (Figure 4.2). North of the river, for East Overton, Lockeridge and Fyfield, cultivation had long been carried out on The Fore Hill, where the land had been divided into furlongs by the early tenth century AD (S449), and on Lockeridge and Fyfield Fields as far as the Valley of Stones (Plate VIII; Figure 2.1). Further north, cultivation of the downs certainly occurred in the tenth, thirteenth and sixteenth centuries (Chapters 6 and 7).

Though some cultivation of the land is likely to have taken place immediately south of the villages in the early medieval period, for example at Rylands in Lockeridge, the expansion of arable on to the large areas of downland south of the villages, that is to say on *scyfling dune* (S449) and White Hill, is likely to be predominantly a thirteenth-century phenomenon. Large-scale reorganisation of the landscape into open-fields divided into strips may have taken place, however, before *Domesday* (cf Hall 1988, 99–122; Costen 1994, 100–1), indeed perhaps at the same time as the insertion of two planned villages (*see below*). From the thirteenth century onwards, this area was the limit of cultivated land south of the settlements, until, that is, the later 1950s when the land on Audley and Boreham Downs, which had until then remained as rough pasture downland, was ploughed up.

The parishes are well endowed with documentary evidence of their southerly field-names. The names are in themselves a roll call of landowners, activities and land-use, and of different landscapes, at various times: eg, *Munkfield* (1280), *Northfield at Scrufeleput* (1312, *scropes pyt* of S449) and *Gravelesputte* (1312). Many of the names are prosaic and other places can easily supply the same or their equivalents; for example, the 1567 Survey notes 'Whithill', 'Baresfeld' and 'Connyfelde'. On the other hand, some names are unusual or particularly eloquent with some speaking of particular activities, others of functions specific to a particular spot in the landscape: eg, again from 1567, 'Coteclose', 'Mylhayes close', 'Puthay', a pasture called 'Sheldford' and a 'Cotagium' called 'Mawdyes'; and from the 1671 Glebe Terrier 'Bittom', 'Blacksmith', 'Paddle Drove' and 'Bum Furlong'. Collectively, they indicate the range of busyness in a working agricultural landscape; they also spell out a perception of a landscape familiar to its inhabitants but perhaps invisible to the passer-by. Overall, they are very much of this place and of its history, landscape and personality quite as much as the pragmatics of tenure and farming.

Equally characteristic of the later landscape in the southern half of the parishes were windmills. West Overton's stood on Windmill Hill, though its precise site is unknown and it may well have disappeared before the 1790s (Figure 8.3); Lockeridge's stood on the spur above Dene, perhaps one of the circular cropmarks (*SL*, colour plate 25).

Plate LVIII West Woods: Victorian estate boundary stone ('HM' stands for 'Henry Meux')

Woodland

Overall, the most striking feature of the woodland in this study area is the stability and consistency of its position and size across the centuries (Plates II, LII, LVII–LXI; *SL*, colour plates 26, 27, figures 14, 54). Here, at the southern edge of the two parishes, is where trees have demonstrably grown since the early Saxon period and very probably since Roman and prehistoric times. Moreover, though the area of the woodland has been reduced since its severance from Savernake, especially along its edges, the acreage covered by trees appears to have been stable since the late eleventh century. That this area was for trees was no doubt reinforced by the construction of high boundary banks, believed to be medieval but not yet sufficiently well dated, which not only demarcated the limits of the arable and pasture with the woodland, but also regulated the flow of animals to and away from the trees.

This woodland was, of course, a precious resource, giving to the landowners and communities of Fyfield

Plate LIX West Woods, Hursley Bottom, looking south west from its north-east corner: this permanent clearing, large enough to provide pasture in the woods, is also a 'nodal point' in the landscape where six tracks of The Ridgeway route meet (cf Figure 16.8)

12.4 *Map showing the permanent woodland (diagonal lines), with eighteenth-century names for the individual plantations that made up the woodland. The areas demarcated by the tree-stump symbol are those with cartographic evidence for active woodland management in the later eighteenth century, namely coppicing, planting and felling. The plantations and their management generally respect Wansdyke*

and Overton something absent from many downland estates (Figure 12.4). Such a resource added significantly to the economic viability of the manorial communities, explaining why all four tithings stretched southwards to reach the woods (Figures 1.2 and 8.1). Fyfield seems to have been least well provided, for it may well not have extended beyond Fyfield Wood in medieval times. Its present (1898) southern extent and shape is the one instance where the modern civil parish differs significantly from earlier arrangements (*see above*). The desirability of having, as of right, access to the woods is obvious: they provided not only wood for buildings, fences and fires, and food in the form of hunted animals, wild fruit and fungi, but also grazing for livestock, pannage for swine and, today more than ever before, a place for recreation (cf Bond 1994).

Hunting was often, and to some extent remains, the *raison d'être* for the creation, management and, ultimately, the survival of much of the woodland in the study area today. In the parishes of West Overton and Fyfield hunting has been a constant feature from prehistoric times through the medieval period to the present day (*WAM* 53, 194–5). In the fifteenth and sixteenth centuries deer were hunted, though by the late nineteenth century the presence of numerous aviaries in West Woods, mostly managed by the Meux estate, indicate that game was then predominantly pheasant. Rabbits, foxes and deer were also hunted. A lone wooden 'game-cabin', perhaps for hunters as much as gamekeeper, survives from palmier days.

The other main crop derived from woodland is, of course, the wood itself. All the woodland in the study area has experienced felling, management (thinning, coppicing, pollarding, lopping) and replanting at some time over the past two millennia, if not for longer. Some hedges and indeed clearings may well be effectively permanent features resulting from woodland management. *Withigmeres hege* (S449, 'Willowpond Hedge'), for example, was perhaps a fairly sizeable hedge in the tenth century, given that it was chosen as a boundary marker. Hursley Bottom (1816 and modern OS maps; Plate LIX) was called *ers lege* (S784) and *hyrs leage* (S449) in the tenth century and 'Hurseley' in 1567. It seems to be a feature of the Anglo-Saxon woodland landscape which has persisted (Figure 12.4, Plate LII). It cuts Wools (or Wolfs) Grove and Wells Coppice and, being a lea and still an open area, may not have been wooded for a millennium (cf FWP 68; Plate LIX). If so, such an open space must have been maintained deliberately. Other documentary sources indicate further woodland felling to create permanent clearances was carried out from the twelfth to the fourteenth centuries (cf Fosbury, Boreham and Breach Cottage, Chapters 10 and 11). The owner in the Tudor period planted hazel, oak, willow and maple on the estate to supply the demands of sixteenth-century England, while modern demands have created a woodland landscape which is above all of beech and fir, with pockets of mixed broad-leaf trees (cf Plate LII). Names such as 'Priest Down Common' and 'Tenants Down' in and around the woodland demonstrate that the grazing and feeding of animals were also an important part of forest activity: cows and sheep grazed the clearings, while pigs foraged under the tree canopy for fungi and acorns.

Though trees have spilt over their plantation boundaries today, much of the woodland in the nineteenth century, and no doubt earlier, was made up of smaller units originally contained by banks, sometimes with ditches. Many of these banks are still visible (Plate LII) and correlate in general with the black enclosure lines on modern OS maps at 1:25,000 and larger scales (*SL*, figure 14).

HEATHLAND

The agricultural viability of the estates was further enhanced by having an area of rough grazing, usually held in common, at their southernmost edges (Plate LIX). At first these areas were wooded or scrub, though felling, grazing and cultivation over the centuries has created land fertile enough today to support three farms in this area. West Overton's common lay *c* 1.5km east of the south-east corner of the tithing and was called Abbess Wood in the Middle Ages. After the insertion of Clatford Park, however, the common, it seems, was displaced a little south and became known as Savernake Park.

East Overton and Fyfield also had areas of heathland, both of which were called Heath Grounds in the late eighteenth century. The origins of these two areas are obscure, though heathland certainly existed hereabouts in the late medieval period (*PNWilts*, 307; Straton 1909, 264), possibly even in the first century AD (*WAM* 61, 98). Some heathland in this area also appears to have belonged to Lockeridge manor before the early nineteenth century (1794 map).

PARKS

The late sixteenth-century creation of two deer parks in the study area south of the Kennet is discussed in Chapter 11 (Figure 10.1).

BRICK-MAKING AND CLAY EXTRACTION

The wooded area between Heath Plantation and Strawberry Ground was called 'Brickkiln Copse' in 1889. The place-name reflects an important though small-scale industry in the area, namely brick-making. Production sites, including kilns, were situated on top of the red clay both here at the southern end of the parish and at the far northern limits (*c* SU 12907218). Wagon routes were created to transport the fired bricks to the major roads and towns. The clay was simply extracted by digging pits, and as deposits are so close to the surface, the pits were seldom very deep. Those in Brickkiln Copse, some in the ditch of Wansdyke, do not exceed *c* 2m.

At the south-east corner of Brickkiln Copse (SU 15576507), bricks were fired but the evidence on the ground is ambiguous about the method. A large pit dug into the north side of the ditch of Wansdyke both contains and is surrounded by fired waste, with indications to its north of the site of one or two buildings shown on the 1889 OS 6-inch map. There may have been an above-ground kiln or kilns, perhaps using an old pit as its flue, or the firing may have been in pits, presumably ones created by clay extraction.

SARSEN STONES

The Kennet valley sarsen industry – the breaking and carrying off of sarsen stones – was at its peak in modern times from the late nineteenth century to the 1930s; but this highly labour-intensive industry did not recover after World War II (Plates III and IV; *SL*, colour plates 8, 9). Its history and methods have been well recorded elsewhere (Free 1948; King 1968), so our coverage of a distinctive local activity is minimal.

Sarsens, worked or roughly cut, were of course one of the main local building materials, along with wood, chalk, clunch and flint. This was so from the fourth millennium BC, as megalithic barrows and other prehistoric stone structures witness.

The excavation of the *polissoir* incidentally showed that sarsen breakers were active on the northern downs in the medieval period (Chapter 5, site OD II; FWP 66; *SL*, colour plate 11, figure 24). Attempts were also made to utilise sarsen technologically, notably as grinding stones (*SL*, colour plate 22, figures 19b, 38), but also to make a bridge.

Stones were also used as boundary markers, especially on the downland areas where an erect stone standing in the treeless landscape would be a distinct feature (cf Saxon charters, FWP 68), and many were given names, such as 'Sadlestone', 'Trippingstoone' and 'three cornered stone' (Straton 1909, 147, 264).

Later, the external demand for kerbs, road-metal and setts for tramways led to relatively large-scale sarsen exploitation and the export of material; yet, in a return to meeting local needs, many of the sarsens from West Woods (Plates V and LIV) were broken by explosives and crushed to produce chippings for the new surface for the A4 road between the Wars. This surface did not last long (*WAM* 52, 338–9).

OVERVIEW

This overview of the lands south of the Kennet may well suggest, despite its sketched and brief nature, a landscape of varied resources under intense exploitation for much of the time in our purview. We see it responding to external and local community need, yet simultaneously influencing the sorts of society acquiring its livelihood in Overton and Fyfield. Natural resource and technology were, however, only part of the dynamic driving – some would say inhibiting – the sorts of lives people lived hereabouts for, from early medieval times onwards, we see two other major factors at work: landlords and tenure. They appear in documents and then in maps too, and therefore late in our time-span; but it is reasonable to envisage their presence earlier.

Chapter 13

The Southern Upland: Shaw and East Wansdyke

SHAW: MANOR, TITHING AND VILLAGE

Shaw lay in the relatively remote south-west corner of our study area (Figures 1.2 and 13.1). It was a peripheral sort of place, now almost completely unoccupied; yet paradoxically its principal settlement was relatively large and on a main route between the downs and the Vale of Pewsey. Almost all its former tithing area is now neatly subsumed within West Overton civil parish, but historically it straddled the boundary between two hundreds, Selkley and Swanborough, and two ecclesiastical parishes, West Overton and a detached part of Alton Barnes (Figure 1.2). Even now, Shaw Copse and part of the adjacent southern area are in Alton parish (Figure 13.1). This current partition is in accord with both facts and ambiguities in the historical record and interpretations of it (*VCH* X, 813; *VCH* XI, 183, 190–2; Bonney *in litt* 1996) though elucidation of the *Domesday* records, at least, will clarify this apparent confusion (*see below*). Behind the uncertainties, however, Shaw exhibits what seems to be a long history of existence in a local boundary zone, topographically, geologically, economically, tenurially and perhaps politically.

The area lies either side of the 225m contour, with qualities partly peculiar to this manor but also with ones that add to the range of geology, soils and resources available in the locality. Its position and altitude put it on a Clay-with-Flints capping. Probably significantly in view of its distinctive history, the area lies on the edge of historically permanent woodland of the sort which exists locally only along these southern reaches of the parishes with, at present, glades of old pasture and clearly ancient hedge-lines, one of which is the Anglo-Saxon estate boundary between West Overton and Shaw (Figure 1.2).

Superficially, then, in topographical and other respects, Shaw is very different from the open downland pasture only 4km to the north (and indeed along the Pewsey escarpment only 2km to the south), yet historically and, to an extent archaeologically, it can be seen in its local context as a southern equivalent to *Raddun* (Chapter 7). Like *Raddun*, hereabouts is a place of long-lived activity; but, unlike the totally abandoned *Raddun*, Shaw is occupied today, at least in the sense that a 'model farm', Shaw House (SU 13156545), exists in the core area of the tithing. The house lies immediately north of Wansdyke and on a slight peninsula from the main plateau immediately to the east (Figure 13.1). It also lies close to the earthworks of the abandoned village and earlier farmhouse. Such conjunction, though, does not 'prove' continuous settlement and indeed the significance of the place for present purposes may well lie in chronological references across the landscape rather than on the spot.

Wansdyke itself clearly indicates that, at least for a time in the fifth/sixth century, any 'Shaw' that may have existed was in 'frontier country'; that it was also in a boundary zone at another time or times, perhaps before and certainly later than the dyke, is further suggested by the pattern of largely ancient parish-cum-ecclesiastical boundaries here. Basically they ignore Wansdyke but jostle for space with each other, giving the impression that the area is at the further limits of economic rather than political units. It is certainly central to Bonney's argument (1972, 174–6) that an organised landscape, eventually represented by parish boundaries, existed before Wansdyke was constructed.

Archaeologically, the core of the Shaw area is indeed a deserted medieval village of which the main earthworks are in a good state of preservation (Beresford and Hurst 1971, 166). The village has not, as yet, been the subject of detailed archaeological ground survey but our Figure 13.1 is the best record currently available. It is based on a MSS OS 1:2,500 'divorced survey' plan now in the NMR, significantly amended and complemented by ground checking and, above all, aerial photography (of which Plate LX is an example). In particular, the extent of the crofts out to the clearly defined western boundary are shown for the first time – they were flattened in the 1960s – and some detail is added on the north east where modern assarting has also destroyed earthworks but revealed soilmarks. Such evidence also suggested that an outer, second ditch might have existed around the curious circular ditched enclosure partly still surviving in the wood on this side. The enclosure itself has been variously identified as a disc barrow or small henge, perhaps unlikely monuments in this location. It is certainly earlier than the medieval village, however, and might more probably be a small settlement enclosure of late prehistoric/Roman date, comparable to, perhaps even part of, the complex just outside our study area but only 0.5km to the south on Draycot Hill/Gopher Wood (*VCH* I, i, 77; partly-shown on OS maps from 1:25,000 scale upwards).

The site of the church, which remains prominent as an earthwork inside a small enclosure, presumably the graveyard, was trenched in 1929, one of only three recorded archaeological excavations of a medieval church site in England before 1939 (FWP 66; Brentnall 1929; Beresford and Hurst 1971, 82). The most prominent earthwork, however, is a length of East Wansdyke, impressive south and south west of Shaw House, but slight and even apparently almost flattened by earthworks at the head of the deserted medieval settlement. This flattening of Wansdyke itself represents the site of 'Shaw Farm' (Figure 13.1; Smith 1885, pl ii, no 3; SU 13546534), at least from the seventeenth century onwards and possibly the manorial farm since the fourteenth century (*VCH* XI, 191). The farm seems to have been deserted by the late eighteenth century, as the site was then referred to as 'Old Shaw' (ibid), presumably as a result of the construction of 'Shaw New Farm' (first edn, 1-inch OS map), present-day Shaw House. The last of its outbuildings, a barn, was demolished *c* 1970 (*VCH* XI, 191).

Immediately north of the site of the church, between it and the parish boundary with West Overton parish and the north-west corner of Shaw Wood (SU 139652), an area of new arable had (in 1995) recently been ploughed over earthworks – presumably of the northern part of the deserted medieval village (Figure 13.1). Indications in the topsoil suggested occupation. A rapid superficial search along the southern and eastern edges of the field produced a scatter of post-medieval material, a Romano-British pot sherd and four Neolithic sherds. Nine worked flints were also of Neolithic type; two small broken blades were of Mesolithic type (FWP 83).

This material clearly hints at phases of activity, probably occupation, somewhat earlier than hitherto imagined in the long story of Shaw. Neolithic activity is not, however, implausible, despite perception of it in the Avebury area as having been largely on the downs and along the Kennet valley. Shaw, despite its marginality to such places then as more recently, nevertheless offers resources characteristic of an interfacial zone, here along a woodland margin, to counterbalance its obvious disadvantages of altitude and clay soil. It would surely have been part of a hunting zone, at the very least for people coming and going at Neolithic Knap Hill, some 2km to the south west. It may too have been more central to Neolithic communities in the area, for the slight evidence here has to be seen in conjunction with other scraps in and around these woodlands (Figure 12.1). There seem to be hints here that Shaw, high on this southern, wooded periphery of the study area, may not only reflect the history and archaeology on the better known downs in the northern part of the parishes, but also complement those areas in making the estates viable as working economic units. The chronological range at this specific spot, as represented by the surface scatter of material, was remarkably similar to that proposed in general for the woodland on the Clay-with-Flints (cf Chapters 11 and 12).

Though it was a manor within the ecclesiastical parish of West Overton during the twelfth century, Shaw appears to have been already a recognisable estate in *Domesday* (*see below*). It possesses a name originally appropriate to its topographical position, that of an area near a small wood (*PNWilts*, 307; Gelling 1993, 208–9). Indeed, the etymological origin of Shaw no doubt arises from the *sceagan* of the West Overton Saxon charter (S784). In fact that charter refers to *langan sceagan*, a 'long, small wood' (Gelling 1993, 208), and this may well be the same wood which measured one league in length by one furlong in width in 1086 (*VCH* II, 154). Though this narrow wood no longer stands now, it may have

Plate LX *Shaw deserted medieval village (*lower left*) on 1 December 1952, showing its position on the Clay-with-Flints plateau, characterised by arable in modern rectilinear fields, between the south-western edge of West Woods (*diagonally from top left to centre right*) and the Vale of Pewsey (*off bottom*). The village site was then marked by earthworks and irregular hedge-lines, now partly removed, and its central hollow-way (*centre left to centre bottom*), part of a through-route (cf Figures 13.1 and 16.7). Three irregular lines in the landscape are early medieval or earlier: Wansdyke runs west–east (*left–right*) across the centre of the photograph; the hundredal/parish boundary between Selkley and Swanborough/Shaw and Alton Priors runs through the village remains; and the estate boundary between Anglo-Saxon West and East Overton runs from* lower right *to the straight 'stoney way' and then north to Edgar's gate on the edge of the woods (cf Figure 13.2). It then passes through the woodland to Hursley Bottom (*top right*) and thence to the 'eye-brows' of the thick hedge (*centre top*); cf Plates LII (where the 'eye-brows' are bottom left) and LIX (NMR 540/958 3078,* © Crown copyright/MoD*)*

done so in 1734 in the shape of Great and East Woods along the northern edge of the estate boundary, south of Pumphrey Wood (SU 139656).

The appearance of the place-name as of a settlement in the historical record, though, is in *Domesday*. By this date, however, it would appear that the Saxon estate had been divided into two roughly equal parts, *Essage* and *Scage* (*VCH* II, 150, 153–4), with both settlement names likely to belong to the late Saxon period (Gelling 1993, 209). In 1086 both estates, according to *Domesday*, supported a similar population and both were very similar in size, if a hide hereabouts can be viewed as comparable in acreage. The later history of these two estates demonstrates that *Scage* was the northern portion, the one attached to Overton, and hence the one which concerns this study (*VCH*, XI, 190–1). The southern part, *Essage*, became attached to Alton Barnes (*VCH* X, 10).

Scage had been held by Cudulf, but in 1086 it was held by a tenant, Hugh, of Robert Fitz Girold. As with *Essage*, *Scage* paid geld for two hides and one and a half virgates, though it had land for only one plough. Though no demesne land is noted, three farmers are, as are 30 acres (*c* 12ha) of pasture and a woodland, following modern and historic computations, of over 80ha (Coleman and Wood 1988). If correct, such a wooded area is likely to have run along the northern edge of the estate, covering the hill adjoining *Chichangles* (Figure 14.2) and along to include Boreham and perhaps on to the slopes of Lurkeley Hill. This estate, like its neighbour *Essage*, was seemingly prospering considering the increase in its 'worth' from 20s to 40s (*VCH* XI, 190–1).

Resources

Shaw still contained a relatively substantial area of woodland in the mid-thirteenth century, including Boreham Wood (*WAM* 49, 424–5) and what later became Great Wood, as it was required to send four representatives to the Savernake Inquisitions (*VCH* IV, 418). A reference in an *IPM* of 1314 to 110 acres (*c* 44.5ha) of arable here suggests an increase in arable similar to that taking place at the same time over large areas of high ground to the north east on Fyfield Down (Chapters 7 and 9). This arable probably lay on the clayey, south-facing slopes west of the village. 'A several pasture', cited in the same document, probably lay east of the Lockeridge to Alton road (SU 127655), much as it did in 1734 when it was called 'Rough Sheep Down' and covered about 20ha.

In 1333, 'Shawe by Savernak' contained at least '8 messuages, 3 virgates and 10 acres of land, 4 acres of meadow, 20 acres of pasture, and 20 acres of wood' (*Feet of Fines WRS* xxix, 38). Another *IPM* reference in 1376 showing Shaw to be held of the manor of Alton Barnes – purchased by the Bishop of Winchester who also owned East Overton and Fyfield – indicates that its position straddling several boundaries continued in the later fourteenth century (*Wilts IPMs*, 1907, 396). Shaw appears by name in the Selkley Hundred for the last time in 1377, when just three poll-tax payers were recorded (*VCH* IV, 310), perhaps a generation before desertion of the village in the early fifteenth century (*VCH* X, 10).

EAST WANSDYKE

Our work on East Wansdyke (hereafter 'Wansdyke') has produced two new observations to add to the authoritative surveys already published (Fox and Fox 1958; Clark 1958), and an interpretation which both reinforces and supplements views already expressed by others. Neither the Foxes nor Clark entertain the notion that Wansdyke may be unfinished; yet the evidence that this is in fact the case is clear on the ground in West Woods. More space is required to describe and expand on the point than is available here. There is much to be said too about ten possible gates through Wansdyke between *Woddes geat* across the Lockeridge–Alton road in Woden's Dene at the south-west corner of our study area and an unnamed gate on the extreme east of Fyfield civil parish where Wansdyke emerges from Short Oak Copse. We also need space to develop the argument that Wansdyke is of military inspiration from Roman precedents, built hastily in the late fifth century to confront potential invaders from the north while trying not to inhibit the daily passage of civil traffic along the various tracks of the 'Ridgeway route' (Chapter 16). We have, therefore, put much of our Wansdyke material into a separate, highly illustrated paper for publication elsewhere (Fowler forthcoming c), while including a copy of it in our infrastructure here (FWP 91), and we content ourselves in this volume with a summary account.

The nature of Wansdyke in West Woods

The dyke becomes quite a slight earthwork as it moves off the downs past Shaw and enters the continuous woodland of West Woods. It diminishes as it approaches Edgar's gate (number iv below) from the west, and east of it becomes little more than a lynchet along the edge of arable south of the wood. On its north, the ditch, which has so far remained more impressive than the bank, peters out as it climbs a gentle slope, ending with what look for all the world like separate slight pits dug separately before being joined up into a ditch proper. Our observation is that here we can see not merely that the dyke was unfinished but exactly how it was being built. No wonder the bank fades away if the ditch was not dug. This ending is on the west side of a newly identified but unfinished probable 'gate' (number v below).

Further east again, along the southern edge of Brickkiln Copse, the earthwork appears impressive; but careful analysis of what actually exists on the ground shows that, though the ditch is as much as 2m deep, the bank above was always slight, no more than a metre high at most. The earthwork is at the crest of a north-falling slope that literally heightens the impression of size, yet the shelf or berm on the inner side of the ditch clearly indicates the ground level on which the slight bank was built. Further, the bank is clearly built in a series of heaps, again indicating the early stages of construction and therefore that it was not completed. Finally, some of the present impression of height is conveyed because a wood or copse bank runs along the top of Wansdyke's bank, in places more upstanding than the original (Plate LXIa).

As Wansdyke approaches Titferth's gate (number vi below), however, the bank becomes larger and more finished in appearance, so that it is once more an impressive earthwork at the gate; eastwards, it is a fine earthwork indeed as it stretches through the woods on a near-straight run to a military-style salient at Daffy Copse where it turns sharply south eastwards towards Clatford Park Farm. Passing north of the farm, and through a probable gate where earthwork and road intersect in Clatford Bottom (number ix below), Wansdyke seems to be missing for *c* 100m on the lower slope before enjoying a fine run up the eastern slope through Short Oak Copse (Plate LXIb). Then, after heavy mutilation by quarries and tracks at another possible gate (x), it continues outside our study area. From the 'salient' at Daffy copse to that point of departure, Wansdyke was used as the northern pale of Clatford Park, which was transferred into Fyfield civil parish in 1896; hence Wansdyke's coincidence with a parish boundary here, a rare happening but of no long-term historical significance.

The gates through Wansdyke in West Woods

Documentary evidence for four *geats* through Wansdyke in the tenth century (numbers i, iv, vi and x below) does not, of course, necessarily make them original, for the earthwork was by then 400 and more years old. Nevertheless, we propose here that they were original gateways. In addition, we bring into consideration six other breaks that might also have been original features of the woodland Wansdyke. The ten possible original gateways through Wansdyke on this stretch are:

- i *Woddes geat*: a 'charter gate' (S272) (SU 127652)
- ii Old Shaw on the way from Boreham to Huish (SU 135653)
- iii 'Triangle gate', a name invented here, following a 1734 field-name, for an otherwise unnamed possible break across a minor coombe (SU 143653)
- iv *Eadgardes gete*: a 'charter gate' (S784) on the way from Hursley Bottom to Huish (SU 14786548)
- v 'Meux gate', a name invented here for a nameless break beside the 'HM' boundary stone on the track from Hursley Bottom to Heath Barn (SU 151655)
- vi *Titferthes geat*: a 'charter gate' (S784) on the way from Hursley Bottom to Oare Hill and Martinsell hillfort (SU 153656)
- vii 'Readdan gate', a name invented here for a nameless break in the earthwork on the line of another old through route, from Fyfield to Oare Hill, using a word from a descriptive phrase in the East Overton charter (S784; SU 156661)
- viii 'Little Wood gate', an invented name, taken from an adjacent part of West Woods, for an original, 70m-gap in the earthwork (SU 159663)
- ix 'Clatford Park gate', a name invented here for a possible gate at the junction between Wansdyke and the track along Clatford Bottom from Clatford to Oare Hill and Martinsell hillfort (SU 165662)
- x 'Short Oak gate', another invented name for a possible 'charter gate' (S424), this time for the way through Wansdyke from Clatford Bottom to Clench Common and Martinsell hillfort (SU 169664)

None of these are certainly original gateways; even if some are original, none are definitely built gateways, in the sense that there were such through Hadrian's Wall; and all may simply be gaps, original or otherwise. Four, however, provide primary field evidence of original structure: numbers iv, v, vi and x. The 'Meux' and 'Short Oak Copse' 'gates' (numbers v and x) have suggestive ditch terminals but little else, whereas the two 'gates' named in the East Overton charter (numbers iv and vi) have more elaborate arrangements, including outworks (Figure 13.2, where they are the only two gates shown, though the break through which passes the track south towards the charter's *readdan sloh* is also arguably the site of a gate, number vii above).

Some 300m south of 'Edgar's gate', the East Overton charter boundary joins the road (SU 148652) from

Plate LXI East Wansdyke in West Woods: (a) in Barrow Copse looking east (above); (b) in Short Oak Copse looking west (below)

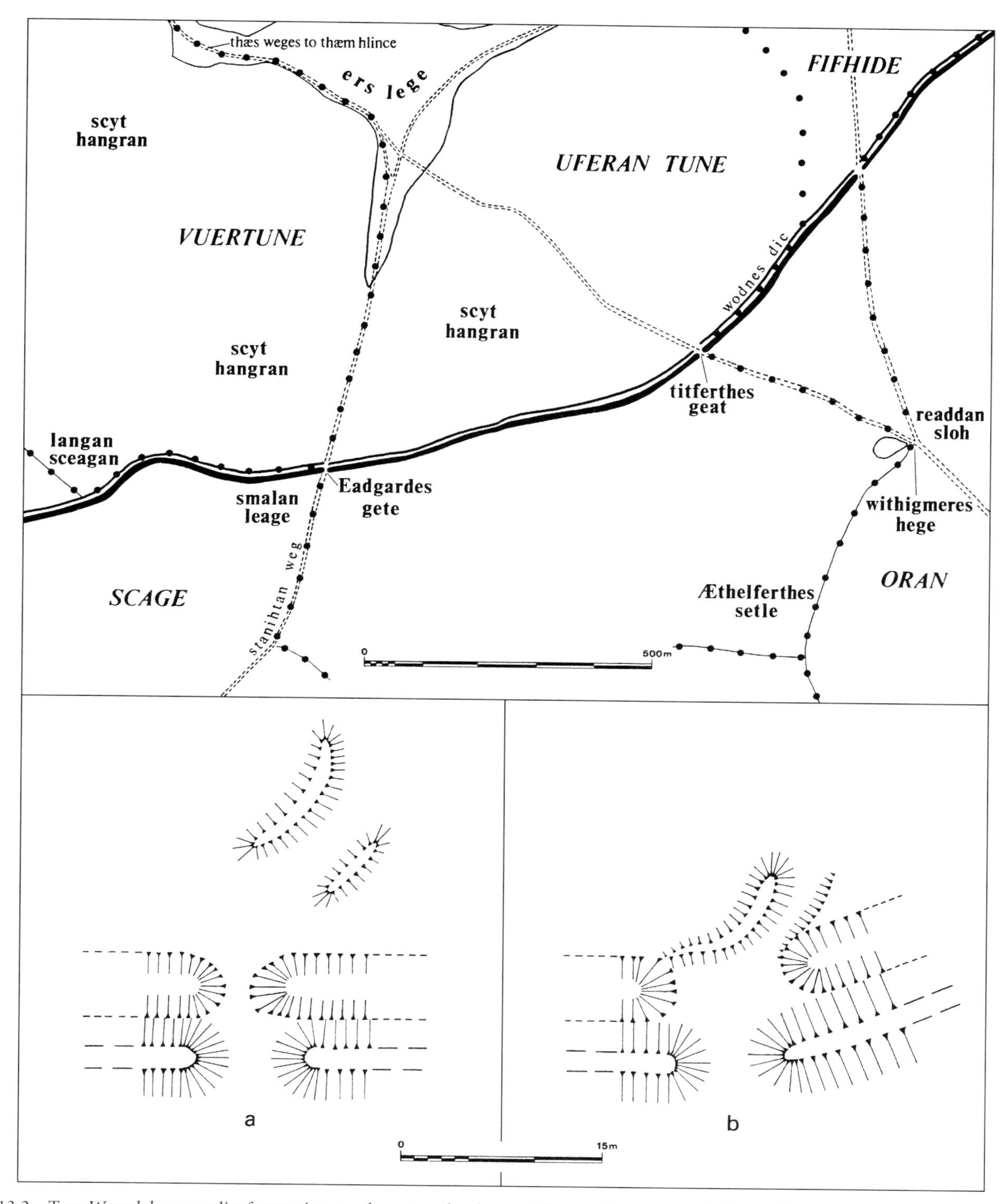

13.2 Top: *Wansdyke as a relict feature in a tenth-century landscape delineated by names and phrases from the late Saxon charters of the two Overtons. The estate boundaries are shown by large dots, which in two places run along Wansdyke itself.* Below are *schematic plans from original field survey of two of the 'gates' through Wansdyke. (a) is* Eadgardes gete *with outer earthworks channelling access from the north on to a narrow causeway; (b) shows* titferthes geat *with a plan also incorporating outer earthworks and apparently attempting a similar effect*

Draycot Farm to Lockeridge, identified as the *stanihtan weg* (Figure 13.2, top). The 'stony way' intersects Wansdyke and the West Overton boundary at *Eadgardes gete*, near *langan sceagan*, long 'shaw' (S784). The East Overton charter (S449), however, only notes a *smalan leage*, a small or narrow lea, presumably a wood like that here in 1816. At the gate itself (Figure 13.2, a), the eastern ditch terminal projected across the entrance to narrow an original causeway to path-width, giving on to a way through between the bank terminals which was itself oblique and narrow. In front two outworks are discernible, with a suspicion that the outer and larger one continued obliquely across the entrance nearer or actually to the outer lip of the western ditch. The shorter, inner bank looks as if it was intended to funnel traffic obliquely into the gateway from the north east; the general effect anyway seems to have been to deflect to one side or both any direct approach up the track head-on to a gateway. The position is right at the head of a narrow coombe, a natural line for a through-track which still exists in use and has arguably been there since late prehistoric times The name, position, context and nature of this gap in Wansdyke make it almost certain that it is original and very probable that it was a built gate of military character.

Titferthes geat (cf Brentnall 1938a, 127) was at the south corner of Wells' Copse and Little Wood and at the west corner of Barrow Copse where a track passes through Wansdyke. Its exact course is shaped now by a modern cut, slightly diagonal to the bank, but the disturbance cannot totally disguise an original causeway between original ditch ends in a pattern strikingly similar to that already recorded at 'Edgar's gate' (Figure 13.2, lower). Here an outer hooked earthwork seems intended to deflect an approach from the north to the east, passing between the outwork and the stepped-back eastern ditch end. It seems to be of some significance that, as modern maps still reflect, two of the major, long-distance 'through-ways' identified in the study area converge on this point, both off the downs, with one passing through the Overtons, the other through Fyfield and Lockeridge (Figure 16.7). This *geat* has good credentials, like Edgar's, to be an original built gateway of military aspect through Wansdyke.

Both gateways, and all the others if they existed, would have been part of a Wansdyke built and abandoned in the woods of our study area (Figure 12.4). It is to environmental matters throughout that area, and throughout several millennia, that we now turn as we move from description to synthesis.

Part III

Synthesis and Discussion

Much of Parts I and II has been descriptive and explanatory, with occasional forays into interpretation. Part III changes to a more discursive mode, with the overt objective of providing an interpretation and synthesis for the whole study area. Chapter 13 does indeed contain summaries of much original data but is also a conscious attempt to begin to embrace generalities of the whole study area rather than the specifics of places and types of evidence within it. Here, we switch from a methodological base rooted in the spatial exploration of the physically existing landscape to a thematic and chronologically led attempt to make sense not just of the study area but also of what we ourselves have done with it, regardless of how the evidence was obtained and utilised.

At a very simple – but in fact highly complex – level, the question facing us can be reduced to 'What does it all mean?' In similar vein, an obvious answer is, interestingly, that we do not know, either in terms of ourselves or of those who have lived here and created the phenomenon that we try to understand. That is clearly an unacceptable answer and cannot be allowed to deter an attempt, however intellectually foolhardy, to 'make sense of', to 'give meaning to', even 'to write the history of' a small, pleasant but undistinguished area of countryside in England.

Interpretation involves subjectivity, though this essay has so far never been other than openly subjective. Part III is simply more so. One of the problems is clearly not knowing whether the interpretative difficulty lies in a deeply complex nature of former times, now impossible to sort out, or in our own personal distance from them – psychologically and chronologically. Going about his business, the writer often silently articulates the thought 'We haven't a clue', which indicates a deep scepticism about 'history' in the sense of its enabling us, or any generation, to 'know' what any period in the past was actually like for those living then. And if that is not its purpose, then the same would apply to history's ability to discern significance except of a *post-hoc facto* nature at least one stage removed from any sort of contemporary reality before the present. The portrayal now, for example, of the 1950s, and the interpretation of them as received by an audience of the late 1990s or the new millennium, bear little similarity to what it was actually like to be living at that time, at least as recalled now by this writer in his memories of his experience then. But who is 'right'?

Before being overwhelmed by doubt as well as by data, the encouraging thing is to ask the question 'Does it matter?' For then we can truly answer with a very positive 'No'. That means, of course, that we are methodologically rigorous and intellectually honest to the best of our ability in attempting to bring some sort of pattern and understanding to the kaleidoscopic chaos that greets our enquiry when we turn to any time before the present. The 'pattern', the 'meaning', as many have previously remarked (eg, Carr 1990; Collingwood 1989; Finley 1986, esp chap 5, Fukuyama 1992; Gardiner 1961; Harvey 1989; *History Today* 1992; Jenkins 1991; Marwick 1989; Plumb 1989; Popper 1986; Tosh 1991), is, however, ours and not that of some independent, objective History, nor that of our subjects of study, whether they be people, events or, as here, landscape. So part of the 'honesty' required in our endeavours is to recognise that 'it does not matter' how accurate or percipient our constructions of the past are in relation to a mythic historical 'truth'. It is essential to recognise that our creations in thoughts, words and graphics are really less about our predecessors and their times than about us and ours.

The following four chapters may, then, bear little or no relation to various pasts that actually occurred in the Fyfield and Overton area, though they are indeed meant to be relevant. Those pasts may concern people's, peoples' or institutional relationships; they may concern what happened as events or processes, they may concern what people saw or thought was happening at the time, or they may, perhaps more probably, involve what was actually happening but was either not perceived as significant or not perceived at all. The mid-fourth and early fourteenth centuries, for example, may well have been such times when events obscured perception of process and any perception of change was unlikely to be illumined by informed understanding of its motors. Similarly, we may well have created a misinformed image of non-significant aspects of the study area in this synthesis. The very chapter heads impose a late-twentieth-century framework, which, if not actually 'wrong' – a pointless word in the world of post-modernist relativism – may well be misdirected and even historically irrelevant; but that does not matter, at least theoretically, provided they express ideas which speak to us rather than being directed solely as probes to find out what happened to others and their surroundings in former times.

Nevertheless, a happy positivism also informs this interpretation, which is based on a countryside that exists and an archaeology that can be seen, walked over and revisited. Countryside and archaeology are also really there in another dimension in that both have provided much pleasure to many and continue to do so. People talk about the area and its local history as if they are real, as they clearly are in both intellectual and popular perception. Old documents and maps also appear fairly convincingly to tell of these same things as if they have been there for some time. That it seems useful to state these truisms makes the point that present concerns shape the history we devise, for these self-evident truths reflect the uncertain cleverness of the 1990s. But we do not doubt that there is something there; even if this is only our view of it.

Chapter 14

Environmental History

Man and the pig between them cleared the Downs, and the sheep carried on the process.

Anderson and Godwin 1982, 75

This chapter presents a summary of the palaeo-environmental evidence from the study area and places it in its local context. Most of the evidence was obtained from the three main site excavations, OD X/XI, OD XII and WC (Chapters 6 and 7); it largely consists of animal bones, Mollusca and charcoal. Some other evidence is also available and is taken into account. Analysis of the material has been carried out by different people at various times since the early 1960s, with major, project-generated reports becoming available in the 1960s, 1970s and 1995–6 (*see also* Chapter 3, methodology).

Our concern now is not with what the environment was like in one place or at one time, though we need to work quite hard at precisely that issue, nor is our synthesis concerned with just one site or a cluster of them through time. Around the theme of 'environment' as inferred from the palaeo-environmental and archaeological evidence, the chapter is actually working towards a story of environmental change. Some documentary evidence is also brought to bear. The evidence is here used primarily as a tool in assessing such change as a factor in the development of the landscape.

THE ARCHAEOLOGICAL EVIDENCE

The discussion which follows draws largely on evidence presented in more detail elsewhere (FWPs 64, 65 and 66). Full details of all the data recovered are available in the archive (FWPs 29, 38c, 39, 40 and 87).

Animal bones

The three main settlement excavations at OD XI, XII and WC (Chapters 6 and 7; FWPs 64–66) provided all of the animal bones discussed here. The material is a mix of butchery waste and structured deposits either found in pits or distributed through the occupation layers. The large proportion of animals identified solely by their teeth and the fragmentary nature of much of the rest of the bones gives some indication of the differential preservation of the assemblage. Barbara Noddle's report, here revised by Michael J Allen, looked at five aspects of the bone evidence:

- proportion of fragments of bone per species;
- the minimum number of individuals represented (MNI);
- the proportion each species made up of the total represented;
- the proportion of certain anatomical fragments;
- where possible, the estimated age of individuals.

Overton Down OD X/XI: Late Bronze Age/Early Iron Age settlement and enclosure

Excavation within the enclosure produced over 1,000 bones of both domestic (*c* 96 per cent) and wild animals. The main species are sheep/goat (42 per cent of bone fragments), cow (38 per cent), pig (8 per cent) and horse (8 per cent). Other animals include red and roe deer, cat, dog, small mammals and amphibians. Of the main domestic species, a minimum number of 30 individuals (= 34 per cent of MNI) are represented by the sheep/goat remains, with MNIs of 22 cattle (25 per cent), 14 pigs (16 per cent) and 8 horses (9 per cent).

Skeletal part analysis revealed specific differences between the species. Sheep/goat are represented by a high proportion of waste parts while most of the skeletons of cattle and pigs are represented. Cattle and sheep/goat are notable in that no phalanges (foot bones) are present. It is likely that these were removed with the skins, though, because of their small size, they may not have survived or were not recovered.

The percentage of mature individuals (over four years old) is similar for sheep, cattle and pigs at 30 to 40 per cent. Horses are unusual in that 70 per cent of the individuals identified were over four years old at the age of death. This discrepancy is significant and it is possible horses were being kept perhaps not even as draught animals, nor for meat, but to ride, or as status symbols. Sheep and pigs, on the other hand, contributed a much larger number of young individuals, so it is more likely they were exploited for their meat. Cattle, of which 40 per cent were mature, may have been kept for their milk and for traction as well as for meat. Evidence for the exploitation of cows' milk has been found at Middle and Late Bronze Age sites on the Marlborough Downs (Maltby in Gingell 1992, 141), and it should not be surprising that this continued into an early phase of the Early Iron Age.

In addition to the major domesticated species, both dogs and cats are represented, all large specimens. The minimum of two dogs were each the size of a modern German Shepherd; the cats were large enough to be from a wild species.

The deer bones presumably represent the exploitation of wild animals. The extent of that exploitation is not clear since there is now no skeletal part information available, though MNI estimates of two red and three roe deer were obtained. Deer may have been hunted for their meat but there is also the possibility that animals were scavenged or that they were killed to protect valuable pasture. It is tempting to envisage hunting parties leaving the downs and crossing the valley to the southern woodlands in order to find the deer in their natural habitat of late prehistoric Savernake Forest, but even in the intensively used and open modern landscape of the downs north of the Kennet valley, deer are not unknown. Roe deer are recorded as present on the National Nature Reserve (NNR).

Small mammals and amphibians were an important component of some pit-fills, though their presence is likely in most cases to have been a result of accidental pit-fall rather than as a source of food (FWP 87). Their structural interest is, obviously, as indicators that some pits were left uncovered and unfilled, at least for a time, but analysis of the spatial patterning of the bone from the pits suggests that some form of deliberate deposition was taking place on site (FWPs 34 and 63). This is not unusual for later prehistoric contexts (cf Hill 1995), so allowance can be made for ceremonial as well as practical reasons to explain the appearance of small mammal bones in some OD XI pits.

Also of interest is the range of animals present in the pits which is typical of open downland pasture and fields. Weasel, short-tailed vole, water vole, harvest mouse, frog and toad are all represented, providing corroborative evidence for the sort of environment inferred from other evidence for the main phases of activity on and around site OD XI (Phase 3; Chapter 6). Given that it is easy to imagine that today's downland environment is roughly similar to that around OD XI *c* 600 BC, it is perhaps curious that the voles, harvest mouse and frog are not on the official record of species present in the NNR, only the weasel (*Mustela nivalis*) and common toad (*bufo bufo*) being common to both lists. The (short-tailed?) vole, however, certainly inhabits the thick red fescue tussocks beside the experimental earthwork today (Bell *et al* 1996, 232).

Overton Down OD XII: Romano-British settlement

A total of 3,133 animal bones were examined. A similar range of species to that at OD X/XI is present and similar calculations were made with the data. Comparisons between the sites are discussed below. Problems of preservation and recovery similar to those on OD XI existed on OD XII. Teeth were by far the most common surviving fragments, particularly from sheep where they represent 70 to 90 per cent of the sample.

The percentage of total individuals is dominated by sheep (50 per cent by MNI, 141 individuals). Cattle and pig are roughly equal (13 per cent), while horses make up a further 8 per cent. Less common domestic animals include dogs and birds.

Skeletal part analysis was carried out on the cattle and sheep remains. The results show a consistent dominance of waste parts. The absence of meatier parts is not unusual and may theoretically be due to the utilised bones being discarded in a separate, unexcavated, area; in practice, however, this is less likely, for excavation, and particularly post-excavation analysis, suggested several dumping areas, notably on the lynchet between Buildings 3 and 4 (Chapter 6, Figure 6.16; FWP 64). Similarly, while the high survival of teeth may simply indicate that conditions for the preservation of bone were poor, in practice this seems most unlikely. Nothing observed suggested that the normal dry, alkaline conditions helpful for bone survival did not pertain on the site.

Of more significance may be a greater degree of

carcass and bone fragmentation carried out on the cattle remains, presumably on-site. One possibility for this may be that cattle were intensively butchered, in contrast to the sheep, which were kept for their secondary products. This interpretation is supported by the contrasting proportions of mature individuals from the cattle and sheep. Only 26 per cent of the cattle were over four years old at death while sheep had a much higher figure of 51 per cent, similar to that of horse. Pigs are represented by 25 per cent of mature individuals, similar to that for cattle. The majority of cattle were killed in the second autumn. This may indicate killing to preserve a milk stock or a desire for younger meat on site, but it could also be indicating a stock-herding response to external market or tenurial demands. Of the sheep which did not reach maturity, some were in their first year while a similar number were in their second or third autumn.

Less common domestic species include the remains of seven dogs, one of which is a puppy. At least one of the dogs is German Shepherd-sized, comparable with the bones found at OD X/XI. Bird bones from domestic fowl represent at least two individuals.

At least six red and five roe deer are represented. Other wild animals are wild pig, hare, rabbit (assumed to be intrusive), small mammals and amphibians. This list is a little closer to that of the current NNR record, though that does not contain wild pig (English Nature 1991). Small mammals and amphibians were again an important component of pit-fills. Their presence is most likely, as on OD XI, to have been a result of accidental pit-fall. The range of animals present is again typical of open downland pasture and fields with short-tailed vole, water vole and frog; though none of those is listed on current NNR records (English Nature 1991).

Marine shells are also part of the faunal assemblage from OD XII. Fifty-four oyster shells were found, representing the remains of at least 51 individuals. They are in a fragmented and worn condition. It seems unlikely that they formed any substantial dietary component or a meaningful part of the local environment.

Wroughton Copse: Raddun *medieval and post-medieval farm*

A total of 2,536 bones were examined from the excavations beside Wroughton Copse on Fyfield Down (Chapter 7; Figure 3.3; FWP 65). The material comes from contexts that span prehistoric, medieval and sixteenth–seventeenth-century occupation. Of these bones, 2,297 are of domestic species. The greater part of the assemblage (some 1,915 bones, excluding goat bones) came from the thirteenth- to early fourteenth-century occupation in and immediately around Enclosures A and B (Figure 7.5). Sheep/goat, cow, pig and horse make up most of the domestic animals. Goat is certainly present but is not individually recorded. Three fish bones were found in a sixteenth-century deposit. The summary of the results below refers only to the thirteenth- to early fourteenth-century assemblage unless otherwise stated.

The assemblage is dominated by sheep bones; a minimum of 255 individuals is represented (60 per cent of domestic animals, including goat, by MNI), with some 48 identified goats (11 per cent). Cattle are the next most numerous (70 individuals, 16 per cent), followed by pig (30 individuals, 7 per cent) and horse (23 individuals, 5 per cent). The number of horses, though few in proportion to the other main species, seems quite high in absolute terms for an outlying, superficially marginal sheep farm. Interestingly, the cattle, sheep and pig are represented by a large percentage of 'first-class joints' (trunk and upper limb). This is particularly clear in the cattle bones, 70 per cent of which come from such joints. Assuming that the meat was locally produced, this proportion strongly suggests that the bovine herd was kept for its meat. It equally strongly suggests that the inhabitants of *Raddun*, wherever they were obtaining their meat, were not dietarily impoverished peasants eking out a poor living on the margins of a subsistence economy. This interpretation is supported by the low percentage of mature individuals represented among the cattle bones. During thirteenth–early fourteenth-century occupation, 22 to 34 per cent were over four years old at death, though it is to such, presumably working, animals to which the documentary evidence refers (Chapters 7 and 9).

In contrast, the sheep demography included a 30 to 56 per cent proportion of mature individuals. The deviation between these results is large and probably economically significant (*see below*), but it does not, of course, preclude the use of cattle for milk or sheep for meat.

Other domestic animals include dog (14 individuals) and domestic fowl (cock, duck and goose). The small number of surviving bird bones may imply that few were kept and eaten or that the small bones were not preserved or recovered. The last seems likely. Among the obligations on the farmstead was one to provide (hen) eggs to the lord of the manor (Kempson 1962).

Wild animals make up a small proportion of the identified remains, probably because of defective recovery techniques on what was the first of the settlement excavations. Survival on a more acidic Clay-with-Flints subsoil was perhaps less good too. Species represented include roe deer (1 individual), red deer (1), fallow deer (1), hare (2) and rabbit (22, many of them likely to be post-occupation). No small mammal or amphibian bones were identified. There is less clear indication there than on OD XI and XII of an open landscape with pasture and arable, and just a possible hint of local woodland. Wroughton Copse itself is partly enclosed by a medieval bank (Figure 7.4) and is suspected on other grounds of having existed, perhaps as woodland rather than a managed copse, in the thirteenth century (Chapter 7).

Discussion

The problems associated with the methodology and interpretation of the animal bone evidence outlined above (Chapter 3) mean that only broad conclusions may be drawn here. However, as the assemblages derive from three settlements, lying close together in similar geographical contexts but spaced temporally at roughly 800-year intervals in different cultural contexts, the opportunity is taken to compare them at a fairly basic level.

The environmental record of the prehistoric site, OD X/XI, is dominated by the remains of cattle and sheep probably some time in the eighth or seventh centuries BC. It is likely that animals were exploited both for their secondary products and for meat. Given the problems outlined in Chapter 3, the composition of the assemblage is not out of place in a Late Bronze Age–Early Iron Age settlement context on either a regional (eg, Maltby 1981) or more local (Locker forthcoming) scale. The presence of pig, generally but not exclusively low on downland sites of this period (Maltby 1981, 163), probably reflects the persistent presence of woodland in the area (Chapters 1 and 12). The nature of butchery practices and deposition on the site is particularly interesting. The proportion of cattle and pig skeletal remains, and their presence as structured deposits in pits, might suggest that these animals may have had a special place in the cosmology associated with the site which may be further highlighted by the contrasting fragmentation of the sheep remains.

The Romano-British assemblage, specifically that of fourth-century date, is, in contrast to the above, dominated by sheep bones. This is quite unusual for a downland site of this period where the trend is generally towards a marked decrease in sheep and increase in cattle (eg, Maltby 1981). Moreover, as Maltby points out (ibid, 163), '... the more "romanised" settlements such as villas, towns and forts tended to have fewer sheep than the native sites which maintained the Iron Age pattern'. Certainly the presence of a main Roman road nearby, connections to urban markets in the region and other local sites, including villas and an apparently large settlement beside Silbury Hill (Powell *et al* 1996, chap 4), can be envisaged as affecting smaller settlements on the downland such as OD XII. The wider range of species present (including wildfowl, wild pig, hare and oyster) suggests much more extensive economic and social contacts than a millennium earlier. We might, therefore, have expected a greater emphasis on cattle at OD XII – evidence of a more 'Romanised' economy. This prompts the question of how Romanised is Romanised? Date as well as location might well be an important element in an answer.

The contrasting bone assemblages from the Iron Age and Romano-British sites emphasise the differing occupation practices that occurred at each. At OD XI evidence suggests a strong symbolic relationship between farmers, domestic animals and the settlement enclosure. At OD XII this relationship is less clear and the assemblage can be understood in purely economic terms. That said, this interpretation would seem naïve, even for farming work within the context of a villa estate – which is what the bones may be reflecting – and it is likely that human/animal relationships were expressed in less tangible ways.

At Wroughton Copse, 800 years later, sheep are again the dominant domesticates represented, a statistic which correlates with the overwhelming thrust of the documentary evidence (Chapter 7). It is typical of rural, medieval bone assemblages. A model of a familiar sort of medieval landscape is the most likely framework for the bone assemblage in general; but perhaps with a reminder that such familiarity was unknown in a downland landscape before this evidence became available. With a touch of old woodland, now perhaps reflected by Wroughton Copse itself, the model here contains large patches of arable strip fields, represented by ridge-and-furrow, surrounded by sheep-pasture grazed in a managed way on the less hospitable downland. The bone evidence also contains some resonances, particularly in relation to the meat represented by the cattle bones and the number of horses, which go beyond environmental matters alone and are discussed elsewhere (Chapter 16).

Interpreting the three sites together in linear sequence, it is tempting to suggest two parallel models through time, respectively environmental and agrarian. We could see, environmentally, through the second

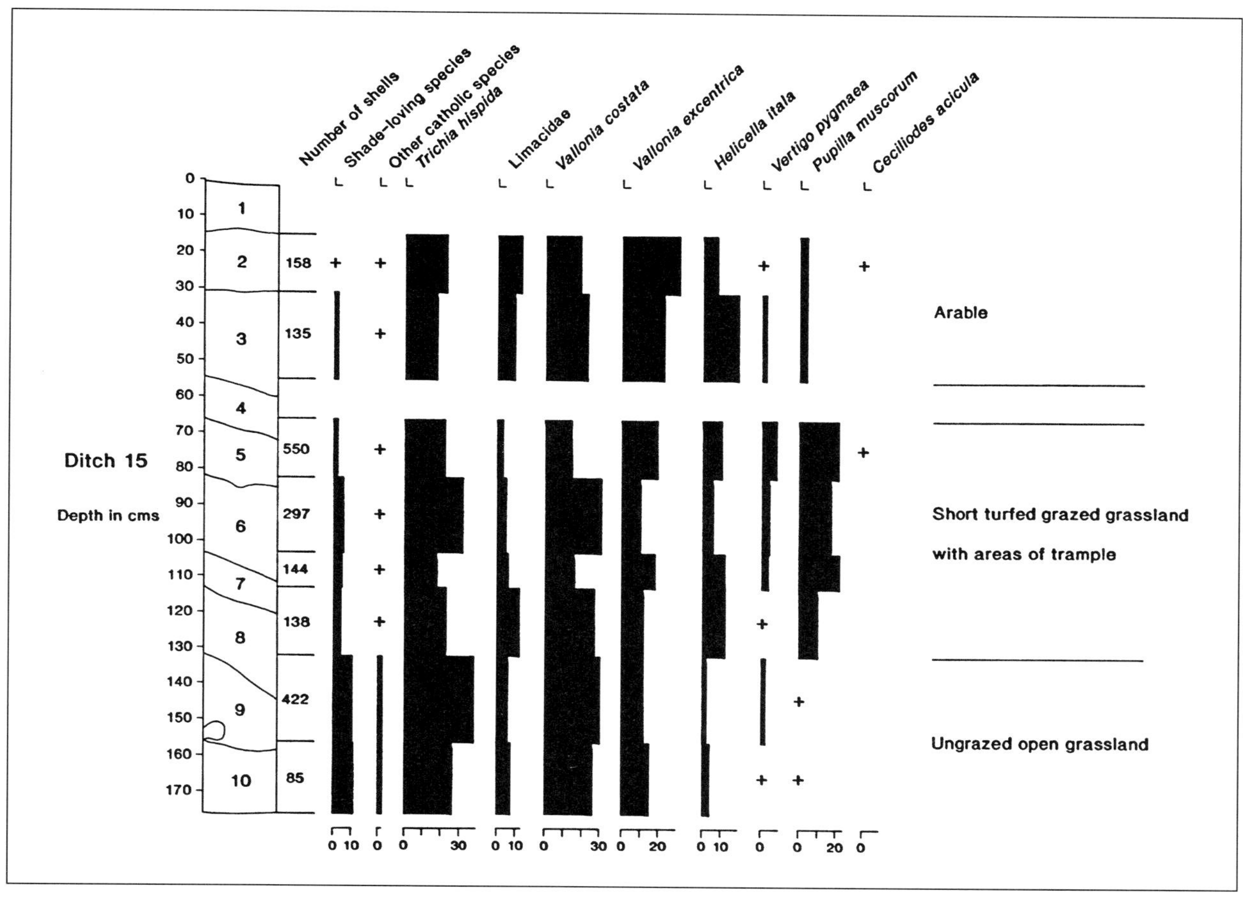

14.1 *Overton Down: molluscan histogram derived from soil samples from the ditch around the enclosed Early Iron Age settlement OD XI (cf interpretative phasing, Figure 6.3). Layer 1 is medieval and modern; layers 9–10 mid-first millennium* BC (see *discussion, Chapter 6). The exact location of the sampling site, cutting X/15, is given in FWP 63, figure FWP 63.31; layers 2–10 on the histogram are the numbers of the field record (in DM: FWP 61, nos 107, 108, 108A) and equate to layers 2, 6b, 6a, 21b, 35b, 2d, 35c, 35a and 46c in Figure FWP 63.31, lower right* (see also *Appendix 2*)

quarter of the first millennium BC the development of a widely opened landscape on the downs, with little macro-flora and, consequently, stabilised with fixed enclosed fields within a pattern of land allotments associated with focal settlements like OD XI. The landscape then remained open, perhaps with patches of scrub and woodland coming and going, the main change through to medieval times being a trend towards more grassland and less arable. The agro-economic model envisages a progressive change through time from mixed farming to one in which sheep became more dominant within a pastoral economy.

We must, however, stress that our sequence, although undeniably chronological, is not necessarily linear in any other respect. And even as a chronological sequence, it is not linear but episodic. So far, in any case, we have only looked at the animal bone evidence, and its import may well be principally economic and cultural rather than environmental. It has nevertheless produced indications of what may well have been happening environmentally in and around three adjacent settlements occupied between *c* 700 BC and AD 1300. Fortunately, some inferences can be matched against those from other evidence from the same sites.

Mollusca

Earlier molluscan work on lynchets from the area has been published elsewhere (Evans 1972, 1975, 1978; Fowler and Evans 1967) and will only be summarised here. New, unpublished primary evidence comes from two sites: OD X and Piggledean (Pickledean) Bottom (Figures 3.3 and 2.4). The samples from OD X were taken at the time of the excavation and, after thirty years in sealed containers, were successfully analysed by Sarah Wyles. Michael J Allen kindly makes the Piggledean data available from his own research. The full reports of both are available in the archive (FWP 38c).

Overton Down OD X: fill from an Early Iron Age enclosure ditch

The soil samples from OD X were taken from cutting 15 on the southern edge of the ditch circuit, down-slope of the main settlement (Chapter 6; FWP 20). The ditch revealed a common stratigraphy through much of its length; at cutting 15, after very close inspection, ten layers were recognised, including deposits on top of the 'normal' sequence (FWP 20, figure FWP 20.31). Samples were taken from eight of these layers with a view to developing an environmental sequence (Figure 14.1).

The deposits were divided into three phases: primary, secondary and tertiary. They showed a clear but apparently gradual sequence of deposition with at least one layer of soil stability when a turf line was able to form (our layer 4). Evans (1972, fig 123) illustrated almost exactly the processes envisaged as represented in the section of cutting 15, but his layer 4, 'tertiary fill', contained at least two, probably three, major phases of cultivation. They occurred after the development of a turf line over the 'natural' primary and secondary fills, and are of Early Iron Age, Romano-British and medieval date.

The primary fill is dominated by open country species, characteristic of established grassland. Shade-loving species are present (10 per cent) and this may indicate that there was wood or scrub close by either when the ditch was dug or during its use when it may well have been kept clean and open for a time. A high mollusc count and a low diversity index at the top of the primary fill indicate a period of stability before secondary erosion took place.

The secondary fill sees a decline in the numbers of shade-loving Mollusca. More specifically, an area of short tufted grassland was present at the time the secondary fill developed. The presence of one species in particular (*Pupilla muscorum*) indicates there were areas of broken ground close by. This may indicate trampling by livestock or disturbance within the ditch itself but, overall, seems best interpreted as representing adjacent arable, fairly unambiguously indicated archaeologically as contemporary with and surrounding the phase (3b) of enclosed settlement. This interpretation only identifies, however, general trends in the mollusc sequence; the actuality is likely to have been much more diverse (as reflected in the higher diversity indices compared with the top of the primary fill and the tertiary fills). The presence of *Trichia hispida*, a species that favours humid closed vegetation, for example, indicates how the local area would almost certainly have included a varied range of micro-environments at any one time.

The sequence also suggests that, as with the primary fill, a period or periods of apparent stability characterised the later history of secondary deposition. A turf line (layer 6), for example, occurred towards the top of the stratification. This may indicate a period of abandonment of the site, or simply that activity did not occur close to the feature. The latter explanation is more likely since the mollusc sequence does not show any evidence for vegetative regeneration. It could well be that this is an indication that the area was being grazed by livestock, an interpretation compatible with other evidence from these downs about their use in the last prehistoric centuries (cf Chapters 6 and 7).

In the tertiary fill, a virtual absence of shade-loving species combined with an increase in Mollusca typical of an arable context. The new land-use does not observe the line of the ditch, the profile of which was lost under the ploughsoil. No dating evidence was present in the tertiary fills but such arable activity would accord well with other, abundant and independent evidence across the landscape for intensive exploitation early in the Roman period. The increase in arable indicators at this point in the stratigraphy, whatever its absolute date, does not, of course, preclude arable activity earlier on. The nature of the evidence is the result of post-depositional processes, for it was only during the accumulation of the tertiary fills that cultivation occurred over the top of the ditch, thereby ensuring the representation of that activity in the molluscan sequence.

Overton Down OD XI/B and Fyfield Down FD 1: lynchets

At approximately the same time as the ditch samples discussed above were taken, further samples were taken from sections through lynchets on both Overton and Fyfield Downs (Chapters 6 and 7, Figures 6.2, 7.3; FWPs 63 and 66). These were collected and analysed by John Evans (Fowler and Evans 1967; Evans 1972, tables 9, 10).

The sequence at OD XI/B was heavily disturbed by later activity on the site, causing mixing of the layers. Molluscs throughout the profile were dominated by open country species. In contrast, FD 1 produced a clear stratigraphy with sufficient molluscs to construct an environmental sequence. The preserved land surface was dominated by shade-loving species, while within the lynchet open country varieties were prevalent. The

Plate LXII The floodplain, West Overton, looking south from the south bank of the River Kennet: the elements of the historic economy on 14 November 1998 – ewes and rams, grass and water management on the meadow, church and manor house on the dry, adjacent knoll

modern turf-line had a similar mollusc content to that found at OD XI/B. Clearly the lynchet was formed in an open country environment that had been cleared just prior to its use. That landscape remained open throughout the lynchet's growth, hardly surprising given that by definition the process involved arable cultivation. Though, again by definition, cultivation ceased and has not been renewed since the lynchet stopped forming, the landscape has remained floristically open until the present day. Interpretation of the archaeological evidence suggests that the old ground surface on the chalk was disturbed, probably by cultivation during prehistoric times, and was buried during intensive agriculture within a field which was finally stone-walled to produce the last stages of the lynchetting effect in the first centuries AD (Chapter 7, Figure 7.3).

Piggledean Bottom: dry valley

A series of auger samples taken from colluvial deposits in the dry valley of Piggledean (or Pickledean) Bottom by Michael J Allen (pers comm) provided molluscan samples which can be related to 'off-site' activity in the area. The results have not been published elsewhere so they are here presented in full.

Cores were taken from four sites along the valley. Two of these reached stone at a depth of 0.4m to 0.6m, while a third revealed a brown rendzina over chalk, a typical chalkland sequence. The fourth core was of more interest in that it revealed a sequence of colluvial deposits overlying a possible buried land surface. A much larger core was then taken from the same site, which produced sherds of pottery from the buried land horizon. The layer sequence is shown in table form on the next page.

The pottery was identified as Beaker, two of the larger sherds having rectangular-toothed comb impressions and a possible chevron. Molluscs were recovered from the Beaker horizon and from the underlying basal deposits, thus giving a sequence before and during anthropogenic influence (Table 3). The basal deposit had a small and poorly preserved assemblage dominated by species which favoured a woodland habitat. In contrast the Beaker horizon had a

Depth	*Layer*	*Description*
0–370mm	*Brown rendzina*	Dark brown (7.5YR 3/2) silty clay loam topsoil, few very small and small chalk pieces
370–580mm	*Upper colluvium*	Yellowish brown (10YR 5/6) silty clay with many small chalk pieces, becoming darker with depth and less calcareous
580–800mm	*Lower colluvium*	Dark yellowish brown (10YR 4/4) fewer chalk pieces, stiffer and becoming moister with depth
800–950mm	*Beaker horizon*	Dark brown (10YR 3/3) silty clay loam, almost stone free, some charcoal flecks and occasional small pieces (*Pomoidea* and *Corylus* identified J Ede), pottery
950–1,380mm	*Basal deposits*	Dark brown (7.5YR 4/4) becoming reddish-brown (5YR 4/4) with depth, silty clay
1,380mm	*Stone*	Sarsen

Layer sequence in core sample from Piggledean Bottom

Table 3 The molluscs from Piggledean Bottom

Mollusca	*Depth (mm)* *Weight (g)*	*Basal deposit* *1,100–1,200* *273*	*Beaker horizon* *800–950* *428*
Pomatius elegans (Müller)		4	–
Carychium spp.		3	–
Vertigo pygmaea (Draparnaud)		–	1
Pupilla muscorum (Linnaeus)		–	4
Vallonia costata (Müller)		7	18
Vallonia excentrica (Sterki)		–	12
Acanthinula aculeata (Müller)		–	1
Discus rotundatus (Müller)		6	+
Clausilia bidentata (Ström)		1	–
Helicella itala (Linnaeus)		–	3
Trichia hispida (Linnaeus)		–	8
Molluscs per kilogram		49	172
Total		21	47
Magnetic susceptibility (x10^{-8} SI/Kg)		13	53

larger and better-preserved assemblage. The species present were characteristic of open country, with dry, short grazed or trampled grassland.

Magnetic susceptibility readings were also taken of the deposits. The Beaker horizon gave a much higher reading, characteristic of a buried soil, possibly supporting short grassland. The deposits that overlay these horizons were made up of colluvium that had sealed the Beaker occupation. The presence of localised colluvium is not unusual in the area and its importance is discussed more fully below. Both in general and in several particulars, these observations and the stratigraphy are similar to those recorded independently in the 1960s and 1995 at the Down Barn enclosure (Chapter 6 and below).

Pound Field, West Overton

Molluscan evidence was also obtained from the site of a round barrow on the south side of the river. The site lay *c* 100m north east of St Michael's Church, West Overton, near the top of the small but prominent knoll at the east end of the village (Chapter 9, Figure 9.2). One sample came from a possible buried ancient soil which unfortunately could also be, or contain elements of, the ground surface buried in the 1960s (layer 16 in Powell *et al* 1996, 18–21, 24, 26, from which publication all the Pound Field evidence used here is taken). It contained 'an open country assemblage with a significant (*c* 15 per cent) shade-loving element ... ' though the generalisation is based on only thirty-two shells and interpretation is therefore tentative. This is a great pity, for evidence unambiguously of 'short grazed grassland' in the Bronze Age on the south side of the valley but closely overlooking the river would be particularly welcome.

Discussion

The molluscan evidence offers only a partial picture of the environmental sequence of the downland. This is a result of the small amount of sampling which took place, a factor compounded by the local scale at which molluscs react to environmental differences (cf Evans *et al* 1993, 159). Consequently, we have only three good micro-environmental sequences that can be supplemented with the data from three other local excavations.

On Fyfield Down, the lynchet at FD 1 provided a sequence that began with a shaded environment at the time of the old ground surface (Neolithic/Early Bronze Age?) which became cleared with the build-up of the lynchet and has remained so until the present time. At Piggledean Bottom the basal deposit (Neolithic/Early Bronze Age?) has Mollusca identified with a woodland habitat, above which the Beaker horizon preserves species of grazed, open country. The sequence from the OD X/15 ditch fill begins with the ditch being dug in a partially shaded environment. Following this the landscape becomes cleared and is possibly maintained that way through grazing until the ditch is ploughed over and the site, or somewhere nearby, is used for arable.

Comparable local samples have come from a prehistoric pit next to The Ridgeway (Smith and Simpson 1964), a round barrow on Overton Hill (Smith and Simpson 1966) and barrow G19 at West Overton (G Swanton pers comm; Swanton 1988). The pit was found while excavations were taking place at the site of three Roman tombs. It was 0.5m deep and just under 1m in diameter. The fill included animal bones from ox, sheep/goat and dog, struck flints and twenty sherds of Peterborough ware. A mollusc sample from the pit-fill showed the presence of *Helicella itala*, *Discus rotundatus* and *Arianta arbustorum*, with the last being the most numerous. The assemblage reflects species found in a diverse range of habitats, including a shaded environment, open country and wet thick vegetation (Evans 1972). This may indicate disturbance of the upper layers of the pit or that the Late Neolithic environment was indeed locally diverse, ranging from dry pasture to woodland.

In contrast, a round barrow excavated on Overton Hill revealed a mollusc sample characteristic of an open dry grassland environment (Smith and Simpson 1966, 142). Barrow G19 at West Overton produced a similar sequence with an environment of cleared grassland existing from the first phase of construction throughout subsequent use of the site. Both barrow sites can be dated from the Early Bronze Age. If the evidence from Pound Field is trustworthy, much the same, especially the established grassland, might have pertained on the other side of the valley and might well reflect in general the clearance sequence indicated across the valley on the northern downs. It might also suggest a pastoral as distinct from an arable land-use on this particular knoll. If so, such use is likely to have been within a complex of environmental resource management which, while perhaps not conceived of in quite such words, would surely have been practised in fact, at the very least on a local territorial basis.

CHARCOAL AND BURNT HAZELNUTS

Charcoal and burnt hazelnuts (identified by Rowena Gale) were not systematically recovered in excavation and make up a small proportion of the recorded finds from the main excavations. The full reports are available from the archive (FWPs 38–40).

Charcoal from OD X/XI (Chapter 6) came from post-holes and pit-fills probably dating to *c* 800–600 BC. Identified species include maple, hazel, ash, oak, *Pomoideae* (hawthorn/apple/pear/rowan/whitebeam/wild service) and *Prunus spp* (wild cherry/bird cherry/blackthorn). The range of species indicates primary woodland (eg, oak) along with thorn-scrub more characteristic of regenerating ground.

A more restricted range of taxa was recorded from *c* AD 400 deposits at Overton Down Site XII (Chapter 6), including hazel, ash, oak, elm and *Prunus spp*, while at Wroughton Copse (*Raddun*; Chapter 7) a similar range, with the addition of elder, *Pomoideae* and charred hazelnuts, were present in a thirteenth–early fourteenth-century context.

A sample from the possible ancient buried soil beneath Pound Field barrow, West Overton (Powell *et al* 1996, 19), produced a small number of charcoal fragments, of which 75 per cent were hazel (including nutshell) or *Pomoideae* with two fragments of oak, from which it was 'only possible to speculate that the environment around the site contained lightly wooded areas' (ibid, 19). These may have been recolonisation in the second millennium of areas cleared in the third. Hazel and hawthorn are common now in the hedges near the site.

Discussion

The charcoal evidence is small and no dating evidence has been obtained from it. Nevertheless, it should provide some indication of the local environment and (so easy to forget with so common a material as wood) any imported material to the site from which it was obtained. A comparison of the results from the sites shows very few significant differences. A stable woodland environment of oak, hazel and ash, and possibly apple or elm, was available. Within what appears to have been a well-cleared landscape it probably lay to the south of the valley or on areas of marginal land, perhaps on the Clay-with-Flints or the wetter valley bottoms. The possible presence of thorn-scrub on all three sites again emphasises the likelihood that land was continually in a process of regeneration through clearance.

This is supported by the charcoal assemblage from a number of early second millennium BC pits containing cremation burials at barrow G19, West Overton (G Swanton pers comm). A mix of woodland species favouring acidic soils or clay overlying chalk is represented (oak, ash, maple, hazel and hawthorn) and species which would have also thrived in a more open habitat on calcareous soils (buckthorn and sloe). This mix of woodland on Clay-with-Flints and scrub on regenerating chalkland is similar to the evidence from the excavated charcoal elsewhere in the locality and, if one accepts the Pound Field barrow evidence, from a similar topographical position on the opposite side of the valley to that of G19 on the north.

Taken together, the charcoal evidence supports that of the molluscan evidence for an essentially wooded environment until at least the later Neolithic/Early Bronze Age (*c* 2000–1800 BC?), when a local 'Beaker horizon' is associated with open grassland. Rapid clearance led to the creation of largely open downland, which included both grazing, occasional localised scrub regeneration and, probably increasingly, arable. This, in turn, led to significant erosion in the second millennium BC.

ENVIRONMENT AND LANDSCAPE IN THE TENTH CENTURY AD: THE CHARTER EVIDENCE

Our study area is fortunate in that much documentary and cartographic evidence bears on its former environment during the historic period. Here we take as exemplar one of the main sources, three pre-Conquest charters, not least because they relate to a time, the tenth century AD, for which we have no other environmental evidence (the medieval and later environment and

Plate LXIII (opposite) Earthworks on the bottomland. (a): the parched agger *of the multi-phase Roman road in 1996, heading west along the north bank of the Kennet towards North Farm and the modern A4, raised on a causeway probably Roman in origin; (b): the main leat inserted to improve Overton's water-meadows under the Parliamentary Act of 1821, here (in 1998) heading down-valley at the foot of the knoll with St Michael's Church towards the near-contemporary, new road north out of East Overton which crossed both it* (middle distance*) and, on the George Bridge, the River Kennet* (to the left*). The leat itself cuts across slight, parallel ridges, aligned south–north towards the river, 9m (30ft) broad and scored down their centres with narrow, infilled ditches, presumably the remains of earlier water-meadows. The stones in the foreground are from a ruinous sluice-gate which formerly controlled the water flow into the meadow on the left where water was collected on its way to the river in the broad channel, at right angles to this leat, shown in Plate LXII*

(a)

(b)

landscape as illustrated by documentary and cartographic sources is specified or implicit in our use of them in Chapters 8 to 13). Two tenth-century charters exist for the modern parish of West Overton: one delineates the bounds of East Overton (S449), the other those of West Overton (S784). An additional description of the cattle farms and downland was appended to the East Overton charter, which covers the northern part of Fyfield (Chapter 7). No charter exists for Fyfield, though one for Oare (S424) describes the south-east corner of the parish.

We use the charters here to try to acquire a better understanding of the environment and ecology of the study area, our approach being through the detail of references to tree species, water, woodland, boundary stones and land-use, followed by extrapolation. As such,

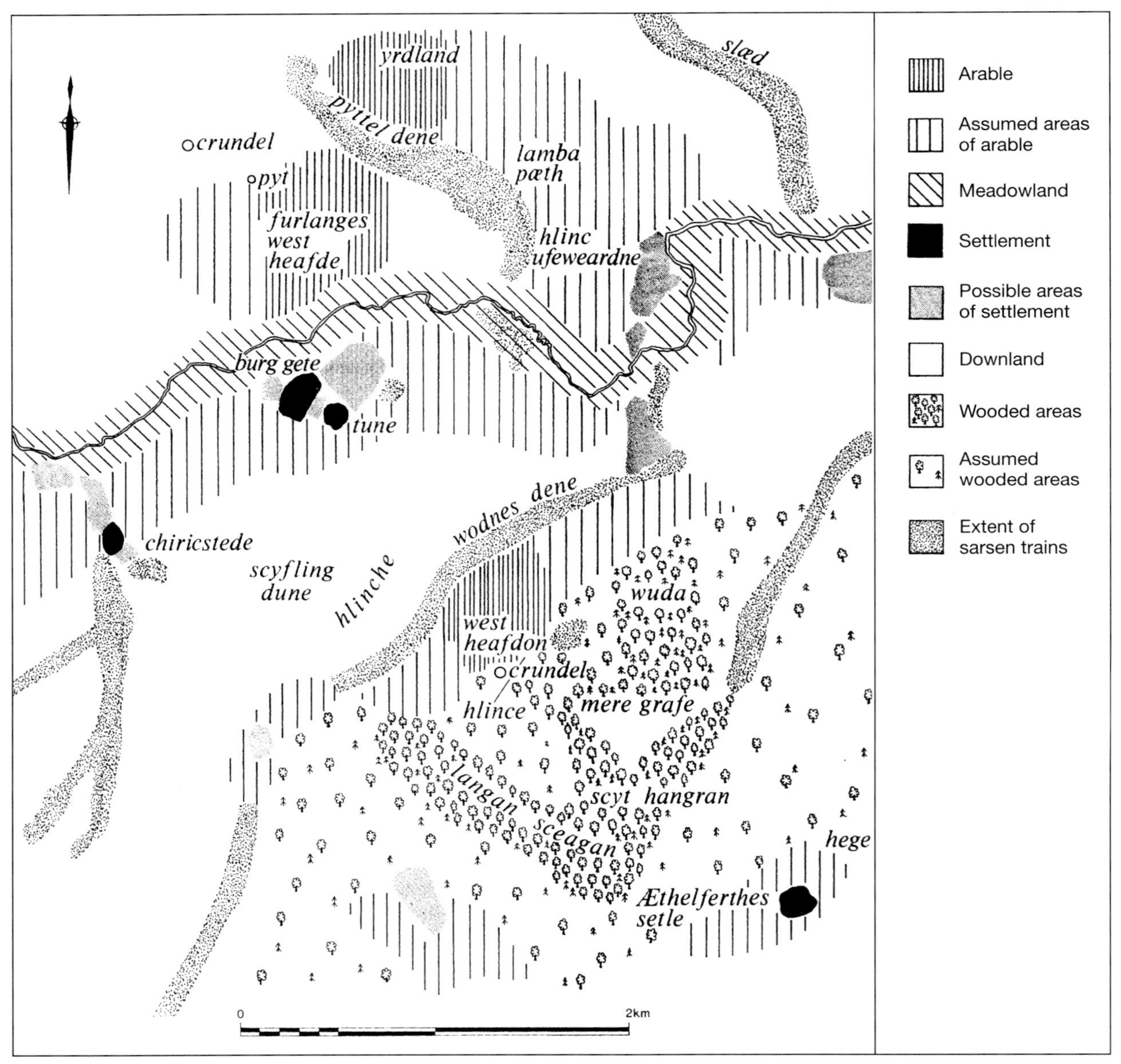

14.2 *Reconstruction of the tenth-century landscape environment of the Kennet valley and an area to its south between 'old' West Overton (*chiricstede*), the south end of the Valley of the Stones (*slæd*) in the north east and* Aethelferthes setle *on the heathlands to the south east. All the words, selected for their environmental/land-use qualities, are from one or other of the two tenth-century land charters for the area, and are correctly located on the ground*

it is the nomenclature which is being examined, not the precise twists and turns of the mere-men (Chapters 8 to 11). So we question the charters here, not about boundaries or social matters, but for any light they can throw on the landscape: of religious and burial practices, the economy, the management of land and resources and the communication and settlement patterns of the period. From such questioning, we have been able to reconstruct a version of the tenth-century landscape (Figure 14.2).

The River Kennet and water

In AD 939, fifteen hides of land at *Uferan tun* (East Overton) were granted to Wulfswyth, a nun, by King Æthelstan. That the settlement created in the Kennet valley was called 'Bank Farm', or 'Upper Town', could suggest that it sat on raised ground. In addition, the 'offtakes' noted in the introduction to the charter may indicate that the settlement was further protected by canalisation work or bank-building, which could counter flooding and drain marshy ground. They may, of course, have been exploitive rather than just protective, perhaps to do with harnessing water power for a mill or with controlling water flow as in an early form of artificial water-meadow (Plate LXIII; *see* Chapter 8).

Two fords are mentioned in the charters, one near the Salt House (SU 131683), the other where today's Ridgeway crosses the Kennet by a bridge (SU 119676). The latter took a metalled road across it, as it is described as a *straetford* (S784). The street was a continuation of the *herpoth* (S449; Costen 1994, 98, 105), The Ridgeway, and a stone bed to the ford suggests the route was a busy one or at least one taking heavy, possibly wheeled traffic. The ford is not on the line of the east–west Roman road (Figure 12.1) nor do we suspect the *straet* name as indicating a previously unrecognised Roman road leading south. On the higher ground, water needed to be contained. The *mere*, or pool, recorded in both the charters of West and East Overton, may have been a forerunner of one called 'Buckpitt' in 1567 (Straton 1909, 147) and possibly the one still visible today at SU 143663.

Vegetation

The charters refer to three species of tree: elder, willow and maple. Elder grew near the modern village of Lockeridge, somewhere between the Kennet (SU 141684) and Lockeridge Dene. After crossing the ford at East Kennet, but before reaching the Seven Barrows, the boundary seeks out another as a marker (SU 119678). As elder trees were certainly not rare in the tenth century (Godwin 1975, 336; Rackham 1996, 187), these ones must have stood out in some way, perhaps because of their size or form, to have been chosen as fixed points. Across the valley, a hedge by 'withy pond', *withigmeres hege*, formed the south-eastern corner of the grant of land at East Overton. The pond lies just beyond the long barrow at the south-east corner of Barrow Copse (SU 15736553; Brentnall 1938a, 128) and willows continue to grow there. Indeed, it has been argued that the relatively rare (for this area at least) species which grew at Withy Pond fifty years ago (*Salix Caprea* [great sallow/goat willow] and *Salix atrocinerea*) are lineal descendants of the Saxon ones (Grose 1946, 576). Presumably willows would have served the basket and hurdle-maker just as osier beds did until recently, and the hedge would have been a useful source of wood, berries and plants as well as a boundary marker. East Overton's charter also refers to *mappeldre lea*, 'the clearing in the maple trees', which was probably situated in the Down Barn area but was possibly a forerunner of Wroughton Copse. Maple was a widely used wood, from bridge uprights to musical instruments, and coppiced well. Other local vegetation included a *méos leage* (S424), 'the mossy clearing', situated near Levetts Farm (SU 173657), and *hacan penne* (Hackpen Hill/Overton Hill), suggesting a fenced enclosure (Field 1972, 270), presumably from a local source of hazel.

Woodland and woodland clearances also feature prominently in the charters. There are references to *ers lege* (S784) and *hyrs leage* (S449; Hursley Bottom, Plate LIX, Chapter 13), *smalan leage* (S449 – 'small or narrow lea'), *lorta lea* (S784; Lurkeley Hill, SU 123663) and *mappledre lea* (S449). As a *lea* or *ley* appears to indicate a permanent clearing in woodland, the number of leas in this study area suggests fairly substantial natural or man-made gaps in the woodland canopy. In addition, the inclusion of *lea* as a second element may imply worship (Yorke 1995, 166–7), thus making Hursley 'the sacred clearing where horses are worshipped'. Would that that reference had been to Overton Down where, in a Late Bronze Age/Early Iron Age context on site OD XI, there is circumstantial evidence of such 'worship' (*see above*; FWP 63).

Grafe, on the other hand, means 'a small, defined, probably managed wood' (Rackham 1996, 46). Thus, *mere grafe* (S784), 'Pond Grove', now part of Wools (Wolfs) Grove (SU 145666), suggests this wooded area was managed in the tenth century just as it was throughout the medieval and post-medieval periods (Figure 12.4). Other descriptions of wooded areas are:

wuda (S449), *scyt hangran* (S449), the forerunner of Pickrudge and Pumphrey Wood (SU 145658), and *langan sceagan*, 'long wood', which was situated in the area of what was to become the deserted village of Shaw (Chapter 13) and the southern part of Pickrudge and Pumphrey Wood. Some settlement had already been established at the edges of this wooded area, such as at *Aethelferthes setle* (S449; ?SU 156651) and five crofts (*quinque cassatos*; S424) possibly lying along the road from Oare to Marlborough. The reference to *Safernoc* in the latter charter, coupled with the general, later name of West Woods for the woodland hereabouts, suggests Savernake Forest stretched right along the southern limits of the estates in the tenth century, possibly as far as Boreham Wood.

Stones, barrows and burials

Tabular sarsen stones were still widespread in this area in the tenth century, with 'brown and rounded' sarsen stones far more limited in number (Chapter 4). It seems that the brown stones were particularly used as markers in an otherwise grey wether landscape. A *dunnan stan*, either a downland stone or, perhaps more plausibly, a dun (brown) one, stood at the entrance to the *burg* of West Overton, with a second north of Pickledean near site OD XII picked out as another boundary feature. *Twegen dunne stanas* (S784; 'two brown stones', ?SU 126692) divided the two Overtons south of Down Barn (Figure 4.1), and on the higher land to the north (?SU 126691), where divisions were less clear, *Aethelferthes stane* (S449) marked the boundary between East Overton, Avebury and Winterbourne Monkton (Figure 5.1 and pp 64–5 *above*). The western edge of the dairy farm was delineated by a *stan ræwe* (which, sadly, we have not so far been able to identify with conviction), and, at the eastern edge of what later became Savernake Park, a *dræg stane* (S424) acted as a boundary marker.

Other 'archaeological' landmarks hint at the antiquity long present in the tenth-century landscape. Some were visually significant then, most obviously the *seofon beorgas* of the West Overton charter, today's Seven Barrows on Overton Hill (Plate XIII; Fowler and Sharp 1990, 187, lower plate; *SL*, colour plate 1, figure 12). Colta's barrow, *colta beorg* (*beorh* in S449), a named barrow common to both charters, lay on The Ridgeway at the north-west corner of West Overton where its boundary with East Overton turns east to the two stones (Figure 4.1). We do not know, however, whether it was a Bronze Age barrow, perhaps with an Anglo-Saxon secondary burial, or a new burial mound of post-Roman times placed on an estate edge or beside a trackway. The *ii beorgas* (*twegan beorgas* in S449) west of North Farm (Figures 4.1 and 4.2) were fairly certainly two quite low-lying Bronze Age round barrows, now revealed by air photography, and interpretable by us as likely to have been an important local point just before one of the main north–south routes met the River Kennet.

Such *beorgas* are presumably to be as carefully distinguished by us as they were by the boundary clerk from the *byrgelas*, burials, encountered high on the downs at the extreme eastern corner of the East Overton dairy farm. This suggests a flat inhumation cemetery. While such could be of pagan Anglo-Saxons, its location might raise the possibility, by analogy with OD XI (Chapter 6, Figure 6.4), of a Beaker burial place. The other such reference is much more specific, though interpretatively as ambivalent: between Shaw and Lurkeley Hill on the West Overton boundary lay *haethene byrgils*, again possibly secondary pagan Saxon burials in 'White Barrow' (Chapter 12).

Agriculture

The charters reflect the general division of the land into arable, pasture and woodland and the location of settlements; the only substantial point to note is that this familiar pattern was already established, long established one suspects, in the tenth century.

It is evident, for example, that the area north of Down Barn, stretching from The Ridgeway in the west to the Valley of Stones in the east, was downland (S449, *dun landes*), used for grazing sheep (S449, *lamba paeth*) and cattle (S449, *feoh wicuna*, the dairy farm). Just north of the valley of Pickledean the land was under cultivation (S449, *yrdland on pyttel dene*), as was land to the south and west (S449, *furlanges west heafde*). It also appears that land east of Wroughton Mead (SU 142706) was also arable (S449, *suth heafod*). The two Overton settlements (S449, *tune*; S784, *burg*) and the *chiricstede* (S784) stood up from, and south of, the Kennet floodplain. South of the settlements the area was described as downland too (S449, *dune*; S784, *scyfling dune*), a point to remember with our present-day tendency primarily to regard those areas north of the A4 as truly downland. There was possibly some cultivation on those southern slopes as well (S449, *riht gemaere*, or 'straight balk'), and west of Boreham Down the land was probably also being ploughed since the reference to the boundary as a *heafnod* (S449) suggests the headland of a field by definition under cultivation. It is also reasonable to assume cultivation, albeit on a small, pioneering scale, around the outlying settlements of *Aethelferthes setle* and the five crofts at *Safernoc*.

The landscape

From the charters we can, then, assemble a picture of detail and some generalities. The landscape was a mosaic, but with a river valley flanked by quite large areas of arable, downland and woodland, all laced by numerous tracks, some metalled or hedged, the whole punctuated by outlying buildings, erect stones, woods, copses and individual trees, pits and ponds, and pre-existing landmarks such as barrows, lynchets and Wansdyke. In functional terms, it was a landscape of a mixed arable and pastoral economy, with local resources such as woodland, the River Kennet and the downland areas also playing a vital part (Plate LXII). Sheep and cattle were kept to provide wool and leather, milk and meat, with the animals grazing the northern and southern downland slopes. The arable land was north and west of Pickledean, with some possibly further south on Boreham Down. Savernake Forest extended this far, but it had been cleared of trees in places and was being managed, at least partially. The river, with its salt house, provided fish, and several ponds, pits and quarries were situated along the boundaries both as recognisable markers and for the benefit of inhabitants on either side.

Downland, fields and settlements were joined by a network of locally and regionally important trackways, many of which also acted as the estate boundary, indicating the bounds represented the limits of property and jurisdiction, but not of communities nor their movement. The principal zone of settlement was on the south side of the Kennet floodplain, much like today we assume, though only an 'old' West Overton with a church by the *herpoth*, a new West Overton (a *burg*) and East Overton (a *tune* or farmstead) are actually attested; Lockeridge and Fyfield, West Kennet and Clatford, implicitly existed as part of that same settlement pattern. An outlying cattle farm lay to the north on Fyfield Down, possibly at *Raddun* (Chapter 7), and six dwellings were inhabited at the southern limits of the study area along the edge of the woods.

SUMMARY AND CONCLUSIONS

Limitations in our use of palaeo-environmental and documentary evidence for environmental interpretation are discussed in Chapter 3. With this in mind, useful data were nevertheless collected, from animal bones and Anglo-Saxon charters, for example, and now offer a different perspective on the landscape to that presented earlier in this book. The following conclusions are based solely on the evidence presented in this chapter; the main points will be discussed in a wider context in Chapters 15 and 16.

The animal bones reflect the varying expression of human/animal relationships from the Late Bronze Age/Early Iron Age to the medieval period. In all cases we can see a diverse and probably complex subsistence pattern with both wild and domestic resources being exploited. Sheep remain one of the key components of the downland environment and human diet. Cattle and pig appear more conspicuous in the Early Iron Age assemblage but the striking expression of their status through structured deposition makes it difficult to make any conclusions on statistics alone. The use of secondary products throughout the periods concerned is very likely, though the bone evidence does not rule out more specialised patterns.

The molluscan samples provide a series of short sequences for the south west of the Marlborough Downs. The initial indications are that clearance was fragmentary across the downland up until at least the later Neolithic. The samples from the basal deposit at Piggledean Bottom and the lynchet at FD 1 have clearance episodes, both apparently around *c* 2000 BC. At the barrows on Overton Hill and West Overton, and at the enclosed settlement OD X/XI, the indications are that a cleared environment existed before their construction, respectively early in the second and first millennia BC. All the sequences agree with the proposal that the downland landscape was a largely open one from a Late Neolithic/Early Bronze Age horizon, most likely as a combination of arable and pastoral activity – probably integral with ceremonial barrow building too (Figure 2.2).

The charred remains can be interpreted as a wider environmental indicator than molluscs but they are more susceptible to anthropogenic influences. All the samples suggest the availability to downland communities of mixed oak woodland and of the presence of regenerating scrub somewhere in their environs. Doubtless the local distribution of woodland and scrub changed through time but the areas of such trees and bushes are most likely to have been on the patches of Clay-with-Flints and in the wetter valley and coombes prior to (possibly Roman? and) medieval drainage.

Overall, the environmental evidence suggests that clearance occurred prior to the Early Bronze Age over parts of the area, creating a patchwork of land in various stages of regeneration and use. The initially widespread woodland eventually became restricted to marginal land on the Clay-with-Flints and in the valleys. Land-use was mixed, and it is difficult from the environmental evidence alone to make any assumptions about the extent to which certain subsistence practices were distributed. Perhaps, however, at any one moment our familiar vegetationally open Bronze Age landscape (Figure 2.1) may have been somewhat more of a mosaic of small shrub and woodland patches dotted around extents of enclosed arable and managed pasture. Scrub over and around barrow ditches, for example, as suggested by West Overton G19, may have been common if the sacredness of a burial ground prevented its being grazed; shrubs may well have grown along lynchets and over abandoned settlement sites. Perhaps, too, the eco-dynamics of such a landscape through time saw more small-scale, local change than has been allowed. Witness the many such changes in the study area's environment in the early nineteenth century and since the 1950s, cumulatively in each case making a lot of alterations, together symptomatic of a general state of change. An understandable interpretation could well be that a general state of local environmental change is, and characteristically has been, normal.

Animals were an integral part of the changing environment, with both domesticates and wild species playing roles in and around places of human settlement. Their status was not one of simple economic necessity, the explicit presence of symbolic deposition at the Early Iron Age settlement enclosure (OD XI) offering a glimpse of the complex web of relationships which would have existed between humans and their environment throughout the history of the area. The hints that a symbolic dimension may also have been present in that relationship in the less familiar animalistic *milieu* of the fourth–fifth centuries AD, and even in the thirteenth century, may well support this. The tradition was unconsciously perpetuated in 1976 (Bell *et al* 1996, 35).

OVERVIEW

Work on the Mesolithic environment has confirmed that by about 8500 BP (*c* 7500 BC) the region was most likely to have been totally wooded (Evans *et al* 1993). People using flint blades at Down Barn around this time were, then, probably doing so in a damp clearing in a relatively low-lying location of the local uplands (not yet 'downland' *sensu* opened environment). The extent of clearance episodes during this period is disputed, with opinions varying between there being no evidence for clearance (Evans *et al* 1993) and episodic clearance being widespread (Smith 1984). Elsewhere in Britain and Ireland Mesolithic clearance is an accepted reality (Simmons 1996), but without evidence for lithic sites on the local northern uplands it is not possible to indicate any precedence for clearance episodes there until the Early Neolithic. In contrast, two sites on the edges of West Woods, one a major one, give clear pointers to the likelihood of woodland management in some form there before *c* 4000 BC. Furthermore, as suggested by the Down Barn Enclosure and Piggledean evidence, there may well be widespread Mesolithic material buried under later prehistoric alluvium and colluvium, if not along the Kennet valley itself then in the miles of coombes which, as with Piggledean, intrude into the Chalk uplands. Mesolithic studies have recently looked more towards human exploitation and manipulation of land as a resource long before the adoption of farming (Simmons 1996). Our data, however, does not allow us to test such a model and it would presently appear that, if there was an environmental impact through Mesolithic land clearance in our study area, then its effect was minimal. We suspect, however, that such was not the case and that further research, especially along the floodplain and through the southern woodlands, will bear this out.

As with the Mesolithic, Early Neolithic evidence is sparse in the study area. Initial clearance in the area may be associated with the silt formation found along the Kennet valley associated with the Avebury soil, which had a radiocarbon date for its earliest formation of 4040±60 BP (OxA-1222; 2870–2460 cal BP at 2 sigma; Evans *et al* 1993, 186). Later mollusc sequences indicating clearance of woodland as a widespread phenomenon in Wessex include those from Marsden, Durrington Walls, the South Dorset Ridgeway, the Dorset Cursus, Maiden Castle and Easton Down, with Windmill Hill, Burderop Down, Horslip, Beckhampton Road and South Street providing local examples (Ashbee *et al* 1979; Entwhistle and Bowden 1991; Evans 1970, 1971, 1972; Evans *et al* 1988; Whittle 1997; Whittle *et al* 1993; Woodward 1991).

It has become a moot point for the modern interpreter as to whether, by the centuries around 2000 BC, the downs bore many clearings, some large, in what was still recognisably woodland or patches of residual woodland in an open landscape. The open Beaker cemetery on Overton Down might have been in one such clearing, the occupation debris on the higher northern part of the down in another, the odd sherds from various places in yet others; but in any case the evidence demonstrates the use of the local upland at a time when such use was apparently common, as suggested by the quantities of Beaker pottery found as surface scatters across much of the Marlborough Downs

(Gingell 1992; Swanton 1987). Such use may well have included arable, as indicated at South Street on lowland and at Red Shore on upland (Ashbee *et al* 1979; Evans 1972; Fowler and Evans 1967; Green 1971), and would surely have included pasture too.

The extent of these clearance episodes appears to have been small since most evidence is from archaeological sites, and off-site analysis has not revealed evidence for major change until the Early Bronze Age (Allen 1988). The significance of this to Fyfield and Overton is possibly best represented by the alluvial deposits along the valley. Woodland soils, once cleared, have a crumb texture which would have eroded very easily (Evans *et al* 1993), so the effect of initial clearance could have been drastic. Large rills appearing in the side of the hillside after heavy rain – which still happens sometimes after ploughing – swept away the loose topsoil. At this early date it is unlikely that the quantity of alluvium reflects major clearance but it is possible to envisage minor clearance producing major erosion locally. The practical and environmental effects are impossible to gauge on present evidence. Regeneration of many non-mortuary sites is, however, known, and the distinction between this and the regeneration of apparently more utilitarian sites where regeneration took place can be explained through differing concepts of spatial importance (Whittle *et al* 1993). The Dorset Cursus, for example, was allowed to return to a shaded environment (Entwhistle and Bowden 1991, 21), the linear, non-natural, form of the monument being lost to encroaching vegetation. Had the monument also lost its significance? Had people directed their energies to conserving their land? Again, Fyfield and Overton do not appear to have been heavily occupied at around this period (Early/mid-Neolithic) but the occurrence of soil in the valleys which may have originated on the uplands could be indicating both activity and a major loss of *in-situ* evidence for such activity.

The evidence here suggests that, within a general story of clearance, floral regeneration was always likely to be a reassertive dynamic within the local environment. That should apply whether or not landscape change was environmentally led. In the second half of the third millennium BC, for example, it is distinctly conceivable that religious or politically driven demands were impacting more on the landscape and its resources than changes induced by the environment; though clearly such demands could themselves have environmental consequences, immediately and longer term (as imagined in *SL*, chapter 11). Were the Avebury stones dragged off Overton Down? Where did all that turf under Silbury come from (Whittle 1997)? Where was the woodland that was so carefully managed and then so severely exploited to produce the materials to put up kilometres of palisading just off the south-western corner of the study area (Whittle 1997)? These are quite important, and practical, questions in a landscape perspective, particularly one of dynamic inter-relationships.

Important too is recognising another aspect of that dynamic: that pasture quite as much as arable became the established land-use within particular patches. There are hints here too that during the second millennium a long-term cultural divide was imposed on the landscape. The uplands to our eyes superficially became mainly arable, but what actually happened was that virtually the whole of the area's extent of enclosed field systems in the Bronze Age came to be spread out across the Downs (Figure 2.1) in what was agriculturally in fact a zone of mixed farming (Figure 15.3). Nevertheless, in contrast, the downland fringes, initially respected for their sanctity but now with their increasingly scrub-infested barrow cemeteries, came to be characterised by rough grazing and permanent pasture as part of a more diverse land-use pattern oriented towards the more varied resources along and in the Kennet valley.

By the later prehistoric centuries, much of the downland seems to have been grass, the permanent arable having presumably already shifted from its upland, not to the valley 'bottomlands' but to that valley-side location which it was to occupy henceforth. If correct, this environmental interpretation has so far provided two key components of the answer to the question of how the landscape came by its present appearance. It became an effectively opened chalkland landscape in the second half of the second millennium BC and by 500 BC the downs were in place as treeless grassland. Their complement, functionally and visually as now, was the location of permanent arable on the south-facing slopes of the valley and coombes, that zone so clearly brought out on Figure 2.1 as a relative 'blank' between the spread of earthworks and the communications corridor later marked by the Roman road and A4.

It was early Roman imperialistic demands of the land that reverted to the earlier, Bronze Age style of upland arable, but unfortunately we produced no environmental evidence specific to this period. Such as there is for Roman times comes from the fourth–fifth centuries, and is economic as much as environmental. Nor is the immediate post-Roman or Anglo-Saxon period represented by environmental evidence from within the study area. However, sections from excavations on the Wansdyke (Green 1971) have yielded both pollen and snail samples from close by to both east and west of the parishes' southern uplands. The Dyke is not closely dated but material beneath it can be taken as having been sealed by *c* AD 500. The samples that were studied come from two locations, Red Shore, and New Buildings, in both cases from the buried land surface. The results from Red Shore (west) produced pollen indicative of rough pasture with some local bracken. While at New Buildings (east) the sample showed evidence for a cleared area with dense woodland nearby, cereal pollen was also present in small numbers.

The contrast in these two samples has been interpreted as evidence for the survival of Savernake Forest, now lying east of the New Buildings section. Ancient forest, however, may also have lain to the west within the Fyfield/Overton area. The presence of forest is not in dispute, particularly locally with the presence of Roman pottery kilns in Savernake Forest (Annable 1962; Swan 1975, 1984). What is important is the extent to which that forest covered the land during the latter half of the first millennium AD. There is no evidence for regeneration of the landscape on the northerly slopes or in the valley so it is likely that the southern slopes were dominated by blocks of woodland, which was potentially heavily managed as a valuable resource. That is certainly the impression from the tenth-century land charters, and may have some bearing on the implications of radiocarbon dates from the formation of a soil layer in the 'bottom-lands'. They bracket AD 890–1280, some four centuries during which the sediment cycle in the valley was apparently stable enough to allow the development of a soil in a dry open environment (Evans *et al* 1993, 190, table 1). A dry open environment was almost certainly also that in which, during the thirteenth and fourteenth centuries, sheep in their thousands, evidenced by bones and documents, grazed their way across by now long-established downland pasture in a pattern which persisted until recently (Plate LXVI; *SL*, figure 4). Theirs was visually a landscape similar to today's, emphasising the permanent nature of the late prehistoric clearances.

Whatever the land-use after agricultural clearance, the important point in trying to understand how the landscape came by its present appearance is that the land in general remained cleared from the second millennium BC onwards up to the present. One reason for that, not clearly brought out until the recent work of Evans *et al* (1993) and Powell *et al* (1996), is that so much of the downland soil, that most precious of environmental resources, washed into the coombe and valley. The downland could not support woodland; without trees, its soils could not be refreshed and could not therefore support arable except on a periodic basis, despite manuring from midden and flock, as in the first–second and thirteenth centuries AD. This remained basically the case between *c* 1000 BC and the day of the artificial fertiliser in the mid-twentieth century AD. The wholesale ploughing up of the downs in our own lifetime falls, therefore, into a historico-environmental perspective, as will the current effects of environmental change.

Throughout this environmentally based narrative it has been possible to observe a number of trends which have given human agencies a solely adaptive role. Such an approach assumes an ever optimising aim for those taking part in its processes, a view we do not fully accept. We must, nevertheless, look realistically at the environment that has nurtured and helped direct the lifeways within it. The environment was a social and practical issue to those who lived in our study area, especially on the downlands, and we may legitimately interpret it as such. Field systems, for example, can be seen as a direct result of this increasing realisation that the land is a resource that must be managed. Husbandry was, for example, an alternative to its exploitation for the building of ostentatious monuments. Land now became more important than metal, more important even than visions of omnipotent gods; it was desirable to return to basics, producing food by working sensibly with the land. Such pragmatism resulted in a landscape of efficient 'units of production', enclosed fields, which helped keep the soil in place. But stones, walls and fences could not trap the nutrients that underpinned arable fertility, and a landscape of pastoralism emerged on the uplands balanced by intensified cultivation on the valley slopes. Despite the interruption of the Romans, 500 years of careful farming, including much pasture, followed by a smaller number of inhabitants and less intensive farming, meant that even in the unstable valley later Anglo-Saxon farming was able to begin to expand in an environment with much woodland and the first stable soil since the 'Avebury formation' in the Bronze Age.

Chapter 15

An Archaeology of a Wessex Landscape

Other new departures like the use of landscape and film as historical sources ... have been only lightly touched on ... because ... their impact has not been so pronounced, nor do they hold such interesting implications for the nature of historical enquiry.

Tosh 1991, 228

THE EVOLUTION OF A LANDSCAPE

The landscape of the parishes of Fyfield and West Overton has now been subject to considerable archaeological and historical analysis. Fieldwork, other research and subsequent study over thirty-nine years have combined to produce a picture of diverse human activity spanning at least eight millennia. The passage of time during this process of investigation has itself been a significant factor in the nature of the study. The quantity and quality of the data collected are so extensive that a volume of this size cannot convey their full scope. Hence, it is important for a student to appreciate the presence and role of each level in the hierarchy of presentation (*see* Editorial Notes) It is hoped, in particular, that the major effort put into the three layers of archive underpinning this volume will be useful for students of human impact on the environment for many years to come.

The following chapter can, meanwhile, only attempt to provide a synthesis of our present understanding. There are many obvious gaps in the data that are beyond the control of the constantly evolving research strategies. The strength of this truism is illustrated by Figure 15.3, which became available after the bulk of this study was conceived and written. We publish it here, courtesy of the (former) RCHME, almost without comment and certainly without the analytical commentary which the comparable map of our study area enjoyed (Chapter 2, Figure 2.1). Despite its small scale, Figure 15.3 clearly asks different as well as similar questions to those asked of the study area embedded within it. The implications are challenging, but daunting. The resources required to investigate these fully and in depth would be prohibitive.

Yet occasionally fate can intervene. For example, the discovery of Mesolithic material at Down Barn in 1995 was the first such recorded instance under controlled conditions. Conspicuous by its absence from the downs until then, it serves to underline the unpredictability of archaeological data collection, a point rammed home in 1997 when a chance meeting with the son of a farmer visited in 1960 led to our being shown significant quantities of Mesolithic material resulting from a family tradition of collecting around Bayardo Farm (which for present purposes we can do no more than make note of; *see* Chapter 14). Similarly, the general absence of artefactual evidence of the later fifth, sixth and seventh centuries AD does not necessarily mean that the landscape was empty during this period. Indeed, one of the successes of the project may well be the demonstration that the landscape itself may often hold the evidence so often sought by excavation. The data as presented here, fairly obviously to us but we would hope suggestively to others, could indicate further avenues of research for future generations. It is fortunate that there are already other studies with which to compare results. Even so, this is only a beginning.

The Neolithic and Early Bronze Age

The study area lies well within the hinterland of the Avebury and Windmill Hill complex. That complex has long been recognised as one of the most important early landscapes in Wessex, a position underscored by its inscription in the World Heritage List (Chapter 17). In addition to the long-known monuments, both earthen and megalithic, the area still has much to yield. The recent discovery of an extensive complex of palisaded enclosures and related features at West Kennet Farm (Whittle 1997) demonstrates the archaeological potential still to be realised, especially on the floor of the Kennet valley. The quantity of timber and other resources required for this complex would have had a major impact on the surrounding landscape and environment. Though more humble in scale, the *polissoir* on Overton Down (Chapter 5), and the sequence revealed by small-scale investigation, suggest that the area was already witnessing activity that required the signalling of points in the landscape.

The most obvious and highly visible monuments of this period are the funerary monuments, long and round barrows. The siting of the long barrows, commencing in the fourth millennium BC, was to have an enduring influence on subsequent attitudes to the landscape. Round barrow groups L and M (Figure 2.4), for example, and the Middle/Late Bronze Age linear ditch F4, lie close to the Manton Down long barrow (Figures 5.4 and 16.3). We may even be able to catch glimpses of possible contemporary boundaries (*see below*).

The siting of round barrows, whether singly or as groups, is clearly important and the location, spacing and aspect are highly suggestive of a non-random placing. The Overton Hill group (H), generally viewed as being sited to present an aspect from the west and the Avebury complex, is equally conspicuous when viewed from the north and east along the river valley. This group (and the adjacent Sanctuary) could, when viewed with barrow groups J, K, L and M, located just above the River Kennet, be taken to underscore the importance of the valley in the Early Bronze Age. Indeed, with the exception of the hilltop location of the Overton Hill group, the 'false crest' locations of these groups reinforces the arguments for contemporary settlement being sought on the valley floor. The West Kennet Farm complex clearly demonstrates the potential of this zone. The possibility of additional complexes of this or similar form further to the east should be given serious consideration. Although heavily modified by water meadows in the vicinity of West Overton and Fyfield, exploratory remote sensing on this stretch of the valley floor should be encouraged.

With the exception of group I, on the high down above Temple Bottom, all of the barrow groups identified are peripheral to the high down and look out over low ground to the west and south. This situation is very similar to that observed on Salisbury Plain where barrow cemeteries cluster around the scarp edges and above the valleys of the Avon, Till, Bourne and Nine Mile rivers (Bradley *et al* 1994; RCHME unpublished fieldwork). Groups L and M are especially striking, being positioned either side of the entrance to Clatford Bottom from the Kennet valley. Some 600m up the coombe lies the megalithic monument of Devil's Den, and the Manton Down long barrow, 2.5km to the north, overlooks the head of the coombe. The higher downland location of the Manton Down long barrow also finds parallels with long barrow distribution on Salisbury Plain. Here many of them are important as boundary markers in the early Anglo-Saxon period (Bonney 1976).

Activity is, however, also attested on the downland and the possibility of land division of this date cannot be discounted even if the physical remains are likely to be masked by subsequent developments. Indeed the general scatter of Neolithic, Beaker and other Early Bronze Age material from the investigations emphasise that this is far from an empty landscape at this period.

The Middle–Late Bronze Age and Early Iron Age

This period marks the first clear episode of large-scale land division and allotment. It was noted in Chapter 2 that extensive areas of field system define themselves as a series of landscape blocks, displaying a restricted range of recurring alignments (Figures 2.1 and 2.2) and that a number of linear ditches emphasise certain boundaries. The deep colluvial deposits under the Down Barn enclosure seal early second millennium BC material and, given the general paucity of Middle Iron Age material, strongly suggests that the horizon is derived from intensive arable farming on the adjacent downland, which commences in the Middle Bronze Age.

This pattern (Figure 15.1) conforms to that observed over other extensive tracts of the Wessex chalk. On the Marlborough Downs north and south west of the project area a major intensification of settlement and land division are associated with the Deverel-Rimbury ceramic tradition and its successors (Gingell 1992). Further south, on Salisbury Plain, a similar pattern is observed, although perhaps more restricted in extent, concentrating on the southern flank (cf Richards 1990) and the area east of the River Avon (Bradley *et al* 1994). Beyond Salisbury Plain, on the chalk of southern and north-eastern Dorset, similar patterns are observed. Fields and settlements appear in the south Dorset Ridgeway area (Woodward 1991) and the work of Barrett *et al* (1991a, 1991b) in Cranborne Chase has produced a detailed sequence of later second millennium BC land division and settlement.

One potential settlement of the mid- to later Bronze Age has been recognised on morphological grounds on

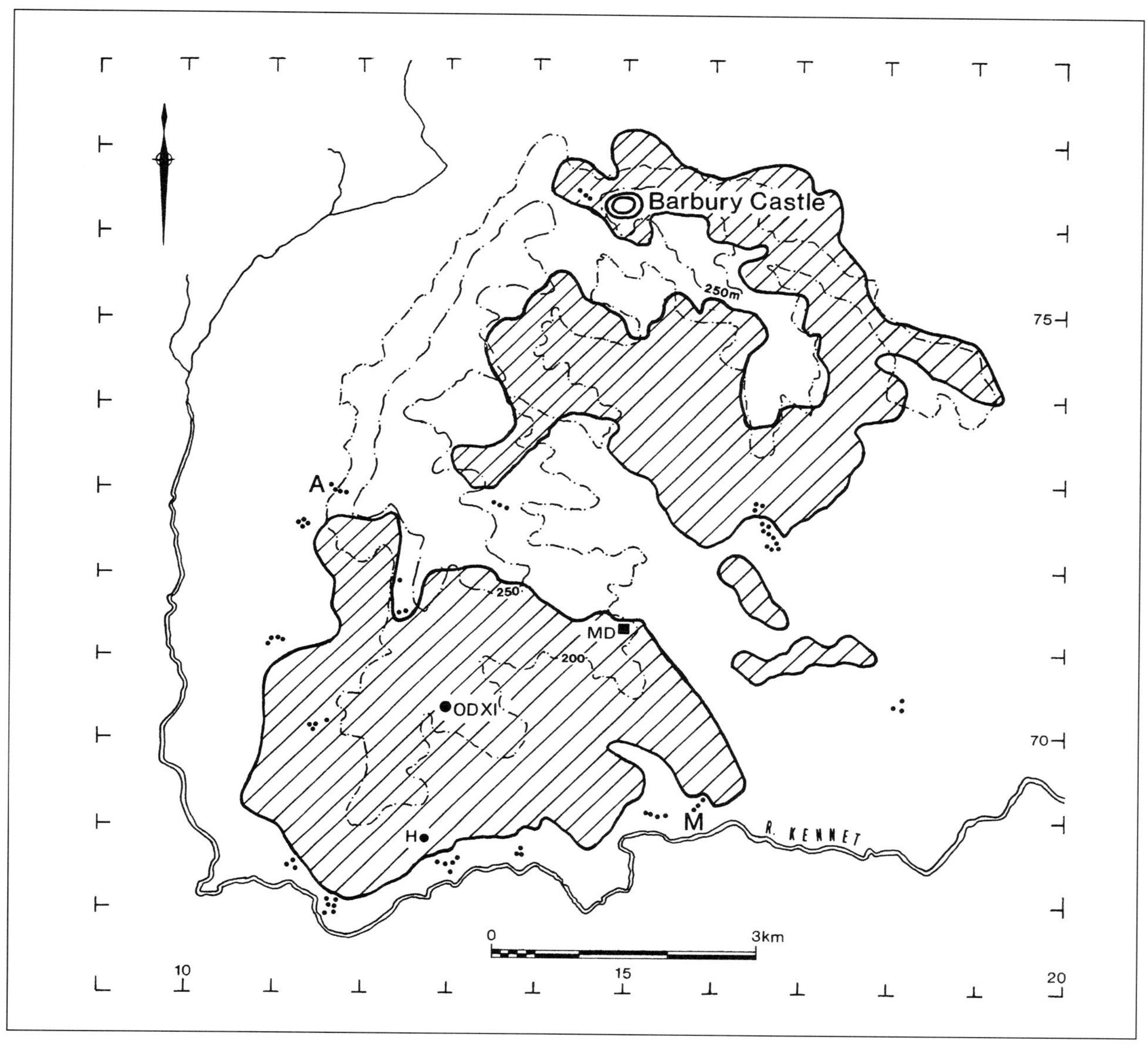

15.1 The Marlborough Downs in later prehistory, putting the study area into a regional context largely by building on the work of Gingell (1992). The hatching indicates the areas of prehistoric fields, here probably all pre-Early Iron Age and likely to be in general earlier than Barbury Castle hillfort. The barrow group on its west is the only one on the northern downs, a major contrast with the density of such groups (A–M) in the study area. There, MD = the Bronze Age enclosure on Manton Down (Figure 5.4), OD XI = the site described here in Chapter 6 and in FWP 63, and H = Headlands (Chapter 4, Figure 4.2)

Manton Down (Figure 5.4); another, on similar grounds but totally lacking in landscape context, lies interestingly well to the south on Windmill Hill Down (Chapter 12). The small square Manton enclosure associated with a distinctive block of small fields on Manton Down (Figure 5.4) has all the characteristics of a Deverel-Rimbury settlement and is accepted as such by Gingell (1992, 156). It is aligned with, and probably overlies, a lynchet, very closely paralleling the sequence at South Lodge, Cranborne Chase (Barrett *et al* 1991a). A linear ditch, bounding the field system and possibly the eastern end of ditch F4 on Lockeridge Down (Figures 2.1, 5.1 and 5.3), passes close by the west side of the enclosure and invites comparison with other Deverel-Rimbury sites such as the much larger Martin Down (ibid) and, closer in size to Manton Down, Boscombe Down East

(Stone 1937). The pattern of small fields around the enclosure form a distinctive block, which could indicate the extent of the arable, associated with a later second millennium BC farm. If the proposed date for the Manton Down enclosure is correct, then we should seek a small cemetery in close proximity. Bradley (1981) has discussed the relationship between Deverel-Rimbury settlements and small barrow cemeteries in detail. Such cemeteries should be within a few hundred metres of the settlement and visible from it (ibid). So far no mound(s) or ring-ditches have been noted close to the Manton Down enclosure and an isolated barrow, some 550m to the south, is perhaps too distant for serious consideration, but this should not necessarily invalidate the proposed date. A Middle to Late Bronze Age landscape with an associated settlement seems a high probability, therefore, on Manton Down, though perhaps the remarkable point about it, given the extent of such landscapes immediately to the north (Gingell 1992), is that it has only been captured for this study by extending the eastern edge of a study area from which such evidence of settlement is otherwise absent.

Other elements of such a landscape are, however, present. Linear ditch F4, rising up the west-facing scarp from the Avebury area and across northern Overton (or Lockeridge) Down and Totterdown (Chapter 5), bounds the north side of an extensive field system on the high down (Plate VII). Excavation has demonstrated a long and complex sequence involving a change of character, but not necessarily its function as a boundary, down to the Romano-British period when part of it became a track. The evidence points to an initial construction date in the late second or early first millennium BC, in happy accordance with evidence from elsewhere in Wessex, notably Salisbury Plain (Bradley *et al* 1994). It is quite possible that the stretch of ditch beyond the change of alignment north of Delling Copse marks an extension associated with a Late Bronze Age ploughing up of Totterdown, an expansion of high downland arable on to Clay-with-Flints (Plates VIII and XVI). At the west end of its course on Avebury Down the ditch passes close by and to the north of barrow groups D and E, possibly emphasising an earlier boundary established by the cemeteries (Figure 2.1). Here, too, unlike the relationship on Totterdown, fields lay both north and south of the ditch, hinting at the vestiges of 'strip' territories similar to those around the valleys of the River Bourne and Nine Mile River (ibid).

By the Late Bronze Age/Early Iron Age the evidence for settlement and fields and associated trackways is even stronger. Aerial photography, reinforced by ground inspection, has identified an enclosure on the western edge of Totterdown Wood (Figure 5.1). This is related to the extension to linear ditch F4 by overlying it; and again, given the total absence of Middle and Late pre-Roman Iron Age evidence from six cuttings on Totterdown, would seem to require a date in a mid-first millennium BC horizon. Its morphological and chronological analogue lies with site OD XI on Overton Down (Chapter 6), and indeed the detail of the land-use sequence on its site may be very similar – fields, settlement, fields, abandonment, within the thousand years 1500–500 BC. On Overton Down the ditched, enclosed settlement (OD XI) is set within an existing field system and contains a number of circular domestic structures, pits and hollows. Associated with furrowed bowl and early All Cannings Cross-type ceramics, its date can be fairly well established between the eighth and sixth centuries BC. The enclosure appears to succeed an unenclosed Late Bronze Age settlement set within an existing field system. The field system continued to function beyond the limits of the enclosure. Although of relatively short duration – three generations or about a century is suggested elsewhere (Chapter 6) – the enclosure was to have an enduring influence on the form of the fields and a track that post-dated its abandonment. The enclosure is of 'tombstone' shape, a type best paralleled in northern Cranborne Chase (Bowen 1991, fig 46c, especially Gussage St Michael 7h). No entrance was located, but it was most probably on the east or south-east side, perhaps giving out on to what subsequently became formalised as a double-lynchet track overlying as well as passing through prehistoric fields (Figures 2.1, 6.1 and 6.11).

Two other morphologically similar enclosures exist in the study area. One, called by us 'Headlands', lies the same distance south west from OD XI as is the Totterdown enclosure to the north east; a less well evidenced but probable one lies across the river to the south west on Lurkeley Hill (Figure 15.3). From there, the other three enclosures are visible and all four occur along the same straight line; three are at 200–250m aOD but 'Headlands' is slightly lower in altitude. Indeed, rather than truly a downland site, it could be regarded as on the side of the Kennet valley adjacent to barrow group J, a situation which may bear on the longevity of the site (Chapters 2 and 6; Figures 4.2 and 4.3). Although unexcavated, the interior of the enclosure has produced material from the surface that shows that its early phase is contemporary with the Overton Down site. Further investigation is required to ascertain whether settlement continued into the Middle and Late

Iron Age. The location was certainly occupied in the Roman period, probably developing into a small villa *(see below)*. 'Headlands' could have become the dominant Iron Age settlement in our study area. Its location, off the high down and close to the river, is likely to have been more attractive than the exposed location on Overton Down. That it also related to the higher landscape is suggested by a north-west facing entrance associated with an 'antenna' ditch that runs north to link with an east–west track approaching from Overton Hill; it is also approached by a track from the higher downland to the north. This may indicate a role central to the control of livestock movement within an economy based on pastoralism and transhumance. Its continued importance in the local landscape, first as a settlement and later as a feature of the tenurial geography, is discussed in Chapter 4.

The general form of this sort of settlement enclosure and its date are easy to parallel elsewhere in Wessex, as has indeed been known since the discovery of OD XI (Bowen 1966, fig 1). The appearance, distribution and nature of such settlements at this period are now, however, much better known, not least as a result of considerable excavation. Relevant examples are Gussage All Saints (Wainwright 1979), Little Woodbury (Bersu 1940), Longbridge Deverill Cow Down (Hawkes 1994), Old Down Farm, Andover (Davies 1981), Pimperne (Harding *et al* 1993) and Winnall Down (Fasham 1985). Some settlements of this period associated with early All Cannings Cross ceramics are also associated with especially large round-houses, although no such structure was located within the limited area explored at OD XI. The economy was based on a mixed farming regime of arable and livestock, with sheep, cattle, pig and horse. The last were mature animals, perhaps used as draught animals or, alternatively, an indication of status. The settlement was abandoned before the end of the sixth century BC and, after a brief interval, its location reverted to arable with field boundaries laid across it. This post-settlement phase of arable was also short-lived and thereafter, ie, from not later than *c* 500 BC, apart from two short phases of cultivation, the site has been the pasture it is today. Of the parallels cited above, those on the higher and more exposed locations like OD XI, Longbridge Deverill Cow Down and Pimperne, also have a relatively short-lived occupation.

Richards (1990) noted of the Stonehenge environs that, apart from a short episode of Late Bronze Age and Early Iron Age settlement, the area is otherwise remarkable for the absence of recognisable Iron Age activity. The proximity of Overton Down to the Avebury complex could additionally suggest that the status of the area was somehow different from other areas of the chalk but like that around Stonehenge. This observation is not incompatible with the proposal below concerning a special boundary or 'neutral' zone status for the Fyfield/Overton area. The study area is also distinctive in being peripheral to the distribution of Middle Iron Age hillforts. The nearest is Barbury Castle, 6km to the north, with Oldbury 8km to the west and the Martinsell/Giant's Grave complex 6km to the south east (Figure 15.2). It should also be remembered here that Cranborne Chase, also close to a major socio-political boundary in the Iron Age, is also peripheral to the neighbouring hillfort zone (Barrett *et al* 1991a, fig 6).

The general lack of Middle Iron Age material from the whole study area strongly suggests a major change in the agrarian economy, probably to pastoralism. Again the best local parallels for this shift in emphasis are to be found on Salisbury Plain. There, evidence for Early Iron Age settlement is strong, especially east of the River Avon (Bradley *et al* 1994; Brown *et al* 1994; RCHME unpublished fieldwork). Apart from hillforts and a small number of other settlements on its periphery, the high central and western downland of Salisbury Plain is noticeably devoid of Middle Iron Age activity. This marked change in the downland economy requires explanation.

Salisbury Plain, like the study area, appears to have experienced a major episode of land allotment in the late second and early first millennia BC. Arable activity only appears to have recommenced at the end of the Iron Age or early Roman period. It is possible that in these two areas we are witnessing a major change in the economic (and possibly social) structure of the region in the last centuries BC. Both were in 'grey' areas in terms of later Iron Age 'tribal' boundaries, a situation that may have originated in the Middle Iron Age and influenced land tenure and use. Our study area, like Salisbury Plain, has not produced many Iron Age coins, again suggestive of a regional boundary location (van Arsdell 1994).

Consideration of the Fyfield/Overton evidence leads to two suggestions. The first, earlier in date, sees much of its area divided up for grazing purposes between possible territories related to hillforts (Figure 15.2); a second sees a similar area forming a sort of 'no man's land', used as common grazing as part of an economy practising transhumance (with or without hillforts).

Quite independently, and more or less synchronously, the RCHME (unpublished MS) inferred the same model from its work over a much larger area on Salisbury Plain, ie, that much of the plain, especially

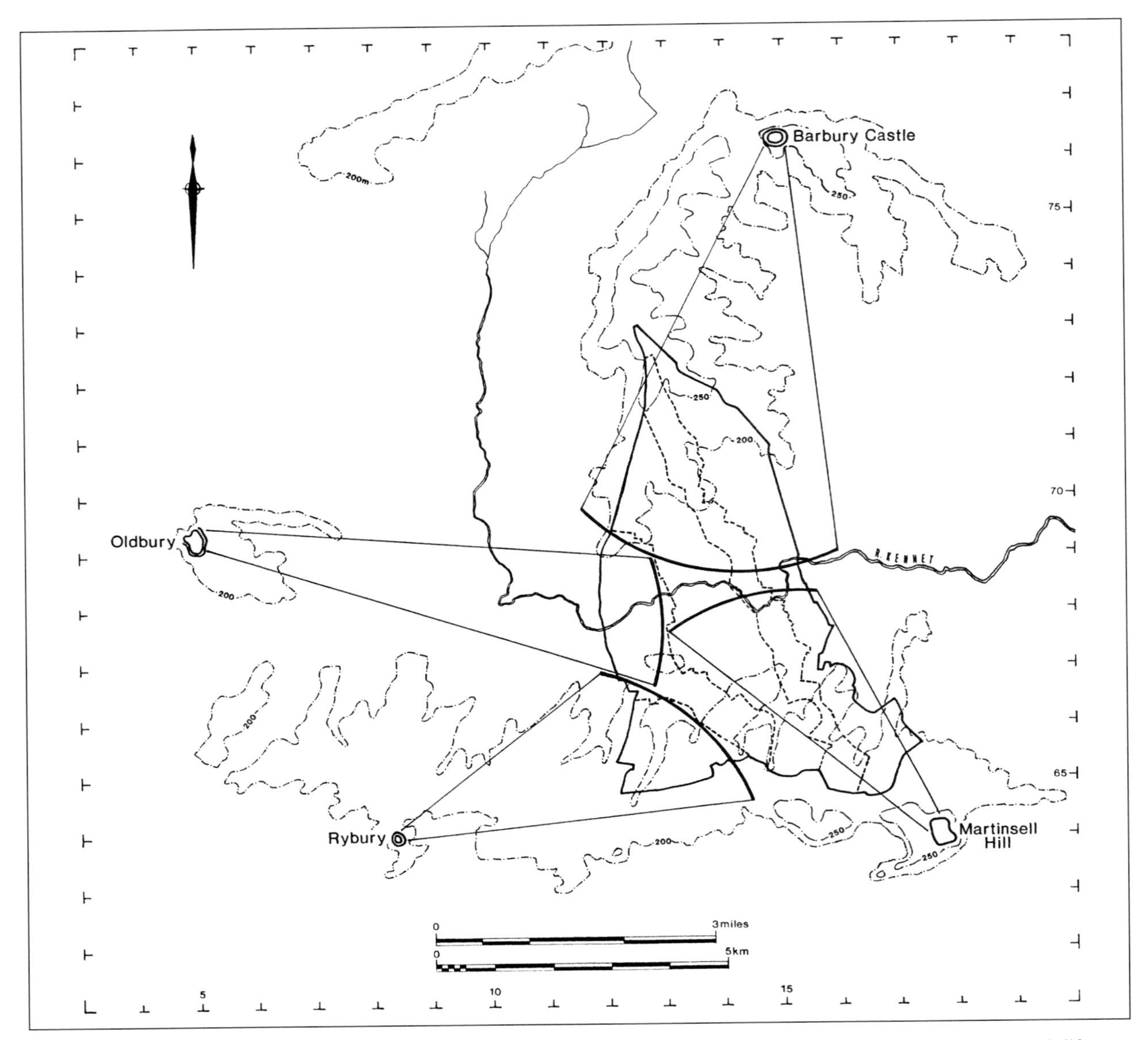

15.2 Diagrammatic map based on a resource-exploitation model to suggest a possible relationship between the four distant hillforts around the study area and the later proprietorial land units within what became West Overton and Fyfield parishes. Barbury's possible 'arc of influence' embraces most of the downs north of the river, later the pastures of the conjoined manors of East Overton and Fyfield; Martinsell could have embraced the equivalent manorial area south of the river, for wood, clay and flints as much as pasture and possibly arable on Boreham Down; Ryburys area could have coincided fairly accurately with the later manor/tithing of Shaw; and Oldbury's territory could well have been what later became the manor of West Overton, on both sides of the river

west of the River Avon, was common grazing for transhumant stock in late prehistoric times. If these suggestions are remotely correct, then the long-established image of the Marlborough Downs and Salisbury Plain teeming with Iron Age cultivators in a busy, inhabited downland landscape must be abandoned (eg, Fowler 1983a, 1983b). Instead, both would have looked more like the plain and, to a lesser extent, Fyfield and Overton Downs today: hectare upon hectare of treeless grassland through which, scarcely visible, sheep, cows and horses nibbled, munched and chewed their unhurried way tended by herdsmen and carolled by skylarks. But, to revert to an earlier model (Piggott 1958), at times there may have been cowboys.

The Late Iron Age and early Roman period

Renewed activity that has left an archaeologically recoverable pattern occurs during the first centuries BC and AD. The changing nature of the political and social patterns of the Late Iron Age is reflected in the region by the development of a major Late Iron Age centre to the east of Marlborough, 8km from Fyfield. Occupying both the high ground on the south and the floor of the Kennet valley, the site, *Cunetio* in the Roman period, suggests the presence of a local tribal centre (Corney 1997). The complex has many of the general characteristics of an *oppidum*, a rich élite represented by well-appointed burial – the 'Marlborough Bucket' – a centre associated with lengths of dyke, a location in a major valley suitable for longer distance exchange and a hinterland which has produced evidence of contacts with the Roman world (Corney 1989, 1997). The emergence of this complex is likely to have had an influence on the surrounding landscape and may provide a context for renewed activity, especially arable farming, within the study area.

This period is marked by a number of significant developments in Fyfield and Overton. After a hiatus of half a millennium, ceramics and other material occur in quantities on the downs. A complex in the Bayardo Farm area on the edge of and in the woodlands, recently brought to our attention and as yet uninvestigated scientifically, has produced much pottery (which we have seen), a bracelet and unconfirmed reports of cremation burials. Associated with this phase of activity, apparently occurring over the length and breadth of the valley, downs and woods, is an extensive reorganisation of the landscape with the laying out of new, or refurbishment of, field systems integrated with a network of tracks. The precise chronology of this is not certain, due in part to the continuing debate over the date of the establishment of the Savernake pottery industry. Swan (1975) sees it as a post-Roman Conquest establishment whilst Timby and Hopkins (pers comm) prefer a pre-Conquest date. The outcome of this debate will be crucial to the dating of the start of this major phase of activity. Here, largely because the landscape implications are so strong, we associate it with immediate post-Conquest Roman land reorganisation.

Whatever the precise date, the impact of Roman rule on the region is clear. It is highly probable that a Roman fort existed at *Cunetio* during the Conquest period although the occupation is certainly short-lived (Corney 1997). A temporary Roman military presence is unlikely to have had a direct long-term dramatic effect on the area, but Roman conquest led to the construction of a major road along the Kennet valley. Heading west from *Cunetio* towards Bath, the road uses Overton Hill as an alignment point where its direction changes to the west, and a new field system was apparently laid out at roughly 90° to the road axis east of the hill (Chapter 2). The establishment of this road, the first formal east–west route recorded in the area (but cf Green Street, Chapter 2), will surely have affected the local economy.

A further result of Roman rule was probably the imposition locally of the *annona militaris*, much of it probably collected in kind. This tax could provide a context for renewed arable on the downs although, as already noted, a Late Iron Age, pre-Conquest date is also possible. Though we are not convinced on present evidence, reinvigorated downland and woodland farming in the first centuries BC/AD, as Caesar may well have been reflecting in his description of the Kentish countryside, is not incompatible with a physical and tenurial reorganisation of a working landscape in the interests of greater productivity after AD 43. The evidence from Salisbury Plain again closely mirrors that from the study area. There a resurgence of activity in the Late Iron Age is seen at a number of sites either side of the Avon Valley, such as Casterley Camp (Corney 1989) and Netheravon (Graham and Newman 1993). At the same time field systems are reactivated and extended (RCHME unpublished fieldwork), a pattern which continues and accelerates into the Roman period (Fulford *et al* 1994; McOmish 1998).

In broader terms the pattern is one seen over much of lowland Britain in the first century AD – the expansion of field systems and an increase in the number of rural settlements. Continuing into the second century, the pattern reflects increasing Roman investment in the rural economy, perhaps accelerating a move already underway before the Conquest. Additionally, a rapidly expanding urban population required produce. Apart from *Cunetio*, other urban sites were developing close enough to the Fyfield/Overton area to have had an influence on demand, notably at Wanborough and Cirencester, probably the local *civitas* capital. Only 2km west of Overton Hill, at the foot of the great mound of Silbury Hill, another Romano-British settlement has only recently begun to be studied in detail (Powell *et al* 1996; Corney 1996; Whittle 1997). Of as yet undetermined character, this covers at least 12ha and, like the fields from Overton Hill to Hackpen (Figure 2.5), is set along and at right angles to the Roman road. It may have had a religious aspect (Corney 1996), perhaps underscoring the continuing importance of a ritual focus in the Avebury region (Dark 1993; Williams 1998).

Within the study area, the settlement complex at 'Headlands' (judging by surface artefact scatters) saw an intensification in activity, perhaps encouraged by its close proximity to the Roman road giving easy access to local market centres such as *Cunetio* (Figure 4.3, 3). The unusual Roman barrows of second-century date on Overton Hill may be related to the 'Headlands' settlement. They would have been clearly visible from 'Headlands' and may be positioned to mark the western limit of a territory whose origins lay in the prehistoric period. Their location, adjacent to a major Roman road also recalls the Roman burial custom of placing cemeteries close to roads (Toynbee 1971). Morphologically distinct from the large Roman barrows of eastern England (Jessup 1962), their form and location, perhaps consciously mimicking earlier monumental burial traditions, suggests an indigenous development. Though their rite is different, their size is similar to that of the post-Conquest but native 'Roman' burial mound at Knob's Crook, Dorset (Fowler 1965).

On Overton Down the site of the Early Iron Age enclosed settlement (OD XI) was again, briefly, put to the plough. Just down-slope, fields were also under contemporaneous cultivation, perhaps not continuing far into the second century, though settlement was probably burgeoning nearby. Most of this down, however, could well have been under grass for 200, perhaps 250, years before a new settlement, not grain seed, was planted out on long-abandoned fields in the vicinity of Down Barn (Chapter 6; Figure 6.11 and below).

The late Roman and early post-Roman period

The fourth century AD witnessed a dramatic intensification of agricultural and settlement activity on much of the Wessex chalk. Villas, many likely to be estate centres, are evident in many places and the majority will have had associated settlements in their hinterland.

At 'Headlands', the air photographic and field evidence strongly suggest that a villa developed adjacent to the Iron Age and earlier Roman settlement (Plate XV). Another villa is suspected under Fyfield Manor House and a further example may be associated with the Silbury Hill settlement. The pattern emerging is one of fairly regularly spaced villa-based settlement along the Kennet valley, on present evidence specifically along its north side west of Fyfield. The implication is clearly of a relationship with the Roman road. This hypothesis would carry greater weight if villas were to be found beneath the medieval villages of Clatford, Manton and Preshute, east of Fyfield, south of the river but still along the Roman road. With regard to placement, it is interesting that the putative villa at 'Headlands' at the very least reused an old site and might well have colonised an existing settlement.

These villas were surely associated with estates which, to be viable economic units, would unavoidably have incorporated downland and woodland. In other words, their shape and size, if not necessarily their position, are likely to have been similar to those of estates visibly well established by the tenth century. The Roman pattern of villas and estates is repeated elsewhere locally, notably around Salisbury Plain (Gaffney *et al* 1998). Furthermore, the evidence from our study area and beyond would support the view that such estates not only reflected pre-Roman land units, but were more assuredly also central to the development of the post-Roman landscape (Chapter 16; cf Bonney 1976; RCHME unpublished fieldwork). Certainly the location of the villa estate-related settlements in Fyfield and Overton strongly suggests continuity of locale, shape and size and land units over a lengthy period of time. The origin of something closely akin to the pattern of the medieval tithings can, on this argument, be envisaged as before *c* AD 900 at latest and before *c* AD 200 at the earliest.

The extensive Romano-British settlement complex on southern Overton Down/Down Barn (Figure 6.11) is likely to have related in some way to the 'Headlands' villa (as was suggested long ago; Fowler 1963a). Its principal track continues south from the settlement to run past the 'Headlands' complex (Figure 16.6). Four Roman settlement areas have been distinguished within the complex – OD XII, OD XIII, ODS and beneath the Down Barn Enclosure – but they may all be part of one whole at the same time (though this is extremely unlikely, *see* Chapter 6) or different parts of the same settlement shuffling around a location in a manner familiar from medieval studies. Nevertheless, certainly the overall plan suggests an element of regulation that must have been long lived (Figure 6.11). The morphology and date of the whole settlement area, and specifically of ODS, which is large enough in itself, show it to be of similar form to the settlements that developed on the downland of Salisbury Plain (Fowler 1966; Frere 1992, figs 20, 21 and 22) and elsewhere in Wessex (Cunliffe 1977). Both the whole complex, and ODS, were integrated with the established local communications network. The regular plan of the settlement, and again specifically of ODS, although partly governed by earlier land use, also suggests an element of control over its layout. This would appear to be so also of OD XII, despite the fact that it did not overlie earlier occupation and did not start until the mid-fourth century; excavation showed its close spatial (tenurial?) relationship to an earlier ditch (Figure 6.16). Such organisation, implying regulations, is closely paralleled on Salisbury Plain at the settlements of Church Pits and Knook Down East (Frere 1992).

The late Roman settlement, OD XII, was laid out over an area of earlier fields last cultivated in the second century at the latest. The evidence from the excavations here suggested a clear division of the settlement based on function and specialised activity. Grain processing was one, but only one, important component of the economy. Faunal remains suggested a strong pastoral element, dominated by sheep. That these were largely mature animals suggests they were kept for secondary products, probably wool. Comparison with the settlements on Salisbury Plain suggests also that many of these Wessex downland settlements of the late Roman period enjoyed a surprisingly high standard of living and had access to the coin-using economy and high quality materials. At the Overton Down settlement some 300 coins of late fourth- and early fifth-century date and fragments of fifth-century glass bottles and tableware indicate a prosperous, well-connected community.

Such an inference leads to the suggestion that the full agrarian potential of the whole area was only now realised in the late Roman period, especially after the middle of the fourth century. Then, generalising from the excavation of OD XII, it is reasonable to envisage a downland reverted to sensible farming, mainly shepherding and probably to meet a strong external demand for wool, a trade likely to have brought in cash. Local downland cultivation, in contrast to an earlier arable régime designed to meet external, cashless demand, was for local need alone. The other side of that model, for which evidence is scant in the extreme, would be of intensive cultivation of the valley sides where, for all we know, arable has been continuous ever since (Figure 16.1).

The context for such an interpretation can be sought in the character of late Roman Britain. Administrative reforms of the early fourth century will have placed the region in the province of *Britannia Prima*, arguably the richest of the provinces of late Roman Britain (Corney in prep). In the second half of the fourth century, state interest in securing tax revenues, in kind as well as cash, is reflected in a major programme of building at a number of key 'small towns'. The nearby site of *Cunetio* forms part of this process. Subject to major redevelopment after *c* AD 360–70, the site received substantial defences and may have acted as a local administrative, fiscal and military centre (Corney 1997). It can convincingly be seen as representing late Roman concerns with security and the collection of raw materials. The general regional pattern of fourth-century agricultural intensification, possibly dominated by pastoral rather than arable products, is well established (*see* Gaffney and Tingle 1989 for a similar pattern on the nearby Berkshire Downs), and accords with the independently generated model from Overton Down. Our study area as a whole, however, lacks comparable evidence, a fact likely to reflect our methodology (Chapter 3) rather than the archaeology of its landscape.

West of the 'Headlands' villa, on Overton Hill, the large polygonal enclosure recorded by Crawford (Chapter 4) is very similar to probable late Roman enclosures elsewhere in Wessex. In Cranborne Chase, notably on Rockbourne Down and Soldier's Ring (Bowen 1991), and on Salisbury Plain at Warden's Down and Wadman's Copse (RCHME unpublished fieldwork), such enclosures may be associated with control of livestock. These are most likely to be components of an economy based on sheep, probably for wool production, as suggested by the faunal remains from OD XII. Britain, of course, possessed at least one state-owned weaving works (*gynaeceum*), placed at *Venta* (possibly Winchester), recorded in the late fourth-century *Notitia Dignitatum*. Further evidence of late Roman wealth and activity in the Fyfield/Overton locality is provided by the large hoard of silver *siliquae* and one gold *solidus* deposited with silver bullion, recently recovered from downland 8km to the west at Bishops Cannings. This assemblage contains a number of other items suggesting the presence of late Roman officials in the region. The hoard may have been deposited as late as AD 420–30 (Guest 1998). South east of the study area, at Castle Copse, Bedwyn, a large villa appears to be occupied well into the fifth century (Hostetter and Howe 1997).

It is uncertain when settlement OD XII was finally abandoned. The quantity of late Roman coins and the fifth-century glass combine, however, to present a convincing picture of activity, and relative wealth, beyond the formal end of Roman Britain. Recent work on two Romano-British settlements on Salisbury Plain, Coombe Down and Chisenbury Warren, has produced convincing evidence (including radiocarbon determinations in the case of Coombe Down) of continued occupation into the sixth century AD (Fulford *et al* 1994). Further south east, at Chalton in Hampshire, Cunliffe (1977) noted that three late Romano-British settlements produced grass-tempered pottery ascribed to the fifth–eighth centuries AD. The only comparable evidence in our study area came, not from a Romano-British settlement but from the site of a medieval farm called *Raddun* (Chapter 7; FWP 65), the sherds possibly indicating the location of a Saxon sheep-cote a long way out on Fyfield Down. Although no firm, equivalent evidence has so far come from OD XII or the Down Barn complex in general, the artefacts recorded from the former make a late fifth- or sixth-century date for the abandonment of that site a possibility. A sub- or early post-Roman date for the Down Barn Enclosure is also a possibility. The probability of other stone-based structures in the vicinity suggests in any case that the end of OD XII was not the end of the Down Barn area as a node of activity (Chapters 6 and 16).

Although specifically sixth-century material has not been identified from the downs it is present, along with fifth-century items, from the secondary burials placed in and around the Roman tombs on Overton Hill (Chapter 4). The burials stand out locally as one of the few tangible traces of immediate post-Roman activity in the area. These are of some significance as they are not only placed around a Roman (and indeed much earlier) cemetery but are also in a location visible from the 'Headlands' complex. There, air photography has suggested the presence of three or four timber buildings of post-Roman form (Figure 4.3, 4). Fieldwalking produced no material from the area of the cropmarks, suggesting they are not Roman in origin. An earlier prehistoric date is also unlikely so a post-Roman date deserves serious consideration. It is tempting to relate the structures to the secondary burials around the tombs on Overton Hill and suggest that we are glimpsing a successor complex to the villa. Even if that were to be so, we could not know at present whether the occupants of such a settlement and its associated cemetery were newcomers of Germanic origin, as grave-goods suggest, or part of the indigenous population using indicators of status from a new material culture.

Whatever the ethnic origin or status of post-Roman occupants, in some or several ways the later history of the 'Headlands' and the southern Overton Down settlement complexes is likely to have been related to the most substantial local monument of the period in our study area, the East Wansdyke. We are persuaded by the arguments for dating it to *c* AD 500 (Eagles 1994). Whether it marks a boundary between Britons and Anglo-Saxons or between two British or two Anglo-Saxon factions is still uncertain, but we argue for the former circumstance elsewhere (Chapter 13; Fowler forthcoming c). That its construction disrupted existing tenurial units (Bonney 1972), we also accept and indeed believe that we have reinforced that idea; we suggest additionally that the construction, particularly if it was operative for some years (*contra SL*, 134–5), may have had a broader effect on an area probably dependent on seasonal transhumance and the free movement of livestock, especially sheep. This may have been an additional factor in the demise of the upland settlement pattern on downs and along the woods, helping to shift habitative emphases to the valleys where the villa centres already were – or at least had been.

By the tenth century the area of the 'Headlands' complex was on a tenurial boundary in the late Saxon landscape; so was the equivalent Early Iron Age enclosure on Lurkeley Hill, also adjacent to a probable Romano-British settlement (G Swanton pers comm), and the probable counterpart at Totterdown Wood was also just inside the Fyfield boundary. The association of Romano-British settlements, especially villas, on or close to documented estate, tithing and parish boundaries is undeniable in the Wessex region. It was first explored by Bonney (1968, 1976), elaborated on by Fowler (1976), statistically demolished by Goodier (1984) and recently illumined by further fieldwork (RCHME unpublished; Gaffney *et al* 1998) in the Avon valley north of Amesbury and the foot of the northern scarp of Salisbury Plain. A number of villas have been identified adjacent to middle to late Saxon estate, and later tithing, boundaries. The association seems too frequent to be coincidental and other explanations need to be sought.

Some possibilities are considered under the 'boundary' theme in Chapter 16, but one is that the Roman sites became peripheral to later land units and were used as convenient markers in subsequent phases of land division. While this may be the case in some instances, at 'Headlands' and examples around Salisbury Plain, in general it seems an unlikely explanation. More promising is the idea that the siting of the boundary at 'Headlands', and probably the Salisbury Plain examples, represents a post-Roman subdivision of a large estate. The creation of smaller, early medieval estates and tenurial units under new political and economic frameworks would be a logical development once a centrally controlled economy designed to generate a large surplus ceased to function. The fragmentation and ultimate demise of Roman political and economic control over Britain would provide the circumstances for such a development, whether or not new peoples moved in and took control.

Logically, this chapter should end with three or more sections similar to those of which it has consisted so far, covering perhaps Anglo-Saxon, medieval and modern times. History from early medieval times onwards has, however, fundamentally determined this volume's structure from Chapters 8 to 13, chapters that are biased throughout to a history of the medieval landscape. Chapter 7 is largely a case study in medieval landscape, a topic also discussed in important passages in Chapters 14 and 16. The volume as a whole has been infused by Ian Blackwell's work on the Anglo-Saxon charters (FWPs 11, 11a and 68) and medieval local history (FWPs 44–48), Simon Yarrow's exploration of the twelfth-century Templar landscape (FWPs 18a and b) and John Hare's analysis of a manorial landscape history (FWP 43). Further chronological treatment would therefore be repetitive, so we change to a thematic approach, embracing many and various times.

Chapter 16

Themes in a Wessex Landscape

I have always been preoccupied with man's position in landscape and his relation to the structure of nature.

Barbara Hepworth, *Some Statements ...*, Barbara Hepworth Museum (St Ives), 1977

Numerous general themes emerged from the detail of the investigations. Ten are selected for general, albeit brief, discussion here. They are:

1 land-use and the basic farming regime: pastoralism
2 land-use: arable
3 a landscape of exploitation: tenure
4 boundaries: boundaries and territories; boundaries and settlements
5 settlement 'shuffle'
6 settlement morphology
7 religion
8 recreation
9 communication: through-routes
10 English countryside, British landscape

1 LAND-USE AND THE BASIC FARMING REGIME: PASTORALISM

The landscape of the downs north of the A4 road, or south beyond the valley bottom towards West Woods, conveys an overwhelming impression of arable farming. Both in terms of the space in our two parishes and in historical terms, this impression is misleading. Even today, the actual state of affairs is that the landscape supports not only mixed farming but, as we discuss further below and in the last chapter, a landscape whose uses extend beyond farming alone. Leaving aside, for the moment, forestry and various recreational uses, there remains a long-term historical truth lurking in the farmed landscape of the downs, which is belied by both their archaeology and their present appearance.

Essentially, the landscape of Fyfield and Overton is a pastoral landscape. Its long-term economic viability rests in animals, not cereal crops, above all in sheep rather than barley and wheat. This may seem a surprising assertion, given the plentiful primary documentary and cartographic evidence of extensive and, in some cases, permanent arable (Figure 16.1). But what was that arable for? Primarily, it was to feed the local population and its stock and was, therefore, in a sense a means to an end. The surplus, and therefore the potential profit that made land-owning so attractive to Church and laity alike, lay in the animal products, not those of arable fields.

Hare (FWP 43) noted a nice example, which illustrates the general point in medieval times:

> ... agriculture seems to have been extensive rather than intensive, with Overton generally at the lower end of yields from the Wiltshire manors of the cathedral priory (Harrison 1995, 13–15). This was despite the livestock figures for the manor which were generally amongst the highest of the priory manors in Wiltshire (ibid, 16). An explanation would seem to lie in the growth of demesne cultivation on poor downland soils (such as *Raddon*), where even large quantities of dung did not make up for the poverty of the soil itself, and the enormous concentration of sheep at one end of the parish, where much of the dung may have been expended on some of the poorest of the soils. The rising demand for land was also reflected in the growing rent totals, particularly between 1267 and 1280.

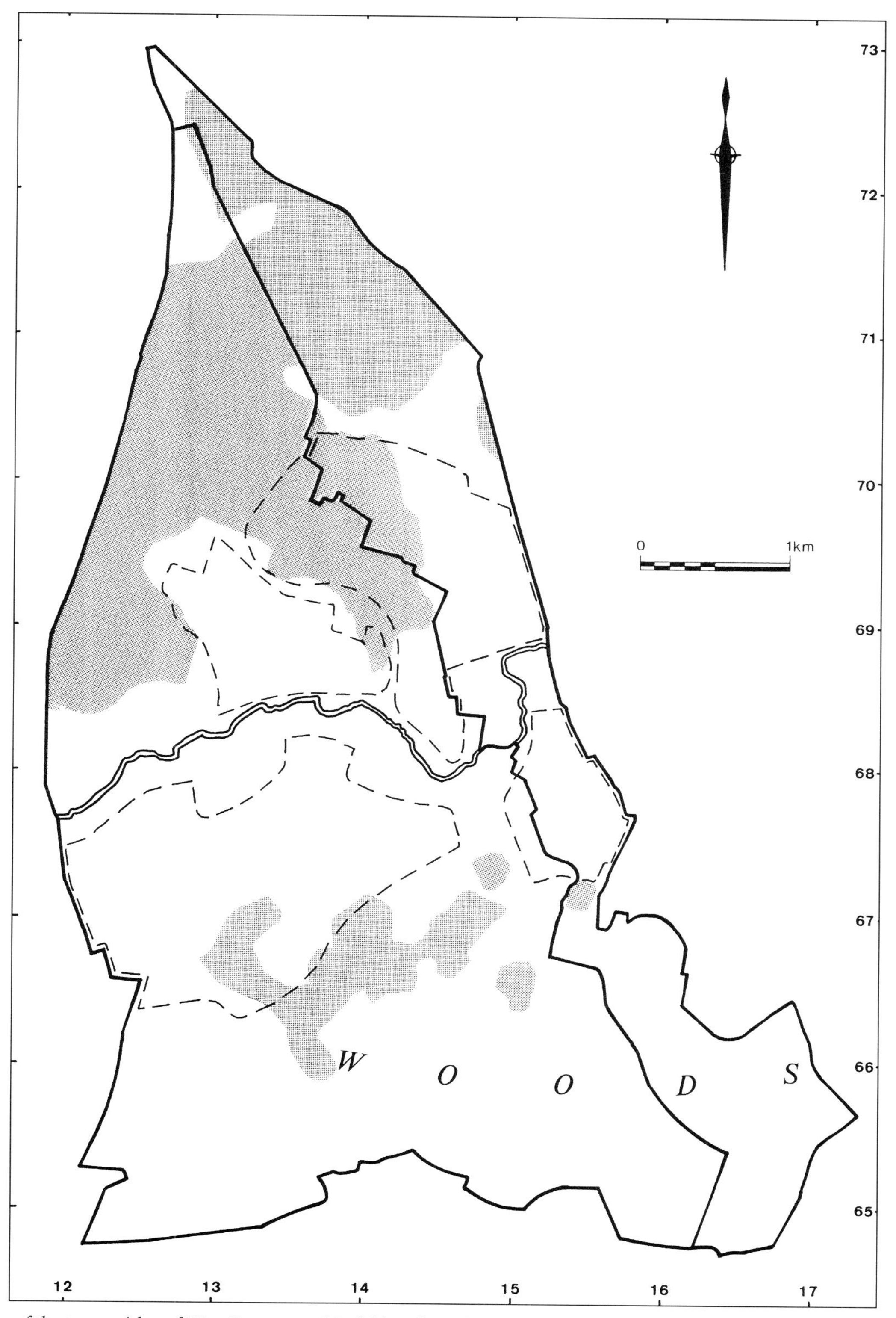

16.1 Map of the two parishes of West Overton and Fyfield to show the total extent of 'permanent' arable land in Roman and prehistoric times (stippled), and in the medieval period (dashed outline)

> ... though it is not clear whether this was the start of a period of growth or a short-term fluctuation. That this early fourteenth century peak coincides with the abandonment of the settlement at *Raddon* (FWP 65), may indicate that land on Overton Down, no longer cultivable due to the cropping demands made on it over the previous decades and the poor weather conditions of the 1310s, was being returned solely to sheep pasture.

That example also illustrates the complement to the assertion above, or the consequence of extensive cultivation. Ultimately, widespread arable did not work; in this context, such land-use was not a viable, long-term strategy. One can well imagine that Hare's interpretation of what happened in the early fourteenth century AD was replaying events of the mid-first millennia BC and AD, reversions to the long-term sustenance provided by pasture reflected in each case by the abandonment of similar farming settlements (OD XI, OD XII and *Raddun*), which, in part at least, depended on extensions of cultivation.

It is in this context that a particular interpretation can be put upon the great extents of prehistoric, arable fields across Avebury, Overton and Fyfield Downs. Taking a longer perspective, the cultivation of the downs in the later second/earlier first millennia BC was as much an intrusion into a grassland landscape as was that of the ecclesiastical landlords, briefly, in the thirteenth century AD. Over that longer perspective, the downs have essentially grown grass and supported stock, mainly sheep; and we can see the essential truth of that interpretation archaeologically on the downs before about 2000 BC and specifically on Overton Down around the seventh century BC and in the later Roman period. The change not so much from arable to pasture but rather back to pasture from arable is as deducible there in pre-medieval times almost as clearly as the documentary evidence indicates for the early fourteenth century.

Nor does this hypothesis end there, for we can continue the story of episodic and temporary downland cultivation, on a much smaller scale apparently, in the sixteenth century and Napoleonic times. And, most significantly, in this context we can see the extensive arable regime inserted into the downs from the 1950s onwards, and still prevailing so destructively in conservation terms, as also temporary. 'Set aside' and other yield-reducing mechanisms were predictable in the broad sweep of landscape history. It is nevertheless of considerable historical significance and conservation interest that, in the autumn of 1998, a 'Countryside Stewardship' proposal for North Farm, West Overton, was approved by the Ministry of Agriculture, Fisheries and Food. As the effects of such constraints begin to be glimpsed in the use of this ancient landscape in the late 1990s, so does the supreme landscape irony come into focus: that Fyfield Down National Nature Reserve, designated in mid-century because it was already becoming a grassland oddity in a sea of arable, actually far better represents traditional land-use in this area than the arable to which we have become accustomed. Yet, in historic terms, the Reserve's principal features include redundant arable field systems that were themselves, in their time, the intruders (frontispiece).

2 LAND-USE: ARABLE

Arable fields have now been put in their local, long-term perspective, but they were nevertheless important both to the economy here at various times and in the development of the landscape. Over the millennia they have occupied a large part of the study area at one time or another, with a core area of arable immediately around the valley villages almost certainly having been in permanent cultivation for at least 2,000 years (Figure 16.1). It was also this, and not pastoral use, which contributed to the erosion which has liberally covered the downs with lynchets, prehistoric (Figure 2.1) and medieval. We have made little of the latter, but strip lynchets exist to the north, just beyond Totterdown, along the southern edge of Fyfield Down (both examples on Plate XXIX) and on the north-facing slopes above Pickledean Barn and round into Fore Hill Fields, as well as across the river around the low spur east of St Michael's Church. In fact, all four examples are on north-facing slopes where there are not known to be earlier fields, so the strip lynchets here might well accord with the conventional wisdom that they sometimes represent the push on to the local limits of arable land in the thirteenth century. A short flight of contour strip lynchets in Foxbury Copse, Clatford SU 165665), facing east above Clatford Bottom, could be another local example of the same phenomenon. If so, though we have no firm dating evidence for any of them, these groups of strip lynchets would be part of the contemporary landscape of patches of downland arable in an extensive sheep run (Figure 2.3); but all of the examples (except the first and last) seem rather to be on the outer limits of permanent arable. Yet attempts to make even such local generalisations are fraught with danger at the detailed level. On Fore Hill, for example, the land had been divided into furlongs by the early tenth century AD (S449).

Here we need not add to the discussion of the prehistoric fields presented in Chapter 2 and elsewhere in Chapters 13 to 15 and above, but we wish to revisit the Roman fields on a comparative basis in company with the one example of a set of 'open' fields which we have looked at in detail (Figure 8.3). We have recorded two good systems of early Roman fields, one unambiguously dated on Totterdown, the other very probably contemporary north along Overton Hill (Figures 5.3 and 2.5); and we have suggested a possible block of late Roman arable fields, by association, on the north side of settlement site OD XII (Figure 6.11). Furthermore, we have suggested that the fragmentary remains of an arable field system on Boreham Down, whatever the date of its origins, might also have continued under cultivation into a post-Roman phase (Figure 12.3). This last is the nearest we have so far come to addressing a central question in this study area, and indeed in various forms in English history. It is, of course, 'How and why did the farmers here move from cultivating the land in field systems of the sort they used in the second millennium BC to those they were using in the first–second centuries AD to those, superficially quite different in form, they were using in the twelfth and thirteenth centuries to those of a, so we believe, similar sort we see them unambiguously using in the eighteenth century?'

Here we offer a contribution to the general question of arable continuity and change by looking in some detail at parts of the local evidence, and then offering a model that suggests one way, perhaps, in which cultivation may have moved from Roman fields to medieval fields. We can only illustrate the physical side of such a change, and make no pretence to illuminate the development of common fields in a tenurial sense. Our model is suggested by the detail on Fyfield Down of how the thirteenth-century ridge-and-furrow was fitted into the framework of much older lynchets that had fossilised patterns of Roman fields modifying prehistoric ones. It is assumed that a similar sequence could have happened in what became the permanent arable of valley settlements, a change that had indeed already happened by the thirteenth century. An expansion (or reclamation?) of arable is likely to have occurred after the establishment of early Saxon settlements just up from the valley floodplains *c* AD 550–650. Initially, for West Overton, this would have been just south of the road between the two settlements of Overton (now the road from East Kennet to West Overton) and probably at the site of 'Headlands'. Also north of the river, Lockeridge and Fyfield's early fields probably stretched as far as the Valley of Stones. Further north, cultivation of the downs certainly occurred in the tenth century. So the insertion of areas of medieval arable into old arable marked by lynchets and field banks may not have been a new experience for Richard of *Raddun*'s contemporaries, though our argument is that such communal experience by then may have lingered only in the folk memory, for in our study area that change would have occurred three to four centuries earlier.

We interpret the evidence of the tenth-century charters as indicating that, already by then but as recognised by the estate boundary jurymen, late Saxon arable was inextricably mixed up with boundaries of fields and estates relating to earlier land shapes and uses. The 'headlands' may well have been describing a function of contemporary strips of land in those tenth-century fields (though of course such strips may well have been long fixed in position) but the *hlincs* (lynchets) were, we suggest, those formed at the edges of prehistoric and/or Roman fields. Should that be so, then, almost by definition, they would also have been the edges of contemporary fields, exactly the sort of situation we see so clearly on Fyfield Down (Plate XXXVIII, Figure 7.3).

We do not, however, have the good state of preservation in the medieval permanent arable that exists beyond their margins on the northern downs, but we can speculate with some actual evidence within the limits of the model the downs suggest. Though virtually nothing of pre-medieval arrangements is mappable in the area covered by the three open fields of West Overton, the air photographic evidence is nevertheless enough to indicate the former presence of such 'cellular' fields as survive on the northern downs (Figure 2.1). So the question is relevant on that hill: how did the pattern of cultivation in 'cellular' 'Celtic' fields change to the pattern of open fields that we see, ultimately, in 1794? Our model suggests that an answer may lie in the pattern of furlongs, not of individual strip fields. The edges of the furlongs are characteristically marked by rectangular changes of directions, assumed to be – and here demonstrably in 1794 – where their boundaries go across the ends of strips. But suppose that pattern was a secondary result, not a cause; suppose that, as we can see was the case on Fyfield Down, strips were fitted into zig-zag lines of former field edges, surviving as lynchets and banks while the rest of the 'cellular' system was overploughed. In such a fashion it would not be too difficult to move physically from a pattern of slightly irregular, conjoined, small, squarish fields to a pattern of 'fields', called furlongs, containing extended strips of ploughland. The reason for making such a change could well have been not tenurial but technological, as the introduction of a plough with coulter and mouldboard quickly replaced the traditional ard.

We can push the model a little further, for we actually have fairly precise local evidence to hand as to exactly how the change may have taken place. We turn away from the generally 'cellular' pattern of the generically pre-medieval field types specifically to the Roman fields, originally picked out on Totterdown and Overton Hill for their morphological distinctiveness. Comparison of their framework – the main linear divisions running through the field systems – shows a close metrical similarity between their field blocks, each containing numerous long, thin fields, and the larger blocks of land defined by runs of boundary along the sides of furlongs in the late eighteenth-century open fields of West Overton (Figure 16.2, a and c). Critical measurements within each field system type are of the same order of numbers, and in some respects quite extraordinarily similar. The 'building blocks' making up the Roman field system on Overton Hill tend to be 150–200m broad and *c* 500m long; the equivalent blocks in the West Overton open fields tend to be *c* 200m broad and 450–500m long. This may be coincidence, of course, but we are suggesting that it is not. Our model sees the similarity as the result of the 'open' type of field system developing from the physical morphology of the Roman system.

It is, of course, a huge leap in time from the field pattern that was on Overton Hill and may have been on West Overton's Windmill Hill in AD 100 and that which was on the latter in 1794; especially as we know that, in detail, field boundaries can change considerably. On the face of it, the time-lapse is too long; 'continuity' in the fields is implausible. Yet the very open fields of our 1794 example are in part defined by an estate boundary of AD 972, and some at least of the trackways helping to outline the fields and furlongs are probably coeval. Furthermore, on the one hand, on the chalk downs in general there is plentiful evidence of prehistoric field systems remaining more or less stable in their macro-patterning over one or two millennia; and on the other, we stress that here our model is based on the framework of the field systems, the primary lines of land-division, not on the size and shape of individual fields. These primary or major divisions are likely to be more long-lasting anyway, both because of what they are and also because other things such as tracks, paths, property boundaries and particular land-uses will tend to relate to them and, by recognising them, help stabilise them as permanent features in the man-worked landscape.

Our proposal here, therefore, based directly on a particular interpretation of locally available evidence, is that rectilinear blocks of land, each containing groups of individual fields, were the 'building blocks' of the local Roman field systems; and that the outline of at least some of these blocks persisted through and after the Roman period to provide at least some parts of the basic physical framework within which furlongs within open fields developed (Figure 16.2). Such a model says nothing about when such a change may have occurred or about the tenurial developments fundamental to the evolution of the English common fields. We would guess, however, that technology drove the physical change, and that that change was occurring in northern Wessex in a post-Viking context, beginning perhaps in the later ninth century and continuing apace through the tenth century. That period could well have seen the conjunction locally of the availability of a new tool – the plough – and proprietorial interests to effect such change.

We must, however, also allow for phases of change. Late Saxon field systems are unlikely to have been changed by central control in a decade, as happened on the Overton manors around AD 1800. And there are of course changes other than morphological, ones of extent, for example. It is possible that the expansion of arable on to the large areas of downland south of the villages, that is to say on *scyfling dune* (S449) and White Hill, is predominantly a thirteenth-century rather than Anglo-Saxon phenomenon; and in any case field development there was perhaps on less-used land with fewer physical constraints. Overall, large-scale reorganisation of the landscape into furlongs arranged in open fields and subdivided into strips was probably well-advanced before *Domesday* (cf Hall 1988, 99–122; Costen 1994, 100–1). Indeed, contrary to the above premise about the nature and speed of change, such change in the management and workings of the arable here may have been centrally controlled and sudden, perhaps at the same time as the proposed creation of one or two planned villages (Chapter 8).

On the East Overton estate, the big change was in the past by AD 939, but perhaps not all that long ago: it was after all apparently worthwhile to comment on the furniture of not just an 'open' landscape, for trees had long been absent on Windmill Hill, but a landscape of heathen burials and ancient lynchets now worked in a different, perhaps even 'common' way.

3 A LANDSCAPE OF EXPLOITATION: TENURE

Hare's example of medieval land-use, quoted above (FWP 43), also illustrates an incident in another long-term characteristic of the study area, the power of the

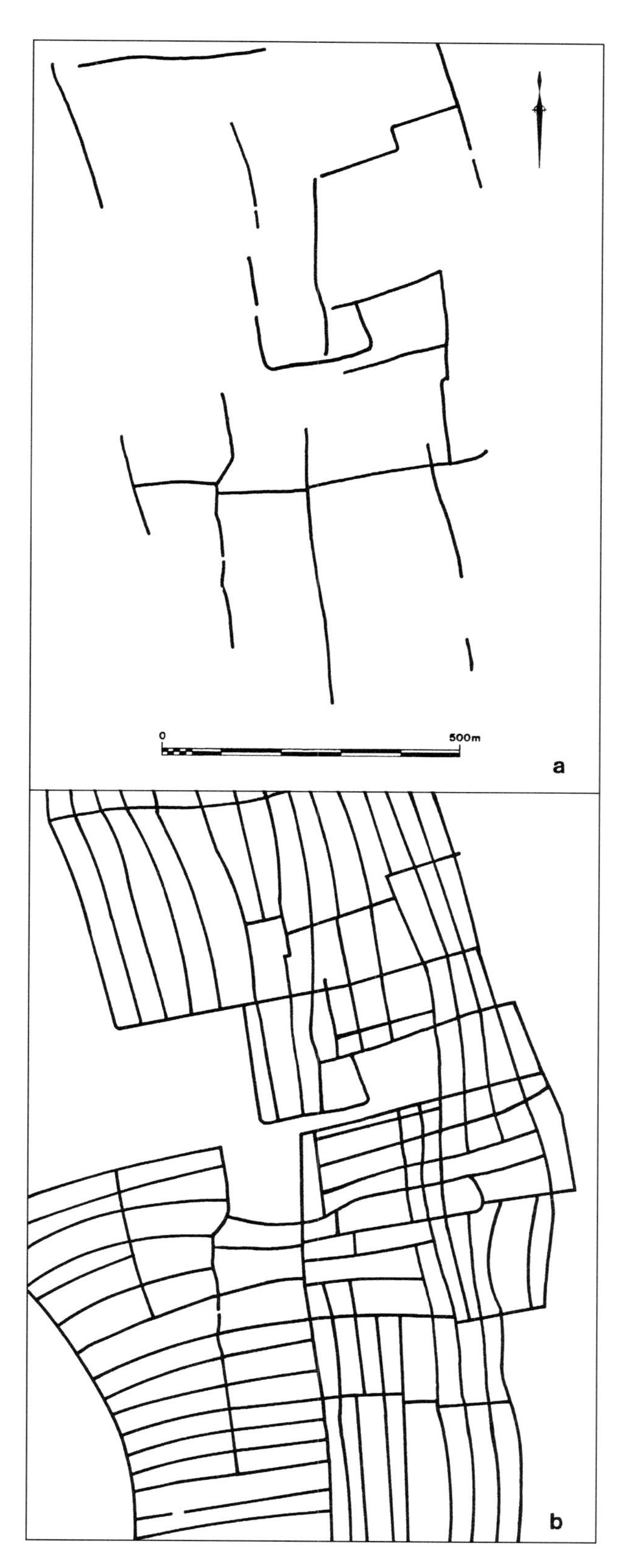

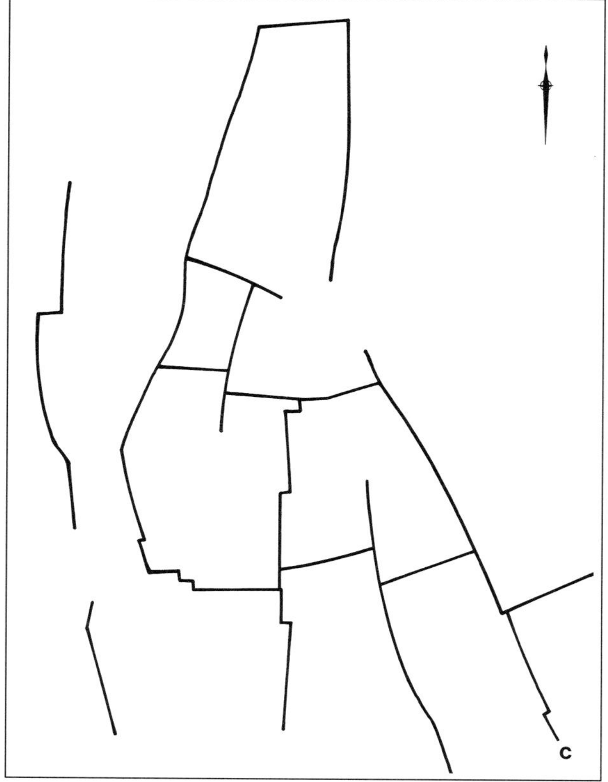

16.2 Diagram exploring a possible metrical relationship between Roman fields and medieval furlongs ((a)–(c) are the same scale)

(a) the main boundaries framing blocks of enclosed, strip-shaped fields in part of the Roman field system, Overton Hill (cf Figure 2.5)

(b) reconstruction of a possible layout of the individual fields within the framework of (a)

(c) the boundaries of the furlongs in part of the open fields of West Overton (cf Figure 8.3, which includes the layout of the individual strip fields within each furlong in a pattern similar to that in (b) above). The frameworks basic to Roman and medieval field systems, (a) and (c), are geometrically similar and, in some respects, metrically of the same order

landlord. The tenurial history of the area is reasonably well documented and researched from the tenth century AD. Over a millennium it shows estates at work under the aegis of a small number of dominant landowners. They are first and foremost ecclesiastical in medieval times, primarily the Bishop at Winchester and the Abbess at Wilton, but also the Knights Templar and Hospitaller and the Prior of St Margaret's at Marlborough. The last was at least local if not actually resident in Fyfield or Overton, but the other three were all distant absentee landowners, a characteristic shared with most of the lesser, secular landowners. This situation continued in post-Dissolution times, when most of the land belonged to the Pembroke and Marlborough estates, and indeed it was not until the later eighteenth/earlier nineteenth centuries that a significant change in the tenurial pattern occurred.

Over a millennium and more, then, we see a landscape dominated by a small number of locally large and characteristically absent landlords. Whether such arrangements prevailed earlier is unknown with any certainty. One favoured inference would be that, if anything, early medieval and earlier times saw even fewer landlords, though their residence locally or otherwise is debatable. It might well have been, however, that the four long, narrow estates, and several smaller parts of estates, detectable in late Anglo-Saxon times were fragmenting from perhaps two larger units, or even just one (*see below*). It is certainly tempting to envisage at least a single 'Overton' unit, perhaps a royal estate, lying behind the ecclesiastical land-holdings delineated in the tenth century as if they were already of some age. So one or more dominant landlords, probably absentee, may well have featured as early as the sixth or seventh centuries AD.

Earlier still, and perhaps precursors, one or two estates in Roman times may plausibly have been based on villas at Overton ('Headlands'; Chapter 4) and Fyfield (on ground later covered by manor house and church; Chapter 11). Landlords of such estates would have been in a general sense absentee, in that the state was the intended beneficiary of local farming arrangements, especially in the first–third centuries AD; but in a more personal sense it is likely that an owner or tenant of a villa at Fyfield, say, left much of the farming to an agent while otherwise devoting time to urban interests in either direction along the Roman version of the A4. Earlier still, it is perhaps only with the appearance of Late Bronze Age/Early Iron Age enclosed farmsteads, like those on Manton and Overton Downs, that we can see people, potentially landowners, actually resident on their land. Such seems likely to have been the case in the second millennium BC too, though not necessarily in the later first millennium BC when ownership may well have lain in the hillforts (Figure 15.2).

The generality emerges, therefore, that the landscape of the study area is the product of ownership by a few, absentee landlords, certainly over a millennium, probably over two millennia, and possibly over two and a half thousand years. A consequence of the recognition of this factor in the history of this landscape is that some of the negatives noted early in the study (Chapter 1) become understandable. Conversely, had we 'read' them for their significance in (more correctly, absence from) the landscape years ago, the hypothesis in these paragraphs could have been adumbrated earlier, and tested. For we are now proposing that the cumulatively impressive lack of monumental evidence in the landscape of Fyfield and Overton is precisely because for most of the time, since at least the later Bronze Age, the area has been used by absentee landlords to support enterprises elsewhere. Candidates would include, on this proposition, the hillforts of Oldbury, Barbury, Rybury and Martinsell; the Roman empire and perhaps specifically and more locally *Verlucio* and *Aquae Sulis* to the west, *Cunetio* and *Calleva Atrebatum* to the east; Winchester Cathedral, St Swithun's Priory and Wilton Abbey; Wilton House and Blenheim, now a World Heritage Site.

Such great places, representing among other things massive and long-term investment to maintain as well as construct, are simply not present in Fyfield and Overton parishes. The reason is now clear: the absence of any substantial structures is not just a quirk of our sample area but a prime piece of evidence, telling us that the parishes and their people were primarily, throughout most of their history, first and foremost but a means to an end. They were a resource to produce a surplus for investment elsewhere. The Fyfield/Overton landscape was farmed and used to produce a profit, and any profit was always intended for use outside the parishes. It follows, therefore, that in looking at the pleasant landscape of the study area we are actually looking at a landscape produced as a result not just of non-investment but of positive economic denudation. It is, *sans* hillfort, castle, abbey, priory or country house, a classic landscape of exploitation.

4 BOUNDARIES

Boundaries exist in many forms and across the whole of the study area. Here we select but two of their aspects, boundaries in relation to the concept of territories and of actual settlements.

Boundaries and territories

If units of land as properties existed, as envisaged in the previous section and in Chapter 15, then there would have been a need to define them spatially. Some of our prime pieces of documentary evidence, notably the tenth-century AD land charters (Chapter 5) and the Pembroke Survey of the sixteenth century (Straton 1909), arise from this need. That such do not exist earlier than the tenth century of itself neither denies nor demands the existence of similar needs and solutions in earlier times, but it is here taken as likely that the landscape, and more particularly its use, was organised in doubtless numerous ways throughout history and much of prehistory. If such a proposition has substance, then 'territories' or 'estates' probably existed in some form and might be recognisable. Such recognition might well come from the physical existence of boundary structures; conversely, all boundary works are likely to have an historical significance, not least in terms of land-use and tenure, for few people are likely to expend effort in marking boundary lines unless such were necessary or advantageous.

Numerous land units and boundaries are attested in our landscape from medieval and post-medieval times. We can also delineate with near-certainty the exact boundaries of two of the main late Anglo-Saxon estates, East and West Overton, and can identify the existence and at least part-location of other land units at that time. Here we extend the concept back in time, which seems entirely reasonable, and propose some theoretical and obviously more contentious territorial arrangements that might have existed in, respectively, the fourth millennium BC, the mid-second millennium BC, the mid-first millennium BC and the early centuries AD. We note them as they might have occurred chronologically.

The study area contains only two long barrows but four others lie immediately outside the historic parish boundaries (Figure 16.3); together they appear to represent that rare thing hereabouts, a relatively large investment of local resources in the locality. They also collectively present a non-random distribution. That of the West Kennet long barrow, East Kennet long barrow and the recently rediscovered White Barrow above Lockeridge is especially striking. All three occupy ridges less than 1km from the Kennet valley and display aspects that appear to relate to the lower lying ground rather than the high downs. The apparently isolated (but *see below*) oval long barrow in West Woods is in size a close companion for White Barrow. North of the river the pattern is less certain, but the location of the Manton Down long barrow and Devil's Den may relate to the now-dry coombe of Clatford Bottom (Valley of Stones) as a boundary zone (*see below*). The four long barrows either side of Fyfield parish are all *c* 2km from their nearest neighbour, with a fifth slightly further away to the north west near Glory Ann Barn. The East Kennet long barrow appears isolated to the south west, though this situation could be remedied by postulating the existence of an as yet unfound long barrow at a position to its south east not only equidistant from it, White Barrow and the West Woods barrow, but some 2km from each. Had this patterning been noted earlier, the position could have been predicted of the Lockeridge example discovered in 1995. As it was, the discovery prompted the spatial observation.

We suggest two inferences from these data. Firstly, it is possible that regularity in the placement of the long barrows was occasioned by the existence of land units, or 'territories', to which they related, rather than, for example, to topography. Secondly, all six lie on or close to (<300m) historic boundaries; the one postulated geometrically to complete the pattern would, if it existed, lie very close to Wansdyke near Shaw House. From this we derive the idea, not to suggest that historical land units originated in the Neolithic, but rather that long barrows, if they were related to 'territories', marked the boundary zones rather than lying at their centres. Renfrew (1973, 132–46) explored both the idea and specific examples on Arran and Rousay, similarly attracted by the thought that he was looking at a complete or near-complete data-set and, therefore, that 'the existence of one tomb inhibited the construction of others very close to it' because each tomb marked a territory.

It is also possible to discern theoretical land units among the mass of evidence plotted from the air (Figures 2.1 and 12.3). A different approach can use groups of round barrows, like the long barrows, as focal points in the landscape for the investment of relatively large amounts of local resources. This only 'works' for the downland north of the Kennet valley, for to the south major concentrations of round barrows do not occur. A dozen groups (A–M, Figure 2.4) occur on the northern downs, however, with perhaps another three to coalesce as agglomerations accrue from further work around already-known barrows between groups B–E, F–G and K–L (around respectively SU 114719, 115691 and 145691, Figure 2.1). Even more obviously than in the case of the long barrows, their distribution is non-

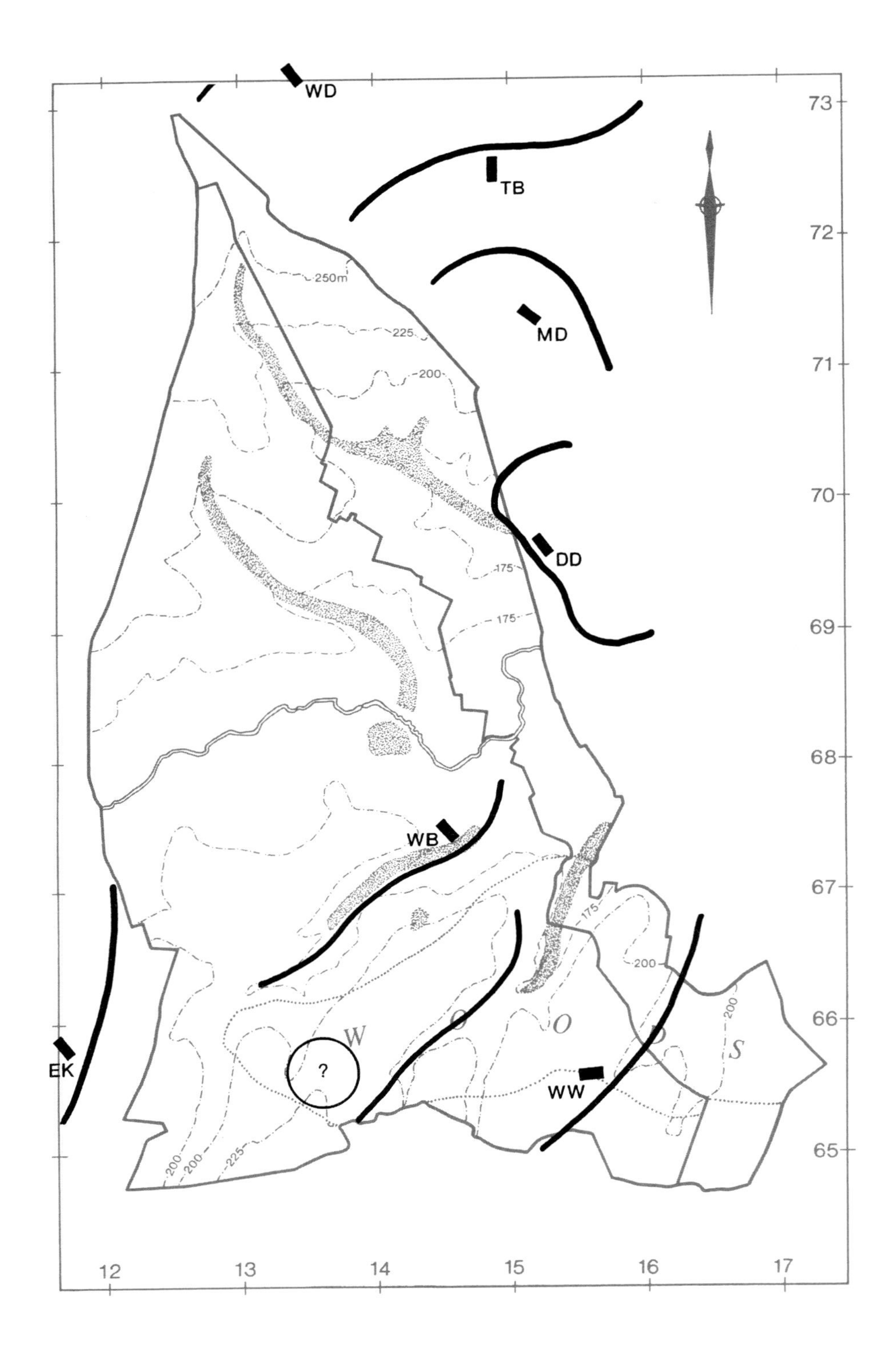

16.3 The distribution of long barrows in and immediately outside the study area, with a suggestion that they may have been sited in the boundary zones of possible topographically defined 'territories'

WD = Wick Down long barrow; TB = Temple Bottom; MD = Manton Down; DD = Devil's Den; WB = White Barrow; WW = West Woods; ? = where a 'Shaw long barrow' should be; EK = East Kennet

random. While in part the siting of groups was topographically influenced, notably on the high ground around the south-western edges of the Marlborough Downs, a metrical patterning based on a unit of roughly 0.5km (x 2, 3, 4 and 8) seems to underlie the distribution. Such deliberation might again indicate land units, in this case perhaps relatively small 'territories' at the family group level. That the burial grounds were also at the edges of the enclosed field systems suggests that they lay, like the long barrows, on the margins of such units; unless of course these familial lands extended down the valley slopes as well as across the downs, in which case the round barrow groups could have been in or towards the middle of possibly long strips of property. The latter interpretation seems more likely in practical, farming terms.

It is possible to envisage such strip-shaped holdings, at a higher tenurial scale perhaps, in the first millennium BC. Two certain, two probable and one possible – five in all – roughly circular settlement enclosures seem, like the barrows, to be distributed carefully across the landscape. They lie more or less in a line from north east to south west from Totterdown to Golden Ball Hill, implying that their associated territories, if laid out symmetrically, may well have tended to stretch from north west to south east in what later became the familiar medieval pattern. If such related to mixed farming with an emphasis on arable, as seems clear from OD XI, then an extension of the 'territory' concept could see a significant change in the later first millennium. Then, so it has been argued (*above*), much of the northern downlands reverted to pasture and stock raising. In the absence of local settlements, a shift of tenurial power to the neighbouring hillforts could allow for the division of what later became the two parishes into four units, one for each hillfort (Figure 15.2).

The south-west corner of modern West Overton civil parish, essentially what we later come to know as the historic tithing of Shaw, might have lain with Rybury. A north-west/south-east strip of land approximating to historic West Overton tithing could have lain with Oldbury; a similar strip, approximating to the later East Overton manor with its lands on the south side of West Woods, plus, perhaps, the land of the later estates of Upper Lockeridge and Clatford in the woods, could have been managed from Martinsell. The northern downs of our study area, approximating to the northern parts of East Overton, Lockeridge and Fyfield, traditionally run together under Overton in medieval times, were perhaps, in 200 BC, part of an even larger estate, the land of Barbury Castle. The ascriptions are, of course, guesswork, though correlations in the number four are attractive, but the general proposition that such 'territories', however organised, existed in later prehistoric times in this area has a certain force.

Similarly, as already mentioned, here as elsewhere the idea of a villa estate is not only attractive but also helpful in understanding both the archaeological evidence and the Roman landscape. If such was the case, perhaps one (Fyfield?) was the principal centre, with tenants living in a smaller villa-like establishment at 'Headlands' which, with the passage of time, might have become the centre of a fragmented estate in the sort of socio-economic changes hinted at on the downs at the sites around Down Barn. Two 'territories' or estates may well have been, then, the Anglo-Saxon inheritance in the sixth/seventh centuries. Then the settlement *foci* shifted permanently off the northern downs and four churches – Shaw, West and East Overton, and Fyfield – appeared in the landscape, probably before the end of the eighth century at latest. This key development suggests the existence of at least four tenurial units, perhaps remarkably similar in size and location to those before the Roman rearrangement and, henceforth, a permanent framework within which the land was worked – and exploited. Lockeridge tithing, itself compounded of parts of several other estates, was somehow squeezed out as a narrow strip of land between East Overton and Fyfield before the Conquest. Now, although land ownership has fragmented and the landscape is no longer dominated by just one or two landlords, paradoxically the five medieval tithings have effectively disappeared and administratively, at least for certain local purposes, arrangements as represented by two civil parishes are back to what they may well have been in late Roman times. Aspects of possible relationships between 'territories' and settlements in the mid-first millennium AD were explored by Fowler (1976, fig 1.9), and have subsequently been examined in much more detail, in particular by Hooke (1988b, 1997, with bibliography, p 83).

Boundaries and settlements

Boundaries often seem to relate significantly to settlements themselves; or it may be that settlement location was either related to a pre-existing boundary or attracted one. Certainly in a number of cases a settlement is close to or actually divided by a long-term boundary.

A specific, documented example can be taken from the East Overton charter (S449). It refers to 'Aethelferthe's house'. The structure and site remain unidentified, though their position within a 200m-diameter circle is fairly clear (*SL*, figure 63), by definition close to the south-eastern boundary of the tenth-century estate. Fortuitously or otherwise, another 'Aethelferthe' boundary marker occurs in the same charter, this time a *stane* at the northern end of the estate (Chapter 5; *SL*, figure 17). The location is as remote on high downland as is the 'house', so perhaps the use of the personal name is topographically-associative rather than proprietorial or tenurial; equally, it might be functional, for example habitative indicating a living place in both cases, or iconographic indicating to those who needed to know the further end of a land unit, as perhaps some long barrows did 4,000 years earlier.

Boundary/settlement relationships apparently range across both more and less significant instances than that tenth-century one. These relationships seem to have been a considerable factor in the landscape's evolution, perhaps modulating it. At the locally major scale, the former West Overton settlement, with its church, was immediately beside The Ridgeway boundary of the eponymous estate (Figure 8.1); the common boundary between the two Anglo-Saxon Overton estates divided West and East Overton villages or, depending on your viewpoint, provided the line up against which they were developed (Figure 8.2). A similar situation, spatially at least, existed at medieval Shaw (Figure 13.1), and apparently the same phenomenon existed at the 'Headlands' settlement complex too (Figures 4.1 and 4.2). There, the tenth-century documentation does not speak of settlement division but, in concentrating on the line of the boundary through arable fields, it allows us to be certain that it physically split a later prehistoric settlement beside which further settlement developed in the first five or six centuries AD, though only on its western side (Plate XV; Figure 4.3).

The extensive 'Crawford settlement complex' on Overton Hill was, like Anglo-Saxon West Overton village, hard up against the western side of the western estate boundary (Plate XIV; Figure 4.2), and similarly the settlement complex on Lurkeley Hill is hard up against the 'inside' of the same West Overton boundary, which, at this spot, seems to respect the settlement by going round it (Figure 12.1). Indeed, the evidence could be construed to be indicating something special about an 'early' West Overton estate, with six settlements adhering to the inside of a boundary certainly older than its tenth-century documentation. Three of them, all existing in Anglo-Saxon times, are 'paired' settlements, with counterparts, all with churches, on the other, outer side of the boundary. The other three settlements have no certain counterparts on the other side, though all three have slight hints that occupation may have existed there in the Roman period. Perhaps significantly, all three are of later prehistoric origin with no habitative existence in early medieval or later times.

A closely comparable situation existed with at least three other cases on Overton Down itself. The settlement complex around Down Barn (Figure 6.11) occupies virtually the whole width of the medieval tithing of East Overton at this point though, perhaps significantly, neither of the boundaries on north east or south west nor the settlements seem to acknowledge a relationship other than by proximity. The tenth-century 'Lamb's path', however, passes through the complex, just as did earlier tracks. One of those trackways was that one referred to in the tenth century which we have tried to distinguish as 'the Overton Ridgeway'; it also served as the east side of the Anglo-Saxon East Overton estate on the downs (Chapter 6; Figure 5.1). Up on the down, a Romano-British settlement lies immediately adjacent to its western side, that is in East Overton, with an earthwork enclosure, tentatively identified as a 'sheep-cote', immediately on its east in Lockeridge tithing (Figure 6.1).

An extension of this track, perhaps still a boundary, passes along the south side of OD XI, cutting through associated lynchets; but the most striking aspect of this site in this context is that the settlement enclosure, originally carved out across field boundaries, was itself crossed in the mid-first millennium BC by a new fenced boundary marking out a field to be cultivated where formerly buildings had stood. At a local boundary level, this mirrors the subdivision of the contemporary 'Headlands' enclosure by a boundary which, whatever its date of origin, had become an estate boundary by the tenth century (Figure 4.2). Both examples are the converse of the comparable probable enclosure on Totterdown, which was laid out across a Bronze Age boundary ditch (Figure 5.1). Whatever the reasons, a certain magnetism seems to be at work within this landscape, pulling boundaries and settlements into close juxtaposition.

5 SETTLEMENT 'SHUFFLE'

Yet one of the major features of this landscape through time is that settlements were characteristically disappearing and appearing, contracting and expanding, and moving short distances, sometimes in an almost stately minuet around a particular place. The Down Barn complex (Figure 6.11) may well represent such a localised shuffle; certainly it can plausibly be interpreted in that way. But all the valley-bottom villages have also 'shuffled' a little in their positioning (Figure 8.1). Probably the best example is Fyfield village itself. This settlement effectively became 'fixed' in the landscape in the Roman period at a central point subsequently occupied by the manor house and church in a pairing which continues to the present day; yet settlement remains extend over an area about half a kilometre square more or less all round that central point (Figure 11.1). These represent, not a large and now deserted settlement, but small clusters of habitation which have formed and then, in part, dissolved as the village has responded to and reflected economic and – notably in this case – transport changes over almost two millennia.

The other medieval valley villages show similar, if less marked, characteristics of local movement. Lockeridge has a long history of 'settlement shuffle', both physically and tenurially, with the habitation places tending to move along the narrow estates rather than, as at adjacent Fyfield, around a central point (Figure 10.1). We argue that Dene was superseded in the twelfth century by a new Templar planned settlement of Lockeridge stretching some 300m northwards almost on to the floodplain of the Kennet towards the possibly *Domesday Book Locherige* across, presumably, a ford (Chapter 10). Much later, the new Lockeridge imploded to an extent with some large houses around its original core at the Dene cross-roads, before again stretching northwards with a new estate village of the mid–late nineteenth century.

A variant on the evidence of 'shuffle' might also be hinted at in place-names, rather than merely archaeological evidence, as the following theoretical example illustrates. A possible settlement sequence, perhaps just a naming sequence, might be suggested by place-name evidence taken with the topography at Lockeridge, Fyfield and Overton. Long before *Domesday*, *Lockerige* may have been the name of the first valley settlement when people came down off the 'ridge with earthworks' above the present village, locating themselves in relation to their former habitation. We do not know the pre-English name of Fyfield but, with its Roman origins, it could well have been called something related to its position on a knoll beside the river – *Avondun*? – or derived from a Romanisation of the name of the River Kennet or even the local River Og; or perhaps, whatever it had been called, it was given a new name for an existing settlement, *tun* with some appropriate prefix in the eyes of arriving settlers, during the seventh century (eg, *Wel-tun*). A new settlement, *Upper tun* (the early West Overton by The Ridgeway), perhaps that of the newcomers, later came into existence further up the River Kennet, an item of history still echoed in AD 972 when it was referred to in precisely those terms. Subsequently, a new Anglo-Saxon settlement developed between the 'mother' and 'daughter' *tuns*. Since it was of the Saxons and not the Britons, it was related to *Uppertun*, not *Wel-tun*, and so was called *East Upper tun*. To protect its identity, *Upper tun* came to be called *West Upper tun*, particularly when the estate centre moved over to a site beside *East Upper tun*. Meanwhile, the original Roman settlement ascribed to the Britons needed to protect its identity in this new landscape of English nomenclature so was simply called 'Five Hides' when it became the centre of an Anglo-Saxon estate in the tenth century.

Illustrating a similar phenomenon but rather different in detail, already explored in some detail (Chapter 8), the compound village of modern West Overton is made up of two late Saxon Overtons either side of a tenth-century, probably much older, boundary. One, the second Anglo-Saxon West Overton, was formed when the first moved 1.5km across the estate from west to east; while the other, East Overton, is likely to have originated on the knoll on which the later church was built (Plate LXII; cf Hare 1994, 58). It then physically expanded south west of the church site but in no other direction, and either shuffled north or simply contracted (Chapters 8 and 9). The large, contemporary village of Shaw high on the edge of the southern woods also formed on either side of a boundary, expanded and decayed, and then shuffled sideways to become only a large farm (Chapter 13).

At a rather larger scale, it is possible to suggest that a fundamental sort of settlement 'shuffle' may have occurred during the critical mid-first millennium AD period which gave us our present-day distribution. Such change may have been in terms of the Lurkeley Hill settlement's functions and population moving 1.5km to Shaw, the 'Crawford settlement complex' moving 1.5km across the river to be replaced by 'early' West Overton, the 'Headlands' complex also crossing the river to the area of the manorial centre in medieval West Overton, and the remnants of the Down Barn settlement moving 1.5km to East Overton on and around the knoll where the church later became established.

It is also possible to suggest yet another type of 'shuffle' on the northern downlands. The idea comes from the three excavated settlements. Each, with varying degrees of certainty, was occupied for only a short period of about a century. *Raddun* is the best evidenced, with habitation of the farm AD 1220–1318 and barely a decade of uncertainty at either end. OD XII was inhabited *c* AD 340–440, and OD XI, the most ambivalent of the three, could on one interpretation have enjoyed a short occupation in the eighth century BC. The proposition is, therefore, that all three small, excavated settlements represent the same phenomenon in closely comparable situations recurring over two millennia: that each enjoyed a life of three (perhaps four) generations – perhaps father, son and grandson – and was then abandoned because that was the length of time that the environmental conditions could sustain. If correct, this interpretation provides a clear indication of a significant constraint in use of this downland for settlement and farming, certainly without major, external inputs. Even where such were made, as was certainly the case at *Raddun*, habitation ceased even though the site continued to function as a significant element of the estate (Chapters 7 and 9). Nevertheless, nothing in the comparative plans of the thirteen buildings excavated denies the inherent unsustainability of downland occupation implicit in such a proposition (Figure 16.4); indeed, if function, furniture and material contents are taken into account too, the life-style is remarkably similar as well as short-lived in three settlements across two thousand years.

We did not find, on the northern downs, successors to the excavated settlements, the farms to which the great-grandchildren moved. We further infer, therefore, that each abandoned farm represents the end of permanent habitation in the locality for the time being. Such a settlement 'shuffle' would be small-scale, involving few people at any one moment, exiting from what were secondary settlements anyway; but the consequential move of those involved could be relatively far, down into the valley, for example, and not just to a house alongside the deserted one.

An internal move, a sort of domestic 'shuffle', is in fact evidenced in each of the excavated settlements individually while they were still in use. The preferred interpretation of OD XI is of its four main buildings forming a succession (Figure 6.8). A similar succession is envisaged at OD XII in late Roman times (Chapter 6), and an internal move sideways of the main residence a few metres to the east is unambiguous at *Raddun* (Figure 7.9).

The significance of the abandonment of such settlements may be far more, however, if we can read it aright, than merely the displacement of a few people. *Raddun*, for example, both typifies and represents a widely detected economic and land-use trend of decline and retraction, not from the Black Death but from early in the fourteenth century, seriously discussed by

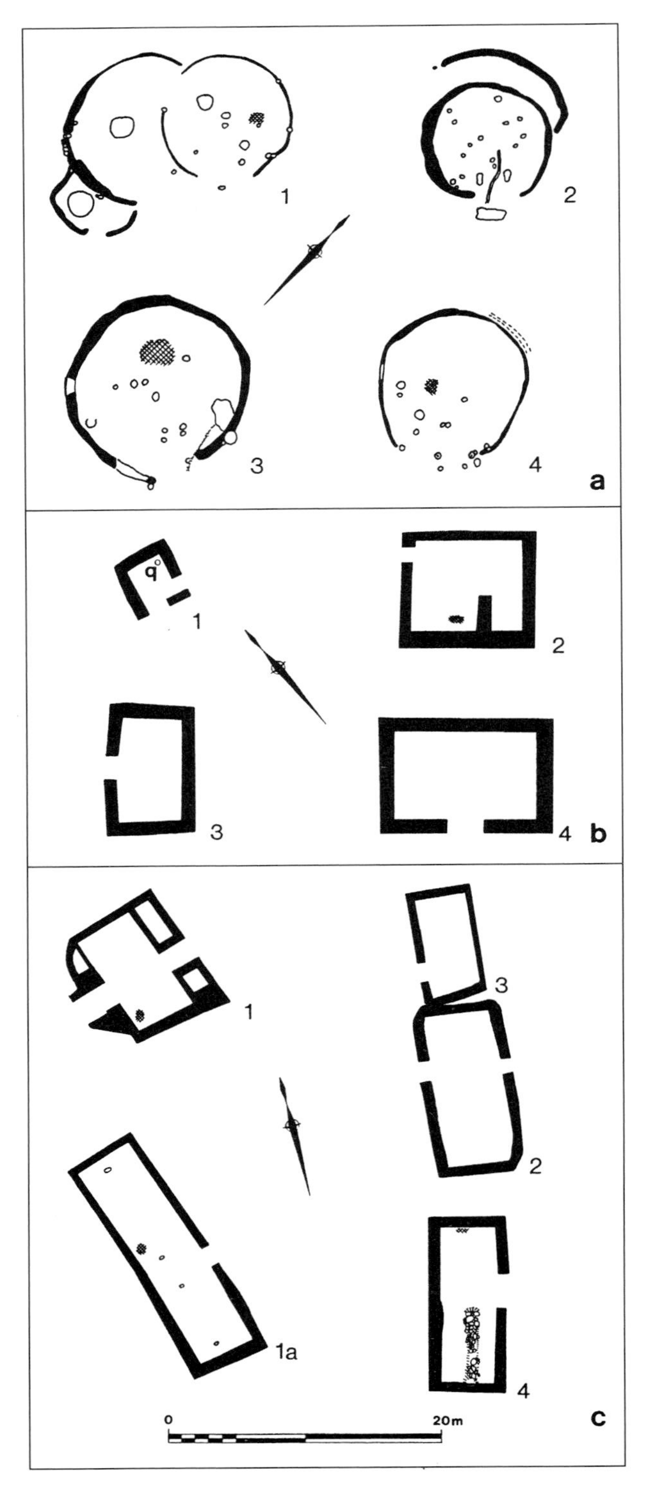

16.4 *Comparative plans of thirteen excavated houses/buildings from three settlements of, respectively, Early Iron Age (site OD XI, top), late Roman (site OD XII,* middle*) and medieval (site WC,* bottom*) date*

historians since Postan (1973), to use his own words, put it 'on the agenda of economic history'. Castigating fellow historians on their failure at that time – the late 1940s – to recognise the significance of the sort of landscape change with which we are here much concerned, he spotted that it was precisely because of their land-use marginality that holdings like that of Fyfield Down were likely to be more sensitive to economic fluctuation, and therefore more 'significant' for historical interpretation, than richer manors elsewhere on the estates of the Bishops of Winchester (Postan 1973, 208–9). It is a distinct possibility that the two similar excavated settlements, OD XI and XII, have a comparable historical significance which we literally cannot read in the absence of documentation.

Tabulating the intangible?

Clearly we have assembled a lot of evidence indicating places of settlement and activity from all over the study area. Much of it is to do with 'settlement shuffle', and other matters develop therefrom. It is tempting to tabulate it in various ways with a view to assessing whether any significant patterns emerge. Here, we plot nineteen places/sites against time, the horizontal dimension in units of 500 years after 1500 BC. The sites/places are selected primarily as those that constantly recur in our considerations. All possess settlement evidence, though in most cases neither their extent nor interest is limited to one settlement alone and at least some of the evidence is of activity and/or land-use rather than specifically of habitation. We accept that the self-selecting sites/places are not all closely comparable in type, size or intensity of investigation; but no set of data from this landscape would be, so if we want to tabulate, then these are the sort of data available.

The list coincidentally includes ten of the eleven 'nodal points' in the landscape selected on other criteria (*see below*). The tabulation did not include Clatford (no. 8 on Figure 16.8), a medieval manorial village beside the intersection of three main routes, because it lies just outside our study area and, unlike the similarly situated Overton Hill and Manton Down (Chapters 4 and 5), has not been critically investigated by us.

In Table 4, the site order in the left-hand column is

Table 4 Selected places/sites in the study area in relation to evidence of activity through time

Column 1	*2*	*3*	*4*	*5*	*6*	*7*	*8*	*9*	*10*	*11*	*12*
Period	BC> 4000	*4000–2500*	*2500–1500*	*1500–1000*	*1000–500*	*500–* AD *1*	AD *1–500*	*500–1000*	*1000–1500*	*1500–2000*	*'hits'*
Area											
West Woods	+	+				+	+	+	+	+	7
Shaw	+	+					+	+	+	+	6
Down Barn	+	+	+				+	+	+	+	7
Lockeridge Down		+	+	+			+	+	+		6
Lockeridge Dene		+	+	+				+	+	+	6
Lockeridge village		+	+					+	+	+	5
Manton Down		+		+					+	+	4
Fyfield Down		+	+	+		+	+	+	+	+	8
Totterdown		+	+	+			+		+	+	6
OD X/XI			+	+	+		+		+	+	6
OD XII			+				+				2
Raddun				+			+	+	+	+	5
East Overton village			+					+	+	+	4
'Headlands'			+		+		+	+			4
North Farm			+	+			+	+		+	5
West Overton village			+					+	+	+	4
Overton Hill			+				+	+		+	4
Delling				+						+	2
Fyfield village							+	+	+	+	4
Total	**3**	**8**	**12**	**9**	**2**	**2**	**13**	**13**	**12**	**16**	

determined by the date of the earliest evidence from the site/place. The 'hits' on the right are merely the total number of times a place/site gains a cross for having produced some evidence of occupation/activity under one of the time-units.

The table brings out immediately that most places in this landscape had already been used by the mid-second millennium BC. Indeed, of our selection only *Raddun*, Delling and Fyfield village have not produced evidence of occupation/activity before 1500 BC. Fyfield village, easily regarded as 'old' because it overlies a probable Roman villa, comes out in fact as one of the 'latest' places in the study area, one moreover with quite a low index of activity (column 12). But then the same 'score' is obtained by Overton Hill, mainly because the time-length embraced by our column 4 means it scores only 1 for its major Late Neolithic/Early Bronze Age phase; yet clearly it was a central area of activity within the sub-region of which our study area is a part. The example emphasises the obvious point that such tabulation has to be looked at very critically: numbers can easily mislead and in any case are by nature inconsistently sensitive to intangibles such as 'continuity' and 'iconography' and both abstract and practical values in the landscape like 'sacred' and 'good drinking water'. Nevertheless, in this presentation the main phases of human activity (within what was surely a continuum within the local landscape overall) were apparently in the fourth millennium; 2500–1500; 1500–1000 BC; AD 1–500; 500–1000 (and, a nuance column 9 cannot bring out, especially 800–1000); and 1000–2000. Conversely, the 'gap' in activity in the Middle and later Iron Age, which we have noted elsewhere (Chapter 6), is clearly highlighted. Similarities in settlement/activity distributions are also brought out, for example, both on the downs and in the valleys in both the later Neolithic/Early Bronze Age and medieval periods (columns 4 and 10), as are patterns of emphasis on the same or similarly-located sites, for example, the similarity between the Roman and post-medieval columns (8 and 11).

In Table 5 the same data are rearranged. While the horizontal time-units remain the same, the running

Table 5 Sites/places in the study area ranked by the number of occurrences of their use ('hits') through time

Column 1 *Period* *Area*	*2* BC> *4000*	*3* *4000–* *2500*	*4* *2500–* *1500*	*5* *1500–* *1000*	*6* *1000–* *500*	*7* *500–* AD *1*	*8* AD *1–* *500*	*9* *500–* *1000*	*10* *1000–* *1500*	*11* *1500–* *2000*	*12* *'hits'*
Fyfield Down*		+	+	+		+	+	+	+	+	8
West Woods*	+	+				+	+	+	+	+	7
Down Barn*	+	+	+				+	+	+	+	7
Shaw	+	+					+	+	+	+	6
Lockeridge Down*		+	+	+			+	+	+		6
Lockeridge Dene*		+	+	+				+	+	+	6
OD X/XI			+	+	+		+		+	+	6
Totterdown*		+	+	+			+		+	+	6
Raddun				+			+	+	+	+	5
North Farm			+	+			+	+		+	5
Lockeridge village		+	+					+	+	+	5
Manton Down		+		+					+	+	4
Overton Hill*			+				+	+		+	4
'Headlands'			+		+		+	+			4
West Overton village			+					+	+	+	4
East Overton village			+					+	+	+	4
Fyfield village							+	+	+	+	4
Delling				+						+	2
OD XII			+				+				2
Total	**3**	**8**	**12**	**9**	**2**	**2**	**13**	**13**	**13**	**15**	

Sites asterisked also feature as 'nodal points' in the landscape, selected by different criteria (Figure 16.8)

order of places/sites in column 1 is determined by the number of 'hits' in column 12. In however crude a way, Table 5 seems to reflect some sort of reality and helps jolt perspectives. Clearly the rearrangement has the effect of representing the visibility of archaeology as much as anything. It certainly reflects intensity and mode of investigation, for the first four sites/areas have all enjoyed at least some excavation; indeed that is true of eight of the first nine places. Conversely, the bottom two sites only creep into their lowly positions as a direct result of excavation. Yet five (starred) of the top seven sites/places were also selected as 'nodal points' in the landscape on quite different criteria which had nothing to do with excavation or indeed with the conventional archaeological and historical evidence which underlies this table (Figure 16.8). We can probably accept that, within this local landscape, seven of the more important places were Fyfield Down, Down Barn, Lockeridge Down, Totterdown and Lockeridge Dene. Apart from Fyfield Down, the list is not perhaps of the names that would come first to mind when looking initially for the significant places in the study area.

On specific places, Lockeridge, which in Chapter 10 is argued to be the latest of the permanent villages, actually evidences earlier and longer-lived activity than Fyfield and the two Overtons. Fyfield, in contrast, can be seen in a perspective reinforcing its status, not as the oldest, most long-lived settlement in the area, but as a place arriving 'late' in the landscape. Another contradistinction suggesting influences underlying such a simple concept as 'settlement shuffle' is that while Fyfield village is located in what would be regarded as a prime settlement situation, the top four places in the table with the most and longest-lived activity are all 'marginal' on a conventional view. Perhaps the question is not so much 'What is marginal land?' as 'When does marginal become marginal?' and 'How often does marginal land change status?'. It is interesting to note in that context that this table indicates the woods to have been almost as active as the downs, and generally with an earlier start to that activity. Overall, the exploitation of resources is early and widespread. The level (or is it spread?) of activity is virtually of the same order in the Late Neolithic/Early Bronze Age (column 4) as it is over the two millennia since Romano-British times (columns 8–11). The consistency of landscape activity, with fifteen 'hits' after AD 1000, bears on our argument (Chapter 11) that medieval population estimates are about, not only absolute figures, but also the carrying capacity of the area in a pre-modern technological and tenurial regime.

In the time dimension, Table 5 shifts any concept of 'dark ages' back from a focus in the mid-first millennium AD to the mid-first millennium BC. Conversely, it exposes the myth of the post-Roman 'dark ages' – they simply do not show in this presentation, with columns 8 and 9 having the same, high number of 'hits', in total nine times the number for the preceding millennium. It seems probable that such a numerical discrepancy does point to some historical 'truth'.

Table 6 ranks the same nineteen sites in relation to the five topographical/land-use zones within the study area defined in Chapter 1. With the notable, and probably prescient, exception of 'Headlands' (Chapter 4), and possibly of the West Overton open fields (*see above*, this chapter), our investigations have brought little new understanding to the landscape history of the permanent arable on the valley slopes; but three of the other zones each contain five sites/places selected here, with three in the remaining 'interface zone', indicating that our effort has eventually been reasonably well distributed across the study area despite an early preoccupation with – and a persistent project image of – the northern downs. Probably more significant historically, however, is that the average 'score' is virtually the same for each zone however many sites/places are included. Numerically, five time-units are represented in each zone whatever other criteria it may be judged by. In other words, it does not seem to matter whether the zone is on good or bad soil, south or north facing, by a river or in the woods: its use through time is fairly consistent and closely comparable to that of other land-use zones in the long term. The numerical exception, 'Headlands' with its 'average' of 4, stays at 4 even if Overton Hill is brought in, as it could well be, as a second 'permanent arable/valley slope' place. Their common factor, keeping their number of 'hits' below the average, is their lack of late prehistoric and medieval activity, a numerical blip which is probably historically significant despite the limitations of a sample of two. At the very least it, and numerous other questions posed by these and similar tables, are a challenge to further research.

6 SETTLEMENT MORPHOLOGY

The discussion above has incidentally included several points about settlement morphology, but perhaps a main inference is that, with change so manifest, from internal domestic level to whole settlement movement, the validity of morphology as a useful criterion for significant interpretation is constrained. What we see is not merely that which has survived and is discoverable but also what happened to be the physical shape of a

Table 6 Sites/places ranked in relation to five land-use zones, giving the total and average number of 'hits' in each zone

Place	*Zone*	*'Hits'/average*
Shaw	High	6
Lockeridge Down*	High	6
Totterdown*	High	6
West Woods*	High	4
Manton Down	High	4
Total	**5 places**	**26/5.2**
Fyfield Down*	Medium	8
OD XI	Medium	6
Raddun	Medium	5
Overton Hill*	Medium	4
Delling	Medium	2
Total	**5 places**	**25/5**
Down Barn	Interface	7
Lockeridge Dene*	Interface	6
OD XII	Interface	2
Total	**3 places**	**15/5**
'Headlands'	Valley-slope	4
Total	**1 place**	**4/4**
North Farm	Bottomland	5
Lockridge village	Bottomland	5
West Overton village	Bottomland	4
East Overton village	Bottomland	4
Fyfield village	Bottomland	4
Total	**5 places**	**22/5.4**

Sites asterisked also feature as 'nodal points' in the landscape, selected by different criteria (Figure 16.8)

settlement at a particular moment. The point, so obvious in the landscape and from its archaeology, is reinforced by the plans of the study area's main settlements on eighteenth- and nineteenth-century maps. Some are inaccurate, and all are partial to some extent; yet overall they show such a degree of change that one has to consider whether the century 1760–1860, particularly the half-century 1780–1830, really was exceptional in what happened on the ground. If not, through the accident of maps becoming historically available as evidence, we can conclude that we are suddenly able to witness the type and pace of change that may have been normal in earlier times too.

Certainly the archaeology suggests that such had always tended to be the case. Nevertheless, three essentially simple generalisations emerge from a morphological consideration of our settlement data. One is that, in a sense, nothing here is academically new. We can see settlements small and large, open and enclosed, dispersed and nucleated, primary and secondary, long-lived and temporary, simple and complex, organic and planned, uni-centred and polyfocal, and prehistoric, Roman, medieval and modern. So much is this the case that perhaps a mildly interesting feature is that so much variety, typical of much wider areas over southern England, is contained within only two parishes. The phenomenon is not, however, unusual, even though not perhaps always looked at in this light (cf RCHME volumes on Dorset and Northamptonshire; Everson *et al* 1991; Taylor 1983).

The second point concerns morphology and size. Recognising that our prehistoric settlement data are sparse in the extreme, nevertheless no evidence exists to suggest that individual settlements were substantial or large before the first century AD. The areas some covered may have been extensive, for example around Bayardo Farm in the Mesolithic and on Overton Hill in the Late Neolithic/Early Bronze Age, but, in the surprising absence of proven Middle Bronze Age settlement enclosures in our study area (that on Manton Down is the most likely, Figure 5.4; cf Gingell 1992), we lack evidence of intensity of structure and occupation until the Late Bronze Age/Early Iron Age at OD XI. Even then, that enclosed settlement is not particularly large in comparison with some later settlements. Indeed, it is not until the (probably early) Roman period that, with ODS (Figure 6.13), we have unambiguous evidence of a large, and in this case organised, settlement. Its striking features are its rectilinear morphology, and its extent which puts it on a par with the large medieval settlements, notably, Shaw, the composite West Overton and the supposedly Templar Lockeridge. Indeed, from ODS onwards, there appears always to be at least one large settlement in the study area, usually two, one in each of what came to be the two parishes. ODS/OD XII, 'Headlands' and Fyfield villa settlements were in some sense contemporary in the first centuries AD, and all were extensive; we can see organisation at the first, and detect a hint of it around or preceding the putative villa at the second (Figure 4.3, 3), but we have no firm evidence to suggest a morphology at Fyfield. There, however, the position of the mosaic pavement, the topography and the positions of both the Roman road and loose Roman material can suggest an extensive area of possibly intensive occupation beside a villa (Figure 11.1).

Roads or tracks play a significant part in the local settlement morphology. All the main Roman and later settlements are related to such lines of communication, notably ODS, early West Overton, composite Anglo-Saxon Overton and Shaw. The factor was probably as important at Fyfield and Templar Lockeridge, but is not so obvious. A consequence is, naturally, that the settlements tend to be, or to become, linear; though, that said, a marked characteristic of Overton is that it did not spread very far, as if strong physical or tenurial constraints existed, as invisibly influential on the west, south and east as was the floodplain on the north.

The third point concerns morphology and location. A few small settlements, some enclosed but most not, existed in most of the study area at any one time throughout its history. They were probably most numerous in Neolithic, Roman and post-medieval times, judging by, respectively, flints, potsherds and maps. Most were probably farmsteads; we know that to have been the case in the eighteenth and nineteenth centuries from cartographic evidence, though then the landscape comes to be dotted also with agricultural barns and yards where no one normally lived (*SL*, figure 69). In the third–fourth millennia BC, 'small' settlements were perhaps the norm and it would be unjustified to label them generically 'farmsteads' when they may have been all-purpose places. Archaeologically, on the record so far, a general characteristic is that they did not involve much in the way of permanent structure though, as the evidence from beneath the Down Barn enclosure suggests (Chapter 6), some sites will emerge which do not conform to that generality. Such small settlements do not appear, at any period, to have been limited to one location in particular; rather is their absence notable from particular areas at particular times. For example, no Bronze Age settlements are known south of the river valley; no Middle and Later Iron Age settlements are known on the northern downs or along the river valley; no Roman farmsteads have been discovered in West Woods.

In complete contrast are some large, planned settlements, all Roman or later and of village size and, apparently, mature. That called ODS is on the northern downs and the earliest; Lockeridge, as recreated by the Meux estate in the later nineteenth century, lies just above the floodplain and is the latest (Figures 6.13 and 10.2). Strictly speaking, that last remark is not correct since the relatively large expansion, mainly infilling, of all three extant valley villages in the second half of the twentieth century is very much the result of planning. The results, visual and morphological, are now just as much part of the archaeology of the villages as are earlier episodes. Such earlier planning is evidenced reasonably clearly in the lineaments, represented by both earthworks and property boundaries, in each of the Overton nucleated villages which became West Overton and, in earthworks and hedgelines, at Shaw (Plate XLVII; Figures 10.2 and 13.1).

Overall, two generalisations stand out. First, even though three of the four largest settlements have been along the valley since the mid-first millennium AD, there does not appear to be any particular correlation between morphology and location. Large and planned settlements occurred on the high ground and valley sides as well as in the valley itself; though no large settlements existed on the northern downs after the Roman period and medieval Shaw is the only example of a large settlement south of the valley at any period. Small settlements, characteristically farmsteads, occur throughout this landscape at all periods.

Secondly, in a study area characteristically a landscape of small settlements, at no time until the late twentieth century has any one settlement, of any shape or anywhere, been significantly large in the sense of providing the residence of, say, more than 500 people. Only two or three large settlements existed at any one time with populations into three figures, and then only from the early centuries AD onwards. Such places were and are nucleated settlements, beside a through-road or drove and, from the seventh century AD onwards, except in the case of Lockeridge, around a church.

7 RELIGION

The two Christian churches still in use, respectively St Michael and All Angels at West (East) Overton and St Nicholas's at Fyfield, are both relatively humble structures as parish churches go but each is a major monument in its local context. Each is at least 700 years old as a structure, and each was significantly restored in the nineteenth century. St Nicholas's nevertheless remains modest in appearance, whereas the tower of St Michael's is magnificently successful in remaining a Kennet valley eye-catcher (Plates LXII and LXIIIb). Each marks a site in use earlier than the main date of the standing structure: St Nicholas's, with nothing obviously earlier than Early English, stands close beside, perhaps even on part of, a probable Roman villa and is certainly within a Roman settlement; St Michael's, with at least three major phases visible in its structure despite the Victorian tidying up, exhibits fragments of structural evidence back to Norman times (Plate L) and is the most likely candidate for the church of, at latest, the tenth-century *burg* of East Overton.

We begin this brief discussion of religion in the landscape with the two parish churches because they are obvious and arguably the two most important buildings in the study area (Figure 16.5). Other aspects of various sorts of religious belief affecting this landscape are generally less obvious, though the churches raise some of the questions. St Michael's and St Nicholas's represent, for example, the interest of the major medieval landowner in the study area; the other is likewise represented but we recognise the longer term effect of the ecclesiastical proprietorial interest (*see above*). Neither religion nor belief have been major concerns throughout this project because they have only periodically offered themselves as influential factors in landscape history; and, while we have never doubted that both have been of the greatest significance in human terms, it is not always easy to see that personal concern carried through into evidence of landscape change in ways comparable to that of, for example, prehistoric field systems and palaeo-botanical evidence. Yet clearly human behaviour has often been led by religious belief and therefore, at the local level as well as nationally, that should be reflected in our parishes whose existence, shape and landscape are, after all, anthropogenic.

Such evidence, of course, exists, contemporaneously with the churches right up to the present day, and earlier than their unattested foundations. In the nineteenth century, for example, the area was well supplied with several manifestations of Nonconformity, subsequently largely eschewed. One chapel, for example, was demolished to make way for a bungalow in the early 1960s (Plate XLIX). A much grander, and perhaps more sacred, place once stood on Overton Hill before it too was abandoned and eventually forgotten. 'The Sanctuary', in use around 2000 BC and so-called when excavated in 1930, was without doubt the major ceremonial monument in the area (strictly, just outside the study area since it lies a few metres inside Avebury parish west of the West Overton boundary) in that it was undoubtedly a sophisticated structure apparently intended for pomp and circumstance and not, primarily, for burial. Whatever the detail of its uses and structure, for us its significance is that, partly capitalising on older use of that hilltop in its placing, thereafter it exercised a most powerful influence on the landscape. It marked and created a nodal place in that landscape, a place of influence over the next four millennia up to the present day (Figure 16.8). There is nothing comparable of its scale and long-lasting effect in Fyfield and West Overton parishes, not even the churches.

It perhaps needs to be emphasised that churches themselves were not immune to disappearance; they, like Nonconformist chapels and Neolithic 'temples', could also be abandoned in medieval times as they are in the twentieth century. Two other churches certainly existed in the study area: one at the earlier West Overton by The Ridgeway, which may already have been abandoned by the tenth century, and the other at Shaw though, strictly speaking and significantly, just outside West Overton parish. The latter church was not just abandoned but was in part removed, its architectural pieces being detached and taken elsewhere, for reuse apparently at Huish church. And there was also the Templars' preceptory at Wick Down Farm (*SL*, figure 47b), again just outside the two parishes and this time far out on the northern downs but representing a signal influence on the landscape of the study area, not in itself but through tenure. The village of Lockeridge is what it is in part because of that site.

That example surely exemplifies the principal significance for this landscape of religion. It lies not so much in specific sites or particular acts as in the fact that all the main landowners from at least the tenth century AD until the sixteenth century were ecclesiastical. Five hundred years, even another century or two, may not be very long in the perspective of this study, but nevertheless it was a critical phase in the development of the landscape and it was one rooted in religion. As landlords, clerics were probably not all that different from secular owners but, in this case, with a crucial difference: with the centres of the ecclesiastical estates already well established at distant places, there was never even the possibility, as there often was with a secular lord, that a principal residence would be founded in the locality or that the local farming would be run other than for the support of a non-local enterprise. So, as we have discussed from an economic angle (*see above*), the first half of the second millennium AD left a real impact on this landscape, not so much by innovation as by the consolidation of a particular way of doing things in this particular landscape. Winchester and Wilton, furthermore, made it difficult for their successors to work this land in other ways so what we see today is not so much a medieval religious landscape as, in a significant sense, very much a landscape of religious institutionalisation.

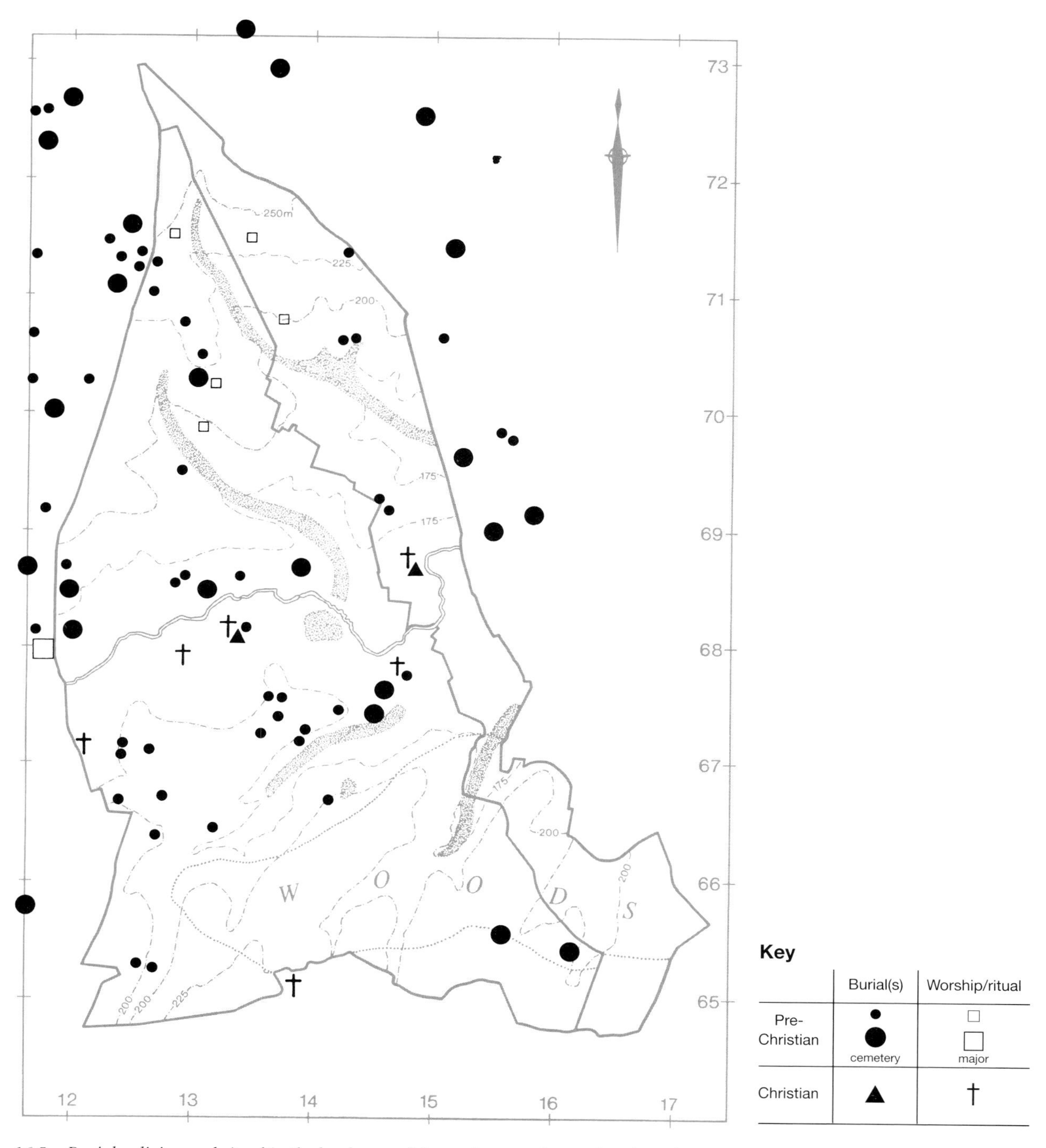

16.5 Burial, religion and ritual in the landscape of the study area: places where burial and religious/ritual activity have occurred, largely identifiable by archaeological evidence. The division of the evidence into only four basic categories is nevertheless self-evidently simplistic: some Christian sites may have pre-Christian origins and some burial sites doubtless witnessed ritual (eg, long barrows). Doubtless, too, most single round barrows, each here shown as a singular 'burial site', were actually the sites of multiple burials and should be shown, like the barrow groups, as a large black circle = 'cemetery'. While much of the landscape has been used for such 'non-productive' activity at one time or another, the overall distribution is clearly non-random, with some patterning apparently reflecting basic topography, such as river valley and northern valley slopes, and land-use/vegetation, such as the sparsity of evidence in the southern woodland

There is another, very different sense in which this landscape is 'religious'. This concerns burial in particular, ceremony in general. If we disregard chronology and archaeological site typology, and instead plot our evidence by function, concentrating on burial and ceremony as recorded over the whole of the study area, we see that activity burying the dead and otherwise expressing belief and ritual need is widespread throughout this landscape (Figure 16.5). Though the individual sites are seldom extensive, collectively the evidence indicates a busy landscape in which human activity concerned with ceremony and belief was widespread over some six millennia. Indeed, in some respects our perception changes as the landscape becomes one of religious expression rather than the one of economic and environmental pressures with which we have been much concerned.

We suspect, of course, that even this evidence is merely picking up the fringes of such activity. Acknowledging that, we nevertheless select for brief comment two aspects of the existing record. Belief, as we have already noted with the churches, can lead to monuments, big public ones and small private ones like the fine (and listed) box tombs outside the south porch of St Nicholas's Church (*SL*, figure 53). In this context, long barrows are monumental predecessors of the churches, as probably are the round barrow groups even if single round barrows are more the equivalent of the family box tombs; but we lack the equivalents of all these structures for the 1,000 years or so centred on 1 BC. We have found no late prehistoric shrine or Roman temple, though the discovery of something relevant to the former in West Woods or to the latter around Down Barn would not be surprising. But if there are not so far field monuments, then excavation gave a hint of belief being expressed in our landscape in a non-monumental form. Indeed all three settlement excavations produced evidence of ritual deposition.

At OD XI, it was of two sorts and two dates. Several of the Early Iron Age pits showed clear evidence of the structured deposition of their contents or of the deliberate deposition of a particular group of material. The practice centred most obviously around cattle and horse skulls but also involved the placement of other materials and objects (table in FWP 34). One such deposit (Pit 20, FWP 63, figure FWP 63.21) suggested from its context that it might have marked the 'killing' of a building, perhaps the last on the settlement. Two instances of similar deposits, though in the tops of already-filled pits, were much later, contextually in the first centuries BC/AD. The other type of deposit concerned post-holes, not pits, and pottery in particular. From the Late Bronze Age structure onwards, potsherds seemed to have been deliberately placed in some post-holes. What had probably been a complete pottery jar had been placed in the most southerly of the post-holes of the post-settlement 'fence' (FWP 63).

Rather less expected were 'ritual deposits' on both sites OD XII and WC (*Raddun*), respectively of the late third century and thirteenth and early fourteenth centuries AD (FWPs 64 and 65). The former was in a pit that almost certainly related to the early Roman fields, having nothing to do with the fourth-century settlement (FWP 64, figure FWP 64.10). Indeed, the deposit and its pit, dug into the negative lynchet at a field corner, again suggested that the 'ritual' it represented may have been marking an end, here presumably something to do with cultivation ceasing and therefore, metaphorically, the 'death of the field'. Though we cannot possibly recognise with certainty what this evidence may actually represent, we may be looking at chronologically unequivocal evidence of about AD 165 for the end of arable on southern Overton Down.

The comparable evidence was different at *Raddun*, though its significance may be similar. An oblong, disturbed and empty pit lay under the eastern wall of Building 4, just south of the doorway. A pit lay in an almost identical position under the eastern wall of the second phase of Building 1 (Figure 7.8). It too had been disturbed; indeed its robbing pit into its western end, from inside the house, was evident (Plate XLV, d). Its curious contents suggested that in it had been a box which had contained various objects; the box had been broken open in the soil and darkness of the pit, leaving behind what we found but presumably other things were taken away. The whole smacked of heirlooms hidden in a family's treasure chest which had been buried as a foundation deposit, first under the original house – Richard of *Raddun*'s on our interpretation – and then under the extension to the new house of the late thirteenth century. Perhaps after only one generation or less, the sentimentally valuable assemblage was partially removed, perhaps in a hurry as the family left their farmstead in or about 1318. They left behind an archaic axehead (FWP 63, figure FWP 63.33).

Belief, religion, and particularly organised religion that required institutions, have all left less tangible marks on this landscape, as elsewhere, in the form of names. Here, two main factors give the study area a fine sprinkling of religious nomenclature: the relatively mighty, and inexplicable, earthwork attributed to Woden (Plate LXI), and the long proprietorial interest of ecclesiastical landlords. From pagan mythology may come 'Pipers Lane' for the Roman road, an oddity, but the better known 'Devil's Den' is, we suggest, not early but a very late corruption of 'Dillion Dene'. Almost similar but not a corruption, 'The Sanctuary' encapsulates a nice Gothic, antiquarian application in the inter-war years of what the name should have been among the pagan Ancient Britons. Woden nevertheless reigns supreme across the southern downs and into West Woods with names such as 'Wodens dene' and Wansdyke. 'Church Ditch' may just reflect an early Christian association, while the various 'Temple' names across the northern reaches of the downs accurately reflect the presence of the Knights Templar in the twelfth and thirteenth centuries (although it is a pity that their influence did not extend to naming their new village 'Lockeridge Templar'). The appearance of 'Monks' in various field names reflects the medieval Winchester interest, but 'Abbess Farm' represents a rare appearance in the landscape of the Wilton nuns, so adversely prominent in documents (*SL*, chapter 10, 114).

8 RECREATION

We discuss recreation as a function of the modern landscape in the next chapter but the concept itself is not new in our study area and is expressed in the landscape archaeology as well as in documents and maps. Hunting has been the traditional recreation, and the training of horses the principal, related activity. The numbers of people involved have always been small, but the effect of these two activities on the landscape has been considerable.

Given the high probability that there has always been woodland across the southern reaches of the study area, hunting is likely to have been normal long before we see it so well documented in medieval times (Brentnall 1938, 1941). It is implied by some of the animal material excavated in the fourth century AD (OD XII; Chapter 6; FWP 64) and around 700 BC (Chapter 6, OD XI; FWP 63). That from *Raddun* in the thirteenth century AD fills out the list of the hunted, suggesting that there may have been some hunting habitats on the northern downs nearer than West Woods across the valley (Chapter 14). Deer was clearly the main quarry, but it is likely that some birds were caught too. The pheasant-shooting and fox-hunting which take place in the area today – though not in the Nature Reserve nor on Forestry Commission land – are not, of course, historic in the proper sense. Nevertheless, some aspects of the present landscape, notably some recent plantations like Totterdown Wood, Delling Copse and Delling Wood between Fyfield and Overton Downs, owe their origin and maintenance to the need for cover for potential human recreational prey.

Earlier that was provided in parks in addition to what was maintained in the managed woodland of the Royal Forest. Curiously perhaps, the Fyfield/Overton landscape does not contain an unequivocally medieval park, but two certainly, three possibly, were created in the later sixteenth century (Plate LII, Figure 4.1). Two, Clatford and Savernake, now partly embraced by the southern end of Fyfield civil parish, are where such enclosures might be expected, on the edge of woodland, but the other, which we have named 'Hackpen Park' in default of an original name, was on the open downland. While it may have represented recognition of a real need to provide cover there for deer, it did not last long and was, presumably, a failure. Yet all three have left an imprint on the landscape in the form of surviving boundaries variously incomplete, though it is only here that their existence is first recognised (cf Watts 1996).

Racehorse training, not a recreation in itself, nevertheless relates to a recreation, however business-like and serious it may be. The downland landscape left by prehistory provides an excellent locale for racehorse gallops and these too have left an imprint on the modern landscape over the last century. They are best seen on air photographs as they change slightly in position and course over the years but on the ground now former gallops are indicated in two ways: by standing stones marking their sides; and by levelled earthworks where the ground has been smoothed to make the running easier. The best example of both is a totally abandoned gallop along the top of Fyfield Down, shown in operation on Allen's air photograph of 1932 (frontispiece). The white streak stretching south east from Totterdown Wood is detectable on the ground today by observation of exactly the same sort of archaeological field evidence relict from much older activities. In the 1990s some other redundant gallops were brought back into use, most notably the 'Derby gallop' along the spine of Overton Down (Plate LXVIII). It also closely follows the line of the 'Overton Ridgeway', some eleven hundred and fifty years older.

Today there are racehorse-training headquarters at Manton House, just outside our study area (Figure 5.4), and the racing interest has only a minor reflection in some of the local buildings, such as Bayardo Farm. The presence of a number of stables, especially in and around the village of Lockeridge, reflects the considerable interest in the area in recreational horse-riding, rather than in racing itself.

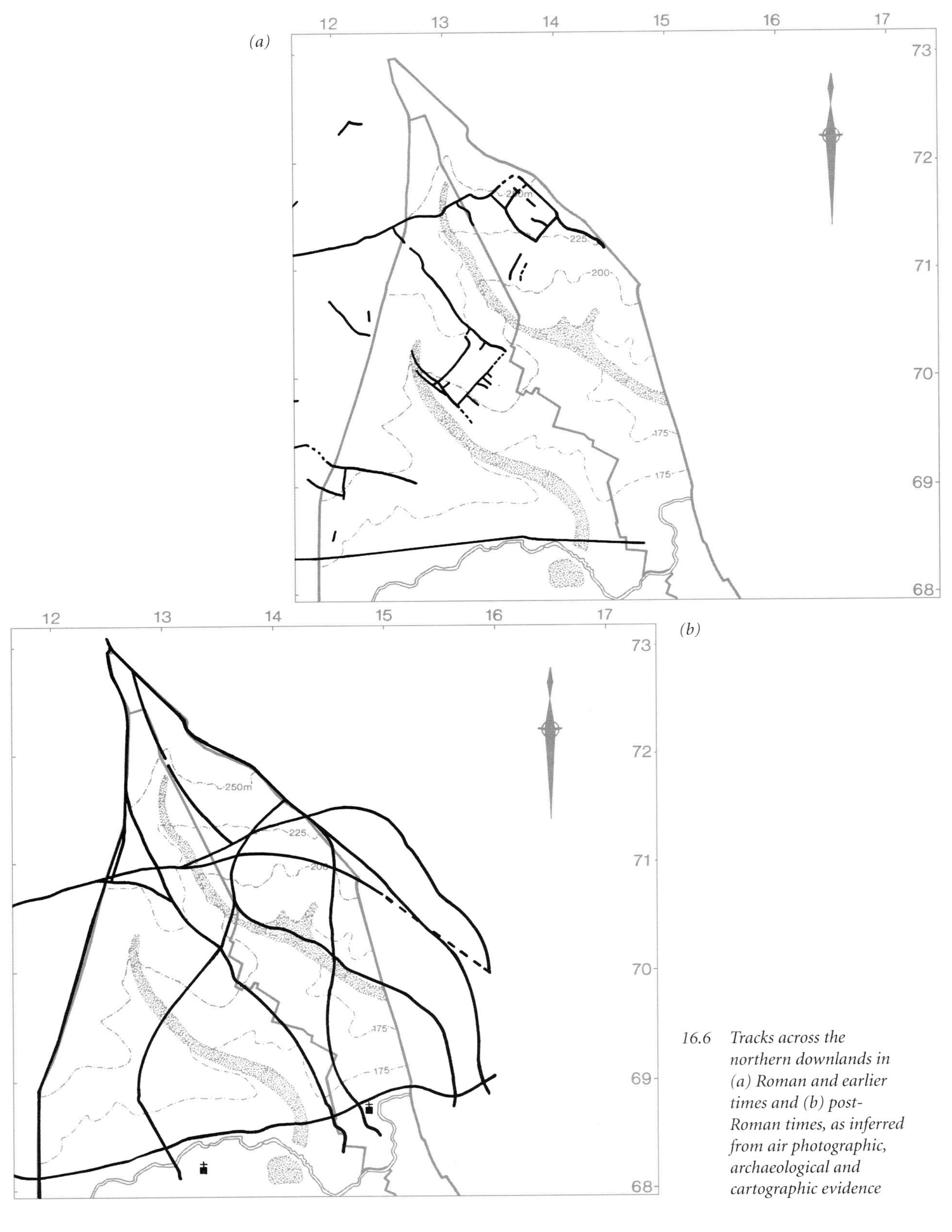

16.6 *Tracks across the northern downlands in (a) Roman and earlier times and (b) post-Roman times, as inferred from air photographic, archaeological and cartographic evidence*

(a)

Plate LXIV *Tracks and roads: (a) The Ridgeway, its western side here forming the Avebury/West Overton parish boundary, heads north from its intersection of the Roman road on Overton Hill past round barrow G.8 (SMR 650) towards Hackpen (cf Figures 4.1 and 16.6); (b) The Ridgeway, along the western skyline, is intersected by the toll road/A4 (from right to centre of view) on Overton Hill and passes The Sanctuary (behind the skyline bushes) and the silhouetted barrows of the Seven Barrow or Overton Hill Group (Group H). The view is from Overton Bridge, across the River Kennet and former water-meadows (cf Figures 8.1 and 16.6)*

(b)

9 COMMUNICATION: THROUGH-ROUTES

Over much (but not all) of Fyfield and Overton Downs is a cover of earthworks, in the main of a landscape which developed between *c* 2000 and *c* 600 BC. This was in its turn selectively overlaid by further workings, which have left earthwork remains in the first–second centuries AD, and in medieval times. Visually, the skein of complexity on the ground seems to be held together by linear features (Figure 2.1). Essentially, these are of three types: field boundaries, tracks and ditches. Here we are concerned with the tracks. Most are integral with, or fitting into, the south-west/north-east and north-west/south-east axial field system (Chapter 2). These lengths of downland track were part of the working Romano-British countryside, forming elements of a network that was already old, and which has persisted. For example, one runs the length of Overton Down on approximately the north-west/south-east axis: we call it the 'Overton Ridgeway', for it traverses the whole length of the medieval tithing of East Overton, partly along the boundary between the two Anglo-Saxon Overton estates. It is also part of a long-distance, north–south route.

But The Ridgeway is commonly perceived as the only major track running north–south in this area. That is a misconception. Wrongly, it is almost invariably considered as a single entity when in fact the line now designated 'The Ridgeway' is but one of a bundle of former track lines. One such line is our 'Overton Ridgeway', actually called 'Ridgeway' in the tenth century AD. All lie within a zone of movement forming what it would be helpful to think of as 'The Ridgeway route' (Figures 16.6 and 16.7). The purpose of this route, it is suggested, was for something crucial in the working of the landscape in mixed or pastoral farming, but an operation now almost forgotten in southern England. The 'Ridgeway route' was a bundle of drove-ways used in transhumance. They are called *drailles* in southern France whence come the working analogies for this suggestion, examined in a little more detail elsewhere (Fowler 1998).

The proposition is, then, that the main function of the local tracks, in part represented by earthworks, was to control access to and the use of pasture by sheep. The purpose of the droves, which included some lengths of local track, was the controlled movement of sheep through the landscape, and specifically an enclosed landscape where crops and properties had to be respected. This function has left its imprint on the landscape, even though a drove-way or 'draille' as such was not built. The Ridgeway, the Overton Ridgeway and the rest of the bundle of north/south downland tracks through Overton and Fyfield reflect, in this model, transhumance; that is, not just movement of flocks from pasture to market but quite specifically the movement of flocks with people in a life-way which involved seasonal living on the summer pastures often far from the winter home. This was until recently a normal mode of livelihood in much of rural Europe (eg, Barker 1995 for southern Italy, Clément 1991 for the Cévennes, France), and it persists in some places; for example, in the Pyrenees (as witnessed 1997, and recorded by Aragnou (1982) and Cavailles (1910/1986), following the mountain tradition documented for medieval times in the Departmental Archives, Tarbes, and further east, for example, in Ladurie (1978)).

An insight into how and why these tracks were probably operating in earlier times in our study area, both for local traffic and as through-routes, is offered by Smith (1885, 24), writing of the 'Ridge Way' in the middle and earlier nineteenth century and arguably just in touch with an older tradition:

> ... [this British road], ... twenty years ago [ie, 1865], ... was the regular route adopted by the thrifty drovers who would avoid the tolls on the high road; and only fifty years since [c 1835] ... was the much-frequented path employed by smugglers for conveying their contraband goods from the south coast to the interior of the country ... They were merely tracks over the turf ... and lying open to the wind in [their] exposed position, are generally firm and hard.

Smith was writing in the 1880s of his countryside one to two generations before his time. He helps us see The Ridgeway 'bundle' of trackways spilling off the Marlborough Downs as droves, used in both directions for through traffic most of the year and for transhumance northwards in spring, southwards in autumn.

If these drove-ways are at all ancient in origin, they would seem to be, as suggested by Figure 2.1 and on the ground, of at least the early centuries AD and very probably of later prehistoric times too. The next step in our argument is to accept such antiquity as premise and therefore be able to see such tracks, individually and collectively, as 'permanent' features of the landscape, even if their precise line wobbled a bit across the countryside from decade to decade and century to century (Plate VI). That being so, they would then have been able to contribute to the shaping of land units, estates, the tithings, the parishes and their predecessors whatever their name or status before the mid-first millennium AD.

Indeed, it would have been as difficult to avoid using them as it clearly was to avoid lesser features such as field edges marked by lynchets; in practical terms, when it came to defining long-term boundaries, it would have been extremely convenient to begin to follow lines already being etched into human consciousness as well as into the land. In part at least, therefore, so the proposition goes, what emerge for us as the tithings of West Overton, East Overton, Lockeridge and Fyfield, are the shape they are, with their boundaries where they are, because of the lie of the drove-ways around which they tended to arrange themselves. Of course the lie of the drove-ways themselves owes much to the lie of the land and the desired line of movement; but such factors do not apply so forcibly to the shapes of estates or the position of boundaries. They could in theory be any shape or be anywhere; in fact, from the time of the Anglo-Saxon estates onwards (and we would argue from much earlier), to a marked degree the main land units shape themselves round the topography and their boundaries lie with the drove-ways (Figure 16.7).

The hypothesis is that this became so because, as land units developed, they did so under the strong influence of what was already there, ie, topography and the main lines of communication across it. So the land units, which we eventually see as Anglo-Saxon estates, draped themselves, as it were, on that structure, flowing with the land and movement through it, and using where convenient the lines of movement – the tracks and drove-ways – as their edges. Such would have been a pragmatic evolution, intended to utilise what everybody knew about this landscape.

The particular point of this hypothesis is that the actual lines of communication that influenced and often became the permanent boundaries were themselves traditional because of their long use in transhumance. While that practice in a general sense goes back to Mesolithic times and could well have been a mainstay of socio-economic life in this area through the fourth and third millennia BC, the spatial patterning of the local archaeology through the third millennium suggests that life-style emphases were more locally concentrated. With the abandonment well before the middle of the first millennium BC of the very landscape patterns which suggest that interpretation and, more positively, with the development of a new emphasis on stock-breeding, the way was literally open for flocks and herds to be led to and through a newly unenclosed grassland along lines selected by their herdsmen from their pastoral perspective, not along trackways designed by arable farmers to secure their field crops. And what suited one herdsman was likely to suit another, along different routes depending on the environmental circumstances; and hence the development of recognised lines of movement through this landscape which, as a matter of convenience and practicality, became drove-ways which in turn, where it suited, were used as the lines of boundaries.

When this could have happened is obviously debatable. It has long been suggested that the early boundaries are at least older than Wansdyke which cuts across the north–south boundaries that it meets. It is significant for this theory that the two original gateways through Wansdyke in our study area are both defended (Figure 13.2) and on the line of the same independently suggested ancient drove-way which only bifurcates in Hursley Bottom below them and to their north (Plate LIX; Figures 16.7 and 16.8). This relationship fairly positively suggests that the drove, a continuation of the track identified as the 'Overton Ridgeway' further north, is the earlier; its line is much used as the boundary between the two Anglo-Saxon Overtons (Figure 16.7). This lends weight to the idea that, since the drove is but one of a bundle of routes forming 'The Ridgeway zone', boundaries generally followed droves, not vice versa. Furthermore, if all this is correctly observed and interpreted, then at least the basics of the tenurial structure were present at latest by late Roman times, related to pre-Roman lines of movement through this landscape which were essentially elements in a late prehistoric transhumant route.

10 ENGLISH COUNTRYSIDE, BRITISH LANDSCAPE

The evolution of such a route is most likely to have occurred in the last few centuries BC. If so, its results, in the form of a tenurial framework defined by boundary lines on the ground, was probably not so much obliterated as adjusted, again for a different set of conveniences, by the Roman administration. For much of its length before meeting Wansdyke, that particular drove – our 'Overton Ridgeway' – is the boundary between the two early medieval Overton estates. Its route is in fact more or less through the centre of a hypothetical single 'Overton estate', perhaps of late prehistoric origin and Roman use around an estate centre at 'Headlands' near a cross-roads formed by old drove and new Roman road (Figure 4.3). Reminding ourselves of earlier arguments, the suggestion would then be that, perhaps in late Roman rather than sub-Roman times – the fourth–fifth rather than the sixth–seventh centuries AD – such a single Roman estate

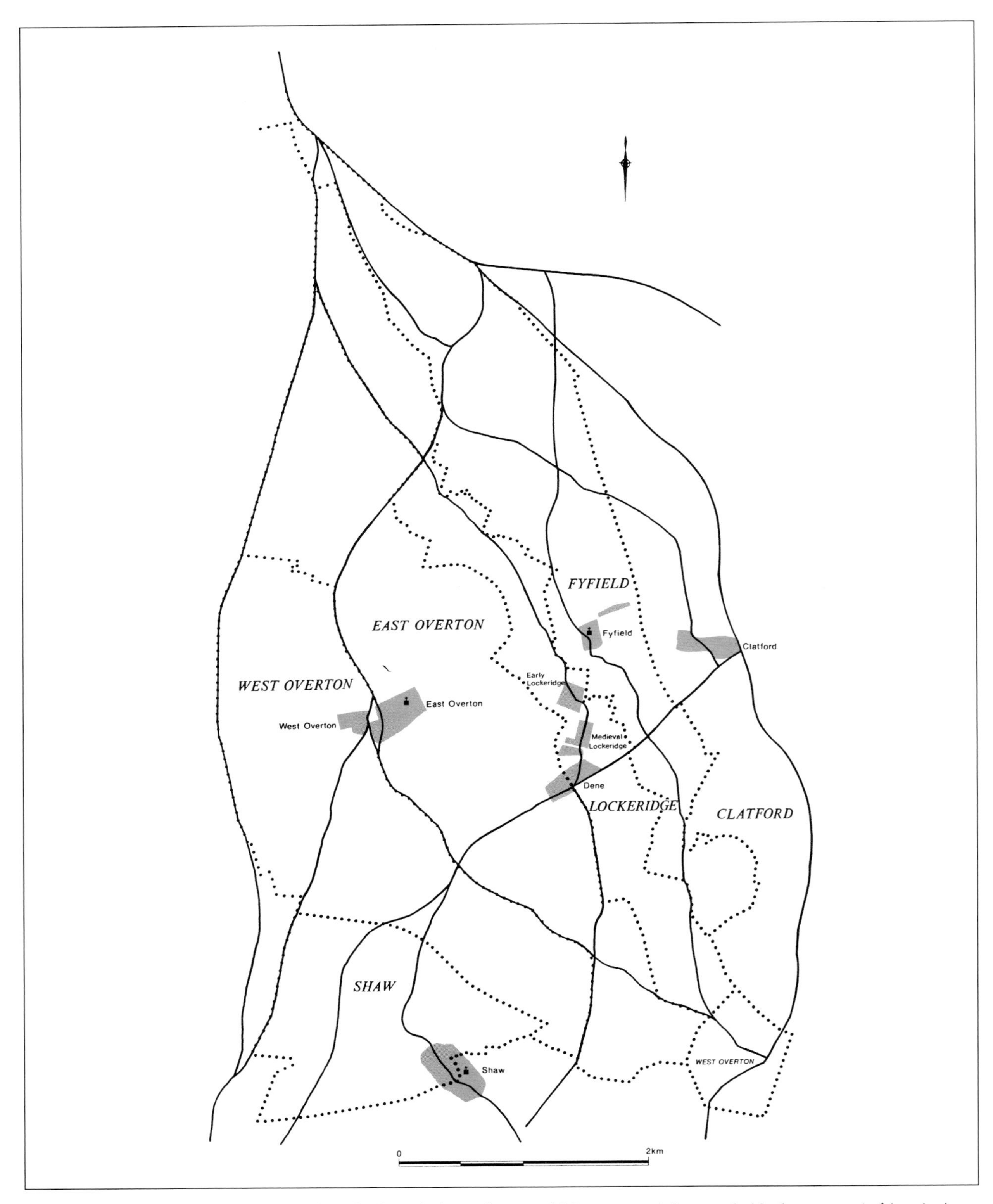

16.7 *Map indicating those well-attested tracks through the study area which were certainly or probably drove-ways in historic times and are here suggested as transhumance routes, part of an all-embracing 'Ridgeway route', with origins in late prehistory*

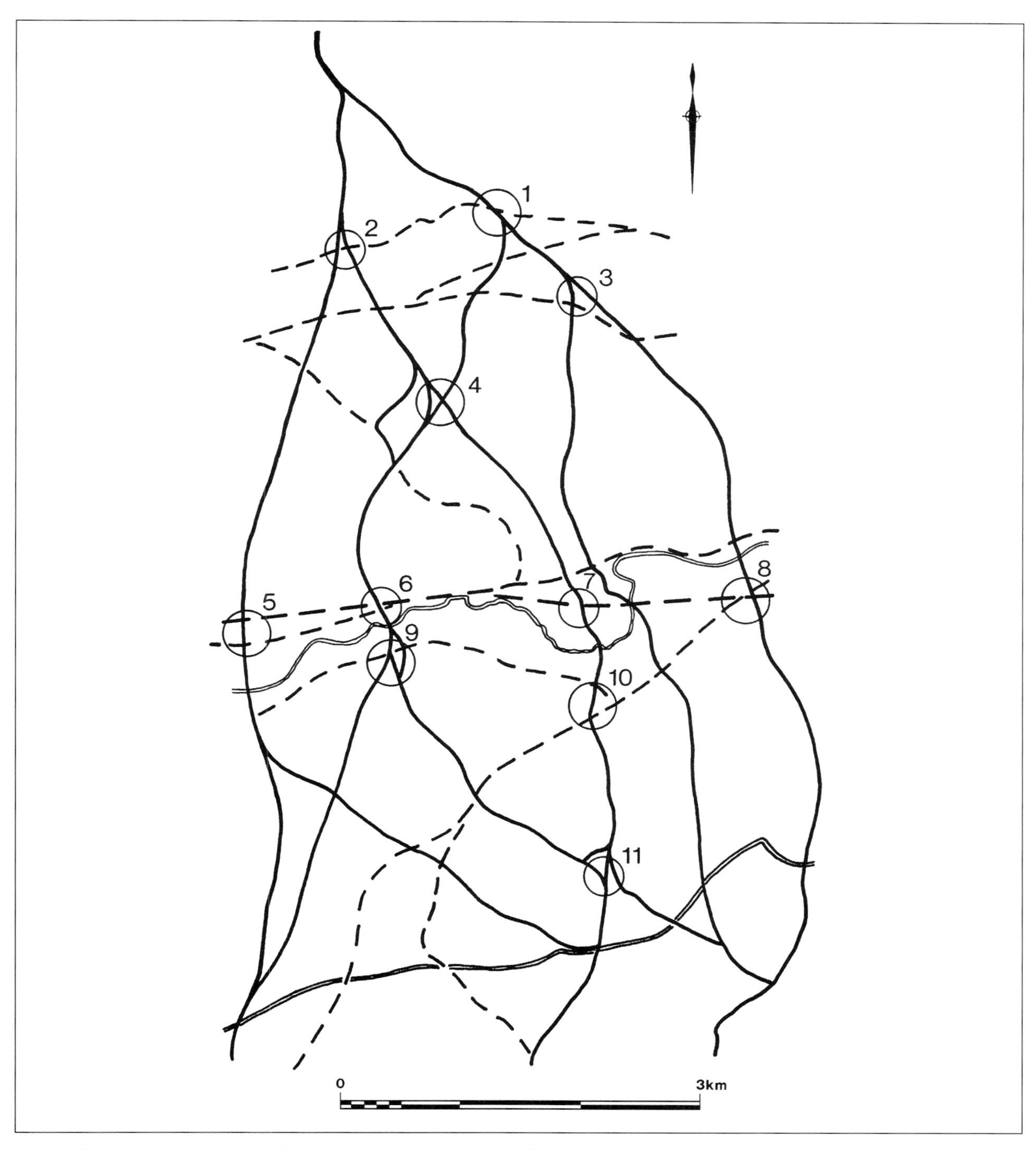

16.8 *Diagrammatic map, in part based on Figure 16.7 but using all sources of evidence, plus inference, attempting to identify significant places, 'nodal points', in the landscape. Eleven are suggested, places that have been consistently 'busy' long term, particularly as habitations and meeting-points of roads, tracks and boundaries. None of the places circled could fairly be described as only 'prehistoric', 'Roman' or 'medieval': they all seem to be present in most landscapes irrespective of period. Solid lines represent north–south routes, broken lines east–west. The double line centre is the River Kennet; south is Wansdyke 1: Totterdown; 2: Aethelferthe's stone; 3: Clatford Down; 4: Derby gallop; 5: Overton Hill; 6: North Farm; 7: Piper's Lane; 8: Clatford Crossways; 9: Overton village; 10: Dene; 11: Hursley Bottom*

split into two smaller estates, forerunners of the two Anglo-Saxon estates of West and East Overton which we pick up as already existing in the tenth century. Such is detail as well as speculation; but much of this later discussion leads to a final proposition of which that speculative detail is but one facet.

The upper Kennet valley, the neighbouring downs and the remnant of the royal forest of Savernake constitute what is understandably regarded as a typically English countryside. The area lies, after all, more or less in the centre of civilised southern England, quite a long way from the wild and unkempt lands of the 'Celtic' west and north, at least as perceived by such English travellers as Fiennes and Defoe until relatively recently (Morris 1949; Rogers 1971). Wiltshire itself, a very English county in many respects, was not around as such until AD 878, but the term 'was probably employed long before its recorded use', perhaps in the eighth century and even as early as the seventh (*VCH* II, 1, 2). So much has happened in the country, on the Marlborough Downs and in our own small study area, since Wansdyke was abandoned that there is little reason to question the perception of an English countryside hereabouts.

Our proposition is that such perception be limited to early medieval times onwards, from, say, possibly the sixth century and certainly from the seventh century with its establishment of Christianity and, central to our concerns, its establishment under ecclesiastical management of countryside estates. King Ine's *Laws* (Whitelock 1955, 364–72) can convincingly be read as applying to a countryside of estates, ecclesiastical and secular. Nevertheless, so we propose, more fundamentally the Fyfield and Overton countryside worked and fashioned by the early English and their successors was in essence a British one; but that assertion is historically self-evident unless we favour a catastrophic interpretation of complete tenurial and land-use breakdown for which no evidence exists in our study area. So we go further. We would argue, because it is those fundamentals of the pre-English landscape which have underpinned the landscape ever since, that what we see today is basically a British landscape.

The proposition should flow from this monograph. The landlords of, say, AD 700 were working a landscape of which the character had been decided long ago – the subsoils and land-forms in geological times, the soils and types and distribution of vegetation in prehistoric times, the land-uses in later prehistoric and Roman times, and the location and perhaps types of the habitations on their estates in the late- and sub-Roman period of the preceding 300 to 400 years. Because all that had happened, it is possible to argue further that the very nature of the communities making up local society was itself also predetermined – not the racial mix but what was required of people seeking to earn their keep in this area and therefore the sort of mix they were likely to be socio-economically and, given other, external constraints, organisationally. We have already suggested a model of acculturated landscape constraints on the northern downs within known climatological and technological parameters (*see above* this chapter, section 5), and we now expand the concept spatially and socially. A shepherd was going to be socially more useful than a silversmith in eighth-century Overton, a ploughman more than a lace-maker in Lockeridge, and a fowler more than a furrier in Fyfield.

Effectively, the stage was set and the cast of characters already written in before the seventh century AD. The new Anglo-Saxon landlords, ecclesiastical and lay, could move some of the props and pick their actors and actresses, even change the words from time to time; but they could not fundamentally change the sort of show that was performed. The fact is that they also chose not to, for they took over the area that became the two parishes as a going concern which, perhaps under-performing a little at the time, nevertheless worked reasonably reliably within its own and contemporary technology's limits. The landlords moved what we might call the 'countryside furniture' around a bit and took their own decisions about what to grow in which field from year to year, so that gradually, over a millennium and a half, particularly with Enclosure and its consequences in the valley and on its southern slopes, the countryside came to look 'English' in the way that the English themselves, under the influence of their painters, writers and poets, came to expect their countryside to look. But the landscape was not of their making, and not even of their ancestors; for the British made it.

Chapter 17

A Once and Future Landscape: Issues and Objectives

England's landscape is its consummate artefact – not merely the locus of the heritage but its mainstay ... It is an English creed that all land requires human supervision.

Lowenthal 1991, 215, 216

Whether consummate or otherwise, the whole of the landscape of Fyfield and Overton, in so many respects so typically English, is also in some degree artefactual. This study has demonstrated much to that effect. The area has been, and continues to be, valued as agricultural land, producing an economic return; that will doubtless continue but another set of values is already being applied to an area that now finds itself increasingly defined as 'landscape' rather than just as 'land'. These other values are primarily based on concepts of conservation, recreation and amenity.

The first concept probably contains elements from the other two, though they do not necessarily return the compliment. Conservation also tends to embrace 'heritage' values, if still rather vaguely, a composite idea promoted more and more at the end of the twentieth century as a form of bait for economic return. It may well be that, in an area so close to visitor-attracting Avebury, the twenty-first century will see a different sort of 'farming' based on such concepts as a 'tourist crop' rather than traditional bushels of wheat or flocks of sheep (Plate LXVI). As this study ends, another phase of historical development in the land-use of our study area is beginning (Fowler and Stabler forthcoming; English Heritage 1998).

The considerable time, effort and resources dedicated (happily) to the attempt to understand the landscape of Overton and Fyfield parishes (and a little beyond) should have more than academic, serendipitous and personal outcomes. The author would not presume to think that he had improved matters but, if this somewhat fuller and different academic perspective can now inform the previous (mid-twentieth-century) understanding of the area, there could be implications for the management of this landscape, with particular reference to the concepts of 'conservation management' and 'World Heritage' (Small 1999).

Saying this is not to challenge the social and economic context that assumes it to be entirely desirable that this land should continue to generate agricultural livelihoods. Any new perspective gained on this landscape must not be used to 'freeze' it as it is now, to fix it in time as it was in the late twentieth century, let alone to justify attempts to return it to fuzzily conceptualised notions of our rustic past.

If this project has illustrated anything, it is that this landscape is what it is because of what it has been, because of the way it has been used and because it has always been changing, often at very different rates and in different ways. The first conclusion to draw is that, as a result of this detailed study, we can now view this landscape with sufficient depth of historical perspective to appreciate that the extensive and environmentally degrading arable regime inserted into the downs from the 1950s onwards (and still prevailing) is likely to be as temporary as any of the farming regimes of the past. 'Set aside' and other yield-reducing mechanisms introduced since the 1980s are predictable in the broad sweep of landscape history. (Thirsk 1997, 223–50, 'The fourth experience, 1980s onwards', sees the same mechanism emerging for a different set of reasons.) As the effects of such measures begin to be glimpsed in the use of this

Plate LXV Old Totterdown Cottage collapsing in 1959 (now completely ruinous and overgrown)

ancient and essentially pastoral landscape, so does a supreme landscape irony come into focus. Fyfield Down National Nature Reserve, designated in the mid-twentieth century because it was already becoming a grassland oddity in a sea of arable, actually far better represents traditional land-use in this area than the arable to which we have become accustomed. The irony is that, among the Reserve's principal features, there are redundant arable field systems (Plates XVI and XXXVIII) that were themselves, in their time, the intruders.

A second inference is that the heritage value that we should pass on, in this case, is not the fossilisation of the landscape but its continuing dynamism (Plate LXV). However, we must manage it, in the sense of looking after it, to produce an agreed effect to common benefit. Posterity may then, like us in our time, be able to read and enjoy the graphic but inarticulate pages of field and down. All of us, whether or not we are 'conservationists' or 'conservation managers', need to ensure that the landscape we hand on continues to have the capacity to stimulate individual experiences, be they aesthetic, emotional, intellectual or physical (not that in real life experiential values can be so neatly packaged).

A cyclist, asked why she and her group had made the effort to ride out along The Ridgeway from Swindon, replied 'It's just the sense of space'. That is very important; free public access to that sense of space, even in a privately owned landscape, is as vital (though no more so) as other sorts of access to the other dimensions that this landscape can retain and offer. The poet, the scientist, the academic, they all want to see their landscapes, like the young cyclist, and their right to do so is as powerful as that of the local resident, the tenant farmer and the legal landowner. Legally, though this *land* may not be their land, assuredly the *landscape* is.

The third conclusion, equally basic, is that we really must bring respect back once more into our relationship with the landscape. It is not just a factory floor, neither is it just a commercial asset, a tourist attraction, a recreational facility, nor even – and we admit it – just an archaeological site. It is, of course, all those things and a sense of respect for the landscape should be the

Plate LXVI Lambing on Overton Down: thatched sheep pens, 1961, as constructed here for the last time by a full-time shepherd

common ground between all such interests. By 'respect' we mean starting not from particularism, not from self-interest, not from ignorance, but from an informed appreciation that the landscape we live with is what it is because others have created it for us. It has the potential to be valued in different ways because our ancestors have put effort into it. Yet that effort has, in so many cases, produced fragility in the nuances that give a landscape its range of values. 'Tread lightly on this earth', the Countryside Commission exhorted on a 1995 poster, and that is sound advice to guide our thoughts and deliberations as well as our actions.

EXPLORING OUR PAST

English Heritage's publication under the above title (1991a) provided helpful guidance at the outset of this project (Chapter 1); it is also helpful at its end. The 1991 edition has been revised (1998), but it was the earlier version that influenced this project, for example with such generalisations as:

> It is through the record left by the surviving remains of past generations that we can interpret most clearly the impact of humankind on the environment of these islands.

The text continues to say that, as a consequence, it is necessary to

> ... identify the surviving individual sites or landscapes which are the most important indicators ... then to ensure that these are properly understood and that their significance is fully recognised ... The most important sites must [then] be managed ...
>
> English Heritage 1991a, 1

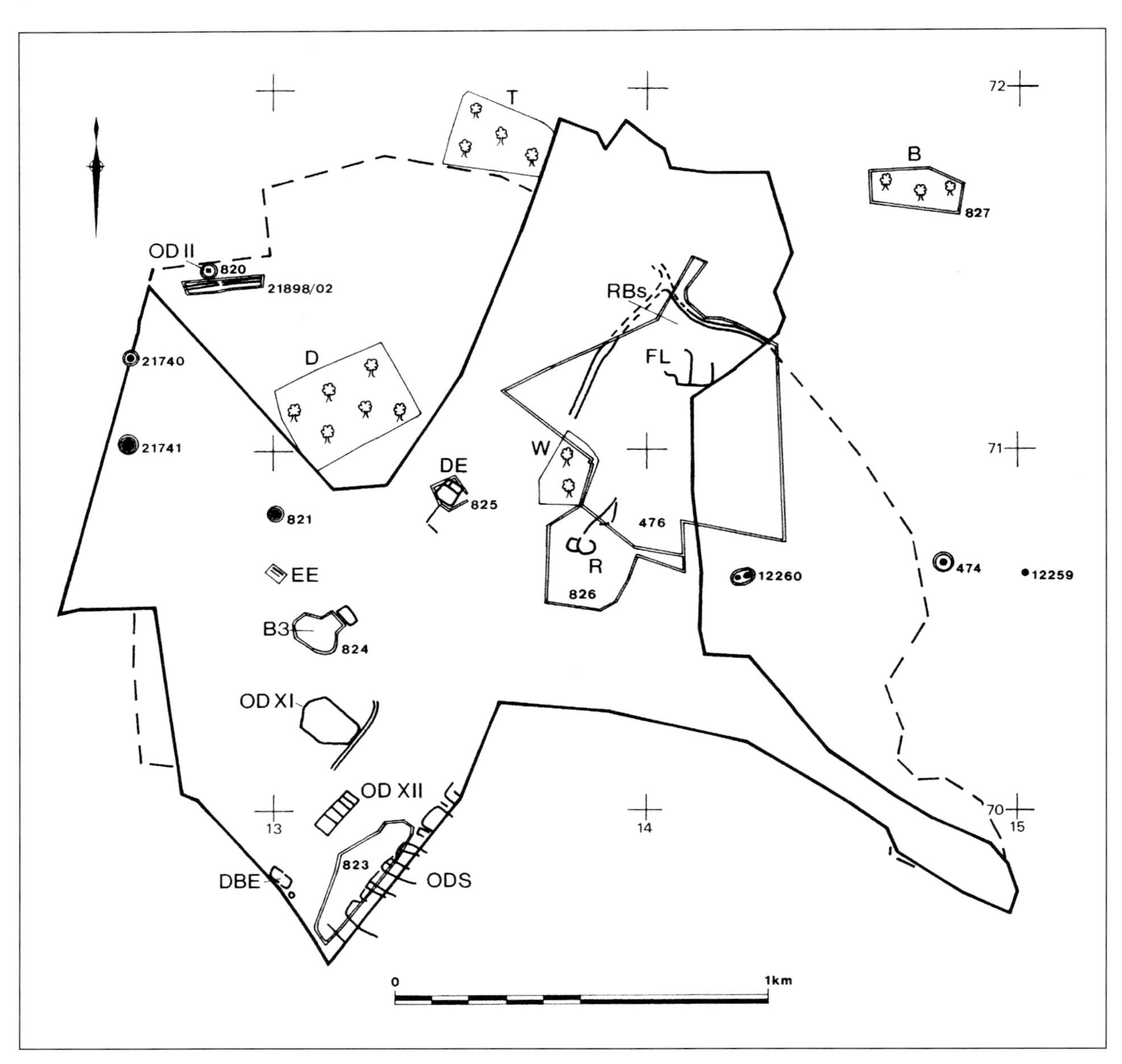

17.1 *Map: a landscape of designation on the downs. The continuous heavy black line defines the Fyfield Down National Nature Reserve, all of which is part of an even larger Site of Special Scientific Interest extending, as indicated by the broken line, to the north, east and west. Scheduled Ancient Monuments are defined by double lines with their AM number, islands in the landscape precisely capturing individual round barrows (eg, 21741), but somewhat inaccurately mapped in relation to some settlements (eg, Overton Down South [ODS], which AM 823 largely misses, and Fyfield Down Romano-British settlement [RBs], which the bold areal scheduling of AM 476 misses altogether. Just scheduling a short length of ditch F.4 on the north [AM 21898/02] also rather misses the significance of the feature). Some major sites are not archaeologically protected (eg, OD XI and XII and the Down Barn Enclosure [DBE]). The scheduling is currently under review. [In 1999/2000, the scheduling of this area was reviewed in the light of data from this project and a proposal for area scheduling has been made, making all the data here redundant and the map itself 'historic' before publication.]*

T = Totterdown Wood; B = The Beeches; FL = Fyfield lynchet excavations;
W = Wroughton Copse; R = Raddun*; DE = Delling Enclosure;*
D = Delling Wood; EE = Experimental Earthwork; B3 = Figure 6.1

There can be few places in the British Isles where 'the record' is more clearly etched on the landscape than on and around Fyfield and Overton Downs. Even though it is less clear in the valley and to the south, in the woods, a record is nevertheless discernible and it has an important complementary story to tell. Although this project and its output have been much concerned with detail and particular sites, the whole is actually about the interaction through time of human and environmental dynamics (Plate LXVII); that is to say, the evolution of a magnificent example of a 'continuing landscape'. Furthermore, our work has been concerned throughout with recognition, understanding and (though the phrase had not been invented in 1959) resource management. The more informed perspective that might now be brought to the study area, and especially to the downs (whether they be archaeologically well preserved or badly damaged), is a potentially significant contribution to the achievement of the official objectives quoted above.

That may seem, and indeed may prove to be, a pious hope, but not necessarily, as three examples of small but significant recent official conservation actions have demonstrated. Firstly, there is the management plan for the Fyfield Down National Nature Reserve (NNR), originally drawn up (and periodically reviewed) by the Nature Conservancy (then English Nature, now within the Environment Agency). Management of the NNR has increasingly reflected not just the growth in the amount of archaeology recognised as existing in the NNR but also, and rather more importantly, the growing awareness of the significance of the archaeology, both in enhancing the quality of the Reserve and in impinging on management responsibilities (English Nature 1991).

Secondly, the number of Scheduled Ancient Monuments on Fyfield and Overton Downs has increased markedly as a direct result of our own early fieldwork and publications (Figure 17.1; SAM numbers 820a, 821, 823, 824, 825 and 826). A revision of the schedule in the 1980s expanded the boundaries of the 'old' SAM 476 to include much of Fyfield Down. That scheduling is now being reviewed again, even as this text is being completed; to that end, Figure 2.1 was made available to English Heritage in advance of publication. All the data displayed on that map were taken into account in the autumn of 1998, during the Monument Protection Programme's review of the study area. Our improved understanding of the area's 'significance' was brought to bear on management judgements, in addition to the quantifiable elements of the location, extent and state of the 'sites'. Our recommendation was not to add to the number of existing schedulings, but to schedule the whole of the NNR out to the SSSI boundaries to the north and east (Figure 17.1).

A third example was an event, not a process, which occurred during the preparation of this volume. In 1995, Ring Close, the field immediately south west of Overton church, came on to the market. Its sale raised the possibility of some form of change from its existing use as grazing land. The field contains the excellently preserved village earthworks illustrated in Figure 9.2, a plan that was already in existence together with an early draft of part of Chapter 9. The draft (FWP 26) described the earthworks and – importantly in the circumstances – brought out their possible historical significance (here proposed in Figure 8.2). The plan and text were sent to English Heritage to help support the case for scheduling the earthworks, which have indeed now become a Scheduled Ancient Monument. The immediate urgency, incidentally, has passed, for the new ownership is content for the field to continue under grass.

Another incident was not so happy, though – curiously – it was also led by conservation considerations, this time linked to amenity. It occurred on The Ridgeway in the long hot summer of 1995, when well-intentioned but misconceived road works, carried out under the aegis of the Countryside Commission and Wiltshire County Council, began on The Ridgeway just south of its intersection with Green Street, between Overton and Avebury Downs (Plate VII; Figures 2.1 and 4.1; FWPs 30 and 81) (had we been certain that the area had previously been called 'Hackpen', we could have brought that name back into circulation at the time).

The 'Ridgeway Incident' – interesting in its own right, particularly in illuminating the interface between research and management – also serves as a metaphor for countryside misunderstanding and heritage tensions. Its lessons are quite clear. At the practical level two requisites can be specified: they are good communication between those involved with an area, together with the involvement and active participation of local people. More generally, divisions between theory and practicality, between research and good management, between academic concerns and the so-called 'real world' are, we believe, false. Indeed, we would go further and assert that such contradistinctions exist only in the minds of their proponents – and mainly in those of managers rather than academics – and actually lead to bad management. Much of what ideally should be involved within a model of conservation management over the south-western corner of the Marlborough Downs seems as yet either too particularistic or too unfocused, leading to flaws in practice and unnecessary threats to the cultural landscape, of which the 'Ridgeway Incident' of 1995 is but one example.

Heritage issues will not disappear from the Avebury landscape in the foreseeable future. Indeed, they have now become part of that landscape, thanks to a wider recognition of the fact that 'heritage issues' are a basic part of the Avebury area's heritage. No longer is that heritage merely a matter of some old stones and a big bank and ditch if we follow, for example, recent discussions of the non-intellectual aspects of Avebury (Pitts 1996) or the sort of highly personal, 'touchy-feely' responses recorded by some visitors (Bender 1998); but then, *plus ça change* if we recall Stukeley's later lucubrations on the site (Piggott 1985, 93–109) and remember similar contrary claims encapsulated from ten years at Stonehenge (Chippindale *et al* 1990). The issues at Avebury are combustible and complex, quite as politically and professionally challenging as the issues of public health and education, and they are far from being the peripheral single issue of an antiquarian clique. And they are now official, too, for henceforth there will be no escaping the *Avebury World Heritage Site Management Plan* as finally promulgated (English Heritage 1998).

A LANDSCAPE OF DESIGNATION

Let us turn, then, to some aspects of landscape conservation and heritage management relating to the area as a whole (in the course of which we will argue that the future of the Avebury area – including the whole of Fyfield and Overton Downs – lies in designation as a World Heritage Site). One of the clearest conclusions of this study is not only that very little of this landscape is 'natural' but that much of it has a very long history of management. From Clatford Bottom to Silbury Hill we are indeed looking at, and have in our stewardship, a 'cultural landscape' (Birks *et al* 1988; Jacques 1995; Fowler forthcoming a and b), a concept argued for as part of the UK's official designations elsewhere (Fowler and Jacques 1995) and apparently now accepted by English Heritage (1998, paras 1.4.1, 5.6.2).

Of course, historically the management that has produced this landscape has varied through time, apparently being strong in some periods, and expressing itself through well-organised landscapes, while being less strong at other times, and therefore not so prominent in the archaeological record. Even so, absence of archaeological evidence does not necessarily mean weak management, nor does positive management necessarily express itself through monuments: one senses, for example, that fairly firm directives were emanating from medieval Winchester to the Overton estate without there being much to show for them on the ground now. At the very least, the constraints and seasonal rhythms of good husbandry would always impose their own management precepts, whether or not there was an estate office managing the affairs of the prehistoric, Roman or medieval landowner.

Management, good or bad, has always been present in the exploitation and sustenance of a landscape's resources. There is therefore nothing new at all in the idea or practice of managing a landscape from a conservation point of view. Husbandry has been practised over our study area for much of its history. Husbandry objectives may embrace different emphases, methods may change, but the process is the same, whether a Bronze Age farmer digs a long ditch across the downs, whether a group of villagers walks its estate boundaries in the tenth century AD, or whether we try to influence the future landscape by designation now. We have merely rediscovered the concept of good husbandry and relabelled it 'sustainability' as some begin to hear the global alarm bells ringing in the sort of way that a few may well have done locally about 900 BC and again, to their horror, about AD 540 (Baillie 1995, chapter 6) and 1313 (Chapters 6 and 16; Fowler and Blackwell 1998, 133–5).

Designation has become the characteristic methodology for managing the Fyfield/Overton landscape over the last forty years. Indeed, the process has gone so far as to justify our calling the landscape a 'landscape of designation'. Designation now characterises the landscape itself, not just the way in which it is managed. A mid/late-twentieth-century way of doing things, in other words, is already leaving its imprint on the landscape in a similar process to that which enables us to pick out the doings of an anonymous Roman *agrimensor* and Richard of *Raddun*.

The principal features in 'the designated landscape' are scientific ones; that is, features and areas that have been ring-fenced in some way for their scientific and academic interest rather than for amenity or recreational use. The main types of designation and the areas they cover over the northern part of our area are shown in Figure 17.1, together with a proposal. The various designations are generally authorised and imprinted on the landscape by bodies in business to conserve or manage conservation; most do not own the land. West Woods (the eastern part of the 'permanent woodland' on Figure 12.1) is, however, managed by the Forest

Authority, in business to produce timber but also with a statutory recreational and conservation responsibility. So, in addition to providing walks through these woods, the Authority also has a policy of archaeological conservation management. In a sense, therefore, the archaeological resource in these woods is being looked after as elsewhere in the study area, but by active management rather than prescriptive designation.

Overall, the landscape between Fyfield and Avebury has become subject to a multiplicity of designations, constraints and policy initiatives, some specifically archaeological and all, whether the originators intended so or not, with archaeological implications (English Heritage 1998, fig 4, appendices B, C, E, I). Even so, it is possible to see at work here one of the great misconceptions of previous generations and one that (understandably but disastrously) came to underpin so much conservation provision nationally: the assumption that the landscape is 'natural'. Here, some of the provisions are overlapping in their objectives while others are partly contradictory. Four main levels or scales of interest exist: international, national, county and local. We will look briefly at the principal national one and then even more briefly at the international one.

The *Ancient Monuments and Archaeological Areas Act 1979*, interpreted and reinforced by the Department of the Environment's *Planning Policy Guideline* 16 (PPG16), entitled *Archaeology and Planning* (DoE 1990), provide the current statutory and non-statutory framework of national policy on archaeological monuments. Doubtless both will be revised from time to time, but their fundamentals are unlikely to depart far from those of a legal situation which has evolved over more than a century. PPG16 in particular contains the following crucial assertion of principle, which seems particularly relevant to archaeological resource management in the Fyfield/Overton/Avebury area:

> Archaeological remains should be seen as a finite and non-renewable resource, in many cases highly fragile and vulnerable to damage and destruction. Appropriate management is therefore essential to ensure that they survive in good condition. In particular, care must be taken to ensure that archaeological remains are not needlessly or thoughtlessly destroyed. They contain irreplaceable information about our past and the potential for an increase in future knowledge. They are part of our sense of national identity and are valuable both for their own sake and for their role in education, leisure and tourism (para 6).

and

> Where nationally important archaeological remains, whether scheduled or not, and their settings, are affected by proposed development there should be a presumption in favour of their physical preservation (para 8).

WORLD HERITAGE

The publication of the *Avebury World Heritage Site Management Plan* (English Heritage 1998) has made redundant a long discussion here, for much of the data relevant to the management and future of the landscape in effect as far as the West Overton/Fyfield parish boundary on the downs is now easily accessible. It is important to stress, however, that (assumptions and impressions in the plan to the contrary) the whole of Fyfield parish, and therefore the whole of Fyfield Down, is currently outside the World Heritage Site (Fowler 1999). Nevertheless, the plan impinges on our study area and uses some of our data, notably Figure 2.1 (the implications of which it ignores), and its objectives have clear implications for land-use in Fyfield and Overton, especially on the downs.

The Avebury area shares with Stonehenge and its environs the distinction of jointly forming a single World Heritage Site (C373: Stonehenge, Avebury and associated sites) (Figure 17.2). This designation was first included on the World Heritage List on 28 November 1986. The World Heritage List was established under the 1972 Unesco *Convention Concerning the Protection of the World Cultural and Natural Heritage* that came into force in 1975 and was ratified by the UK in 1984. The concept and practice of World Heritage is concerned with the identification, protection, conservation and presentation of those parts of the heritage all over the world which, because of their exceptional characteristics and qualities, are considered to be of outstanding universal value. The concept embraces both 'natural' and 'cultural' heritage as two parallel classes of designation, and now, for precisely the reasons we can see operating on our own local area, the idea of 'cultural landscapes' is being pioneered in World Heritage terms. The move recognises that some of the world's most interesting places are actually those resulting from interaction between people and their environment (von Droste *et al* 1995).

In World Heritage terms, cultural landscapes fall into three main categories (World Heritage Committee 1995, paras 35–42, reproduced in von Droste *et al*, 431–2, annex II, and also discussed with reference in part to Britain in Fowler forthcoming a and b; FWP 88):

i 'landscapes designed and created intentionally by man';

ii 'the organically evolved landscape', a category that subdivides into two types, a 'relict or fossil landscape' and a 'continuing landscape';

iii 'the associative cultural landscape'.

Neither (i) nor (iii) concerns us here, but clearly, in the light of this and other studies, the landscape of the Avebury region is well qualified in category (ii).

The 'relict landscape' of category (ii) is a familiar concept, exemplified in many parts of upland Britain and notably on Dartmoor and Bodmin Moor (Fleming 1988; Johnson and Rose 1994). The phenomenon is something of a rarity on a world scale in terms of the time-scale of such examples (Fowler and Jacques 1995). The 'continuing landscape' type is both obvious and more subtle, though the definition is quite clear: 'it retains an active social role in contemporary society closely associated with the traditional way of life, and in which the evolutionary process is still in progress'. At first glance, the downs over the north of our Fyfield and Overton Downs study area might well be thought merely a relict landscape, an excellent archaeological landscape of high-quality preservation, but now 'dead', its development expired. This study has surely dispelled such a view. The downs are alive, still evolving as landscape in their use and perception. Particularly when linked spatially and conceptually to their parochial and sub-regional context, as we have attempted to show,

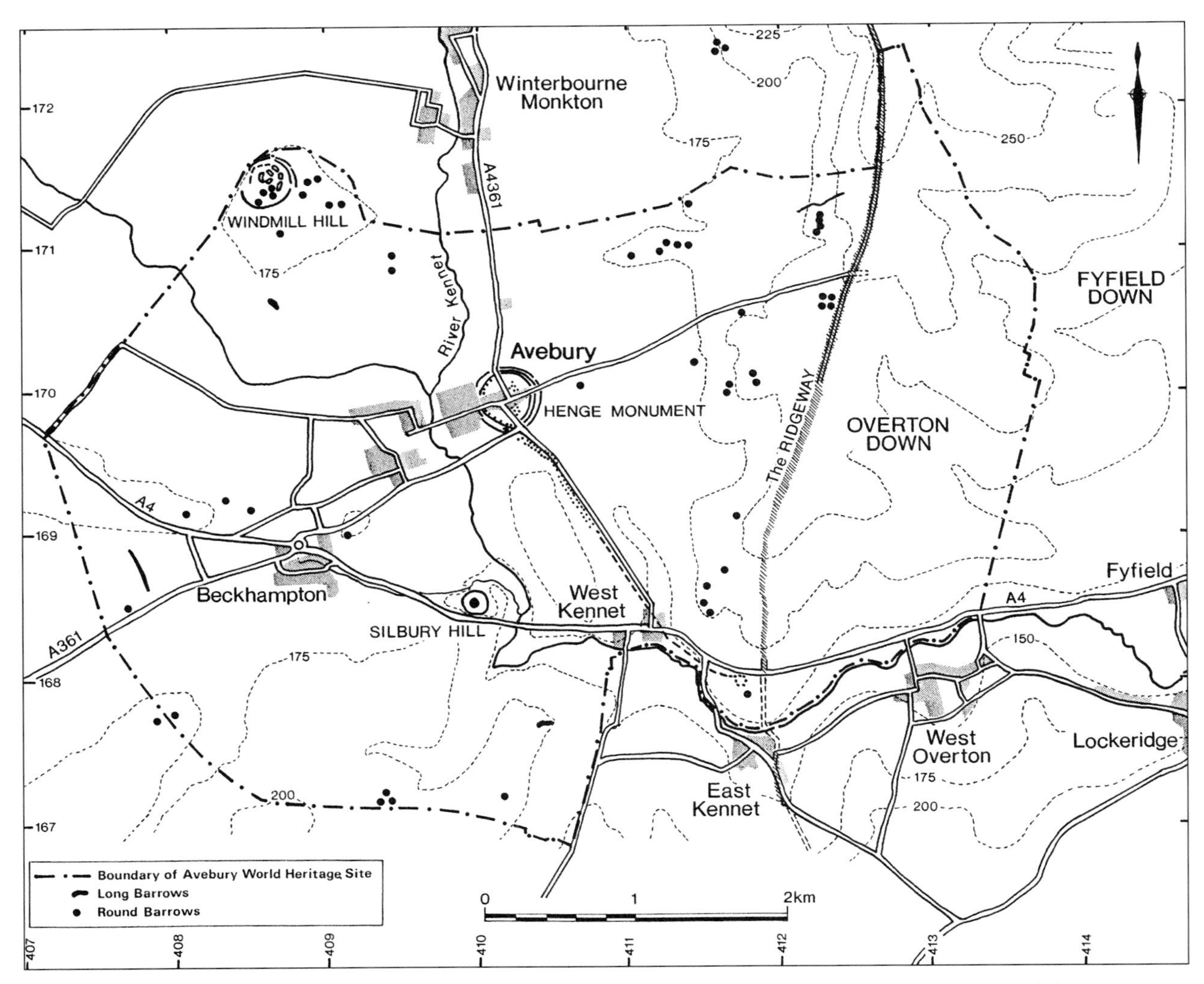

17.2 Map of Avebury World Heritage Site, embracing Overton Down but excluding Fyfield Down, Totterdown and the greater part of the study area

Fyfield and Overton Downs clearly retain 'an active and social role in contemporary society because "the evolutionary process" is still very much in progress'. They represent 'a continuing landscape'. That landscape may not still be being ploughed into prehistoric fields, but it cannot be dismissed as not 'continuing' for that reason, especially when such land-use moved into the past tense two and a half thousand years ago. The downs continue, as they have done since later prehistory, with their truly traditional land-uses, which are pasture, recreation and racehorse training (Plate LXVIII).

Inscription on the World Heritage List endows a place with a significance which raises it above purely local considerations, yet at the same time charges local interests with a considerable responsibility on behalf of the world community (Boniface and Fowler 1993; Boniface 1995). Everybody, in other words, who deals with the site should do so with respect, as indeed the management plan now emphasises (English Heritage 1998, para 10, objective B). Obviously this applies in particular to public bodies, visitors and local residents. The importance of World Heritage status is conveniently outlined in the *Operational Guidelines* (World Heritage Committee 1995). Two of its paragraphs are particularly important for our area:

> The cultural heritage and the natural heritage are among the priceless and irreplaceable possessions, not only of each nation, but of mankind as a whole. The loss, through deterioration or disappearance, of any of these most-prized possessions constitutes an impoverishment of the heritage of all the peoples in the world (para 1).

and

> The Convention provides for the protection of those cultural and natural properties deemed to be of outstanding universal value. It is not intended to provide for the protection of all properties of great interest, importance or value, but only for a select list of the most outstanding of those from an international viewpoint (para 6 (i)).

The World Heritage guidelines are placed in a wider and more practical context by Feilden and Jokilehto (1993).

Plate LXVII Overton Down experimental earthwork, July 1960, on completion of its construction at the start of the experiment to study change through time, planned to last 128 years until 2088 (see page 81)

Meanwhile, it flies in the face of logic and intellectual rigour that one half of the northern part of our study area, Overton Down, should carry the status of World Heritage, while its culturally integral eastern neighbour, Fyfield Down, does not.

Understandably, reservations exist about the argument that this status and its consequences should take priority over other policies within the Avebury World Heritage Site; yet the problem is common to most such sites. It can be overcome, given time. Hadrian's Wall, for example, another World Heritage Site, already has a management plan which looks at time frames over the next five and thirty years (English Heritage 1996), a model followed at Avebury. For such attitudinal developments to occur in managing the Avebury landscape, three changes are highly desirable:

- i Avebury should be separated from Stonehenge and become a World Heritage Site in its own right;
- ii the Avebury World Heritage Site should be reclassified as a 'cultural landscape';
- iii the boundaries of the Avebury World Heritage Site should be very carefully redrawn.

The case for such changes is outlined in Fowler and Stabler (forthcoming) and discussed in Fowler (1999). The third change proposed here concerns the spatial definition of the Avebury World Heritage Site, an issue postponed by English Heritage (1998, objective H). Fyfield and Overton are vital to considerations of any redefinition of the area of the World Heritage Site, specifically its limits on the east along the boundaries of Fyfield and Clatford. To omit Totterdown and Fyfield and Manton Downs (Figure 2.1; Plates VII, VIII, IX, XVI, XIX, XXII and XXXVIII) at this stage seems perverse. How, after all, can management decisions be made unless your boundaries are secure? On the east, they are neither secure on the ground nor in terms of intellectual justification.

The World Heritage Site's present eastern marches include quite large areas of the two parishes (Figure 17.1). Despite the use of the base map commissioned for this project (English Heritage 1998, fig 2, here Figure 2.1), a blind eye is turned without informed comment to the fact that that boundary is now known to cut across a recorded cultural landscape, excluding huge areas of landscape that are physically and contemporaneously part of the landscape that is included. The present boundary is now shown to make no sense either archaeologically or topographically, tenurially or managerially. For most of its way it does not follow a line which is readily identifiable on the ground. It runs right across the ancient landscapes and the straightforward archaeology of Fyfield and Overton Downs in a way that is both impractical and intellectually unacceptable. It splits the area of the Scheduled Ancient Monument and of the National Nature Reserve, as well as ignoring farm boundaries and therefore the opportunity to work with the occupiers of the land.

A new eastern boundary is, therefore, required, either pulling right back to somewhere much closer to Avebury or facing the landscape reality adduced in this volume. Neither the World Heritage Committee nor some considerable body of public opinion would now be likely to accept the first option which, in any case, retreats from the original idea of including some of the area of sarsens from which the Avebury megaliths presumably came.

One obvious conclusion to draw from this study is that the second option is preferable: in other words, that the eastern extent of the World Heritage Site should be enlarged to a culturally meaningful boundary (Figure 17.3). The World Heritage Site so redefined on its east would be justifiable academically, defensible politically and workable managerially. It would also make a big step forward towards meeting one of the objectives of World Heritage: it would make sense in terms of presentation and interpretation. The site – over and above its closely related sarsen stone and megalithic interest – would include an extent of landscape that was explicable as an entity in terms of its visible and known archaeology and history.

FINALE

World Heritage Sites, now numbering well over 500, have become (perhaps elsewhere more than in Britain) one of the icons of the late twentieth century. We create our own versions, not so much of the several hundred Wonders of the Modern World as of 'sacred places' and even 'sacred landscapes'. The core of our study area was in, or on the edge of, an undoubtedly sacred landscape some 4,000 years ago, though we can but wonder about the sort of sanctity. We can also wonder what was really going on in the late twentieth century AD when, beside one of the nodal points in that landscape, the world's largest conservation organisation bought a transport café to demolish it and restore its site and lorry park to grass.

A landscape of designation is clearly not just lines on maps. It can, unfortunately, be a proliferation of

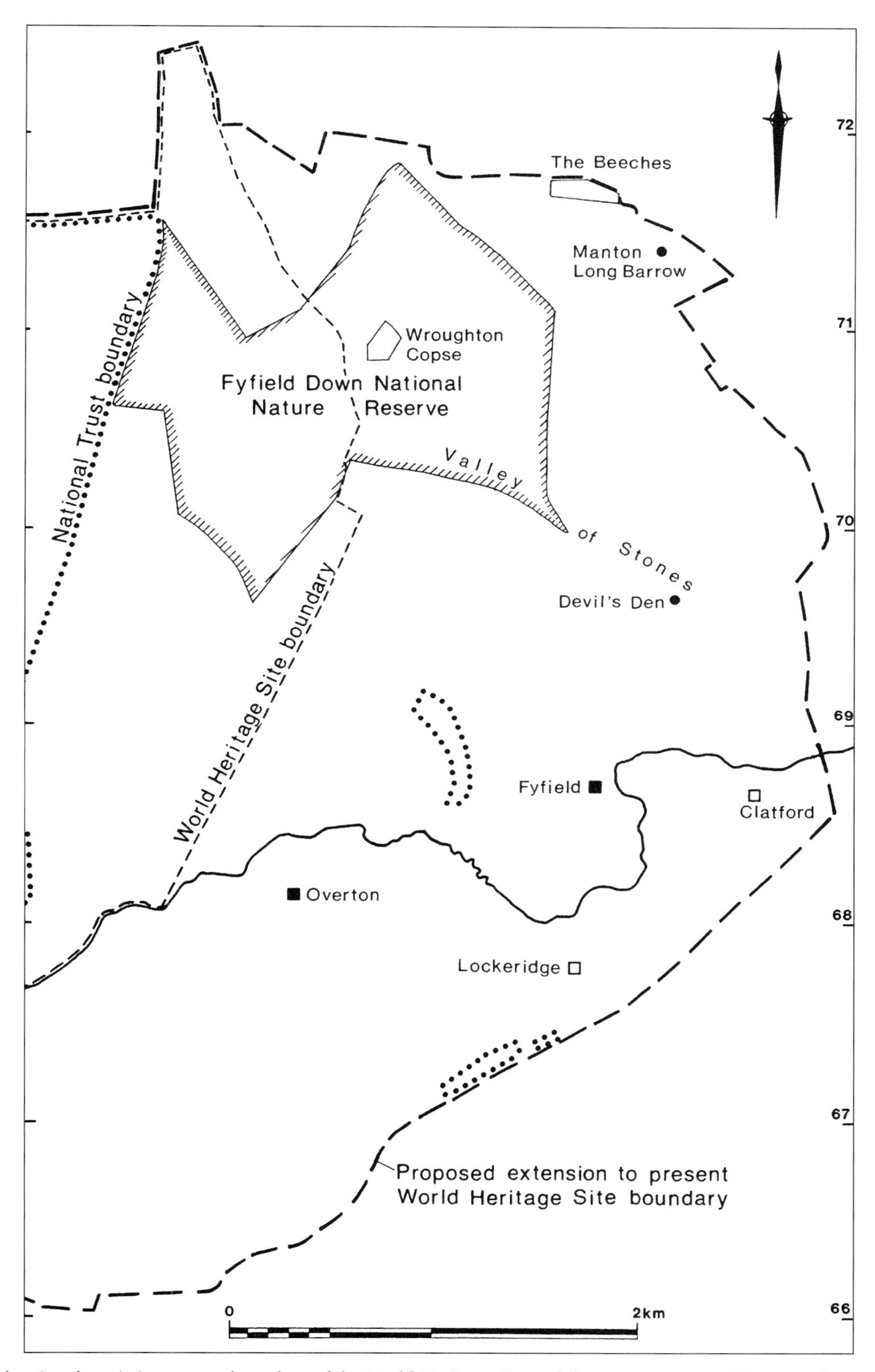

17.3 *Map showing the existing eastern boundary of the World Heritage Site and the new one proposed here to take account of the matter in this volume*

countryside labels, but it can also, and more significantly, be something not there: it is grassland where there was ploughland; it is derelict buildings that no one dare touch; it is downland so precious you hardly dare breathe. Perhaps it was previously so in another twentieth century. Perhaps the idea of a 'sacred geography' – for that is what we are creating in our 'landscapes of designation' – has quietly persisted, to re-emerge after 4,000 years. Is that just conceivably one interpretation of what was and is on Overton Hill? The site of a transport café 'killed' in pursuit of a contemporary ideology beside a Neolithic 'Sanctuary' reconstituted in concrete in the 1930s AD; near burial mounds of prehistoric, Roman and Anglo-Saxon date and a battle site, cemetery and boundary of a thousand years ago; near the crossings of Roman road, modern trunk road and The Ridgeway, and at a place which is now the starting point, or the end, of a national trail intended for anyone to journey across landscapes?

We end (echoing T S Eliot) where we began, in this case high on the downs. We asked a long time ago: 'To what extent do Fyfield and Overton Downs typify the chalk downlands of southern England?' It is an important question, for in the answer to it probably lies some justification, if such there has to be, for this increasingly personal landscape journey. Geologically, geomorphologically and, in some respects, botanically, these downs are unusual, even remarkable, as their NNR and SSSI status indicates (English Nature 1991; English Heritage 1998, appendix E, 160–1, a formal description of Fyfield Down SSSI): so that is one answer. Archaeologically, they are outstanding in their state of preservation: so that is another. These downs are therefore not typical on at least two counts.

Yet one initial premise of this project was that the study was worth carrying out precisely because – and despite the strikingly visual and extensive nature of the earthworks – the archaeology was indeed typical of what had once been common over other downland areas in Wessex. Nothing has happened to change that view; rather has it been reinforced, not least by relating Fyfield and Overton Downs to their environmental context and the dynamics of manorial history.

Minor caveats exist. These downs do not contain, as far as is known, examples of certain types of site recognised elsewhere on other downs, such as a henge, a cursus or a 'banjo' enclosure. Nor, in general, do they contain unique examples of types of sites unknown elsewhere, though it is difficult at the moment to think of exact parallels for the Roman barrows by The Ridgeway (Smith and Simpson 1964), the settlement sites OD XII and by Wroughton Copse, or the Down Barn and Delling Enclosures. Those exceptions are all, it will be noted, post-prehistoric, suggesting that the prehistoric field archaeology of these downs, though without some monument types, is to a useful degree typical. Rather than further suggesting that the area is especially significant for its Roman and later field archaeology, that list probably indicates that the study of downland field archaeology of the last two millennia is less well advanced than for earlier times. Overall, though, these downs are a very good display case of Wessex field archaeology, and excavation has shown their buried archaeology to be both typical and a valuable scientific resource.

Whether the history is typical is a different matter. In detail, these downs doubtless enjoy their own idiosyncrasies of periodical developments, as does anywhere. The late prehistoric grassland, Roman field systems and medieval settlement witnessed here do not all occur on every piece of Wiltshire downland and certainly not in stratified succession. But each is known individually elsewhere. Generally, it can only be guessed that, overall, the downland history here reflects a common experience. That seems to be the case from 1086 onwards (*VCH passim*; Hare in Chapter 9 and FWP 43), and it is to be expected that a similar ebb and flow in the extent and nature of land-use occurred elsewhere in earlier millennia. In pursuit of that point, it is hoped that the interpretation of the history of Fyfield and Overton Downs proposed here can act as a model, a test-bed, for histories of downland elsewhere.

The fundamental point, however, is the existence not just of lots of old earthworks spread out across the landscape but of their existence on the ground in a matrix of chronological, functional and cultural relationships. The surface of the downs is like a palimpsest, impressed with the evidence of how people have been using this land for thousands of years. This project may not have come up with the correct answers nor with the right interpretations but it has demonstrated that we can do more than just look and wonder: we can actually sort out in a reasonably rational way the sort of stories that such a palimpsest can tell. That is a very important piece of knowledge in itself, grasped at by many before we began and amplified many times over since.

The Fyfield and Overton palimpsest and the range of sites and features that are its components are indeed typical of what was once common on the Wessex Downs, though it would require a separate discussion as to whether that 'once' was before 1660, when Aubrey was complaining about archaeological destruction, before the 1870s and 1880s when Long and Smith were horrified at what was happening to the downs, or in the 1920s and 1930s in field archaeology's 'golden age' of agricultural depression when Crawford viewed and Grinsell strode the turf-rich Wessex Downs (Ashbee 1972, 64–5, 69; Crawford 1955; Fowler 1997; Grinsell 1989, 9; Long 1862; Smith 1885; Ucko *et al* 1991). Nevertheless, it is a typicality now demonstrated not just by what is known of these downs but also by a plethora of other landscape-oriented archaeologies throughout the generation spanned by the project. These include the RCHME's air photographic exposition of the Danebury area, Hampshire (Palmer 1984), and its voluminous coverage of Dorset (1952–75), followed up on the south Dorset Ridgeway by Woodward (1991) and in Cranborne Chase by Bowen (1990) and Barrett *et al* (1991a, 1991b), and by the RCHME's similar fieldwork (partly forthcoming) in Wiltshire, where a considerable amount of recent and current work makes the same point (*WAM passim*).

In general, then, Fyfield and Overton Downs are not archaeologically unusual. It is that typicality which makes them so precious now, not just in the fact of their own archaeological survival but also because so much of that which they typify is no longer accessible. Sadly, much of the field archaeology of the Danebury area, for example, or the south Dorset Ridgeway, or Cranborne Chase or whole swathes of Wiltshire's downland, especially the Marlborough Downs, no longer exists. A lot of it has been ploughed over during the very same decades that have seen Fyfield and Overton Downs become a less-common habitat and consequently more valuable as representing typical former land-use. They now possess a rare archaeological value, of national, and probably at least of European, significance, in that they are both unique unto themselves and typical of former extents of chalk downland, not only locally but, more generally, of Wessex. Fyfield and Overton Downs, now viewable in at least facets of their parochial context, not only preserve the typical but also make available a telescoped version, an encapsulation, of the landscape archaeology of the Wessex Downs (*see* frontispiece). With the rest of their parishes, they represent, illuminate and, thankfully, question, the archaeology and history of the whole of the Wessex landscape, both on and off the downland, plotted and pieced.

Plate LXVIII 'The Derby Gallop': the south end of the racehorse-training course on Overton Down in 1996, looking north towards Delling Wood

Appendix 1

The Archive

PRINCIPLES AND PRACTICE

The idea of an archive was built into the project from the start. Although a private initiative, the project (1959–65) was initially carried out from a base in the Salisbury office of the Royal Commission on the Historical Monuments of England (RCHME), an organisation with its own high standards and practices of research, recording and archiving which were consciously carried over into the Fyfield and Overton work. The fieldwork was also supported to some degree by official resources so that, for example, some of the photography at the Wroughton Copse excavation was carried out by Commission staff and photographs were therefore automatically lodged in the National Monuments Record (NMR; now the National Monuments Record Centre in Swindon) from an early stage of the project.

When the programme to publish the project was begun in 1995, preparing a publicly accessible archive was quite as much a part of the brief from English Heritage as preparing a printed volume. Indeed, not only has preparing the former taken quite as much time and effort as the latter, but both were early accepted, in concept and practice, as part of the same process. The aim was defined as producing an entity, the whole of which, irrespective of its media, made the project, its data and materials as accessible as seemed appropriate to its various potential users. In December 1995 (FWP 80), the proposal was made that, such was the bulk of material, some should be published electronically; but at that stage the cost was too great. Advances in technology, and therefore in lowering the cost, were such, however, that by 1998 it was possible to implement as well as conceive the idea of making more material more widely available than would be the case with only conventional printed publication and a static archive. This development brought nearer both making an archive relatively easily accessible and seeing an archive existing in several media as part of the publication of the project in a wider sense than is sometimes attempted.

The result is an output arranged hierarchically by its nature, and accessible in different media, partly as appropriate to the material but also as designed to meet different needs:

1. The primary archive at the Museum of the Wiltshire Archaeological and Natural History Society (WANHS), at Devizes, Wiltshire (DM)
2. Material available electronically at *www.ads.ahds.ac.uk/catalogue/*
3. Material available in print

EXPLANATION

The primary archive

The primary archive consists of a fairly comprehensive paper record from 1959 onwards, all the existing excavated material and a select photographic collection.

(a) The *paper records* are themselves subdivided into:
   - i a graphic component in a plan chest containing 526 field drawings and subsequent working drawings – a collection which has been catalogued (FWP 61); to which has been added the artwork for the illustrations in this volume and in FWPs 63–66 (*see* below for explanation);

ii a largely verbal component arranged in box files which, when finalised, will be catalogued as an addition to FWP 61.

(b) The *excavated material* is systematically arranged in numbered boxes, currently on racking in the museum basement, apart from the material on display. The contents of the boxes have been catalogued (FWP 89).

(c) In addition, a *photographic archive*, also catalogued (FWP 97), is deposited at DM but storage requirements may mean that negatives and transparencies finally come to rest in the NMR. Details will be documented at both places. Some photographs taken 1959–65 are already in the NMR, coded with an ISF prefix when taken by the author, archived under parish name otherwise.

The electronic archive

The electronic archive is managed by the Archaeological Data Service (ADS), York University (*see* note by Robinson and Richards below), and is available at *www.ads.ahds.ac.uk/catalogue/*. It consists of:

(a) Conventional excavation reports

Four volumes of conventional, illustrated excavation reports covering the three main excavations and, in a fourth volume, all the smaller excavations:

Fowler, P J 2000. 'Excavation within a later prehistoric field system on Overton Down, West Overton, Wilts: land-use over 4,000 years' (FWP 63, *c* 50,000 words)

Fowler, P J 2000. 'Excavation of a settlement of the fourth and fifth centuries AD on Overton Down, West Overton, Wilts' (FWP 64, *c* 70,000 words)

Fowler, P J 2000. 'Excavation of the medieval settlement of *Raddun*, Wroughton Mead, Fyfield Down, near Marlborough, Wilts' (FWP 65, *c* 45,000 words)

Fowler, P J 2000. 'Seven small excavations on Fyfield and Overton Downs, with summaries of eighteen other excavations in and near Fyfield and West Overton parishes, Wiltshire' (FWP 66, *c* 15,000 words)

Much, but not all, of this material is in FWP 75. The above four reports were edited from that volume as 'stand-alone' conventional excavation reports when it became apparent that the volume was too bulky to print. They include all the illustrations as prepared for that draft volume, plus some new ones, all now in digitised form. With present technology, any one of them will take only seconds to download. Paper copies of any one drawing can be obtained from the original artwork or copies in DM. Bound paper copies of the four volumes have been deposited in various libraries (*see below*).

(b) 'Fyfod Working Papers' (FWPs)

Beginning as the drawing together of data and drafts of parts of early versions of a volume in 1995, these in-house papers quickly became, first, 'building blocks' in working towards a single major volume and then a device for both handling ideas and material and keeping an electronic record of the development of the publication and archival programme and spin-offs from it. There are, or will be, one hundred FWPs in all.

Many of the FWPs were revisited several times and went through numerous revisions; they were never intended for wide publication, though it was envisaged that floppy discs of such material would be part of the archive deposited at DM and in the NMR. With the development of appropriate technology, however, an accessible electronic archive became not only possible but, warts and all, even attractive. Although we retain the original numbering of FWPs, which reflects spurts of activity in particular directions, the later FWPs represent a conscious attempt to round off this part of the archive in a structured way – and indeed in a way which allows for the archive to be added to.

A full catalogue of the FWPs, with bibliographic and other information, is in FWP 33. A select subject index to ninety-nine FWPs is in FWP 100. Four early FWPs (1, 5, 8 and 9), prepared before the project had use of a personal computer, are available only in paper in the primary archive at 1/a/ii above. FWPs 30, 31, 38, 39, 40 and 41 now exist only in hard copy in the same archive because they were deconstructed on disc and their components redistributed appropriately as elements of several different chapters during the numerous reorganisations of the monograph's structure 1995–8 (*see* FWP 12).

Otherwise, the contents of one hundred FWPs, some with various additions and revisions, are available at *www.ads.ahds.ac.uk/catalogue/*. A complete list of papers follows.

1 OD X and OD XI: an overview
2 The lynchet and 'fence' posts of OD XI/A and OD XI/B
2a OD XIA/B/C: post-holes under the lynchet
2b OD XIC: features analysis

2c OD XI pits
3a Areas OD XI/A South 1
3b OD XI/A South 2
4a OD XI: towards a text of the excavation report
4b OD X: towards a text of the excavation report
5 OD XII: preliminary analysis of general finds numbers
6 A potted history of Fyfield, Wiltshire
7 *Raddun*: the documentary evidence
7a *Raddun*: the documentary evidence
8 Wroughton Copse medieval site (WC): *Raddun*, some archaeological implications
9 Draft text of miscellaneous excavations (FWP 66)
10 OD XII: finds distributional analysis
11 East Overton Saxon boundary S449
11a West Kennet Saxon boundary S547
12 Monograph structure
12a Monograph structure
12b Monograph structure
12c Monograph structure
12d Monograph structure
13 OD X and OD XI Interpretation: three models
14 Intersection of gullies 6 and 8 OD XI/A/East 3
15 List of illustrations for publication
15a List of illustrations for the monograph
15b Final list of illustrations for the monograph
16 OD XI/Area East 1: analysis of features and finds
17 OD XI: summary of small finds and pottery results from TWA Finds Report (FWP 38a–c)
18a Sandford charters
18b Report on the Templars in Rockley and Lockeridge
19 OD XII: features inventory
20 OD X/15: cutting 15
21 Thoughts on OD XII
22 WC: initial analysis of finds and features (*see* FWP 92 for final data)
23 Field archaeology: its data and place
23a The field archaeology of the Down Barn area
24 The villages
24a Village houses and their dates
25 Bibliographic work
25a Bibliography
25b Bibliography
25c Bibliography (as in *LPP*)
26 Morphology of West Overton village
26a Morphology of villages
27a LBA/EIA abstracts
27b Important themes, LBA/EIA
28 OD XI East 2 and 3
29 Environmental evidence
29a Environmental evidence
30 Ridgeway assessment report (plus SMR inventory)
31 Flint report
31a Summary of flint report and interpretation
32 OD XII: initial text and interpretation
32a OD XII: draft text
33 List of Fyfod Working Papers
34 OD XI: table of pits
34a OD XI: table of pits
35 Aerial cartography: draft discussion of RCHME AP map, 1995
36 Early Chapter 14 (environment)
37 Early Chapter 16 (general discussion)
38a TWA Finds Report OD XI: tables
38b TWA Finds Report OD XI: archive catalogue
38c TWA Finds Report OD XI: text
39a TWA Finds Report ODXII: tables
39b TWA Finds Report OD XII: archive catalogue
39c TWA Finds Report OD XII: text
39d TWA Finds Report OD XII: comments
40a TWA Finds Report WC: tables
40b TWA Finds Report WC: archive catalogue
40c TWA Finds Report WC: text
41 TWA Finds Report: miscellaneous sites
42a OD XI ard-marks: record note
42b OD XI ard-marks: description and analysis
43 Medieval landscape history: with commentary
44 Documentary analysis of West Overton, East Overton, Lockeridge and Fyfield
45 Documentary analysis: Lockeridge
46 Documentary analysis: East Overton
47 Documentary analysis: Fyfield
48 Documentary analysis: West Overton
49 Miscellaneous excavations: early/draft accounts
50 List of photographs for the monograph
51 Window 1: 'The High Downland' (*see* FWP 75)
52 Windows 2, 4 and 5 (*see* FWP 75)
53 Window 3 (*see* FWP 75)
54 Windows 6 and 7 (*see* FWP 75)
55 Window 8 (*see* FWP 75)
56 FWPs related to miscellaneous excavations
57 FWPs related to OD X and OD XI
58 FWPs related to OD XII
59 FWPs related to WC
60 The archive: objectives and explanations
61 Archive of original drawings deposited in Devizes Museum
62 The Beaker accompanying Burial 1a, OD XI
62a The Beaker skeletons
63 Excavation report OD XI
64 Excavation report OD XII
65 Excavation report WC

66 Reports on smaller excavations
67 Society of Antiquaries lecture, 27 November 1997
68 Analysis of Saxon charters: the tenth-century Saxon landscape
69 OD XI and OD XII: layer descriptions for illustrations in monograph
69a Layer numbers and descriptions for OD XII illustrations
70 Layer numbers and descriptions for illustrations used in Chapters 3–7
71 Pottery analysis for OD XII
72 SMR data for West Overton and Fyfield parishes south of the Roman road (*LPP* figure 12.1; cf FWP 78)
73 References in *Wilts Archaeol Mag* for study area
74 July 1996 monograph text
75 July 1997 monograph text
76 Translation of 1567 Pembroke survey
77 WC pottery database (*per* TWA 1995: a wdb file)
78 SMR data for area (incl parts of Preshute, Savernake, Marlborough and Mildenhall parishes)
79 Captions to the plates (for FWP 75)
80 Proposal to archive an 'Accessible public archive'
81 'The Ridgeway incident'
82 The pottery finds from TD VIII and IX, and TD I–III
83 The finds from Shaw
84 'Crimean War' cultivation on Overton Down
85 AP catalogue, incl RCHME database
86 AP report (RCHME, rev edn, 12/98)
87 Small mammals report, OD XI and XII
88 Cultural landscapes
89 Catalogue of finds … as boxed in Devizes Museum
90 'Moving through the [Fyfield] landscape'
91 'Wansdyke in the woods'
92 WC archive
93 *The Land of Lettice Sweetapple. An English Countryside Explored* (published August 1998, Tempus, Stroud): a list of the contents and all the illustrations contained in the book, with captions of the colour plates
94 Note on the Bayardo Farm flint and pottery collection
95 Note on the coins from OD XII
96 The glass from OD XII: report and catalogue
96a Note on the glass from OD XII
97 Catalogue of the photographic archive
98 Bibliography of Fyfield and West Overton: the primary publications
99 Summary of the project's 'total product' and its archival deposition (Appendix to LPP)
100 A select, subject index to the preceding 99 FWPs

PRINTED MATERIAL

(a) 'Popular'

Fowler and Blackwell 1998 = P Fowler and I Blackwell, *The Land of Lettice Sweetapple. An English Countryside Explored*, Tempus, Stroud

(b) 'Academic'

Fowler 2000 (this volume) = P J Fowler, *Landscape Plotted and Pieced. Landscape History and Local Archaeology in Fyfield and Overton, Wiltshire*, Society of Antiquaries, London

(c) 'Research/archival'

P J Fowler assisted by I W Blackwell, *Landscape Plotted and Pieced. Landscape History and Local Archaeology in Fyfield and Overton, Wiltshire*, July 1997 (though with '1998' optimistically but misleadingly on its front cover), produced in-house by the Computer Centre, Newcastle University.

This was an unedited draft of the then proposed monograph in six hard copies circulated to readers. A copy will be deposited with the paper archive in DM, and in the library of the Society of Antiquaries, London. The present monograph is essentially a reordered, partly rewritten and significantly reduced version of this draft. Its contents were as follows (with word count in thousands in brackets; total word count for main text: 194.6):

Frontispiece
List of illustrations
Acknowledgements
Editorial notes
A note on the archive
Summary

Part I A landscape and its investigation
1 Landscape in a locality (9.9)
2 Aerial cartography: Avebury, Overton and Fyfield Downs (12.7)
3 Old land-use on old grassland: the high northern downs (7.6)
4 Old grassland, former fields: Fyfield Down (4.6)
5 A medieval settlement: *Raddon* in Wroughton Mead (23.1)
6 Preceding old pasture: four millennia on Overton Down (23)
7 A dene in the downland: *pytteldene* and Down Barn (18.2)
8 Old arable, pasture and parkland: from The Ridgeway to Headlands (8.6)
9 Lands to the South: of valley, wood and heath (6.2)
10 Shaw and West Overton: landscapes of two small estates (10.2)
11 The manor of East Overton: landscape with people (10.6)
12 Lockeridge and Fyfield: landscape and landlords (13.4)

Part II Landscapes and interpretations
Introduction
13 Environmental history (14.1)
14 Archaeology and the landscape (14.6)
15 Time and themes in a local landscape (11.4)
16 A once and future landscape: issues and objectives (6.4)

Location of the project archive
Bound paper copies of the four excavation volumes have been deposited in the following libraries: the Wiltshire Archaeological and Natural History Society, Devizes Museum, Wiltshire; the Trust for Wessex Archaeology, Salisbury; the Ashmolean Museum, Oxford; the Society of Antiquaries of London at Burlington House, Piccadilly, London; the NMR, Swindon; the Institute of Archaeology, University College London; and the Universities of Bristol, Reading and Southampton.

Individually made paper copies of any or all of these excavation reports can be obtained from either the NMR or the ADS at York University, as single orders at prices to be negotiated with the supplier. (The approximate size of the volumes is indicated by the wordage as noted, and by the fact that all except FWP 66 contain some thirty-five to forty line drawings.)

Other printed material
The study area has also generated a considerable bibliography quite apart from this study. Primary printed sources and significant studies are listed in FWP 98, and most of the useful authors are listed alphabetically with their publications in the Bibliography in this volume. In particular, *see* Bonney, Bowen, Evans, Featherstone *et al*, Fowler, Fox and Fox, Free, Gingell, Greatrex, Green, Grinsell, Kempson, King, Lacaille, Meyrick, Smith and Simpson, and Swanton.

This appendix is also available as FWP 99.

The Digital Archive

Damian Robinson and Julian Richards
Archaeology Data Service, Department of Archaeology, University of York

A major product of the Fyfield and Overton project is an integrated monograph and Internet publication. In combination with the World Wide Web, readers are able to move from the high-level interpretations contained in this monograph to the minutiae of the data held in the digital archive. The project digital archive has been deposited with the Archaeology Data Service (ADS) and can be remotely accessed via its catalogue, ArcHSearch (online: *www.ads.ahds.ac.uk/catalogue/*).

There are several pathways into this digital part of the Fyfield and Overton archive. For example, ArcHSearch contains many thousands of index records describing individual archaeological projects and interventions. These may be searched via the Keyword search button, which is located in the frame on the left-hand side of the screen (Figure A). A Keyword search on Fyfield, for example, would return several different resources; one of these is entitled 'Fyfield and Overton Project'. Selection of this resource would take the user to the Project page from which elements of the digital archive may be downloaded or viewed online.

The archive can be accessed most directly via the Excavation and Fieldwork Archives section of ArcHSearch, by clicking on the Project Archives button. This is also located in the frame on the left-hand side of the screen (Figure A). Once in the Excavation and Fieldwork Archives section, the Fyfield and Overton Project Archive may be selected, which takes the user to the Project page.

The digital archive facilitates the long-term secure storage and dissemination of the one hundred Fyfod Working Papers (FWPs, catalogued above). The four volumes of excavation reports (FWPs 63–66) are available in html format. This allows the text to be viewed over the Internet and enables links to be made between text, illustrations and other FWPs. The remainder of the working papers are available simultaneously as Microsoft Word, plain text and html file formats.

The Fyfield and Overton digital archive is an integral component of the project's dissemination strategy (*see* the introduction to this Appendix, above). The digital archive provides the opportunity to remove the detailed technical data from the main body of the published monograph and locate it in a universally accessible environment in order to facilitate its scholarly reuse, both for research and teaching purposes. This encounter between the text of the monograph and the data contained in the archive will ensure that the interpretative process is not fossilised at the moment of the monograph's publication but will become dynamic and recursive in nature.

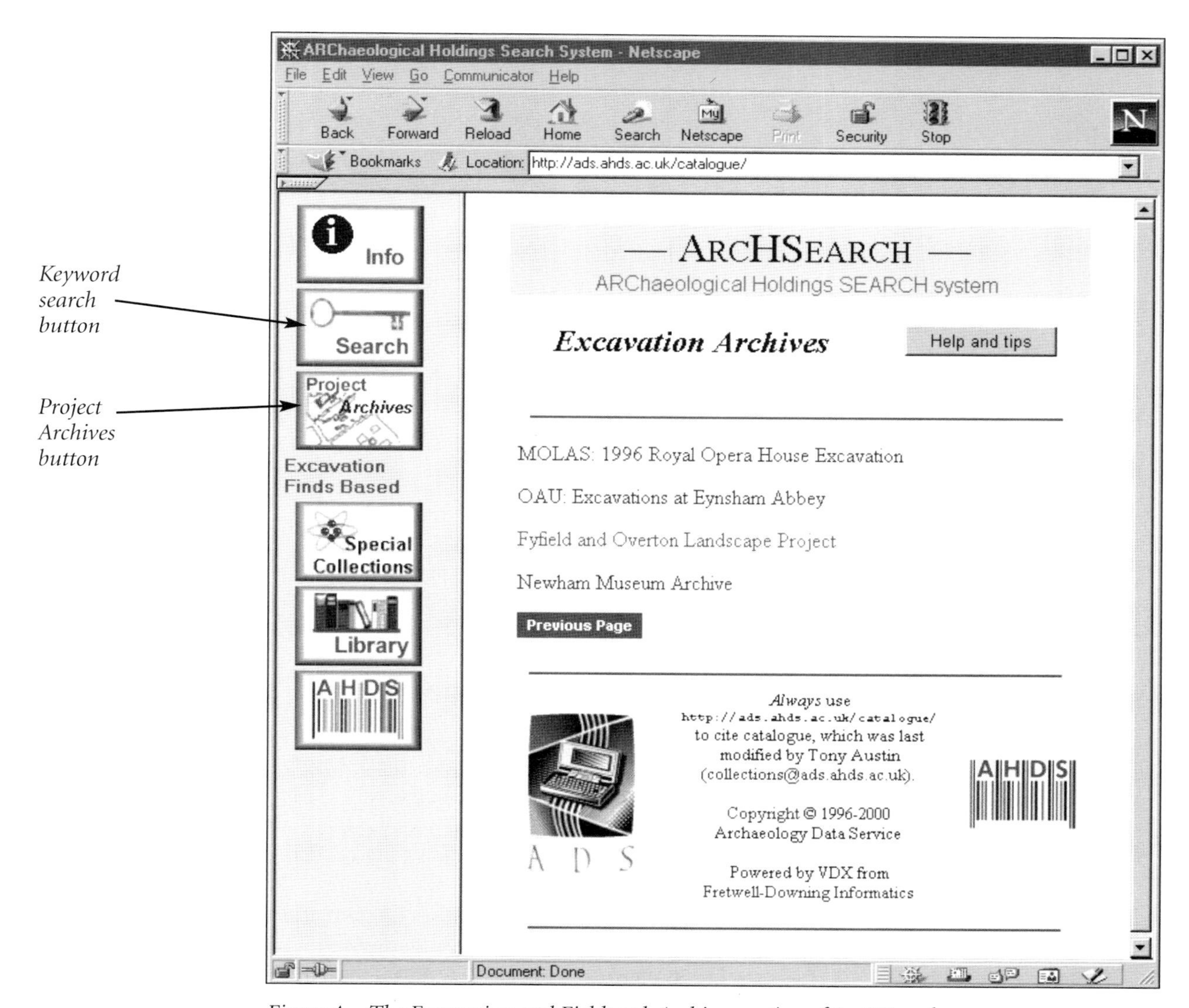

Figure A The Excavation and Fieldwork Archives section of ArcHSearch

Appendix 2

Layer Numbers and Descriptions used in Chapters 5 to 7

The following layer numbers and descriptions are used on the section drawings relating to excavations undertaken during the project. They are to be used in conjunction with the standardised graphic symbols provided in the Key (*see* page 2).

EXCAVATIONS AT TOTTERDOWN, FYFIELD DOWN AND WROUGHTON COPSE

1	Turf and topsoil
1a	Humus
1b	Turf and clayey humus
1c	Buried turf
2	Humus and small flints
3	Blackened soil
4	Brown clay and flints
5	Brown soil with many flints
5a	Flints in light brown loam
5b	Light brown soil with chalk lumps
6	Blackened soil with flints
7	Dark earth with flints
8	Brown loam with small flints
9	Reddish clay with small flints
10	Chalk lumps with clay
11	Gingery brown loam
11a	Gingery loam with flints
11b	Flints with some gingery loam
12	Dark earth with flints and charcoal
13	Clayey loam with a few stones
13a	Chalk lumps in clayey soil
13b	Dirty chalk with clay, soil and some flint
14	Brown soil
14a	Brown soil with burnt chalk nodules
14b	Light brown soil with chalk fragments
15	Humus, flint and small chalk rocks
16	Chalk rubble: (a) small; (b) medium; (c) large; (d) with traces of clay
17	Cob
18	Black burnt soil, stones and charcoal
19	Burnt flints
20	Burnt cob
21	Charcoal
22	Brown soil with chalk and charcoal flecks
23	Flints: (a) small; (b) medium; (c) large
24	Black soil
24a	Pale grey soil
24b	Dark grey soil
25	Black earth, chalk, charcoal, sarsen and burnt flint
26	Brown-red clay with chalk and flints
26a	Orangey, silty soil with chalk lumps
26b	Silt with chalk lumps
27	Cob with dark humus
28	Red clay
29	Dark brown soil with flints and chalk lumps
29a	Primary silt and dark humus with chalk flecks
30	Brown soil with flints and chalk lumps
31	Brown soil with charcoal
32	Weathered chalk with some clay and sarsens
32a	Chalk with clay and charcoal
32b	Chalky soil
32c	Flints and chalk lumps
33	Clayey soil with flints

33a Grey spill, large flints and chalk
33b Grey clay with charcoal and small flints
34 Red clay with chalk lumps
35 Black earth with charcoal and some clay
36 Clay with chalk lumps and flint
36a Silty clay with chalk flecks
37 Brown soil with flints and charcoal
37a Brown soil with flints, chalk and charcoal
38 Dark soil with charcoal
38a Dark soil with chalk
38b Soil with chalk flecks
38c Dark brown soil
39 Brown clay
39a Brown clay lens
40 Clayey soil
40a Light brown clayey soil with flints and some charcoal
40b Clayey soil with a few flints
40c Clayey soil with chalk lumps
41 Dark brown stones with some flints and chalk
42 Light brown clay with chalk and some large flints
43 Brown soil with flints
44 Clay and flints
44a Clay and large flints
45 Clay with flint chips
46 Brown clay with flint chips and chalk tip lines
47 Blackened clay
48 Light brown soil with a few flints
49 Clay-with-Flints (subsoil)
49a Rotten chalk
50 Chalk (subsoil)

EXCAVATIONS AT OD X AND OD XI

1 Topsoil
2 Flints
2a Small flints
2b Small flints in chalky light brown soil
2c Flints and sarsens
2d Flints and chalk lumps
3 Chalky humus
4 Humus with flints, sarsens and chalk lumps
5 Humic layer
6 Grey-brown soil
6a Grey-brown soil with large chalk lumps and small flints
6b Grey-brown soil with flints and chalk flecks
7 Grey-brown soil with small chalk lumps
8 Brown soil
8a Brown soil with flints
9 Brown soil with chalk lumps
9a Brown soil with burnt chalk
9b Brown soil with chalk flecks
10 Dark soil
11 Dark soil with flints
12 Fine dark brown soil with chalk lumps and large flints
13 Grey-brown soil with chalk lumps and flint chips
14 Clean loose chalk lumps
15 Loose chalk lumps
16 Chalk rubble
16a Large chalk rubble
16b Medium chalk rubble
16c Small chalk rubble
16d Chalk powder
16e Chalk rubble with flints
17 Dark grey-brown soil with lines of chalk
18 Chalk soil
19 Dark soil with flints and charcoal
20 Dark humus
21 Dark soil with chalk lumps
21b Dark soil with flints and chalk lumps
22 Silt and chalk lumps
22a Silt and charcoal
23 Dark brown soil with chalk lumps and flint chips
24 Charcoal
25 Grey chalk
26 Clayey soil, round flints and sarsens
26a Clayey loam and flints
27 Chalk lumps with some soil
28 Compact chalk
29 Soil with chalk lumps
30 Soil with flints and chalk
31 Dark soil with chalk lumps
32 Grey soil with chalk and flint
33 Chalk lumps in silt
34 Brown soil with chalk and charcoal flecks
35 Light brown soil
35a Light brown soil with chalk lumps
35b Light brown soil with chalk and flints
35c Light brown soil with chalk flecks
36 Soft grey brown soil
37 Disturbed soil with chalk lumps
38 Flint in soil
39 Light grey soil with chalk flecks
39a Chalk lumps and light grey soil
39b Light grey soil with flints
39c Dark grey soil with chalk flecks
40 Clay
40a Yellow clay
40b Orangey clay
40c Chalk and yellowy clay

40d	Weathered chalk and light brown clay
40e	Burnt clay
41	Brown soil with chalk grains
41a	Brown soil with chalk grains and charcoal
42	Dark grey soil
42a	Grey soil with chalk grains
42b	Grey soil with chalk lumps
42c	Grey soil with chalk and flints
43	Brown loam
43a	Brown loam with small flints
43b	Brown loam with chalk and flints
43c	Brown loam with flints and sarsens
43d	Brown loam with chalk flecks and lumps
43e	Brown loam with chalk lumps
43f	Brown loam with large sarsens
43g	Brown loam with charcoal flecks
43h	Brown loam with chalk lumps and flint packing
44	Fine grey soil
44a	Grey soil
45	Brown soil chalk and flints
45a	Pinky brown soil with chalk lumps and flints
46a	Clay silt
46b	Brown silt
46c	Brown silt with chalk lumps
46d	Grey silt
46e	Chalky silt
47	Lens of black soil and charcoal
47a	Black soil and burnt sarsens
48	Red-brown soil with chalk lumps
49	Rotten chalk
49a	White decomposed chalk
49b	Orangey decomposed chalk
50	Chalk subsoil

EXCAVATIONS AT OD XII

1	Topsoil (brown humus)
2	Dark soil with flints
3	Large flints with earth
4	Flints
5	Flints and soil
6	Brown soil and large flints
7	Grey soil with chalk lumps
8	Brown soil
9	Brown soil and chalk
10	Soil with chalk nodules
11	Rotten chalk and light brown soil
12	Light grey soil with chalk flecks
13	Flinty layer
14	Flinty soil
15	Ploughsoil
16	OGS? ploughsoil
17	Disturbed dark soil with chalk lumps
18	Disturbed dark soil
19	Chalk lumps
20	Chalky clay
21	Chalk silt
22	Light brown soil with rotten chalk
23	Dark layer
24	Clay
25	Burnt clay
26	Light brown clay
27	Humus with chalk flecks
28	Brown clay with chalk and flint
29	Brown clay and chalk
30	Silt
31	Brown silt
32	Dark soil with flint and flecks of chalk
33	Dark soil with flints and chalk lumps
34	Flints packed with clay
35	Stiff yellowish soil with flints and chalk lumps
36	Soil with flecks of chalk and some flint
37	Soil with chalk lumps
38	Loose gritty soil
48	Clay-with-Flints
49	Rotten chalk
50	Chalk subsoil

Bibliography

Allen, M J 1988. 'Archaeological and environmental aspects of colluviation in south-east England', in *Man-Made Soils* (eds G Groenman-van Waateringe and M Robinson), 67–92

Anderson, A S, Wacher, J S and Fitzpatrick, A P forthcoming. *The Romano-British 'Small Town' at Wanborough, Wiltshire: Excavations 1966–76*

Anderson, J R L and Godwin, F 1982. *The Oldest Road: an exploration of the Ridgeway*, Guildford

Andrews and Dury 1773. *Map of Wiltshire* (reproduced by Wiltshire Archaeol Natur Hist Soc, Devizes, 1952)

Annable, F K 1962. 'A Romano-British pottery in Savernake Forest Kilns 1–2', *Wiltshire Archaeol Natur Hist Mag*, 58, 142–55

Annable, F K and Simpson, D D A 1964. *A Guide Catalogue of the Neolithic and Bronze Age Collections in Devizes Museum*, Devizes

Anon 1997. 'Three Mesolithic houses discovered near Avebury', *Brit Archaeol*, 28, 4

Anon nd. *The Church of St Michael, West Overton, Wiltshire* (undated pamphlet available in the church)

Aragnou, C (ed) 1982. *Hommes et troupeaux des Pyrénées. Images des Hautes-Pyrénées* (illustrated catalogue to an exhibition, with bibliography), Tarbes, Hautes-Pyrénées

Arsdell van, R 1994. *The Coinage of the Dobunni*, Oxford

Ashbee, P 1972. 'Field archaeology: its origins and development', in Fowler 1972b, 38–74

Ashbee, P 1984. *The Earthen Long Barrow in Britain*, Norwich

Ashbee, P, Smith, I F and Evans, J G 1979. 'Excavation of three long barrows in Wiltshire', *Proc Prehist Soc*, 45, 207–300

Aston, M and Lewis, C (eds) 1994. *The Medieval Landscape of Wessex*, Oxbow Monogr 46, Oxford

Atkinson, R J C 1953. *Field Archaeology*, London

Atwood, G 1964. 'A study of the Wiltshire water meadows', *Wiltshire Archaeol Natur Hist Mag*, 58, 403–13

Baillie, M G L 1995. *A Slice Through Time: dendrochronology and precision dating*, London

Barber, K E 1981. 'Pollen analytical palaeoecology in Hampshire: problems and potential', in *The Archaeology of Hampshire* (eds S Shennan and R T Schadla-Hall), 91–4

Barber, M 1995. *The New Knighthood: a history of the Order of the Temple*, Cambridge

Barker, C T 1984. 'The long-mounds of the Avebury region', *Wiltshire Archaeol Natur Hist Mag*, 79 [pub 1985], 7–36

Barker, G 1995. *A Mediterranean Valley. Landscape Archaeology and* Annales *History in the Biferno Valley*, London

Barrett, J C 1994. *Fragments from Antiquity: an archaeology of social life in Britain, 2900–1200 BC*, Oxford

Barrett, J C, Bradley, R and Green, M 1991a. *Landscape, Monuments and Society: the prehistory of Cranborne Chase*, Cambridge

Barrett, J C, Bradley, R and Hall, M 1991b. *Papers on the Prehistoric Archaeology of Cranborne Chase*, Oxbow Monogr 11, Oxford

Bell, M G 1981. 'Valley sediments and environmental change', in *The Environment of Man: the Iron Age to the Anglo-Saxon Period* (eds M Jones and G W Dimbleby), 75–91

Bell, M G 1982. 'The effects of landuse and climate on valley sedimentation', in *Climatic Changes in Later Prehistory* (ed A F Harding), 127–42, Edinburgh

Bell, M G 1983. 'Valley sediments as evidence of prehistoric landuse of the South Downs', *Proc Prehist Soc*, 49, 119–50

Bell, M G and Boardman, J (eds) 1992. *Past and Present Soil Erosion*, Oxbow Monogr 22, Oxford

Bell, M G, Fowler, P J and Hillson, S W (eds) 1996. *The Experimental Earthwork Project 1960–92*, Counc Brit Archaeol Res Rep 100, York

Bender, B 1998. *Stonehenge: making space*, Oxford

Beresford, G 1975. *The Medieval Clay-land Village: excavations at Goltho and Barton Blount*, Soc Medieval Archaeol Monogr 6, London

Beresford, G 1979. 'Three deserted medieval settlements on Dartmoor: a report on the late E Marie Minter's excavations', *Medieval Archaeol*, 23, 98–158

Beresford, G 1987. *Goltho, the Development of an Early Medieval Manor c 850–1150*, London

Beresford, G 1988. 'Three deserted medieval settlements on Dartmoor: a comment on David Austin's reinterpretations', *Medieval Archaeol*, 32, 175–83

Beresford, M W 1959. 'Fifteenths and tenths: quotas of 133'; 'Poll tax payers of 1377'; 'Poor parishes of 1428', in *Victoria County History of Wiltshire*, IV, 294–314

Beresford, M W and Hurst, J G (eds) 1971. *Deserted Medieval Villages*, London

Beresford, M W and Hurst, J G (eds) 1989. *Deserted Medieval Villages*, Gloucester

Bersu, G 1940. 'Excavations at Little Woodbury, Wiltshire', *Proc Prehist Soc*, 6, 30–111

Betton, F 1991. 'The Anglo-Saxon foot: a computerised assessment', *Medieval Archaeol*, 35, 44–50

Biddle, M 1976. 'Hampshire and the origins of Wessex', in *Problems in Economic and Social History* (eds G de G Sieveking, I Longworth and K Wilson), 323–41

Biddle, M and Hill, D 1971. 'Late Saxon planned towns', *Antiq J*, 51, 70–85

Birks, H H, Birks, H J B, Kaland, P E and Moe, D (eds) 1988. *The Cultural Landscape – Past, Present and Future*, Cambridge

Boast, R 1995. 'Fine pots, pure pots, Beaker pots', in *'Unbaked Urns of Rudely Shape': essays on British and Irish pottery for Ian Longworth* (eds I Kinnes and G Varndell), 69–80, Oxford

Bond, J 1994. 'Forests, chases, warrens and parks in medieval Wessex', in Aston and Lewis 1994, 115–58

Boniface, P 1995. *Management of Cultural Tourism*, London
Boniface, P and Fowler, P J 1993. *Heritage and Tourism in 'the global village'*, London
Bonney, D 1966. 'Pagan Saxon burials and boundaries in Wiltshire', *Wiltshire Archaeol Natur Hist Mag*, 61, 25–30
Bonney, D 1968. 'Iron Age and Romano-British settlement sites in Wiltshire: some geographical considerations', *Wiltshire Archaeol Natur Hist Mag*, 63, 27–38
Bonney, D 1972. 'Early boundaries in Wessex', in Fowler 1972b, 168–86
Bonney, D 1976. 'Early boundaries and estates in southern England', in *Medieval Settlement, Continuity and Change* (ed P H Sawyer), 72–82
Bowen, H C 1961. *Ancient Fields*, London
Bowen, H C 1966. 'Origins and types of settlement', in Bowen and Fowler 1966, 43–53
Bowen, H C 1978. 'Celtic fields and ranch boundaries in Wessex', in *The Effect of Man on the Landscape: the lowland zone* (eds S Limbrey and J G Evans), 115–23
Bowen, H C 1990. *The Archaeology of Bokerley Dyke*, London
Bowen, H C and Fowler, P J 1962. 'The archaeology of Fyfield and Overton Downs, Wiltshire (interim report)', *Wiltshire Archaeol Natur Hist Mag*, 58, 98–115
Bowen, H C and Fowler, P J 1966. 'Romano-British settlements in Dorset and Wiltshire', in Thomas (ed) 1966, 43–67
Bowen, H C and Fowler, P J (eds) 1978. *Early Land Allotment in the British Isles*, Brit Archaeol Rep 48, Oxford
Bradbury, J 1996. *Stephen and Matilda: the Civil War of 1139–53*, Frome
Bradley, R 1978a. 'Colonisation and land use in the later Neolithic and earlier Bronze Age', in *The Effect of Man on the Landscape: the lowland zone* (eds S Limbrey and J G Evans), 95–103, London
Bradley, R 1978b. 'Prehistoric field systems in Britain and northwest Europe: a review of some recent work', *World Archaeol*, 9, 265–80
Bradley, R 1981. '"Various styles of urns": cemeteries and settlements in southern England *c* 1400–1000 BC', in *The Archaeology of Death* (eds R Chapman, I Kinnes and K Randsborg), 93–104
Bradley, R 1996. 'Ancient landscapes and the modern public', in '*The Remains of Distant Times*': *archaeology and the National Trust* (eds D M Evans, P Salway and D Thackray), 38–46
Bradley, R and Ellison, A 1975. *Rams Hill: a Bronze Age defended enclosure and its landscape*, Brit Archaeol Rep 19, Oxford
Bradley, R, Entwhistle, R and Raymond, F 1994. *Prehistoric Land Divisions on Salisbury Plain*, London
Bradley, R, Lobb, S J, Richards, J C and Robinson, M 1980. 'Two Late Bronze Age settlements on the Kennet Gravels: excavations at Aldermaston Wharf and Knight's Farm, Burghfield, Berkshire', *Proc Prehist Soc*, 46, 217–95
Branigan, K 1977. *Gatcombe Roman Villa*, Brit Archaeol Rep 44, Oxford
Branigan, K and Fowler, P J (eds) 1976. *The Roman West Country*, Newton Abbot/London
Branigan, K and Miles, D (eds) 1989. *The Economies of Romano-British Villas*, Sheffield
Brentnall, H C 1929. 'Shaw Church', *Rep Marlborough Coll Natur Hist Soc*, 78, 95–6
Brentnall, H C 1930. 'The church of Shaw-in-Alton', *Wiltshire Archaeol Natur Hist Mag*, 45, 156–65
Brentnall, H C 1938a. 'The Saxon Bounds of Overton', *Rep Marlborough Coll Natur Hist Soc*, 87, 116–36
Brentnall, H C 1938b. 'Savernake Forest in the Middle Ages', *Wiltshire Archaeol Natur Hist Mag*, 48, 371–86
Brentnall, H C 1941. 'The metes and bounds of Savernake Forest', *Wiltshire Archaeol Natur Hist Mag*, 49, 391–434
Brentnall, H C 1949. 'Venison trespasses in the reign of Henry VII', *Wiltshire Archaeol Natur Hist Mag*, 53, 191–212
Brentnall, H C 1950. 'The origins of the parish of Preshute', *Wiltshire Archaeol Natur Hist Mag*, 53, 294–310
Brown, G, Field, D and McOmish, D 1994. 'East Chisenbury midden complex', in *The Iron Age in Wessex: recent research* (eds A P Fitzpatrick and E Morris), 46–9
Brown, J C 1997. 'Flints from Bayado Farm, West Overton.' Unpublished A-level student submission, Chippenham College (copy in Devizes Museum)
Burchard, A 1966. 'Ancient scrub clearance? at Bayardo Farm, West Overton', *Wiltshire Archaeol Natur Hist Mag*, 61, 98
Burchard, A 1973. 'Some Beaker habitation sites in north Wiltshire', *Wiltshire Archaeol Natur Hist Mag*, 68, 116–19
Burgess, C B and Miket, R (eds) 1976. *Settlement and Economy in the Third and Second Millennia BC*, Brit Archaeol Rep 33, Oxford
Burl, A 1979. *Prehistoric Avebury*, London
Burl, A 1993. *From Carnac to Callanish, The Prehistoric Stone Rows and Avenues of Britain, Ireland and Brittany*, London
Campbell, B M S 1991. *Before the Black Death*, Manchester
Canham, R 1981. 'Aerial photography in Wiltshire 1975–81', *Wiltshire Archaeol Natur Hist Mag*, 76, 3–20
Carr, E H 1990. *What is History?* 2nd edn, London
Case, H J 1993. 'Beakers: deconstruction and after', *Proc Prehist Soc*, 59, 241–68
Case, H J 1995. 'Some Wiltshire Beakers and their contexts', *Wiltshire Archaeol Natur Hist Mag*, 88, 1–17
Casey, P J 1979. *The End of Roman Britain*, Brit Archaeol Rep 71, Oxford
Cavailles, H 1910/1986. *Lies et passeries dans les Pyrénées (1910) et Actes de la 3ème journée de recherches de la Société d'Études des Sept Vallées Luz-Saint-Sauveur*
Chadburn, A and Swanton, G (eds) forthcoming. *Proceedings of the World Heritage Sites Conference, Devizes, 1995*, Devizes
Champion, T C, 1987. 'The European Iron Age: assessing the state of the art', *Scott Archaeol Rev*, 4, 98–107
Chappell, H G, Ainsworth, J F, Cameron, R A D and Redfern, M 1971. 'The effect of trampling on a chalk grassland ecosystem', *J Applied Ecol*, 8, 869–82

Chippindale, C, Devereux, P, Fowler, P J, Jones, R and Sebastian, T 1990. *Who Owns Stonehenge?*, London

Clark, A 1958. 'The nature of Wansdyke', *Antiquity*, 32, 89–96

Clark, M J, Levin, J and Small, R J 1967. 'The sarsen stones of the Marlborough Downs and their geomorphological implications', *Southampton Res Ser Geog*, 4, 3–40

Clarke, D L 1968. *Analytical Archaeology*, London

Clarke, D L 1970. *Beaker Pottery in Great Britain and Ireland*, Cambridge

Clarke, D L 1976. 'The Beaker network – social and economic models', in *Glockenbecher Symposium: Oberried 1974* (eds J N Lanting and J D van der Waals), 459–76

Clarke, H 1984. *The Archaeology of Medieval England*, Oxford

Clément, P A 1991. *En Cévennes avec les bergers. Récits de transhumance*, Montpellier

Clutton-Brock, J and Legge, A J 1981. 'Contributions to discussion', in *Farming Practice in British Prehistory* (ed R Mercer), 218–22, Edinburgh

Coleman, S and Wood, J 1988. *Historic Landscape and Archaeology: glossary of terms*, Bedford

Coles, J and Minnitt, S 1995. *Industrious and Fairly Civilised, The Glastonbury Lake Village*, Somerset

Collingwood, R G 1989. *The Idea of History*, Oxford

Conzen, M P (ed) 1990. *The Making of the American Landscape*, London

Corney, M C 1989. 'Multiple ditch systems and Late Iron Age settlement in central Wessex', in *From Cornwall to Caithness: some aspects of British field archaeology* (eds M Bowden, D Mackay and P Topping), 111–28

Corney, M C 1996. 'New evidence for the Romano-British settlement by Silbury Hill', *Wiltshire Archaeol Natur Hist Mag*, 90, 139–41

Corney, M C 1997. 'The origins and development of the "small town" of *Cunetio*, Mildenhall, Wiltshire', *Britannia*, 28, 337–50

Cornwall, I 1956. *Bones for the Archaeologist*, London

Cornwall, I 1958. *Soils for the Archaeologist*, London

Costen, M 1994. 'Settlement in Wessex in the tenth century: the charter evidence', in Aston and Lewis 1994, 97–113

Countryside Commission 1996. *Historic Landscape*, Countryside Commission, Cheltenham

Crawford, O G S 1922. 'Notes on fieldwork round Avebury, December 1921', *Wiltshire Archaeol Natur Hist Mag*, 42, 52–63

Crawford, O G S 1924. *Air Survey and Archaeology*, Ord Surv Prof Pap, ns 7, London

Crawford, O G S 1929. *Air Photography for Archaeologists*, Ord Surv Prof Pap, ns 12, London

Crawford, O G S 1953. *Archaeology in the Field*, London

Crawford, O G S 1955. *Said and Done. The Autobiography of an Archaeologist*, London

Crawford, O G S and Keiller, A 1928. *Wessex from the Air*, Oxford

Crowley, D A (ed) 1988. *The Wiltshire Tax List of 1332*, Trowbridge

Cunliffe, B W 1977. 'The Romano-British village of Chalton, Hants', *Hampshire Fld Club Archaeol Soc*, 33, 45–67

Cunliffe, B W 1987. *Hengistbury Head Dorset, vol 1: the Prehistoric and Roman settlement 3500 BC–AD 500*, Oxford Univ Comm Archaeol Monogr 13, Oxford

Cunliffe, B W 1991. *Iron Age Communities in Britain, an Account of England, Scotland and Wales from the Seventh Century BC until the Roman conquest*, 3rd edn, London

Cunnington, M E 1909. 'Notes on a late Celtic rubbish heap near Oare', *Wiltshire Archaeol Natur Hist Mag*, 36, 125–39

Cunnington, M E 1923. *The Early Iron Age Inhabited Site at All Cannings Cross*, Devizes

Cunnington, M E 1931. '"The "Sanctuary" on Overton Hill, Avebury; being an account of the excavations carried out by Mr and Mrs B H Cunnington in 1930', *Wiltshire Archaeol Natur Hist Mag*, 45, 300–35

Darby, H C and Finn, R W 1967. *The Domesday Geography of South-west England*, Cambridge

Dark, K 1993. 'Roman period activity at prehistoric ritual monuments in Britain and the Armorican peninsula', *Proceedings First Theoretical Roman Archaeology Conference* (ed E Scott), 133–46

Darvill, T 1984. 'Value systems and the archaeological resource', *Int J Heritage Stud*, 1, 52–64

Darvill, T 1987. *Ancient Monuments in the Countryside: an archaeological management review*, London

Davies, S M 1981. 'Excavations at Old Down Farm, Andover, Part II: prehistoric and Roman', *Hampshire Fld Club Archaeol Soc*, 37, 81–163

Department of the Environment 1990. *Archaeology and Planning*, Planning Policy Guidance 16, London

Dillon, P, Skeggs, S and Goody, C 1992. 'Some investigations on habitats of lichens on sarsen stones at Fyfield Down, Wiltshire', *Wiltshire Archaeol Natur Hist Mag*, 85, 128–39

Dimbleby, G W and Evans, J G 1974. 'Pollen and land snail analysis of calcareous soils', *J Archaeol Sci*, 1, 117–33

Drewett, P 1982a. 'Later Bronze Age downland economy and excavations at Black Patch, East Sussex', *Proc Prehist Soc*, 48, 321–400

Drewett, P 1982b. *The Archaeology of Bullock Down, Eastbourne, East Sussex: the development of a landscape*, Sussex Archaeol Soc Monogr 1, Lewes

Drewett, P 1986. 'The excavation of a Neolithic oval barrow at North Marden, West Sussex, 1982', *Proc Prehist Soc*, 52, 31–51

Droste, B von, Plachter, H and Rössler, M (eds) 1995. *Cultural Landscapes of Universal Value, Components of a Global Strategy*, Jena

Dury, G H 1984. 'Crop failures on the Winchester manors 1232–1349', *Trans Inst Brit Geog*, ns 9, 401–18

Dyer, C 1986. 'English peasant buildings in the later Middle Ages (1200–1500)', *Medieval Archaeol*, 30, 19–45

Eagles, B N W 1986. 'Pagan Anglo-Saxon burials at West Overton', *Wiltshire Archaeol Natur Hist Mag*, 80, 103–20

Eagles, B N W 1994. 'The archaeological evidence for settlement in the fifth to seventh centuries AD', in *The Medieval Landscape of Wessex* (eds M Aston and C Lewis), 13–32

Ellison, A B 1973. *Village Survey: south-east Somerset*, Bristol

Ellison, A B 1981. 'Towards a socio-economic model for the Middle Bronze Age in southern England', in *Pattern of the Past: studies in honour of David Clarke* (eds I Hodder, G Isaac and N Hammond), 413–38

Ellison, A B 1983. *Medieval Villages in South-east Somerset*, Bristol

Ellison, A B 1987. 'The Bronze Age settlement at Thorney Down: pots, post-holes and patterning', *Proc Prehist Soc*, 53, 385–92

English Heritage, 1991a. *Exploring our Past: strategies for the archaeology of England*, London

English Heritage, 1991b. *Management of Archaeological Projects*, London

English Heritage, 1996. *Hadrian's Wall World Heritage Site Management Plan*, London

English Heritage, 1998. *Avebury World Heritage Site Management Plan*, London

English Nature, 1991. 'Management Plan for Fyfield Down National Nature Reserve and Site of Special Scientific Interest.' Unpublished MS, Devizes office

Entwistle, R and Bowden, M 1991. 'Cranborne Chase: the molluscan evidence', in *Papers on the Prehistoric Archaeology of Cranborne Chase* (eds J Barrett, R Bradley and M Hall), Oxbow Monogr 11, 20–48, Oxford

Evans, D M, Salway, P and Thackray, D (eds) 1996. *'The Remains of Distant Times': archaeology and the National Trust*, Woodbridge

Evans, J G 1968. 'Periglacial deposits on the chalk of Wiltshire', *Wiltshire Archaeol Natur Hist Mag*, 63, 12–26

Evans, J G 1970. 'Interpretation of land snail faunas', *Univ London Inst Archaeol Bull*, 8/9, 109–116

Evans, J G 1971. 'The pre-henge environment', in *Durrington Walls, 1966–1968* (eds G Wainwright and I Longworth), 329–37, London

Evans, J G 1972. *Land Snails in Archaeology*, London

Evans, J G 1975. *The Environment of Early Man in the British Isles*, London

Evans, J G 1978. *An Introduction to Environmental Archaeology*, London

Evans, J G, Atkinson, R J C, O'Connor, T and Green, H S 1983. 'The environment in the Late Neolithic and Early Bronze Age and a Beaker burial', *Wiltshire Archaeol Natur Hist Mag*, 78, 7–30

Evans, J G and Vaughan, M P 1985. 'An investigation into the environment and archaeology of the Wessex linear ditch system', *Antiq J*, 65, 11–38

Evans, J G, Rouse, A and Sharples, N 1988. 'The landscape setting of causewayed enclosures; some recent work on the Maiden Castle enclosure', in *The Archaeology of Context in the Neolithic and Bronze Age: recent trends* (eds J Barrett and I Kinnes), 73–8

Evans, J G, Limbrey, S, Máté, A and Mount, R 1993. 'An environmental history of the Upper Kennet Valley, Wiltshire, for the last 10,000 years', *Proc Prehist Soc*, 59, 139–95

Everson, P L, Taylor, C C and Dunn, C J 1991. *Change and Continuity. Rural Settlement in North-west Lincolnshire*, London

Fasham, P J 1985. *The Prehistoric Settlement at Winnall Down, Winchester*, Hampshire Fld Club Archaeol Soc Monogr 2, Winchester

Featherstone, R, Horne, P, Macleod, D and Bewley, R 1995. 'Aerial reconnaissance in England, summer 1995', *Antiquity*, 69, 981–8

Feilden, B M and Jokilehto, J 1993. *Management Guidelines for World Cultural Heritage Sites*, Rome

Fernie, E C 1991. 'Anglo-Saxon lengths and the evidence of the buildings', *Medieval Archaeol*, 35, 1–5

Field, J 1972. *English Field Names: a dictionary*, Newton Abbot

Finberg, H P R 1964. *The Early Charters of Wessex*, Leicester

Finley, M I 1986. *The Use and Abuse of History*, London

Fleming, A 1971. 'Territorial patterns in Bronze Age Wessex', *Proc Prehist Soc*, 37, 138–66

Fleming, A 1978. 'The prehistoric landscape of Dartmoor, Part 1, South Dartmoor', *Proc Prehist Soc*, 44, 97–123

Fleming, A 1983. 'The prehistoric landscape of Dartmoor, Part 2: North and East Dartmoor', *Proc Prehist Soc*, 49, 195–241

Fleming, A 1987. 'Coaxial field systems: some questions of time and space', *Antiquity*, 61, 188–203

Fleming, A 1988. *The Dartmoor Reaves*, London

Fowler, P J 1963a. 'The archaeology of Fyfield and Overton Downs, Wiltshire (second interim report)', *Wiltshire Archaeol Natur Hist Mag*, 58, 342–50

Fowler, P J 1963b. 'Archaeology', in Jewell 1963b, 64–6

Fowler, P J 1965. 'A Roman barrow at Knob's Crook, Woodlands, and a reconsideration of the evidence for Roman barrows in Wessex', *Antiq J*, 45, 22–52

Fowler, P J 1966. 'The distribution of settlement', in Bowen and Fowler 1966, 54–67

Fowler, P J 1967. 'The archaeology of Fyfield and Overton Downs, Wiltshire (third interim report)', *Wiltshire Archaeol Natur Hist Mag*, 62, 16–33

Fowler, P J 1969. 'Fyfield Down 1959–68', *Current Archaeol*, 16,124–9

Fowler, P J 1972a. 'Field archaeology in future', in Fowler (ed) 1972b, 96–126

Fowler, P J (ed) 1972b. *Archaeology and the Landscape, Essays for L V Grinsell*, London

Fowler, P J 1975a. 'Continuity in the landscape? Some local archaeology in Wiltshire, Somerset and Gloucestershire', in Fowler (ed) 1975b, 121–36

Fowler, P J (ed) 1975b. *Recent Work in Rural Archaeology*, Bradford-on-Avon

Fowler, P J 1976. 'Agriculture and rural settlement', in *The Archaeology of Anglo-Saxon England* (ed D Wilson), 23–48
Fowler, P J 1977. *Approaches to Archaeology*, London
Fowler, P J 1981a. 'Later prehistory', in *The Agrarian History of England and Wales, vol I:1 Prehistory* (ed S Piggott), 61–299
Fowler, P J 1981b. 'Wildscape to landscape: enclosure in prehistoric farming?', in Mercer (ed) 1981, 9–54
Fowler, P J 1981c. 'The Royal Commission on Historical Monuments (England)', *Antiquity*, 55, 106–14
Fowler, P J 1983a. *Farming in Prehistoric Britain*, Cambridge
Fowler, P J 1983b. *Farming in England*, London
Fowler, P J 1987. 'The contemporary past' in *Landscape and Culture: Geographical and Archaeological Perspectives* (ed J M Wagstaff), 173–91
Fowler, P J 1995a. 'Avebury', *History Today*, 45, 10–15
Fowler, P J 1995b. 'Archaeology in Trust' in Newby (ed) 1995b, 104–16
Fowler, P J 1996. 'Landscape: personality, perception and perspective', in *'The Remains of Distant Times': archaeology and the National Trust* (eds D M Evans, P Salway and D Thackray), 10–20
Fowler, P J 1997. 'Writing on the countryside', in Hodder *et al* (eds) 1997, 100–9
Fowler, P J 1998. 'Moving through the landscape', in *The Archaeology of Landscape* (eds P Everson and T Williamson), 25–41
Fowler, P J 1999. Review of *Avebury World Heritage Site Management Plan*, *Antiquity*, 73, 719–20
Fowler, P J forthcoming a. 'Cultural landscape: dreadful phrase, great concept' in *Proceedings of the Cultural Landscape Conference, Oxford, May 2000*, ICOMOS UK, London (*see also* FWP 88)
Fowler, P J forthcoming b. 'Cultural landscape: archaeology, ancestors and archive' in *Monument-Groups of Buildings – Cultural Landscape exemplified by the Wachau* (ed G Hajós), 52–9, Vienna (*see also* FWP 88)
Fowler, P J forthcoming c. 'Wansdyke in the woods' in *Roman Wiltshire* (ed P Ellis), Wiltshire Archaeol Natur Hist Soc, Devizes (*see also* FWP 91)
Fowler, P J and Blackwell, I 1998. *The Land of Lettice Sweetapple. An English Countryside Explored*, Stroud (*see also* FWP 93) (paperback 2000)
Fowler, P J and Evans, J G 1967. 'Plough-marks, lynchets and early fields', *Antiquity*, 61, 289–301
Fowler, P J and Jacques, D 1995. 'Cultural landscapes in Britain', in von Droste *et al* (eds) 1995, 350–63
Fowler, P J, Lucas, C J M and Blackwell, I W 1995. 'Fyfod: Fyfield and West Overton, Wiltshire. An assessment report and project design for publication and archive deposition.' Unpublished MS, Newcastle (in Archive, Devizes Museum)
Fowler, P J, Musty, J W G and Taylor, C C 1965. 'Some earthwork enclosures in Wiltshire', *Wiltshire Archaeol Natur Hist Mag*, 60, 52–74
Fowler, P J and Sharp, M 1990. *Images of Prehistory*, Cambridge
Fowler, P J and Stabler, M J forthcoming. 'World Heritage designation: expectations and consequences' in Chadburn and Swanton (eds)
Fox, A and Fox, C 1958. 'Wansdyke reconsidered', *Archaeol J*, 115, 1–48
Free, D W 1948. 'Sarsen stones and their origins', *Wiltshire Archaeol Natur Hist Mag*, 52, 338–44
Free, D W 1950. *Marlborough and District. An Original Peep into the Past*, Marlborough
Frere, S 1977. 'Roman Britain in 1976. 1. Sites explored', *Britannia*, 8, 356–425
Frere, S 1992. 'Roman Britain in 1991', *Britannia*, 23, 255–308
Fukuyama, F 1992. *The End of History and the Last Man*, New York
Fulford, M, Entwistle, R and Raymond, F 1994. *Salisbury Plain Project. 1993–4 Interim Report*, Reading
Gaffney, C, Gaffney, V and Corney, M 1998. 'Changing the Roman landscape: the role of geophysics and remote sensing', in *Science and Archaeology: an agenda for the future* (ed J Bayley), 145–56
Gaffney, V and Tingle, M 1989. *The Maddle Farm Project: an integrated survey of prehistoric and Roman landscapes on the Berkshire Downs*, Brit Archaeol Rep S200, Oxford
Gardiner, P 1961. *The Nature of Historical Explanation*, Oxford
Garmonsway, G N (ed) 1990. *The Anglo-Saxon Chronicle*, London
Gelling, M 1978. *Signposts to the Past, Place-Names and the History of England*, London
Gelling, M 1993. *Place-Names in the Landscape*, London
Gibson, A M 1982. *Beaker Domestic Sites*, Brit Archaeol Rep 107, Oxford
Gingell, C 1982. 'Excavation of an Iron Age enclosure at Groundwell Farm, Blunsdon St Andrew, 1976–77', *Wiltshire Archaeol Natur Hist Mag*, 76, 33–75
Gingell, C 1992. *The Marlborough Downs: a Later Bronze Age Landscape and its Origins*, Wiltshire Archaeol Natur Hist Soc Monogr 1, Devizes
Gingell, C and Lawson, A J 1985. 'Excavations at Potterne 1984', *Wiltshire Archaeol Natur Hist Mag*, 79, 101–8
Godwin, H 1975. *The History of the British Flora: a factual basis for phytogeography*, 2nd edn, Cambridge
Goodier, A 1984. 'The formation of boundaries in Anglo-Saxon England: a statistical study', *Medieval Archaeol*, 28, 1–24
Gover, J E B, Mawer, A and Stenton, F M 1939. *The Place-Names of Wiltshire*, Cambridge
Graham, A and Newman, C 1993. 'Recent excavations of Iron Age and Romano-British enclosures in the Avon Valley, Wiltshire', *Wiltshire Archaeol Natur Hist Mag*, 86, 8–59
Gray Birch, W de 1885–99. *Cartularium Saxonicum*, 3 vols
Greatrex, J (ed) 1978. *The Register of the Common Seal of the Priory of St Swithun, Winchester 1345–1497*, Hampshire Rec Ser 2, Winchester
Green, H S 1971. 'Wansdyke: excavations 1966–1970', *Wiltshire Archaeol Natur Hist Mag*, 66, 129–46
Grinsell, L V 1957. 'Archaeological gazetteer', in *Victoria County History of Wiltshire I.1* (eds R B Pugh and E Crittall), 21–279

Grinsell, L V 1958. *The Archaeology of Wessex*, London

Grinsell, L V 1989. *An Archaeological Autobiography*, Gloucester

Grose, J D 1946. 'Botanical references in the Saxon charters of Wiltshire', *Wiltshire Archaeol Natur Hist Mag*, 186, 555–83

Grundy, G B 1917. 'The ancient highways and tracks of Wiltshire, Berkshire and Hampshire, and the Saxon battlefields of Wiltshire', *Archaeol J*, 2nd ser, 24, 69–194

Grundy, G B 1919. 'The Saxon land charters of Wiltshire', *Archaeol J*, 2nd ser, 26, 143–301

Grundy, G B 1920. 'The Saxon land charters of Wiltshire', *Archaeol J*, 2nd ser, 27, 8–126

Guest, P 1998. 'The Bishops Cannings Hoard', in *Coin Hoards from Roman Britain*, 10, London

Hall, D 1988. 'The late Saxon countryside: villages and their fields', in Hooke (ed) 1988a, 99–122

Hamilton, M A, Dennis, I and Swanton, G 1998. *A Geophysical Study of North Farm, Wiltshire, 1998*, Cardiff Stud Archaeol Specialist Rep 8, Cardiff

Harding, D W 1973. 'Round and rectangular: Iron Age houses, British and foreign', in *Archaeology into History 1, Greeks, Celts and Romans* (eds C F Hawkes and S Hawkes), 43–62

Harding, D W 1974. *The Iron Age in Lowland Britain*, London

Harding, D W, Blake, I M and Reynolds, P J 1993. *Iron Age Settlement in Dorset*, Univ Edinburgh Dept Archaeol Monogr 1, Edinburgh

Hare, J N 1976. 'Lord and tenant in Wiltshire, *c* 1380–1520, with particular reference to regional and seigneurial variations.' Unpublished PhD thesis, London

Hare, J N 1980. 'Durrington, a chalkland village in the later Middle Ages', *Wiltshire Archaeol Natur Hist Mag*, 74/75, 137–47

Hare, J N 1981a. 'Change and continuity in Wiltshire agriculture in the later Middle Ages', in *Agricultural Improvement: medieval and modern* (ed W Minchinton), Exeter Pap Econ Hist 14, 1–18

Hare, J N 1981b. 'The demesne lessees of fifteenth-century Wiltshire', *Agri Hist Rev*, 29, 1–15

Hare, J N 1985. 'The monks as landlords: the leasing of the monastic demesnes in southern England', in *The Church in Pre-Reformation Society* (eds C Barron and C Harper-Bill), 82–94

Hare, J N 1992. 'The lords and their tenants: conflict and stability in fifteenth-century Wiltshire', in *Conflict and Community in Southern England* (ed B Stapleton), 16–34

Hare, J N 1994. 'Agriculture and rural settlement in the chalklands of Wessex and Hampshire from *c* 1200–*c* 1500', in Aston and Lewis (eds) 1994, 159–93

Harrison, B 1995. 'Field systems and demesne farming on the Wiltshire estate of St Swithun's Priory, Winchester, 1248–1348', *Agric Hist Rev*, 43, 1–18

Harvey, D 1989. *The Condition of Post-modernity. An Enquiry into the Origins of Cultural Change*, Oxford

Hase, P H 1994. 'The Church in the Wessex heartlands', in Aston and Lewis (eds) 1994, 47–81

Hawkes, S C 1994. 'Longbridge Deverill Cow Down, Wiltshire, house 3: a major roundhouse of the Early Iron Age', *Oxford J Archaeol*, 13 (1), 49–69

Hill, J D 1993. 'Can we recognise a different European past? A contrastive archaeology of later prehistoric settlements in southern England', *J European Archaeol*, 1, 57–75

Hill, J D 1995. *Ritual and Rubbish in the Iron Age of Wessex. A study in the formation of a specific archaeological record*, Brit Archaeol Rep 242, Oxford

Hill, P 1961. 'Air photograph of Delling Enclosure', *Geog J*, 127, opposite p 489

History Today 1992. *After the End of History*, London

Hoare, R Colt 1821. *The Ancient History of North Wiltshire*, London

Hodder, I 1976. 'The distribution of Savernake Ware', *Wiltshire Archaeol Natur Hist Mag*, 69, 67–84

Hodder, I, Shanks, M, Alexandra, A, Buhli, V, Carman, J, Last, J and Lucas, G (eds) 1997. *Interpreting Archaeology. Finding Meaning in the Past*, London

Holgate, R 1987. 'Neolithic settlement patterns at Avebury, Wiltshire', *Antiquity*, 61, 259–63

Holt, J C 1987. *Domesday Studies*, Oxford

Hooke, D 1987. 'Two documented pre-Conquest Christian sites', in *Medieval Archaeol*, 31, 96–101

Hooke, D (ed) 1988a. *Anglo-Saxon Settlements*, Oxford

Hooke, D 1988b. 'Regional variation in southern and central England in the Anglo-Saxon period and its relationship to land units and settlement', in Hooke (ed) 1988a, 123–51

Hooke, D 1989. 'Pre-Conquest woodland: its distribution and usage', *Agri Hist Rev*, 37, 113–29

Hooke, D 1997. 'The Anglo-Saxons in England in the seventh and eighth centuries: aspects of location in space', in *The Anglo-Saxons from the Migration Period to the Eighth Century: an ethnographic perspective* (ed D Hines), 65–85

Hoskins, W G 1955. *The Making of the English Landscape*, London

Hostetter, E and Howe, T N 1997. *The Romano-British Villa at Castle Copse, Great Bedwyn*, Indiana

Hudson, J C 1990. 'Settlement of the American grassland' in Conzen 1990, 169–85

Huggins, P J 1991. 'Anglo-Saxon timber building measurements: recent results', *Medieval Archaeol*, 35, 6–28

Hunt, T 1996. 'The National Trust and World Heritage Sites', in *'The Remains of Distant Times': archaeology and the National Trust* (eds D M Evans, P Salway and D Thackray), 211–16

Hurst, J G, Neal, D S and Beuningen, H J E van 1986. *Pottery Produced and Traded in North-west Europe 1350–1650*, Rotterdam Pap VI, Rotterdam

Ingold, T 1993. 'The temporality of the landscape', *World Archaeol*, 25, 152–74

Jackson, R H 1985. 'The Tisbury landholdings granted to Shaftesbury monastery by the Saxon Kings', *Wiltshire Archaeol Natur Hist Mag*, 79, 164–77

Jacques, D 1995. 'The rise of cultural landscape', *Int J Heritage Stud*, 1, 91–101

James, S, Marshall, A and Millett, M 1984. 'An early medieval building tradition', *Archaeol J*, 140, 182–215

James, T B and Robinson, A M 1988. *Clarendon Palace*, London

Jarrett, M G and Wrathmell, S 1981. *Whitton, an Iron Age and Roman farmstead in South Glamorgan*, Cardiff

Jenkins, K 1991. *Re-thinking History*, London

Jessup, R F 1962. 'Roman barrows in Britain', *Collection Latomus* 58, 853–67

Jewell, P A 1963a. 'Cattle from British archaeological sites', in *Man and Cattle. Proceedings of a Symposium on Domestication* (eds A E Mourant and F E Zeuner), Royal Anthropological Institute Occas Pap 18, 80–91

Jewell, P A (ed) 1963b. *The Experimental Earthwork on Overton Down Wiltshire 1960*, London

Jewell, P A and Dimbleby, G W (eds) 1966. 'The experimental earthwork on Overton Down, Wiltshire, England: the first four years', *Proc Prehist Soc*, 32, 313–42

Johnson, N and Rose, P 1994. *Bodmin Moor. An Archaeological Survey, Vol 1: the human landscape to c 1800*, London

Jones, M 1984. 'Regional patterns in crop production', in *Aspects of the Iron Age in Central Southern Britain* (eds B Cunliffe and D Miles), 120–5

Jones, M 1986. *England before Domesday*, London

Kempson, E G H 1953. 'The Devil's den', *Wiltshire Archaeol Natur Hist Mag*, 55, 71

Kempson, E G H 1962. 'Wroughton Mead: a note on the documentary evidence' in Bowen and Fowler (eds) 1962, 113–15

Kerney, M P 1976. *Atlas of the Non-marine Mollusca of the British Isles*, Monks Wood

Kerridge, E 1953. 'The floating of the Wiltshire water-meadows', *Wiltshire Archaeol Natur Hist Mag*, 55, 105–18

King, N E 1968. 'The Kennet valley sarsen industry', *Wiltshire Archaeol Natur Hist Mag*, 63, 83–93

Kinnes, I 1977/8. 'The Beaker grave group from East Kennet', *Wiltshire Archaeol Natur Hist Mag*, 72/73, 167–70

Kinnes, I, Gibson, A, Ambers, J, Bowman, S, Leese, M and Boast, R 1991. 'Radiocarbon dating and British Beakers: the British Museum programme', *Scott Archaeol Rev*, 8, 35–68

Lacaille, A D 1962. 'A cup-marked sarsen near Marlborough, Wiltshire', *Archaeol Newslett*, 8, 123–9

Lacaille, A D 1963. 'Three grinding stones', *Antiq J*, 43, 190–6

Ladurie, E le Roy 1978. *Montaillou. Cathars and Catholics in a French Village 1294–1324*, London

Lamb, H H 1981. 'Climate from 1000 BC to AD 100', in *The Environment of Man: the Iron Age to the Anglo-Saxon Period* (eds M Jones and G W Dimbleby), 95–127

Lawson, A J 1983. *The Archaeology of Witton, near North Walsham*, E Anglian Archaeol 18, Dereham

Lawson, A J forthcoming. *Potterne 1982–5: animal husbandry in later prehistoric Wiltshire*, Wessex Archaeol Rep 17, Salisbury

Lawson, A J, Martin, E A, Priddy, D with Taylor, A 1981. *The Barrows of East Anglia*, E Anglian Archaeol 12, Dereham

Leach, P 1982. *Ilchester Volume 1 Excavations 1974–5*, Western Archaeol Trust Monogr 3, Bristol

Leech, R 1981. 'The excavation of a Romano-British farmstead and cemetery on Bradley Hill, Somerton, Somerset', *Britannia*, 12, 177–252

Leech, R 1982. *Excavations at Catsgore 1970–1973*, Western Archaeol Trust Monogr 2, Bristol

Leeds, E T 1934. 'Recent Bronze Age discoveries in Berkshire and Oxfordshire', *Antiq J*, 14, 264–76

Leeds, E T 1938. 'Beakers of the Upper Thames district', *Oxoniensia 3*, 7–30

Lees, B A (ed) 1935. *Records of the Templars in England in the Twelfth Century*, London

Lewis, C 1994. 'Patterns and processes in the medieval settlement of Wiltshire', in Aston and Lewis (eds) 1994, 171–93

Locker, A forthcoming. 'Animal bone', in *Potterne 1982–5: animal husbandry in later prehistoric Wiltshire* (A J Lawson), 101–19

Long, W 1862. *Aubry Illustrated*, Devizes

Lowe, P 1995. 'The countryside' in Newby (ed) 1995, 87–103

Lowenthal, D 1991. 'British national identity and the English landscape', *Rural Hist*, 2, 205–30

Lyne, M A B and Jeffries, R S 1979. *The Alice Holt/Farnham Roman Pottery Industry*, Counc Brit Archaeol Res Rep 30, London

Macinnes, C M and Whittard, W F 1955. *Bristol and its Adjoining Counties*, Bristol

Major, A and Burrow, E J 1929. *The Mystery of Wansdyke*, Cheltenham

Malone, C 1989. *Avebury*, London

Maltby, J M 1981. 'Iron Age, Romano-British and Anglo-Saxon animal husbandry – a review of the faunal evidence', in *The Environment of Man: the Iron Age to the Anglo-Saxon Period* (eds M Jones and G W Dimbleby), 155–203

Manning, W H 1985. *Catalogue of Romano-British Iron Tools, Fittings and Weapons in the British Museum*, London

Margary, I D 1967. *Roman Roads in Britain*, London

Martin, E A nd. *Dew Ponds. History, Observation and Experiment*, London

Marwick, A 1989. *The Nature of History*, 3rd edn, London

McCarthy, M R and Brooks, C M 1988. *Medieval Pottery in Britain AD 900–1600*, Leicester

McOmish, D 1998. 'Landscape preserved by the men of war', *Brit Archaeol*, 34, 12–13

Mellor, M 1994. *Medieval Ceramic Studies in England*, London

Mercer, R (ed) 1981. *Farming Practice in British Prehistory*, Edinburgh

Meyrick, O 1950. 'An early medieval site on Manton Down', *Wiltshire Archaeol Natur Hist Mag*, 53, 328–31

Meyrick, O 1973. 'Some Beaker habitations in north Wiltshire', *Wiltshire Archaeol Natur Hist Mag*, 68, 116–18

Millett, M 1990. *The Romanisation of Britain: an essay in archaeological interpretation*, Cambridge

Millett, M 1995. *Roman Britain*, London

Millett, M and Graham, D 1986. *Excavations on the Romano-British Small Town at Neatham, Hampshire, 1969–1979*, Hampshire Fld Club Archaeol Soc Monogr 3, Winchester

Mizoguchi, K 1995. 'The "materiality" of Wessex Beakers', *Scott Archaeol Rev*, 9/10, 175–85

Monkhouse, F J (ed) 1964. *A Survey of Southampton and its Region*, London

Moore, J and Jennings, D 1992. *Reading Business Park: a Bronze Age Landscape*, Oxford

Morris, C (ed) 1949. *The Journeys of Celia Fiennes*, London

Morris, E L 1991. 'Ceramic analysis and the pottery from Potterne', in *Recent Developments in Ceramic Petrology* (eds A Middleton and I Freestone), 277–87

Morris, P 1979. *Agricultural Buildings in Roman Britain*, Brit Archaeol Rep 70, Oxford

Morris, R 1989. *Churches in the Landscape*, London

Musty, J W G 1959. 'A Romano-British building at Highpost, Middle Woodford', *Wiltshire Archaeol Natur Hist Mag*, 57, 173–5

Musty, J and Algar, D J 1986. 'Excavations at the deserted medieval village of Gomeldon, near Salisbury', *Wiltshire Archaeol Natur Hist Mag*, 80, 127–69

Musty, J, Algar, D J and Ewence, P F 1969. 'The medieval pottery kilns at Laverstock, near Salisbury, Wiltshire', *Archaeologia*, 102, 83–150

National Trust 1997. *Avebury Management Plan*, 2nd draft, Avebury

Needham, S and Ambers, J 1994. 'Redating Rams Hill and reconsidering the Bronze Age Enclosure', *Proc Prehist Soc*, 60, 225–44

Newby, H 1995. 'The next one hundred years', in *The National Trust, The Next Hundred Years* (ed H Newby), 150–63

Ordnance Survey, 1951. *Field Archaeology Notes for Beginners*, 3rd edn, Chesham

Ordnance Survey, 1978. *Map of Roman Britain*, 4th edn, Southampton

Ormrod, W M 1990. *The Reign of Edward III*, Yale

Oschinsky, D 1971. *Walter of Henley and other Treatises on Estate Management and Accounting*, Oxford

Palliser, D M 1992. *The Age of Elizabeth*, Harlow

Palmer, R 1984. *Danebury: an Iron Age hillfort in Hampshire. An aerial photographic interpretation of its environs*, London

Passmore, A D 1923. 'Chambered long barrow in West Woods', *Wiltshire Archaeol Natur Hist Soc*, 42, 49–51

Pearson, M P 1997. 'Tombs and territories' in Hodder *et al* (eds) 1997, 205–9

Pevsner, N 1963. *The Buildings of England. Wiltshire*, Harmondsworth

Phillips, A 1995. 'Conservation', in Newby (ed) 1995, 32–52

Piggott, C M 1942. 'Five late Bronze Age enclosures in north Wiltshire', *Proc Prehist Soc*, 8, 48–61

Piggott, C M 1950. 'Late Bronze Age enclosures in Sussex and Wessex', *Proc Prehist Soc*, 16, 193–5

Piggott, S 1958. 'Native economies and the Roman occupation of north Britain', in *Roman and Native in North Britain* (ed I A Richmond), 1–27

Piggott, S 1963. *The West Kennet Long Barrow: excavations, 1955–56*, London

Piggott, S 1971. 'An archaeological survey and policy for Wiltshire, part 3: Neolithic and Bronze Age', *Wiltshire Archaeol Natur Hist Mag*, 66, 47–57

Piggott, S 1985. *William Stukeley. An Eighteenth-century Antiquary*, London

Pitt Rivers, A L F 1887–98. *Excavations in Cranborne Chase I–IV*, privately printed

Pitts, M 1996. 'The vicar's dewpond, the National Trust shop and the rise of paganism', in *'The Remains of Distant Times': archaeology and the National Trust* (eds D M Evans, P Salway and D Thackray), 116–31

Pitts, M and Whittle, A 1992. 'The development and date of Avebury', *Proc Prehist Soc*, 58, 203–12

Plumb, J H 1989. *The Death of the Past*, London

Pollard, J 1992. 'The Sanctuary, Overton Hill, Wiltshire: a re-examination', *Proc Prehist Soc*, 58, 213–26

Popper, K 1986. *The Poverty of Historicism*, London

Postan, M M 1972. *The Medieval Economy and Society*, London

Postan, M M 1973. 'Some agrarian evidence of declining population in the later Middle Ages', in *Essays on Medieval Agriculture and General Problems of the Medieval Economy* (ed M Postan), 186–213

Powell, A, Allen, M J and Barnes, I 1996. *Archaeology in the Avebury Area*, Wessex Archaeol Rep 8, Salisbury

Pugsley, A J 1939. *Dewponds in Fact and Fable*, London

Rackham, O 1986. *The History of the Countryside*, London

Rackham, O 1996. *Trees and Woodland in the British Landscape*, London

Rahtz, P (ed) 1974. *Rescue Archaeology*, Harmondsworth

RCHME: Royal Commission on Historical Monuments (England) 1952–75. *Inventory … Dorset I–V*, London

RCHME 1979a. *Stonehenge and its Environs: monuments and land-use*, Edinburgh

RCHME 1979b. *Long Barrows in Hampshire and the Isle of Wight*, London

Rees, S E 1979. *Agricultural Implements in Prehistoric and Roman Britain*, Brit Archaeol Rep 69, Oxford

Renfrew, C 1973. *Before Civilization. The Radio-carbon Revolution and Prehistoric Europe*, London

Reynolds, P J 1979. *Iron Age Farm. The Butser Experiment*, London

Richards, J C 1990. *The Stonehenge Environs Project*, Hist Build Monuments Comm Engl Archaeol Rep 16, London

Rivet, A L F 1970. 'The British section of the *Antonine Itinerary*', *Britannia*, 1, 34–82

Roberts, B K 1996. *Landscapes of Settlement. Prehistory to the present*, London

Robinson, M 1984. 'Landscape and environment of central southern Britain in the Iron Age', in *Aspects of the Iron Age in Central Southern Britain* (eds B W Cunliffe and D Miles), 1–11

Rodgers, B and Roddham, D 1991. 'The excavation at Wellhead, Westbury 1959–1966', *Wiltshire Archaeol Natur Hist Mag*, 84, 51–60

Rodwell, W J and Rodwell, K A 1985. *Rivenhall: investigations of a villa, church and village, 1950–1977*, Counc Brit Archaeol Res Rep 55, London

Rogers P (ed) 1971. *Daniel Defoe. A Tour through the Whole Island of Great Britain*, Harmondsworth

Rowley, T 1994. *Villages in the Landscape*, London

Russett, V nd. *Marshfield. An Archaeological Survey of a Southern Cotswold Parish*, Bristol

Saville, A 1977/8. 'Five flint assemblages from excavated sites in Wiltshire', *Wiltshire Archaeol Natur Hist Mag*, 72/73, 1–28

Sawyer, P H 1968. *Anglo-Saxon Charters: an annotated list and bibliography*, London

Scott, R 1959. 'Medieval Agriculture', in *VCH Wiltshire, Volume IV*, 7–42

Sharples, N M 1991. *Maiden Castle: excavations and field survey 1985–6*, London

Sheail, J 1968. 'The regional distribution of wealth in England as indicated in the 1524/5 lay subsidy returns.' Unpublished PhD thesis, London

Shennan, S 1985. *Experiments in the Collection and Analysis of Archaeological Survey Data: the East Hampshire survey*, Sheffield

Shortt, H 1969. 'A Denier of Charles the Bald from Preshute', *Wiltshire Archaeol Natur Hist Mag*, 64, 51–5

Simmons, I G 1996. *The Environmental Impact of Later Mesolithic Cultures*, Edinburgh

Small, F 1999. *Avebury World Heritage Site Mapping Project, Wiltshire*, English Heritage Survey Report, Swindon

Smith, A C 1885. *British and Roman Antiquities of the North Wiltshire Downs*, 2nd edn, Wiltshire Archaeol Soc

Smith, A H 1970. *English Place-Name Elements*, 2 vols, English Place-Name Soc, 25, 26, Cambridge

Smith, I F 1965. *Windmill Hill and Avebury: excavations by Alexander Keiller, 1925–1939*, Oxford

Smith, I F 1967. 'The Beaker pottery from site OD XI', in Fowler 1967, 30–1

Smith, I F 1991. 'Round barrows Wilsford cum Lake G51–G54: excavations by Ernest Greenfield in 1958', *Wiltshire Archaeol Natur Hist Mag*, 84, 11–39

Smith, I F and Simpson, D D A 1964. 'Excavation of three Roman tombs and a prehistoric pit on Overton Down', *Wiltshire Archaeol Natur Hist Mag*, 59, 68–87

Smith, I F and Simpson, D D A 1966. 'Excavation of a round barrow on Overton Hill, North Wiltshire', *Proc Prehist Soc*, 32, 122–55

Smith, R W 1984. 'The ecology of Neolithic farming communities as exemplified by the Avebury region of Wiltshire', *Proc Prehist Soc*, 50, 99–120

Stead, I M and Rigby, V 1986. *Baldock: the excavation of a Roman and pre-Roman settlement 1968–72*, Britannia Monogr 7, London

Stevens, C E 1966. 'The social and economic aspects of rural settlement', in Thomas (ed) 1966, 108–28

Stone, J F S 1937. 'A Late Bronze Age habitation site on Thorny Down, Winterbourne Gunner, S Wilts', *Wiltshire Archaeol Natur Hist Mag*, 47, 640–59

Stone, J F S 1941. 'The Deverel-Rimbury settlement on Thorney Down, Winterbourne Gunner, S Wilts', *Proc Prehist Soc*, 4, 114–33

Straton, C R (ed) 1909. *Survey of the Lands of William First Earl of Pembroke*, vol 1, Oxford

Stuart, D 1992. *Manorial Records*, Cambridge

Summerfield, M A and Goudie, A S 1980. 'The sarsens of southern England: their palaeoenvironmental interpretation with reference to other silcretes', in *The Shaping of Southern England* (ed D K C Jones), 71–100

Swan, V 1973. 'Aspects of the New Forest late Roman pottery industry', in *Current Research into Romano-British Coarse Pottery* (ed A Detsicas), 117–34

Swan, V 1975. 'Oare reconsidered and the origins of Savernake Ware in Wiltshire', *Britannia*, 6, 36–51

Swan, V 1984. *The Pottery Kilns of Roman Britain*, London

Swanton, G R 1987. 'The Owen Meyrick Collection', *Wiltshire Archaeol Natur Hist Mag*, 81, 7–18

Swanton, G 1988. 'Interim report on the excavation of barrow G19, W Overton', *Wiltshire Archaeol Natur Hist Mag*, 82, 181–2

Taylor, C C 1968. 'Three deserted medieval settlements in Whiteparish', *Wiltshire Archaeol Natur Hist Mag*, 63, 39–45

Taylor, C C 1972. 'The study of settlement patterns in pre-Saxon England', in *Man, Settlement and Urbanism* (eds P J Ucko, R Tringham and G W Dimbleby), 109–13

Taylor, C C 1983. *Village and Farmstead*, London

Thirsk, J 1997. *Alternative Agriculture. A History from the Black Death to the Present Day*, Oxford

Thomas, A S 1960. 'Chalk, heather and Man', *Agric Hist Rev*, 8, 57–65

Thomas, C (ed) 1966. *Rural Settlement in Roman Britain*, Counc Brit Archaeol Res Rep 7, London

Thomas, R 1989. 'The bronze–iron transition in southern England', in *The Bronze Age–Iron Age Transition in Europe: aspects of continuity and change in European societies c 1200 to 500 BC* (eds M L Stig Sørensen and R Thomas), 263–86

Thomson, T R and Sandell, R E 1963. 'The Saxon land-charters of Wiltshire', *Wiltshire Archaeol Natur Hist Mag*, 58, 442–6

Thornes, J B 1987. 'The palaeo-ecology of erosion', in *Landscape and Culture. Geographical and Archaeological Perspectives* (ed J M Wagstaff), 26–55

Tilley, C 1994. *A Phenomenology of Landscape. Places, Paths and Monuments*, Oxford

Topping, P 1989. 'Early cultivation in Northumberland and the Borders', *Proc Prehist Soc*, 55, 161–79

Tosh, J 1991. *The Pursuit of History. Aims, Methods and New Directions in the Study of Modern History*, 2nd edn, London

Toynbee, J M C 1971. *Death and Burial in the Roman World*, London

Turner, J 1981. 'The Iron Age', in *The Environment in British Prehistory* (eds I G Simmons and M J Tooley), 250–81

Ucko, P J, Hunter, M, Clark, A J and David, A (eds) 1991. *Avebury Reconsidered from the 1660s to the 1990s*, London

Victoria County History 1955. *A History of Wiltshire, Volume II*, London

Victoria County History 1957. *A History of Wiltshire, Volume I, Part 1 and 2*, London

Victoria County History 1970. *A History of Wiltshire, Volume IX*, London

Victoria County History 1973a. *A History of Wiltshire, Volume IV*, London

Victoria County History 1973b. *A History of Wiltshire, Volume XI*, London

Victoria County History 1975. *A History of Wiltshire, Volume X*, London

Vince, A G 1981. 'The medieval pottery industry in southern England: 10th to 13th centuries', in *Production and Distribution: a ceramic viewpoint* (eds H Howard and E Morris), 309–22

Wacher, J 1974. *The Towns of Roman Britain*, London

Wainwright, G L 1979. *Gussage All Saints: an Iron Age settlement in Dorset*, London

Wainwright, G L, Donaldson, P, Longworth, I H and Swan, V 1971. 'The excavation of prehistoric and Romano-British settlements near Durrington Walls, Wiltshire', *Wiltshire Archaeol Natur Hist Mag*, 66, 76–128

Ward Perkins, J B 1940. *London Museum Medieval Catalogue*, London

Watts, K 1993. *The Marlborough Downs*, Bradford-on-Avon

Watts, K 1996. 'Wiltshire deer parks: an introductory survey', *Wiltshire Archaeol Natur Hist Mag*, 89, 88–98

Wells, T C E, Sheail, J, Ball, D F and Ward, L K 1976. 'Ecological studies on the Porton ranges. Relationships between vegetation, soils and land-use history', *J Ecol*, 64, 589–626

Williams, H M R 1998. 'The ancient monument in Romano-British ritual practices', in TRAC 97: *Proceedings of the Seventh Annual Theoretical Roman Archaeology Conference, Nottingham 1997* (eds C Forcey, J Hawthorne and R Witcher), 71–86

Williams, R J and Zeepvat, R J 1994. *Bancroft: a Late Bronze Age/Iron Age settlement, Roman villa and temple-mausoleum*, Buckinghamshire Archaeol Soc Monogr 7

Whimster, R 1989. *The Emerging Past: air photography and the buried landscape*, London

Whitelock, D (ed) 1955. *English Historical Documents c 500–1042*, London

Whittle, A W R 1993. 'The Neolithic of the Avebury area: sequence, environment, settlement and monuments', *Oxford J Archaeol*, 12, 29–53

Whittle, A W R 1994. 'Excavation at Millbarrow Chambered Tomb, Winterbourne Monkton, north Wiltshire', *Wiltshire Archaeol Natur Hist Mag*, 87, 1–53

Whittle, A W R 1997. *Sacred Mound, Holy Rings, Silbury Hill and the West Kennet Palisade Enclosures: a later Neolithic complex in north Wiltshire*, Oxbow Monogr 74, Oxford

Whittle, A W R and Thomas, J 1986. 'Anatomy of a tomb – West Kennet revisited', *Oxford J Archaeol*, 5, 127–56

Wilson, A R and Tucker, J H 1982. 'The Langley Charter and its boundaries', *Wiltshire Archaeol Natur Hist Mag*, 77, 67–70

Wilson, D M (ed) 1976. *The Archaeology of Anglo-Saxon England*, Cambridge

Wilson, D R 1982. *Air Photo Interpretation for Archaeologists*, London

Woodward, P J 1991. *The South Dorset Ridgeway Survey and Excavations 1977–84*, Dorset Natur Hist Archaeol Soc Monogr 8, Dorchester

Woodward, A and Woodward, P 1996. 'The topography of some barrow cemeteries in Bronze Age Wessex', *Proc Prehist Soc*, 62, 275–92

World Heritage Committee 1995. Intergovernmental Committee for the Protection of the World Cultural and Natural Heritage, *Operational Guidelines for the Implementation of the World Heritage Convention*, Paris

Wormald, F 1973. *The Winchester Psalter*, London

Yorke, B 1995. *Wessex in the Early Middle Ages*, London

INDEX

NOTES TO INDEX

There are three ways of using this index: by archaeological and historical periods, by subjects and by place. Civil parish names are attached if not obvious. Some Anglo-Saxon place-names and words seen as being significant in the charters are included, though 'Saxon', rather than 'Anglo-Saxon', is used throughout the Index. Under the locators, buildings, coins and pottery, the sub-heads are arranged chronologically, not alphabetically. Plates and figures are indexed in **bold** type in their page sequence, with the figures in the end pocket placed at the end of an entry. '*and see*' indicates that there is relevant information under other named locators.
The place-names West Overton, East Overton, Lockeridge and Fyfield are listed consistently west to east, not alphabetically, when '*and see*' is indicated. The names describe variously modern civil parishes, ecclesiastical parishes, villages, medieval manorial estates and tithings, and Saxon estates, and these are indexed as separate locators. The manorial estate of Overton is also indexed separately but, inevitably, the locators of its component parts, East Overton and Fyfield, duplicate in part some of its entries. This is particularly so of entries under 'boundaries'. Consultation of the principal maps of the study area, set out in Chapter 1, is recommended.